Advances in Solid State Circuits Technologies

Advances in Solid State Circuits Technologies

Edited by

Paul K. Chu

Advances in Solid State Circuits Technologies
Edited by Paul K. Chu

CBS, Edition **2016**

Published by InTech
Janeza Trdine 9, 51000 Rijeka, Croatia

Technical Editor Teodora Smiljanic

Cover Designer Dino Smrekar

Additional hard copies can be obtained from orders@intechopen.com

Advances in Solid State Circuits Technologies, Edited by Paul K. Chu
p. cm.
ISBN 978-953-307-086-5

Preface

Invention of solid-state transistors and integrated circuits has spawned the information age and the growth in the past 50 years has been phenomenal and unrivaled. Nowadays, information is at people's fingertips and communications take seconds rather than days like 20 years ago. Such rapid development stems from tremendous developments in both hardware and software such as solid-state circuits. The field of integrated circuits has obeyed Moore's Law for 40 years but as materials are being pushed to the limit, scientists and engineers are finding it harder to continue on the trend predicted by Gordon Moore. Approaches such as parallel processing, new circuit design, and particularly novel materials are necessary. This book brings together contributions from experts in the fields to describe the current status of important topics in solid-state circuit technologies. It consists of 20 chapters which are grouped under the following categories: general information, circuits and devices, materials, and characterization techniques.

The first two categories consist of chapters about CMOS nonlinear signal processing circuits, transconductors, dynamically reconfigurable devices, new unified random access memory devices, low-voltage fully differential CMOS switched-capacitor amplifiers, low-voltage, high linear, tunable, and multi-band active RC filters, multi-clad single mode optical fibers for broadband optical networks, continuous-time analog filters for CMOS and VHF applications, CMOS low noise amplifiers, PCM performance, ESD protection elements, directional tuning control of wireless / contactless power pickup for inductive power transfer systems, regulated gate drivers in CMOS, millimeter-wave CMOS, CMOS integrated switched-mode transmitters, and metal-oxide-semiconductor memories and transistors. The chapters covering materials science and engineering include hafnium-based high-k gate dielectrics, liquid phase oxidation on InGaP and applications, as well as germanium-doped Czochralski silicon. The final two chapters pertain to miniature dual-axes confocal miscroscopy for real time *in vivo* imaging and scanning near-field Raman spectroscopic microscope.

These chapters have been written by renowned experts in the respective fields making this book valuable to the integrated circuits and materials science communities. It is intended for a diverse readership including electrical engineers and material scientists in the industry and academic institutions. Readers will be able to familiarize themselves with the latest technologies in the various fields. In addition, each chapter is accompanied by an extensive list of references for those who want to obtain more detailed information and perform more in-depth research.

The tremendous cooperation from contributing authors who devoted their valuable time to write these excellent chapters and meticulous assistance provided by the editorial staff to make this book a reality are highly appreciated.

Editor

Paul K. Chu
City University of Hong Kong

Paul K. Chu is chair professor of materials engineering in City University of Hong Kong. He received his BS in mathematics from The Ohio State University in 1977 and MS and PhD in chemistry from Cornell University in 1979 and 1982, respectively. Paul's research activities are quite diverse encompassing plasma surface engineering and various types of materials, biomedical engineering, and nanotechnology. Prof. Chu has published over 800 journal papers and been granted 8 US, 5 Chinese, and 1 European patents. He is Fellow of the American Physical Society (APS), American Vacuum Society (AVS), Institute of Electrical and Electronic Engineers (IEEE), and Hong Kong Institution of Engineers (HKIE). He is senior editor of IEEE Transactions on Plasma Science, associate editor of Materials Science and Engineering Reports and International Journal of Plasma Science and Engineering, as well as a member of the editorial board of 6 journals including Biomaterials. He has received many awards such as the IEEE Nuclear and Plasma Sciences Society Merit Award in 2007 and Materials Research Society (MRS-Taiwan) JW Mayer Lectureship in 2008.

Contents

1

CMOS Nonlinear Signal Processing Circuits

Hung, Yu-Cherng
National Chin-Yi University of Technology
Taiwan, R.O.C.

1. Introduction

In VLSI circuit design, nonlinear signals processing circuits such as minimum (MIN), maximum (MAX), median (MED), winner-take-all (WTA), loser-take-all (LTA), *k*-WTA, and arbitrary rank-order extraction are useful functions (Lippmann, 1987; Lazzaro et al., 1989). In general, median filter is used to filtering impulse noise so as to suppress the impulsive distortions. The MAX and MIN circuits are important elements in fuzzy logic design. With regard to WTA application, it is the major function in pattern classification and artificial neural networks. Thus, design of these nonlinear signal-processing circuits to integrate smoothly within SoC (System-on-a-chip) applications becomes an important research. Recently, complementary metal-oxide-semiconductor (CMOS) technology is widely used to fabricate various chips. In this chapter, the designs of all circuits are realized by using CMOS process. However, since CMOS transistor is continuously scaled down via thinner gate oxides and reduced device size, supply voltage is necessary to reduce in order to improve device reliability. Therefore, a high reliable WTA/LTA circuit, a simple MED circuit, and a low-voltage rank-order extractor are addressed in the chapter. The organization of this chapter is as follows. Section 1 introduces the background of these nonlinear functions, including definitions and applications. Section 2 describes conventional WTA/LTA architectures and presents a high reliable winner-take-all/loser-take-all circuit. Section 3 shows an analog median circuit, with advantage of simple circuit. Section 4 describes a CMOS circuit design for arbitrary rank order extraction. Restrictions and design techniques of low voltage CMOS circuit are also addressed. Section 5 will briefly conclude this chapter.

Given a set of external input n variables a_1, …, a_n, the operation of MAX (or MIN) circuit determines the maximum (or minimum) value. A median filter puts out the median variable among a window of input samples. The function of a WTA network is to select and identify the largest variable from a specified set of variables. A counter part of WTA, LTA identifies the smallest input variable and inhibits remain ones. Instead of choosing only one winner, the k-WTA network selects the largest k numbers among n competing variables ($k \le n$), which allows for more flexibility in applications. For arbitrary rank order identification, a rank-order filter (extractor) is designed to select the k-th largest element a_k among n variables a_1, …, a_n. Depending on application requirements, these input variables are either voltage, or current signals.

In order to clearly describe these nonlinear functions, taking one example indicates these definitions. Two output responses of a circuit corresponding to a set of input currents I_{in1},

I_{in2}, …, and I_{inN} : one is analog output current I_o, the other one is digital outputs set $V_{o1(\text{rank})}$, $V_{o2(\text{rank})}$, …, and $V_{oN(\text{rank})}$. Assuming five external input currents are 9, 7, 10, 5, and 3 μA. Depending on various functions requirement, the output current I_o and the corresponding digital outputs responses are as follows.

1. **MAX:** I_o = Maximum(I_{in1}, I_{in2}, …, I_{inN})= I_{in3}= 10 μA
2. **MIN:** I_o = Minimum(I_{in1}, I_{in2}, …, I_{inN})= I_{in5}= 3 μA
3. **MED:** I_o = Median(I_{in1}, I_{in2}, …, I_{inN})= I_{in2}= 7 μA
4. **WTA:** Output voltages $V_{o1(\text{rank})}$, $V_{o2(\text{rank})}$, …, and $V_{o5(\text{rank})}$ respond to logic high to identify which one is the maximum value among I_{in1}, I_{in2}, …, and I_{inN}. In this case, ($V_{o1(\text{rank})}$, $V_{o2(\text{rank})}$, …, $V_{o5(\text{rank})}$)= (0, 0, 1, 0, 0), where "0" and "1" are the logic low and logic high, respectively.
5. **LTA**: A reverse operation of WTA function, and outputs set is (0, 0, 0, 0, 1) for this case.
6. ***k*-WTA:** Depending on *k* value, *k* winners are selected. The function has more flexible in application than WTA. For example, the outputs of 2-WTA is ($V_{o1(\text{rank})}$, $V_{o2(\text{rank})}$, …, $V_{o5(\text{rank})}$)= (1, 0, 1, 0, 0) in this case.
7. **Rank order:** The function of the *r*th rank-order extraction identifies the *r*th largest magnitude among I_{in1}, I_{in2}, …, and I_{inN}. For example, outputs of the 2nd and 3rd rank order are (1, 0, 0, 0, 0) and (0, 1, 0, 0, 0) in this case, respectively.

Rule 1: IF *x* is PL and *y* is ZR, then *z* is NS.
Rule 2: IF *x* is ZR and *y* is NL, then *z* is ZR.

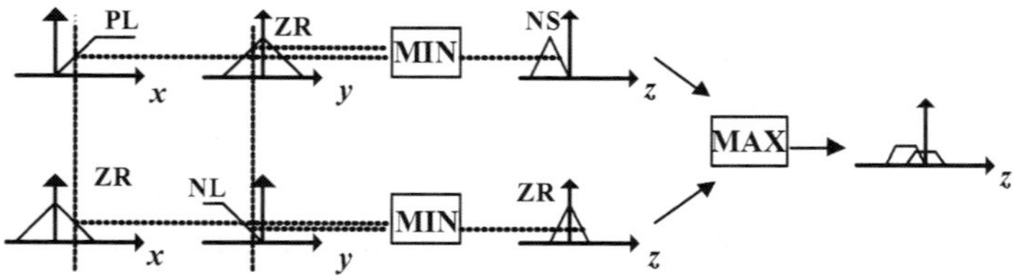

Fig. 1. Applications of MIN and MAX operations in fuzzy inference.

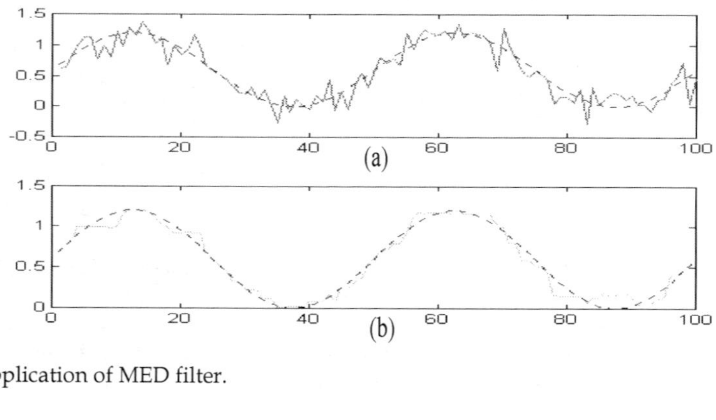

Fig. 2. Application of MED filter.

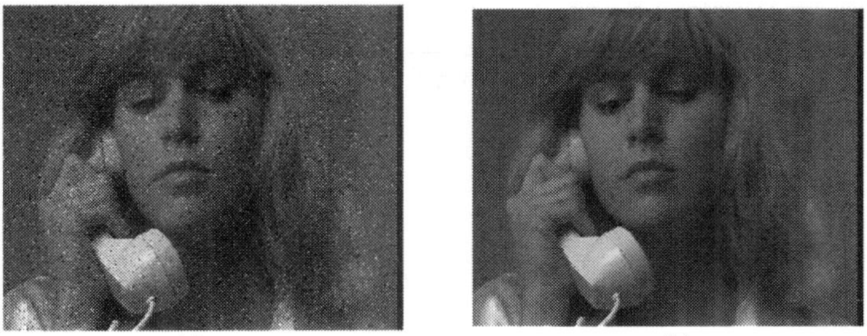

(a) (b)

Fig. 3. Two-dimension application of MED filter.

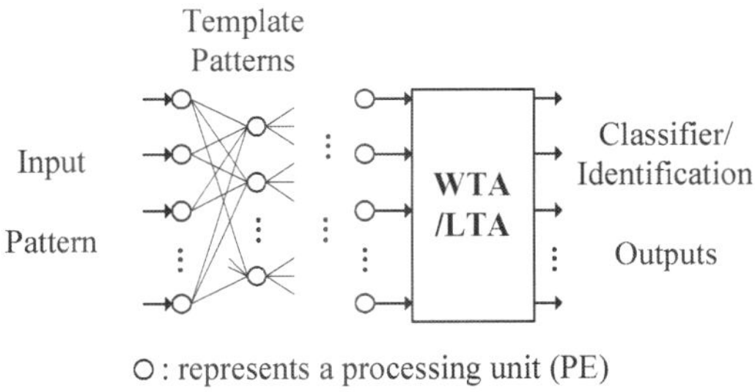

Fig. 4. Applications of WTA/LTA function in artificial neural network.

Various applications for these nonlinear functions are described as follows. The MAX and MIN circuits are important elements in fuzzy logic design (Yamakawa, 1993). Fig. 1 shows the MAX and MIN operations in fuzzy inference. Variables "x" and "y" are inputs; variable "z" is the corresponding output response. In a specific status, either rule 1 or rule 2 is satisfied. MIN function realizes the "and" operation in fuzzy rules, and MAX function realizes the "or" operation. In image signal processing, MED function in general is used to filtering impulse noise so as to suppress the impulsive distortions. Figure 2 shows a one-dimension application for noise cancellation. Fig. 2(a) shows a V_{pp} 1.2 V sinusoidal signal corrupted by noise, and Fig. 2(b) shows the processed signal after MED filtering with a window of size five. In addition, Figure 3 shows a two-dimension application also for noise cancellation of image. With regard to WTA application, it is the major function in pattern classification, vector quantization, data compression, and self-organization neural networks. Figure 4 shows WTA application for pattern identification. Commonly, an analogue rank order filter is widely used in signals sorting and classification.

In general, these nonlinear functions are achieved either by using digital or analog implementations. Under digital implementation, since most of signals obtained from the real world are continuous forms, the continuous inputs must first be transferred to digital type by using one-or-multiple analog-to-digital converter (A/D). As a result, the circuit complexity, chip area, and power consumption are increased due to the extra data converters in digital realization. Whereas for analog implementation, the circuit accuracy is slightly lost than digital operation and there is weaker tolerance to fabricate process variation. However, without extra data transfer, the analog operation is with many advantages such as saving time, bandwidth, and computation at the system level. Considering the practicality and flexibility, design issues of a CMOS analog signal processing circuit therefore must include 1) precision; 2) speed; 3) high tolerance to fabrication process variation; 4) wide range of supply voltage; 5) wide input range; 6) low circuit complexity; 7) low power consumption; 8) scalability; 9) programmability, and so forth, to allow these functions easily integration within various system-embedded chips. Additionally, when the device size of CMOS transistor is shrunk thinner and smaller, supply voltage is necessary to scale down in order to improve device reliability. A forecast of high-performance CMOS circuit operated within low voltage had been reported (Semiconductor Industry Association, 2008). Figure 5 shows the trend of CMOS supply voltage and physical gate length. Moreover, portable equipments such as biomedical electronics, computer, and portable telecommunication equipments are common used recently. Battery operation and low-power consumption are also important design requirements for these circuits.

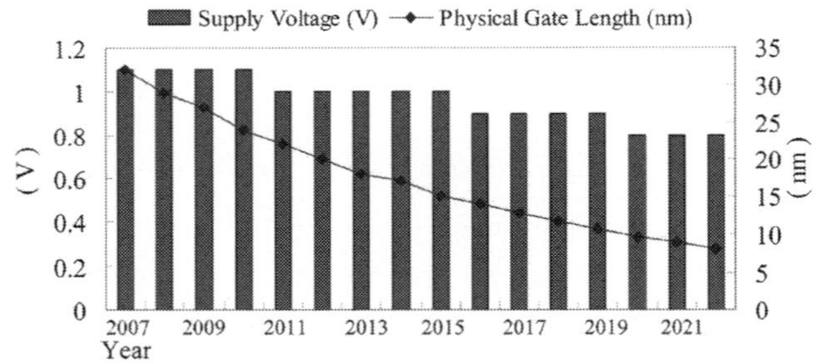

Fig. 5. Trend for supply voltage and physical gate length by ITRS 2008 update.

2. Winner-Take-All and Loser-Take-All circuit

2.1 Architectures of WTA/LTA circuits

Based on different circuit structures, conventional WTA/LTA circuits are roughly cataloged into four types: 1) global-inhibition structure, in which the connectivity increases linearly with the number of inputs (Lazzaro et al., 1989; Starzyk & Fang, 1993); 2) cell-based tree-topology (Smedley et al., 1995; Demosthenous et al., 1998); 3) excitatory/inhibitory connection (He & Sanchez-Sinencio, 1993); and 4) serial cascade structure (Aksin, 2002). Figure 6(a-d) shows the conceptual diagrams of these topologies. In Fig. 6(a), each cell receives the same global inhibition, and a common current I_{comn} or voltage V_{comn} is shared by all the competing cells. The cells represented in a square block are nonlinear signal processing elements. Therefore, the precision of the circuit is degraded as the number of inputs increases. Since the operation of this circuit relies on the cells matching, a stable fabrication process is required for manufacturing a high-precision system. The complexity of the connectivity of the circuit is $O(N)$, where N is the number of inputs. Figure 6(b) shows a cell-based tree-topology, with N-1 cells arranged in a tree topology for N inputs. Each cell receives two input variables to compare and outputs the larger (or smaller) of the two input signals. The backward digits in the bottom cell are then successive feedback to 1st-layer cells to identify the maximum (or minimum) input. The precision of this circuit is also sensitive to cell matching. With this circuit design, the device sizes must be rescaled when the supply voltage is modified.

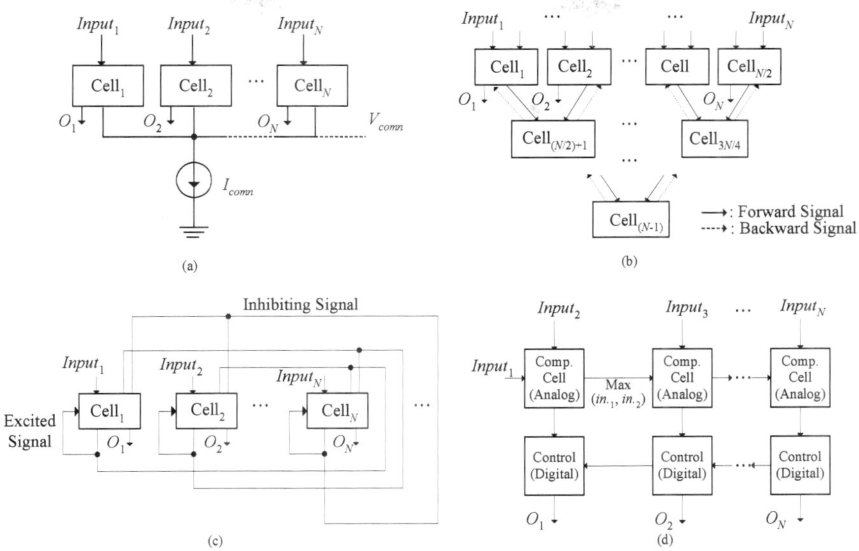

Fig. 6. Conventional architectures. (a) Global-inhibition structure. (b) Cell-based tree topology. (c) Excitatory/inhibitory connection. (d) Serial cascade.

Figure 6(c) shows an excitatory/inhibitory connection with an $O(N^2)$ connectivity complexity. Each cell receives the inhibited signals from other cells and an excitatory signal from itself. With this design, chip area increases with the square of the number of inputs. Based on comparators operation, Figure 6(d) shows an N-1 analog comparison blocks and N-1 digital blocks cascaded in serial. Within a comparison time T_{comp}, the larger magnitude of inputs in each analog block is sent to next stage to compare with other inputs. The result of the each comparison is then sent to the corresponding digital block, and a decision digit is feedback from right block to left block to identify the maximum input. As a result, the response time of the circuit is approximated to $(N-1)\cdot T_{comp} + T_{dig}$, where T_{dig} is the total propagation time of the digital part. The offset voltage of each comparator dominates the precision of the architecture. Circuit implementation of Fig. 6(d) is also sensitive to process variation. For a high precision application, identical internal circuit blocks shown in Figs. 6(a-d) are necessary. The primary limitations of accuracy for the conventional architectures are fabricated process variations and matching requirement of internal cells. The variations of CMOS fabricated process include transistor threshold voltage, actual device size, thinness of the gate oxide, and other variety of factors. In a common process, threshold voltage in general varies from -10% to +10% of its nominal value. Due to the non-uniform etch and diffusion procedures, actual device sizes are also varied. In a real CMOS process, these variations are hard to eliminate completely. How can we improve the accuracy of analog circuit in a conventional process?

2.2 A high reliable WTA/LTA circuit

In the section, a highly reliable CMOS signal processing circuit with a programmable capability for WTA function and LTA function is described (Hung & Liu, 2004). A symbol $COMP_t^i(V_{inj}, V_{ink})$ ($1 \leqq j,\ k \leqq N$ and N is the number of inputs) is defined such that the ith comparator cell receives two input variables (V_{inj} and V_{ink}) to compare in magnitude at time t, and the output Z_t^i of the cell is the larger variable or a binary value. For a $COMP_t^i(V_{inj}, V_{ink})$ operation, Z_t^i is defined as

$$z_t^i = \begin{cases} 1 \text{ or } V_{inj}, & \text{when } V_{inj} > V_{ink} \\ 0 \text{ or } V_{ink}, & \text{otherwise.} \end{cases}$$

Therefore, returning to the conventional architecture the tree topology of Fig. 6(b), WTA mode, is represented as:

$$t_1\text{: } COMP_{t1}^1(V_{in1}, V_{in2}),\ COMP_{t1}^2(V_{in3}, V_{in4}),\ \ldots,\ COMP_{t1}^{N/2}(V_{in(N-1)}, V_{inN})$$

$$t_2\text{: } COMP_{t2}^{(N/2)+1}(Z_{t1}^1, Z_{t1}^2),\ COMP_{t2}^{(N/2)+2}(Z_{t1}^3, Z_{t1}^4),\ \ldots$$

$$\ldots$$

$$t_{(\log_2 N)} : COMP_{t(\log N)}^{(N-1)}(Z_{t(\log N)-1}^{N-3}, Z_{t(\log N)-1}^{N-2}).$$

After time $O(\log_2 N)$, the maximum (or the minimum) input variable is obtained. Total N-1 identical comparators are necessary for this operation.

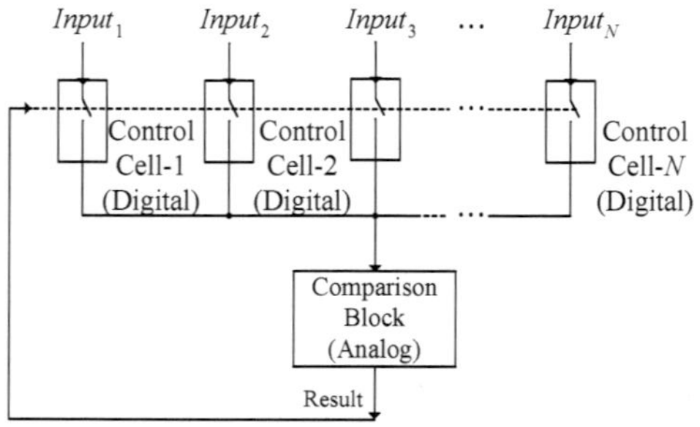

Fig. 7. A high reliable WTA/LTA architecture.

To reduce the matching requirement of internal cell, Figure 7 shows a conceptual diagram of high reliable circuit. In the scheme, there are N identical 'digital' control cells and a single comparator for N input variables. A single comparator block multiplexes in time to achieve all inputs comparisons. The operating procedures are described as follows:

$$t_1\text{: } COMP_{t1}^1(V_{in1}, V_{in2})$$

$$t_2\text{: } COMP_{t2}^1(Z_{t1}^1, V_{in3})$$

...

$$t_{(N-1)} : COMP^1_{t(N-1)}(Z^1_{t(N-2)}, V_{inN}) .$$

The strategy adopted to find the maximum/minimum among a set of variables is that two variables are first compared; then the result of this comparison is compared with the next input variable using the same comparator. The procedure continues until the comparisons of all input variables are completed. Conceptually, circuit operation is similar to a serial comparison. Unlike the traditional architectures that require *N*-1 analogue comparators; this architecture requires only a single comparator to eliminate sensitivity to component matching requirements. Using the same algorithm, the LTA function is easily obtained by only reversing the output state Z_t^i in the same architecture.

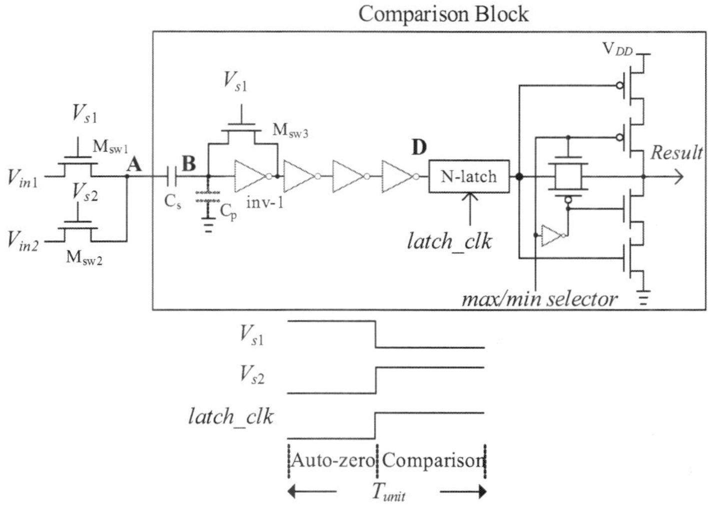

Fig. 8. Comparison block and control signals.

The key block in this architecture is the comparator cell. Comparator performance is a crucial factor for realizing high-speed data conversion systems and telecommunication interfaces. The precision of a comparator is usually defined as the minimum identifiable differential voltage (or current) between inputs, that is, the comparator's resolution capability. A comparator design from (Hosotani et al., 1990) is used herein; the schematic diagram is shown in Fig. 8. Transistors M_{sw1}, M_{sw2}, M_{sw3} are used as switches. The circuit operates on two phases, auto-zero phase and comparison phase. Assuming the voltage at node B is V_x. Based on charge conservation, after the comparison phase, V_x arrives at the following:

$$V_x = V_b + (V_{in2} - V_{in1}) \cdot \frac{C_s}{C_s + C_p + C_{in}} . \quad (1)$$

The effect of the $C_s/(C_s+C_p+C_{in})$ term in (1) represents a degrading factor. To reduce the decision time, the succeeding inverters amplify the different voltage (V_{in2} - V_{in1}) to pull node D up to high (logic 1) or push it down to 0 V (logic 0). The functions of the N-latch are to sample the voltage at node D as *latch_clk* turns high and to hold the comparison result as *latch_clk* turns low. Ultimately, the output polarity of the N-latch will be changed according to the *max/min selector* setting. The *max/min selector* signal modifies the polarity of the compared result; therefore, without the need for structural modification, this circuit possesses win/lose configurable capability. The comparison block shown in Fig. 8 is reused during all comparison procedures. The architecture of *N*-inputs circuit is shown in Fig. 9, in which Control_Cell$_n$ ($1 \leqq n \leqq N$) are identical. *N* cells are required for *N* input variables. Each cell contains a status block, a control_switch block, and two latch blocks.

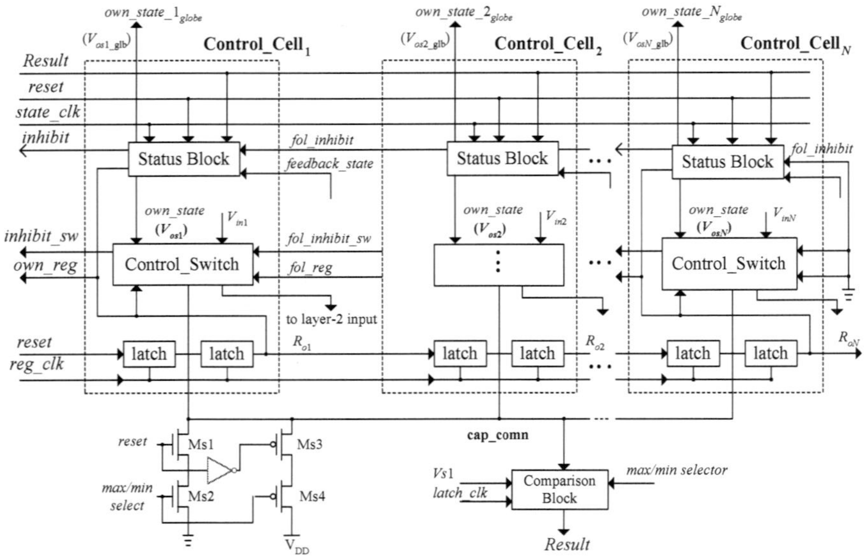

Fig. 9. The block diagram of the high reliable WTA/LTA.

Figure 10 shows the clocks for the whole circuit. Signal *reset* and clock *reg_clk* must be generated externally; other clocks are produced by *reg_clk* and some logic gates.

To describe the operations of the entire circuit, the circuit architecture in Fig. 9 and the clock waveform in Fig. 10 are referred. First, at *t*1, *reset* signal is used to initiate the status blocks, control_switch blocks and latch blocks. The N-latch in the status block and R_{o1}, R_{o2}, ..., R_{oN} are reset to zero by *reset* signal. Based on *max/min selector* signal, the MOS transistors Ms1, Ms2, Ms3 and Ms4 preset the initial sampling voltage (0 V or V_{DD}) at node cap_comn. Despite the magnitude of input-1 variable, the input-1 variable must be a winner during an initial interval for a serial comparison. The initial sampling voltage at node cap_comn is thus set as 0 V when the *max/min selector* signal is set to logic 1 for WTA operation, and vice versa.

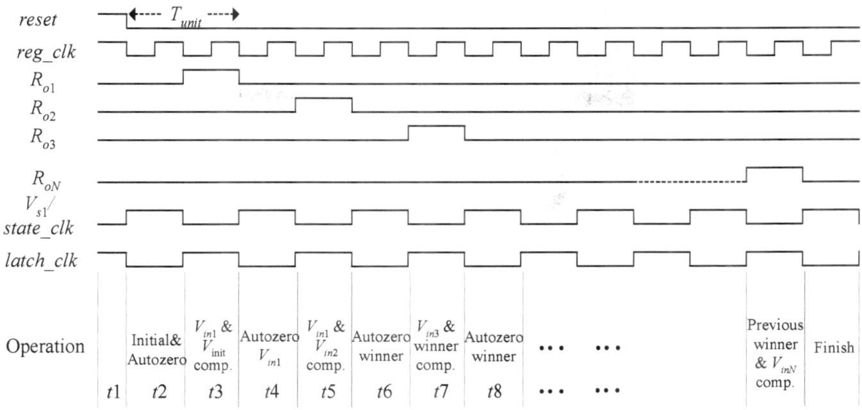

Fig. 10. Clock waveforms.

Then, at *t*2, the V_{s1} clock turns high (auto-zero phase) to sample the initial voltage (0 V or V_{DD}) at node cap_comn. Next, at *t*3, R_{o1} turns high to sample voltage V_{in1}. At this time, the clock V_{s1} turns low (comparison phase) to compare the V_{in1} with the initial sampling voltage, and the compared result is stored in the N-latch of the first status block. The state of the N-latch is logic 1 if the variable is the winner. At *t*4, the present winner V_{in1} is sampled again. At *t*5, a new comparison between previous winner V_{in1} and V_{in2} is performed. At *t*6, the winner (the result for the V_{in1} and V_{in2} comparison) is sampled again. After this procedure, a new comparison between the present winner and V_{in3} is performed. The procedure continues until comparison of all the input voltages is completed. Ultimately, only one state V_{osn} (**n**=1, ..., *N*) in these cells is logic 1 for WTA/LTA indication; others are logic 0. Therefore, a WTA or a LTA operation has been accomplished.

Figure 11 shows the status block. Figure 12 shows the control_switch block. It receives an input variable and controls the transmission gate to sample input level. A true single-phase latch composed of an N-latch and a P-latch is used to reduce the clock skew issue (Yuan & Stensson, 1989).

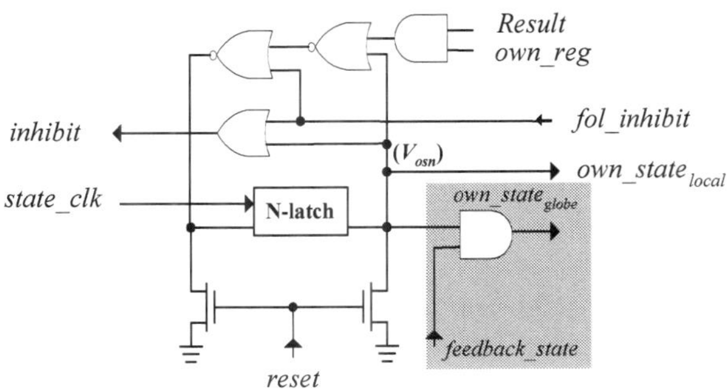

Fig. 11. Status block.

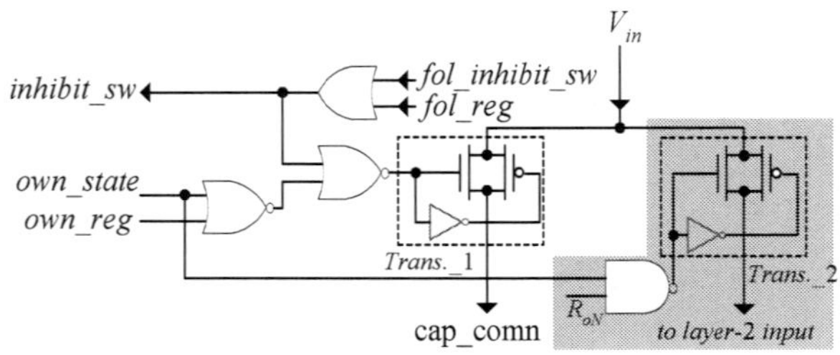

Fig. 12. Control_switch block.

2.3 Simulation results and reliability test

With regard to the high reliable WTA/LTA circuit, an experimental chip with six inputs was also fabricated using a 0.5-μm CMOS technology. The sampling capacitance C_s implemented by using two-layer polysilicon is set to be 3 pF. The period of *reg_clk* clock is 100 ns with a 50% duty cycle. WTA/LTA functions, supply-voltage range, and Monte Carlo analysis of transistor variation by simulation were also tested.

1) WTA/LTA functions

To test the function of the circuit, each example takes ten input voltages for the WTA/LTA operation. For supply voltage V_{DD}=3.3 V, the input variables V_{in1}, V_{in2}, …, and V_{in10} are 0.003, 0.006, 1.000, 0.997, 2.000, 2.003, 2.000, 3.297, 3.300, and 3.297 V for testing WTA function, respectively, and 3.297, 3.294, 2.000, 1.997, 2.000, 1.000, 0.997, 0.006, 0.009, and 0.003 V for testing LTA function. During the WTA operation, the logic state V_{osn} of each cell at each time slice becomes:

V_{os1}= 1,0,0,0,0,0,0,0,0,0 V_{os2}= 0,1,0,0,0,0,0,0,0,0 V_{os3}= 0,0,1,1,0,0,0,0,0,0 V_{os4}= 0,0,0,0,0,0,0,0,0,0
V_{os5}= 0,0,0,0,1,0,0,0,0,0 V_{os6}= 0,0,0,0,0,1,1,0,0,0 V_{os7}= 0,0,0,0,0,0,0,0,0,0 V_{os8}= 0,0,0,0,0,0,0,1,0,0
V_{os9}= 0,0,0,0,0,0,0,0,1,1 V_{os10}=0,0,0,0,0,0,0,0,0,0.

When all comparisons are finished, the outputs V_{os1}, V_{os2}, V_{os3}, ..., and V_{os10} respond as logic 0, 0, 0, 0, 0, 0, 0, 0, 1, and 0, respectively. Therefore, among these ten inputs, input variable V_{in9} is the maximum. Figure 13 shows the results of HSPICE simulation for the WTA operation. The time period of the latch clock (top trace) is 100 ns. In the same operation, Fig. 14 shows the results for the LTA operation. The final outputs V_{os1}, V_{os2}, V_{os3}, …, and V_{os10} are logic 0, 0, 0, 0, 0, 0, 0, 0, 0, and 1, respectively, and the input variable V_{in10} is the minimum one. Choice for the above tested voltages was based on the followings: 1) input voltages of neighbor cells should be as close as possible to test discrimination capabilities; 2) input voltages are distributed from 0 V to 3.3 V to test for wide dynamic range.

2) Supply voltage range

All circuit parameters such as transistor dimensions, clock periods and sampling capacitance C_s are held constant. A supply voltage V_{DD} varies from 2 V to 5 V, and the logic high of these clocks are also modified when the supply voltage alters. The supply voltage V_{DD} for each iteration increases in 0.1 V steps. The simulation results show that the circuit operates successfully within 3-mV discrimination when the supply voltage ranges from 2.7 V to 5 V. Without any procedure for rescaling the device size, the circuit works under various commonly used supply voltages.

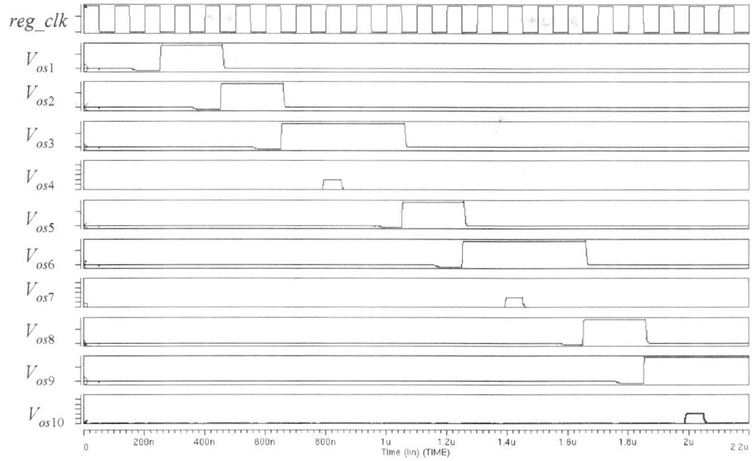

Fig. 13. Simulation results of the WTA operation.

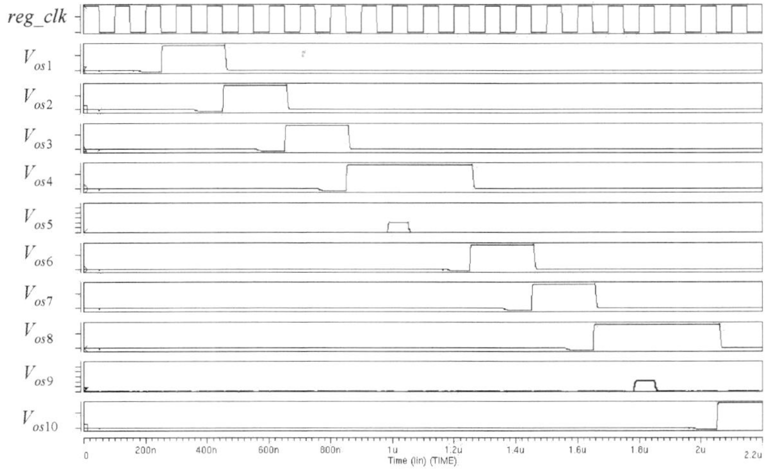

Fig. 14. Simulation results of the LTA operation.

3) Process variations

A statistical distribution of manufacturing parameters often occurs during CMOS fabrication. Wafer-to-wafer, run-to-run and transistor-to-transistor process variations determine the electrical yield and critical second-order effects. Threshold voltage, channel widths, and channel lengths of all MOS transistors were set to nominal values with ±5 % variation at the 3 sigma level, and each transistor was given an independent random Gaussian distribution. After 30 Monte Carlo iterations, HSPICE results indicate that circuit precision and speed are not degraded over this range. In addition, to verify the circuit with multi-technology support capability, using various CMOS fabrication parameters also simulates the circuit performance. The results show that the performance of the circuit under various fabrication processes is functional work, without needing to tune any device dimension. The following reasons contribute to the robustness of this circuit: 1) the circuit is designed with only a single analog cell (comparator), while the other active components are digital; 2) the comparator itself is designed with a auto-zero property, therefore, the operation of the comparator is more tolerant to manufacturing process variation.

4) Circuit precision

The accuracy of the comparator cell dominates the identified precision. The comparator accuracy is dependent on two factors. One is the clock feed-through error and charge-injection error in transistor M_{sw3}, shown in Fig. 8; the other is the degrading factor in Eq. (1). Charge-injection error is a complicated function of substrate doping concentration, load capacitor, input level, clock voltage, clock falling rate, MOS channel dimension, and the threshold voltage. Therefore, this error is difficult to be completely eliminated. In general, complementary clock, transmission gates, and dummy transistor are adopted for a switch realization to reduce the error.

3. CMOS analogue median cell

Median (MED) filter is a useful function in image processing application to eliminate pulse noise. Given a set of external input n variables $a_1, \ldots, a_n$, the operation of MED circuit determines the median value. The extracted median operation is a nonlinear function. The MED circuit realizations can be classified as analog filtering and digital filtering depending upon what type of input signals are. The digital filtering architecture has a variety of sophisticated algorithms to support the circuit realization so as with advantages of higher flexible and higher reliability. For power consumption and chip area considerations, however, it is costly expensive than analog architecture. In 1994, without using an operational amplifier, an analogue median extractor with simple structure and high sharp DC transfer characteristic was presented (Opris & Kovacs, 1994). The circuit expects to reduce the errors in the transition region. In 1997, for the same authors, an improved version with high speed operation was proposed. The median circuit has transient recovery less than 200 ns by using 2-um CMOS process (Opris & Kovacs, 1997). In 1999, a current-input analog median filter composed of absolute value and minimum circuits was proposed (Vlassis & Siskos, 1999). The operational amplifier and transconductor are also not needed in design of the circuit. Based on transconductance comparators and analog delay elements, a fully continuous-time analog median filter is presented in 2004 (Diaz-Sanchez et al., 2004). By using the median filter cells, an image of 91×80 pixels can be processed in less than 8 μs to remove salt and pepper noise. In the section, an intuitional and simple CMOS analog median cell is described (Hung et al., 2007). Based on current-mirror, current comparison, and some basic digital logics, a simple analog median filter cell is achieved. By using TSMC 0.35 μm CMOS technology, simulation shows that the median filter provides a 0.4-μA discriminability and well tracked the median value among input currents.

Figure 15 shows a basic one-input current cell composed of current mirror and control logic circuits. The cell has one signal input (i_s), a current source (i_{s_src}) output and a current sink (i_{s_sink}) output, a control signal $V_{ctr.}$, and an output current (i_{out}). Transistors M_1-M_{12} are cascode current mirrors. M_{swp} and M_{swn} constitute transmission gate for analog switch function. M_{dummy} is designed to compensate the M_{swn} and M_{swp} loading to improve the accuracy of output current. M_{iso} is used to isolate the clock noise from transmission gate. M_{dis1-2} and M_{res} are used to speedup transmission operation and control the discharge timing. Corresponding to Fig. 15(a), Fig. 15(b) is a symbol representation, which is named as *current signal control unit* and is abbreviated as *CSCU*.

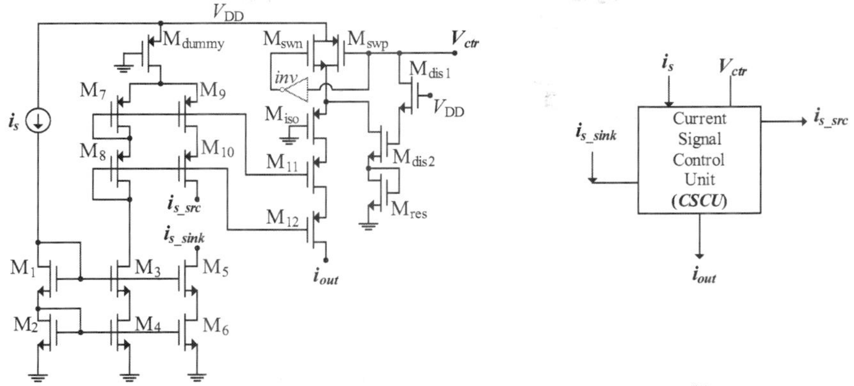

(a) (b)

Fig. 15. Current signal control unit (*CSCU*): (a) circuit and (b) symbol representation.

Three input signals i_{s1}, i_{s2}, and i_{s3}, how can circuit extract the median value? Assuming i_{s2} is a median current. The criteria must be satisfied.

$$\mathrm{MED}(i_{s1}, i_{s2}, i_{s3}) = i_{s2} = \begin{cases} ((i_{s2} > i_{s3}) \text{ and } (i_{s2} < i_{s1})) \\ \text{or} \\ ((i_{s2} < i_{s3}) \text{ and } (i_{s2} > i_{s1})) \end{cases} \quad (2)$$

As a result, current level comparison and logic decision are required to realize the function. Figure 16 shows a three-input median circuit composed of three *CSCU* cells and three decision logic blocks. The decision logic circuit is simply realized by AND-OR gate circuit to perform

$$\overline{V_{ctr}} = ① \cdot ② + ③ \cdot ④ \quad (3)$$

where ①, ②, ③, and ④ represent the corresponding the logic inputs, that is, these signals come from comparison results Ⓐ-Ⓕ signals. Depending on the output status of each decision logic, Eq. (3) determines V_{ctr} a low level or a high level, respectively. A low V_{ctr} will turn on the transmission gate of corresponding *CSCU* cell to switch on the input current; otherwise, the input current is prohibited. As a result, three-input MED filter cell is successfully arrived. Due to the transition pulse noise, a capacitor C_{filter} is used to suppress the switch noise.

In the circuit, NMOS transistor size $(W/L)_N$=5μ/1μ and PMOS transistor size $(W/L)_P$=10μ/1μ are used for M_1-M_{12}. The sizes of inverters are $(W/L)_N$=5μ/0.35μ and $(W/L)_P$=20μ/0.35μ. The device site of switch transistors M_{swn} and M_{swp} are equal to $(W/L)_{N\text{-}P}$=20μ/0.35μ. All transistors in decision logic block are sizing $(W/L)_N$=5μ/0.35μ and $(W/L)_P$=10μ/0.35μ. The filter capacitance C_{filter} is designed as 10 pF. The supply voltage V_{DD}

is commonly used as 3.3 V. Input current signals i_{s1}, i_{s2}, i_{s3} have 10 μA peak value at different 5 μs, 10 μs, and 15 μs time slot, respectively. Figure 17 shows three triangle waves and the corresponding median output. The red line represents the MED output. The output is tracked well with the median value of the three inputs current. By observing Fig. 17, when two input values are closed to each other, the minimum difference must be larger than 0.4 μA. That is the discriminability of the MED filter. However, there are some little spike occurs in the transition point.

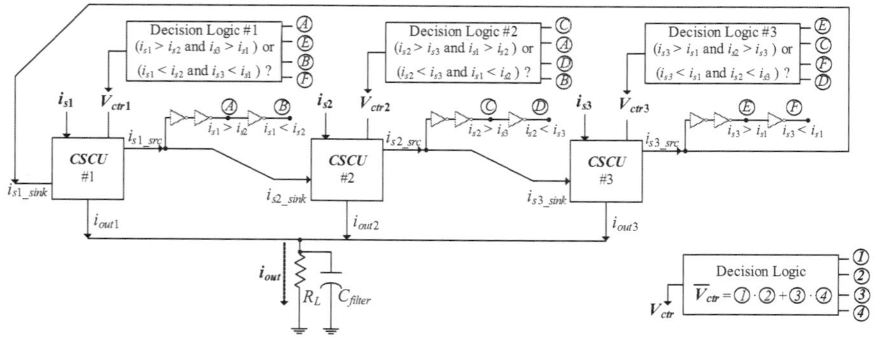

Fig. 16. Three-input median cell.

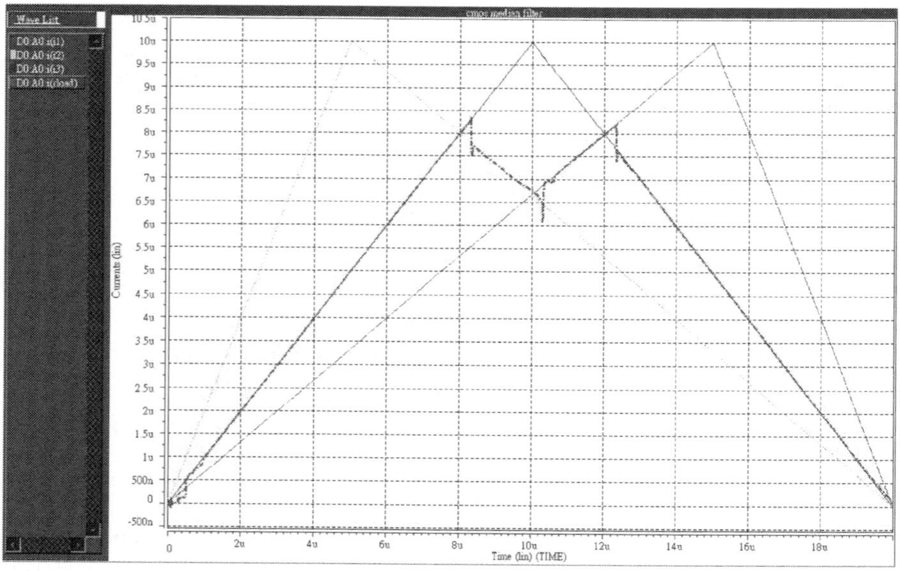

Fig. 17. The output response of the median filter for triangle waveforms.

Inspecting Fig. 16, the proposed three-input median cell has three input pins (i_{s1}, i_{s2}, and i_{s3}) and a common output pin (i_{out}). By modifying the switch transistors and decision logic, the

MED cell can be easily modified as three inputs and three outputs. The modified MED cell will have maximum value i_{maxmin}, median value i_{median}, and minimum value i_{minmum} outputs, simultaneously. As a result, the multiple modified MED cells can be organized cooperation to perform the 'sorting' function. In the design, no critical components such as operational amplifier and precise voltage reference are required in the MED cell. These properties are useful for the MED cell simply embedded into a larger system.

4. Low-voltage arbitrary rank order extraction

4.1 Principle of rank-order extraction

Ether WTA, LTA, or MED function, however, is only a single order operation. In 2002, a low-voltage rank-order filter with compact structure was designed (Cilingiroglu & Dake, 2002). The filter is based on a pair of multiple-winners-take-all and a set of logic gates. In the section, a new architecture for with both arbitrary rank-order extraction and k-WTA functionalities is described (Hung & Liu, 2002). An rth rank-order extraction is defined that identifies the rth largest magnitude of input variables. In the design, the circuit locates an arbitrary rank order among a set of input voltages by setting different binary signals. A set of output voltages V_{o_1}, V_{o_2}, ..., and V_{o_M} corresponds to the output voltages of a rank-order extractor for inputting of a set of variables V_1, V_2, ..., and V_M. The output status D_{ij} of a comparator with two-input terminals is defined as

$$D_{ij} = \begin{cases} 1 & if\ V_i > V_j \\ 0 & otherwise \end{cases} \quad 1 \le i, j \le \mathrm{M},\ j \ne i \tag{4}$$

where M is the number of the input variables. For convenience of description, a temporal index S_i defines the total number of winners for the ith input variable compared with the others. Thus, S_i is represented as

$$S_i = \sum_{j=1,\, j \ne i}^{\mathrm{M}} D_{ij} \qquad 1 \le i \le \mathrm{M}\,. \tag{5}$$

Based on the definition of (5), S_i is expanded as follows

$$S_1 = D_{12} + D_{13} + \ldots + D_{1M} \tag{6a}$$

$$S_2 = D_{21} + D_{23} + \ldots + D_{2M} = \overline{D_{12}} + D_{23} + \ldots + D_{2M} \tag{6b}$$

$$S_3 = D_{31} + D_{32} + \ldots + D_{3M} = \overline{D_{13}} + \overline{D_{23}} + \ldots + D_{3M} \tag{6c}$$

$$\ldots$$

$$S_M = D_{M1} + D_{M2} + \ldots + D_{M(M-1)} = \overline{D_{1M}} + \overline{D_{2M}} + \ldots + \overline{D_{(M-1)M}}\,. \tag{6d}$$

Thus, from the left-hand side of (6), M(M-1) comparators' cooperation is required for M input variables to identify the rank order. Since D_{ji} is the complementary of D_{ij} ($D_{ji} = \overline{D_{ij}}$), the expression is replaced by $\overline{D_{ij}}$ in the right-hand side of (6). The physical meaning is that if both the output of the comparator and its complementary are given, the total number of comparators can be reduced from M(M-1) to M(M-1)/2.

In this section, the comparator generates a unit current I_{unit} when input variable V_i is larger than V_j. Thus, the index S_i in (5) is rewritten as

$$S_i^* = \sum_{j=1,\ j\neq i}^{M} D_{ij} I_{unit},\ 1\leq i \leq M \quad = nI_{unit},\ 0\leq n \leq (M-1) \tag{7}$$

where n is the number of the winner in comparison. If the inputs are arranged in ascending order of magnitude, V_1, V_2, …, V_M, which satisfy $V_1<V_2<\ \ldots\ <V_M$, then $S_1^* = 0$, $S_2^* = I_{unit}$, …, $S_M^* = (M-1)I_{unit}$. Obviously, the minimum, next minimum, …, maximum input variables can be found by checking the index S_i^*. The k-WTA function is defined so that the outputs must be logic high when

$$S_i^* \geq (M-k)I_{unit}. \tag{8}$$

For example, if the input variables are (0.5, 0.6, 0.9, 0.2, 0.4), the first variable 0.5 is larger than variables 0.2 and 0.4. Thus, the index S_1^* is $2I_{unit}$; the meaning is that the variable wins two other input variables among all comparisons. For the same reason, the $S_2^* = 3I_{unit}$, $S_3^* = 4I_{unit}$, $S_4^* = 0$, $S_5^* = I_{unit}$. Therefore, the rank order is found among the input variables by checking the index S_i^*. In this example, the output voltages (V_{o_1}, V_{o_2}, …, V_{o_5}) of the extractor respond to be (0, 0, 1, 0, 0), (0, 1, 0, 0, 0), (1, 0, 0, 0, 0), (0, 0, 0, 1, 0) for the maximum operation, next maximum operation, median operation, and the minimum operation, respectively. The "0" and "1" are the logic low and high. Similarity, if the extractor is configured as k-WTA function, the output voltages (V_{o_1}, V_{o_2}, …, V_{o_5}) of the circuit respond to be (1, 1, 1, 1, 1), (1, 1, 1, 0, 1), (1, 1, 1, 0, 0), …, and (0, 0, 1, 0, 0) for 5-WTA, 4-WTA, 3-WTA, …, and 1-WTA operations, respectively.

4.2 Architecture of rank-order extraction

The structure of the extractor is shown in Fig. 18 for five input variables (Hung & Liu, 2002). There are a total of M(M - 1)/2 comparators and M evaluation cells for M input variables. Each comparator cell accepts two input signals, and the results of each comparison are fed into the individual evaluation cell. In the first row of Fig. 18, the input V_1 is compared with other input variables. In addition, the results of the comparison will generate the proper unit currents I_{unit}. Then, these currents will be summed up in *Eval*-1 cell if V_1 is larger than the other samples; otherwise, the result of the comparison will be fed into the corresponding evaluation cell. The connecting strategy is the same for other input variables. Therefore, equation (7) have been realized in this architecture.

The signal V_{choice} in Fig. 18 is used to decide the function of the circuit. V_{choice} is preset at logic high to allow the rank-order operation; otherwise, the k-WTA function is enabled. The binary signals *sel*_1, *sel*_2, and *sel*_3 are used to determine which rank-order/k-WTA will be located. Based on the select signals (*sel*_1-3) setting, the logic states of the evaluating cells indicate which input variable belongs to this rank order. For example, in the seven inputs rank-order operation, the (*sel*_1, *sel*_2, *sel*_3) signals are set to logic (0, 0, 0) to find the minimum variable; the logic (0, 1, 1) and (1, 1, 0) setting are the median and maximum functions, respectively. Similarity, in the k-WTA operation, the (*sel*_1, *sel*_2, *sel*_3) is set as (0, 0, 1) and (1, 1, 0); therefore, the 6-WTA and 1-WTA are obtained, respectively.

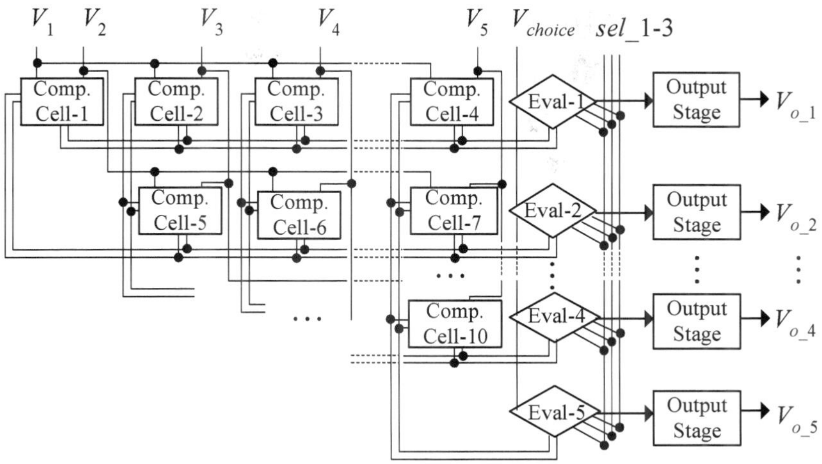

Fig. 18. The architecture of arbitrary rank-order extractor for five input variables.

4.3 Circuit design

4.3.1 1.2-V comparator

Comparator is a key element in Fig. 18. An auto-zero comparator shown in Fig. 19 is designed to operate at low voltage supply. To improve the speed of the comparator, the succeeding gain stage is designed to operate in dynamic mode. First, in the auto-zero phase, the input V_1 is sampled at the top plate of the capacitor C_s, and the MOS transistor M11 is biased at V_{bias} voltage. In next phase, the voltage at node E is $V_{bias}+(V_2-V_1)(C_s/C_s+C_p)$ during the comparison phase. Then, a deviation voltage is amplified by transistors M11 and M12. To reduce the power dissipation, the adjustable biasing voltage V_{bias} is chosen simply to overcome the threshold voltage of a MOS transistor, and the biasing voltage is also adjusted for the comparator operation in different voltage supplies. The succeeding transistors M13 and M14 provide the current to generate the proper voltage at node F. Depending on which input voltage is larger, either the voltage at node H or node G will be at logic high. The output node G of the comparator and its complementary node H are fed into next stage to generate unit currents I_{large_1}, I_{large_2}, I_{small_1}, and I_{small_2}. During the evaluation phase, the unit currents I_{large_1} and I_{large_2} will be presented when V_1 is larger than V_2. Otherwise, the I_{small_1}, I_{small_2} are generated. The symbol representation of the comparator cell is shown in the right-bottom of Fig. 19.

The function of the comparator shown in Fig. 19 is summarized as

$$V_1 > V_2, \quad \begin{cases} I_{large\_1} = I_{large\_2} = I_{unit} \\ I_{small\_1} = I_{small\_2} = 0 \end{cases},$$

$$V_1 < V_2\,, \quad \begin{cases} I_{large\_1} = I_{large\_2} = 0 \,, \\ I_{small\_1} = I_{small\_2} = I_{unit} \end{cases}$$

where I_{unit} is the unit current of the PMOS transistor M_{base}.

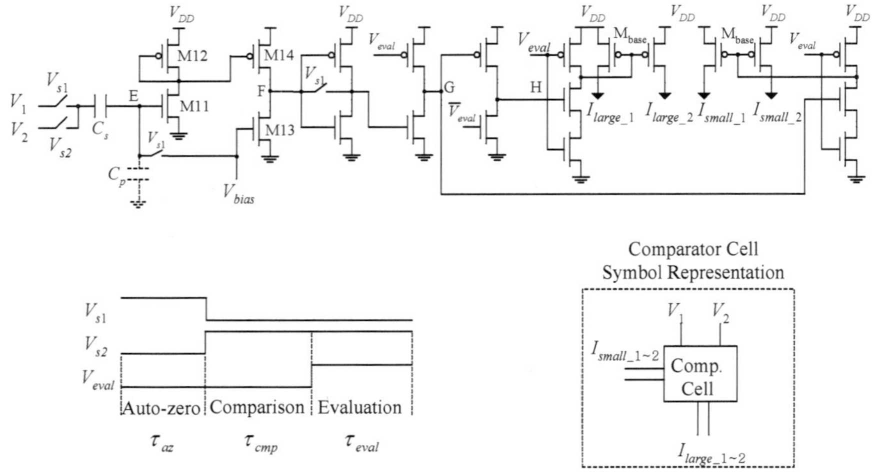

Fig. 19. 1.2-V auto-zero comparator, clock, and symbol representation.

4.3.2 Evaluation cell

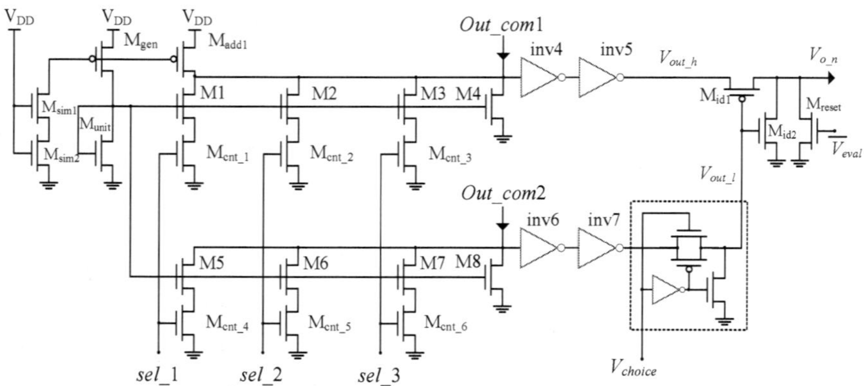

Fig. 20. Evaluation cell.

The circuit of the evaluation cell is shown in Fig. 20. The MOS transistors M_{gen} and M_{unit} reproduce the same unit current. The unit current is equal to the I_{large_1}, I_{large_2}, I_{small_1}, and I_{small_2} in Fig. 19. In order to find the various rank orders for all input signals, the cell must identify that the unit-current summation in (7) comes from *Out_com*1 and *Out_com*2 terminals. It is not easy to identify the exact current value in the VLSI circuit. However, whether the summation current S_i^* lies inside a valid range or not can be checked by the criterion,

$$nI_{unit} - \delta_1 < S_i^* < nI_{unit} + \delta_2 \,. \tag{9}$$

It is a reasonable and safe design to choose $\delta_1 = \delta_2 = I_{unit}/2$. Therefore, the dimensions of these MOS transistors are designed as

$$(\frac{W}{L})_{M_1} = (\frac{W}{L})_{M_5} = 4(\frac{W}{L})_{M_{unit}}, \quad (\frac{W}{L})_{M_2} = (\frac{W}{L})_{M_6} = 2(\frac{W}{L})_{M_{unit}}$$

$$(\frac{W}{L})_{M_3} = (\frac{W}{L})_{M_7} = (\frac{W}{L})_{M_{unit}}, \quad (\frac{W}{L})_{M_4} = (\frac{W}{L})_{M_8} = \frac{1}{2}(\frac{W}{L})_{M_{unit}}$$

where W is a channel width and L is a channel length. MOS transistors M_{add1} and M4 realize the δ_2 effect, and the M8 realizes the $-\delta_1$ one. Depending on the *sel*_1-3 signals setting, the transistors $M_{cnt\_1\text{-}6}$ enable the corresponding binary-weight current. The inverters *inv*4-7 support sufficient gain to amplify the current difference between the currents which come from *Out_com*1-2 terminals and the binary-weight currents. This mechanism is similar to a current comparator. In the upper row of Fig. 20, the extra PMOS transistor M_{add1} generates an extra unit current; therefore, the voltage $V_{out\text{-}h}$ is always larger or equal to $V_{out\text{-}l}$. If the V_{choice} is preset to 0, the dash block in Fig. 20 resets the $V_{out\text{-}l}$ to 0. Then the effect of lower row in Fig. 20 is disabled. At this time, the function of the cell resembles performing only the

$$S_i^* < nI_{unit} + \delta_2 \,. \tag{10}$$

Thus, this is a *k*-WTA criterion.

Take an example to describe the function of the evaluation cell. The number of input variables is seven, and the *sel*_1-3 signals are set as (0, 0, 1) to find the next minimum input variable. Since the next minimum is only larger than the minimum one, only a single unit current comes from *Out_com*1-2 terminals of the corresponding evaluation cell. In the upper row of Fig. 20, the summation of one unit current and the extra unit current (M_{add1}) is larger than binary weight current $1.5I_{unit}$; therefore, V_{out_h} is logic 1. In contrast with the upper row, in the lower row the unit current I_{unit} (which comes from *Out_com*1-2 terminals) is smaller than the binary weight current $1.5I_{unit}$; therefore, V_{out_l} is logic 0. Thus, the transistors M_{id1} and M_{id2} only allow the situation (V_{out_h}, V_{out_l})= (1, 0) to pull up the corresponding output (V_{o_n}, n=1, ..., 7) to logic 1. Otherwise, the status of V_{o_n} will be logic 0 or open state for other cases. Therefore, by inspecting the logic state of V_{o_n}, it is found which input variable belongs to this desired rank order.

4.4 Measured results and design consideration

A seven-input experimental chip was fabricated using a 0.5 μm CMOS technology. Bias voltage V_{bias} is set to 0.9 V in this design. The sampling capacitor C_s is 0.8 pF, and these analog switches in this circuit are implemented by CMOS transmission gates. The micrograph of the experimental chip is shown in Fig. 21, and the active area is 610 × 780 μm². An individual comparator cell was built in this chip for measuring the accuracy. The supply voltages of the core circuit and the input/output pads were all set as 1.2 V. The accuracy of the individual comparator was measured roughly as 40 mV, that is, the resolution of the comparator was near five bits under a 1.2 V supply voltage. Figure 22(a)

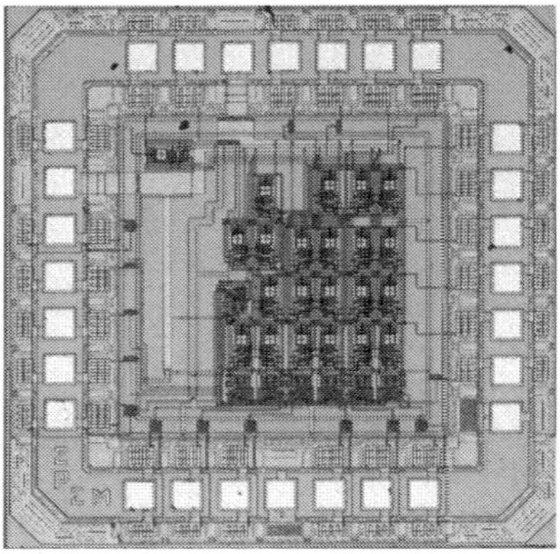

Fig. 21. Micrograph of the 1.2-V rank-order chip.

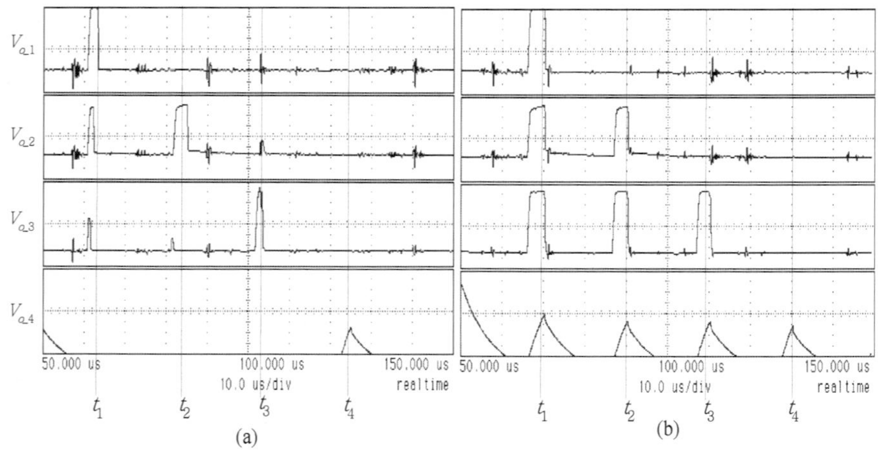

Fig. 22. The measurement results of (a) rank-order (b) *k*-WTA operations.

shows the rank-order function, whereas Fig. 22(b) shows the function of the *k*-WTA. On the average, the accuracy of whole circuit was approximated 150 mV. The performance of the chip was degraded by many factors such as the mismatch in comparator cells, the different capacitance at input terminals of the evaluation cells, and the clock feed-through error. Due to these non-ideal effects, each rank-order function was finished in 20 μs. After increasing supply voltage up to 1.5 V and proper biasing voltage V_{bias} adjusting, the performance of the circuit can be improved. Including power consumption of the input/output pads, the static power consumption of the chip was 1.4 mW.

Many factors such as precision, speed, process variation, and chip area must be considered for design of a low-power low-voltage rank order extractor.

1. Limitations of low voltage and low power

The average power consumption of the circuit is expressed by

$$P = P_{dynamic} + P_{static} + P_{short\_current}$$

$$= f\,C\,V_{DD}^2 + (I_o + I_{leakage})V_{DD} + Q_{sc}\,f\,V_{DD} \quad (11)$$

where f is the frequency, C is the capacitance in the circuit, V_{DD} is the voltage supply, I_o is the standby current, $I_{leakage}$ is the leakage current, and the Q_{sc} is the short-current charge during the clock transient period. In order to reduce the power consumption, the voltage supply V_{DD} must be reduced, and the standby current in the comparator and evaluation cell must be designed as small as possible. In mask layout, the clock and its complementary are generated locally to reduce delay and mismatch. Thus, the probability of a short current occurring in the circuit is minimized.

2. Speed and precision

The accuracy of the comparators determines the resolution of the circuit. For the comparator design, the smallest differential voltage, that is, distinguished correctly is influenced by two factors. One is the charge-injection error in analog switches, and the other is the parasitic capacitor C_p effect. The effect is reduced by enlarging the sampling capacitor C_s and making the switches dimension as small as possible. In the design, the response time τ of the extractor is the summation of the auto-zero time τ_{az}, the comparison time τ_{cmp}, and the evaluation time τ_{eval}.

$$\tau = \tau_{az} + \tau_{cmp} + \tau_{eval} \quad (12)$$

Reducing τ_{az}, τ_{cmp} and τ_{eval} will improve the response time τ. The minimum auto-zero time τ_{az} is required to sample the input voltage correctly at sampling capacitor C_s and to bias the inverter properly at high gain region. The switches shown in Fig. 19 with larger dimension reduce auto-zero time τ_{az}. However, the clock feed-through error and charge injection error will also be enlarged during the clock transition. In the same situation, the smaller sample capacitor C_s will reduce the time τ_{az}. Unfortunately, it will reduce the effective magnitude of the difference voltage; thus, the comparator accuracy is degraded. The comparison time τ_{cmp} dominates the response time τ, especially when the input levels are close each other. Since the amplification in the transition region of a CMOS inverter operated at low voltage supply is not high enough, the comparator must take a long time to identify which input variable has a larger level. The evaluation time τ_{eval} is defined so that the time interval between the comparator cells generates the proper currents and the extractor has finished finding the desired rank order. Time τ_{eval} is a function of the current I_{unit}. The maximum number M of input variables is also influenced by the current I_{unit}. Although reducing the magnitude of the current I_{unit} is able to reduce the power consumption, however, the relationship among τ_{eval}, I_{unit}, and M in this architecture is a complicated function.

3. Process variation analysis

With contemporary technology, process variation during fabrication cannot be completely eliminated; as a result, mismatch error must be noticed in VLSI circuit design. The match in dimension of the binary-weight MOS in the evaluation cell (M1 - M8 in Fig. 20) is an important factor for the circuit operation. If the mismatch error induces an error current I_{err} larger (or smaller) than half of the unit current I_{unit}, decision of the evaluation cell fails. Thus, a rough estimated constraint for I_{err} is

$$I_{err} < I_{unit} / 2 . \tag{13}$$

5. Conclusion

The chapter describes various nonlinear signal processing CMOS circuits, including a high reliable WTA/LTA, simple MED cell, and low-voltage arbitrary order extractor. We focus the discussion on CMOS analog circuit design with reliable, programmable capability, and low voltage operation. It is a practical problem when the multiple identical cells are required to match and realized within a single chip using a conventional process. Thus, the design of high-reliable circuit is indeed needed. The low-voltage operation is also an important design issue when the CMOS process scale-down further. In the chapter, Section 1 introduces various CMOS nonlinear function and related applications. Section 2 describes design of highly reliable WTA/LTA circuit by using single analog comparator. The analog comparator itself has auto-zero characteristic to improve the overall reliability. Section 3 describes a simple analog MED cell. Section 4 presents a low-voltage rank order extractor with *k*-WTA function. The flexible and programmable functions are useful features when the nonlinear circuit will integrate with other systems. Depend on various application requirements, we must have different design strategies for design of these nonlinear signal process circuits to achieve the optimum performance. In state-of-the-art process, small chip area, low-voltage operation, low-power consumption, high reliable concern, and programmable capability still have been important factors for these circuit realizations.

6. References

Aksin, D. Y. (2002). A high-precision high-resolution WTA-MAX circuit of O(N) complexity. *IEEE Trans. Circuits Syst. II, Analog Digit. Signal Process.*, vol. 49, no. 1, 2002, pp. 48–53.

Cilingiroglu, U. & Dake, L. E. (2002). Rank-order filter design with a sampled-analog multiple-winners-take-all core. *IEEE J. Solid-State Circuits,* vol. 37, Aug. 2002, pp. 978 – 984.

Demosthenous, A.; Smedley, S. & Taylor, J. (1998). A CMOS analog winner-take-all network for large-scale applications. *IEEE Trans. Circuits Syst. I, Fundam. Theory Appl.*, vol. 45, no. 3, 1998, pp. 300–304.

Diaz-Sanchez, A.; Jaime Ramirez-Angulo; Lopez-Martin, A. & Sanchez-Sinencio, E. (2004). A fully parallel CMOS analog median filter. *IEEE Trans. Circuits Syst. II*, vol. 51, March 2004, pp. 116 - 123.

He, Y. & Sanchez-Sinencio, E. (1993). Min-net winner-take-all CMOS implementation. *Electron. Lett.*, vol. 29, no. 14, 1993, pp. 1237–1239.

Hosotani, S.; Miki, T.; Maeda, A. & Yazawa, N. (1990). An 8-bit 20-MS/s CMOS A/D converter with 50-mW power consumption. *IEEE J. Solid-State Circuits*, vol. 25, no. 1, Feb. 1990, pp. 167-172.

Hung, Y.-C. & Liu, B.-D. (2002). A 1.2-V rail-to-rail analog CMOS rank-order filter with *k*-WTA capability. *Analog Integr. Circuits Signal Process.*, vol. 32, no. 3, Sept. 2002, pp. 219-230.

Hung, Y.-C. & Liu, B.-D. (2004). A high-reliability programmable CMOS WTA/LTA circuit of *O(N)* complexity using a single comparator. *IEE Proc. – Circuits Devices and Syst.*, vol. 151, Dec. 2004, pp. 579-586.

Hung, Y.-C.; Shieh, S.-H. & Tung, C.-K. (2007). A real-time current-mode CMOS analog median filtering cell for system-on-chip applications. *Proceedings of IEEE Conference on Electron Devices and Solid-State Circuits* (*EDSSC*), pp. 361 - 364, Dec. 2007, Tainan, Taiwan.

Lazzaro, J.; Ryckebusch, R.; Mahowald, M. A. & Mead, C. A. (1989). Winner-take-all networks of *O(N)* complexity. *Advances in Neural Inform. Processing Syst.*, vol. 1, 1989, pp. 703-711.

Lippmann, R. (1987). An introduction to computing with neural nets. *IEEE Acoust., Speech, Signal Processing Mag.*, vol. 4, no. 2, Apr. 1987, pp. 4-22.

Opris, I. E. & Kovacs, G. T. A. (1994). Analogue median circuit. *Electron. Lett.*, vol. 30, no. 17, Aug. 1994, pp. 1369-1370.

Opris, I. E. & Kovacs, G. T. A. (1997). A high-speed median circuit. *IEEE J. Solid-State Circuits*, vol. 32, June 1997, pp. 905-908.

Semiconductor Industry Association. (2008). International technology roadmap for semiconductors 2008 update. [Online]. Available: http://public.itrs.net/.

Smedley, S.; Taylor, J. & Wilby, M. (1995). A scalable high-speed current mode winner-take-all network for VLSI neural applications. *IEEE Trans. Circuits Syst. I, Fundam. Theory Appl.*, vol. 42, no. 5, 1995, pp. 289–291.

Starzyk, J.A. & Fang, X. (1993). CMOS current mode winner-take-all circuit with both excitatory and inhibitory feedback. *Electron. Lett.*, vol. 29, no. 10, 1993, pp. 908–910.

Vlassis, S. & Siskos, S. (1999). CMOS analogue median circuit. *Electron. Lett.*, vol. 35, no. 13, June 1999, pp. 1038-1040.

Yamakawa, T. (1993). A fuzzy inference engine in nonlinear analog mode and its applications to a fuzzy logic control. *IEEE Trans. Neural Netw.*, vol. 4, no. 3, May 1993, pp. 496–522.

Yuan, J. & Stensson, C. (1989). High - speed CMOS circuit technique. *IEEE J. Solid-State Circuits*, vol. 24, no. 1, Feb. 1989, pp. 62-69.

2

Transconductor

Ko-Chi Kuo
Department of Computer Science and Engineering, National Sun Yat-sen University Kaohsiung, Taiwan

1. Introduction

The transconductor is a versatile building block employed in many analog and mixed-signal circuit applications, such as continuous-time filters, delta-sigma modulators, variable gain-amplifier or data converter. The transconductor is to perform voltage-to-current conversion. Linearity is one of most critical requirements in designing transconductor. Especially in designing delta-sigma modulators for high resolution Analog/Digital converters, it needs high linearity transconductors to accomplish the required signal-to-(noise+distortions) ratio. The tuning ability of transconductor is also mandated to adjust center frequency and quality factor in filter applications.

The portable electronic equipments are the trend in comsumer markets. Therefore, the low power consumption and low supply voltage becomes the major challenge in designing CMOS VLSI circuitry. However, designing for low-voltage and highly linear transconductor, it requires to consider many factors. The first factor is the linear input range. The range of linear input is justified by the constant transconductance, G_m. Since the distortion of transconductor is determined by the ratio of output currents versus input voltage. The second factor is the control voltage of transconductor. This voltage can greatly impact the value of transconductance, linear range, and power consumption. For example, when the control voltage increases, the transconductance also increase but the linear input range of transconductor is reduced and power consumption is increased. Hence it is critical in designing transconducotr operated at low supply voltage. The third factor is the symmetry of the two differential outputs. If the transconductance of the positive and negative output is $G_{m+}=I_{O+}/V_i$ and $G_{m-}=I_{O-}/V_i$, then how close G_{m+} and G_{m-} should be is a critical issue, where I_{O+} is the positive output current, I_{O-} is the negative output current, and V_i is the input differential voltage. This factor is the major cause of common-mode distortion of transconductor which occurs at outputs.

In general, the design of differential transconductor can be classified into triode-mode and saturation-mode methods depending on operation regions of input transistors. Triode-mode transconductor has a better linearity as well as single-ended performance. On the other hand, saturation-mode transconductor has better speed performance. However, it only exhibits moderate linearity performance. Furthermore, the single-ended transconductor of saturation-mode suffers from significant degradation of linearity. Several circuit design techniques for improving the linearity of transconductors have been reported in literatures. The linearization methods include: source degeneration using resistors or MOS transistors

[Krummenacher & Joeh, 1988; Leuciuc & Zhang, 2002; Leuciuc, 2003; Furth & Andreou, 1995], crossing-coupling of multiple differential pairs [Nedungadi & Viswanathan, 1984; Seevinck & Wassenaar, 1987] class-AB configuration [Laguna et al., 2004; Elwan et al., 2000; Galan et al., 2002], adaptive biasing [Degrauwe et al., 1982; Ismail & Soliman, 2000; Sengupta, 2005], constant drain-source voltages [Kim et al., 2004; Fayed & Ismail, 2005; Mahattanakul & Toumazou, 1998; Zeki, 1999; Torralba et al., 2002; Lee et al., 1994; Likittanapong et al., 1998], pseudo differential stages [Gharbiya & Syrzycki, 2002], and shift level biasing [Wang & Guggenbuhl, 1990].

Source degeneration using resistors or MOS transistors is the simplest method to linearize transconductor. However, it requires a large resistor to achieve a wide linear input range. In addition, MOS used as resistor exhibits considerable varitions affected by process and temperture and results in the linearity degradation. Crossing-coupling with multiple differential pairs is designed only for the balanced input signals. The Class-AB configuration can achieve low power consumption. On the other hand, the linearity is the worst due to the inherited Class-AB structure. The adaptive biasing method generates a tail current which is proportional to the square of input differential voltage to compensate the distortion caused by input devices. However, the complication of square circuitry makes this technique hard to implement. The constant drain-source voltage of input devices is a simple structure. It can achieve a better linearity with tuning ability. However, it needs to maintain V_{DS} of input devices in low voltage and triode region. Therefore, this technique is difficult to implement in low supply voltage. Hence, a new transconductor using constant drain-source voltage in low voltage application is proposed to achieve low-voltage, highly linear, and large tuning range abilities.

In section 2, basic operatrion and disadvantage of the linerization techniques are described. The proposed new transconductor is presented in section 3. The simulation results and conclusion are given in section 4 and 5.

2. Linearization techniques

In this section, reviews of common linearization techniques reported in literatures are presented. The first one is the transconductor using constant drain-source voltage. The second one is using regulated cascode to replace the auxiliary amplifier. The third one is transconductor with source degeneration by using resistors and MOS transistors. The last one is the linear MOS transconductor with a adaptive biasing scheme. Besides introducing their theories and analyses, the advantages and disadvantages of these linearization techniques are also discussed.

2.1 Transconductor using constant drain-source voltage

The idea of transconductors using constant drain-source voltages is to keep the input devices in triode region such that the output current is linearized. The schematic of this method is shown in Fig. 1. Considering that transistors M_1, M_2 operate at triode region, M_3, M_4 are biased at saturation region, channel length modulation, body effect, and other second-order effects are ignored, the drain current of M_1 and M_2 is given by

$$I_D = \beta\left[(V_{GS} - V_T)V_{DS} - \frac{V_{DS}^2}{2}\right] \tag{1}$$

where $\beta = \mu_n C_{OX}(W/L)$, V_{GS} is the gate-to-source voltage, V_T is the threshold voltage, and V_{DS} is the drain-to-source voltage. If the two amplifiers in Fig. 1 are ideal amplifiers, then

$$V_{DS1} = V_{DS2} = V_C \tag{2}$$

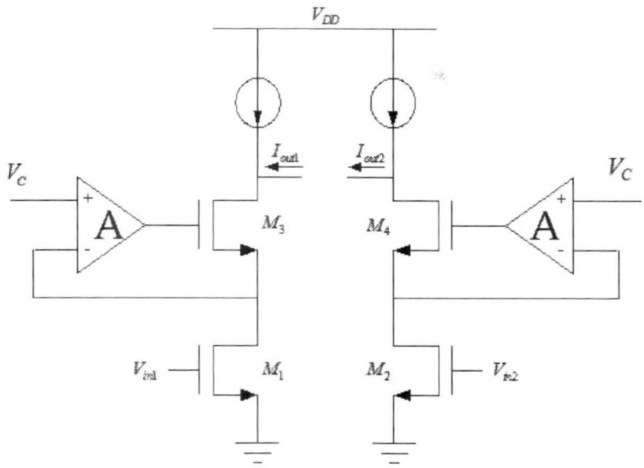

Fig. 1. Transconductor using constant drain-source voltage

The transfer characteristic of this transconductor is given by

$$I_{out1} = \beta\left[(V_{GS1} - V_T)V_{DS1} - \frac{V_{DS1}^2}{2}\right] = \beta\left[(V_{GS1} - V_T)V_C - \frac{V_C^2}{2}\right]$$

$$I_{out2} = \beta\left[(V_{GS2} - V_T)V_{DS2} - \frac{V_{DS2}^2}{2}\right] = \beta\left[(V_{GS2} - V_T)V_C - \frac{V_C^2}{2}\right]$$

$$I_{out} = I_{out1} - I_{out2} = \beta V_C (V_{in1} - V_{in2}) \tag{3}$$

The transconductance value is

$$G_m = \beta V_C \tag{4}$$

In fact, it is difficult to design an ideal amplifier implemented in this circuits. However, it can force $V_{DS1} = V_{DS2} = V_{DS}$ by using two auxiliary amplifiers controlled with the same V_C to keep V_{DS} at the constant value. Therefore, the transfer characteristic of this transconductor is changed as follows:

$$I_{out1} = \beta\left[(V_{GS1} - V_T)V_{DS1} - \frac{V_{DS1}^2}{2}\right] = \beta\left[(V_{GS1} - V_T)V_{DS} - \frac{V_{DS}^2}{2}\right]$$

$$I_{out2} = \beta\left[(V_{GS2} - V_T)V_{DS2} - \frac{V_{DS2}^2}{2}\right] = \beta\left[(V_{GS2} - V_T)V_{DS} - \frac{V_{DS}^2}{2}\right]$$

$$I_{out} = I_{out1} - I_{out2} = \beta V_{DS}(V_{in1} - V_{in2}) \tag{5}$$

, where V_{GS1}= V_{in1} and V_{GS2}= V_{in2}.
Therefore, the new transconductance value is

$$G_m = \beta V_{DS} \tag{6}$$

The linearity of this transconductor is moderated. It is also easy to implement in circuit. However, V_{DS} of the input devices must be small enough to keep transistors in triode region. The following condition has to be satisfied:

$$V_{DS} < V_{GS} - V_T \tag{7}$$

On the other hand, the auxiliary amplifiers need to design carefully to reduce the overhead of extra area and power.

2.2 Transconductor using regulated cascode to replace auxiliary amplifier

In Fig. 2(a) regulating amplifier keeps V_{DS} of M_1 at a constant value determined by V_C. It is less than the overdrive voltage of M_1. The voltage can be controlled from V_C so as to place M_3 in current-voltage feedback, thereby increasing output impedance. The concept is to drive the gate of M_3 by an amplifier that forces V_{DS1} to be equal to V_C. Therefore, the voltage variations at the drain of M_3 affect V_{DS1} to a lesser extent because amplifiers "regulate" this voltage. With the smaller variations at V_{DS1}, the current through M_1 and hence output current remains more constant, yielding a higher output impedance [Razavi, 2001]

$$R_{out} \approx A g_{m3} r_{O3} r_{O1} \tag{8}$$

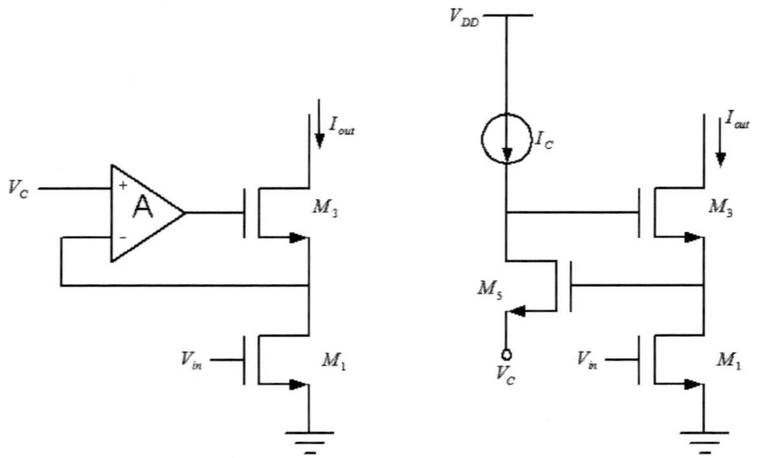

(a) (b)

Fig. 2. (a)Basic triode transconductor structure (b) Simple RGC triode transconductor

It is one of solutions using regulated cascode to replace the auxiliary amplifier in order to overcome restrictions on Fig. 1. The circuit in Fig. 2(b) proposed in [Mahattanakul & Toumazou, 1998] uses a single transistor, M_5, to replace the amplifier in Fig. 2(a). This circuit called regulated cascode which is abbreviated to RGC. The RGC uses M_5 to achieve the gain boosting by increasing the output impedance without adding more cascode devices. V_{DS1} is calculated by follows: Assuming M_5 is in saturation region in Fig. 2(b). It can be shown that

$$I_C = \frac{1}{2}\beta_5\left(V_{GS5} - V_T\right)^2$$

$$\Rightarrow V_{GS5} = V_{DS1} - V_C = \sqrt{\frac{2I_C}{\beta_5}} + V_{T5}$$

$$\Rightarrow V_{DS1} = V_C + \sqrt{\frac{2I_C}{\beta_5}} + V_{T5} \tag{9}$$

From (6) $G_m = \beta_1 V_{DS1} = \beta_1\left(V_C + \sqrt{\frac{2I_C}{\beta_5}} + V_{T5}\right)$. Thus, G_m can be tuned by using a controllable voltage source V_C or current source I_C. However, it is preferable in practice to use a controllable voltage source V_C for lowering power consumption since V_{DS1} only varies as a square root function of I_C.

Simple RGC transconductor using a single transistor to achieve gain boosting can reduce area and power wasted by the auxiliary amplifiers. However, it still has some disadvantages. First, it will cause an excessively high supply-voltage requirement and also produce an additional parasitic pole at the source of transistors. Therefore, it can not apply to the low-supply voltage design. Second, the tuning range of V_{DS1} is restricted. The smallest value of V_{DS1} is $\sqrt{\frac{2I_C}{\beta_5}} + V_T$ when $V_C = 0$. In other words, V_{DS1} can not be set to zero. Owing to the restriction of (7), V_{DS} is as low as possible and the best value is zero. Third, V_T dependent G_m may be a disadvantage due to the substrate noise and V_T mismatch problems [Lee et al., 1994].

In Fig. 3, another RGC transconductor that can apply to the low-voltages applications is proposed in [Likittanapong et al., 1998]. The circuit overcomes the disadvantages mentioned above is to utilize PMOS transistor that can operate in saturation region as gain boosting. The use of this PMOS gain boosting in the feedback path can result in a circuit with a wide transconductance tuning range even at the low supply voltage. In [Likittanapong et al., 1998], it mentions that at the maximum input voltage, M_3 may be forced to enter triode region, especially if the dimension of M_2 is not properly selected, resulting in a lower dynamic range. Besides, β_2 may be chosen to be larger for a very low distortion transconductor. It means that the tradeoff between linearity and bandwidth of transconductor is controlled by β_2. Therefore, β_2 should be selected to compromise these two characteristics for a given application.

V_{DS1} is calculated by follows. Assuming M_3 is in saturation region in Fig. 3.

$$I_C = \frac{1}{2}\beta_3 (V_{GS3} - V_{T3})^2$$

$$=> V_{GS3} = V_C - V_{DS1} = \sqrt{\frac{2I_C}{\beta_3}} + V_{T3}$$

$$=> V_{DS1} = V_C - \left(\sqrt{\frac{2I_C}{\beta_3}} + V_{T3}\right) \tag{10}$$

From (6) $G_m = \beta_1 V_{DS1} = \beta_1 \left[V_C - \left(\sqrt{\frac{2I_C}{\beta_3}} + V_{T3}\right)\right]$. It shows that V_{DS1} can be set to zero when $V_C = \sqrt{\frac{2I_C}{\beta_3}} + V_{T3}$. Therefore, this transconductor has a wider tuning range compared to that of RGC transconductor and is capable of working in low-supply voltage (3V). However, this transconductor still has some drawbacks. The major drawback is the tuning ability. For example, it is difficult to control $V_C = \sqrt{\frac{2I_C}{\beta_3}} + V_{T3}$ if V_{DS1} is set to zero. The minor drawback is that V_T depends on the G_m. It also may cause substrate noise and V_T mismatch problems [Lee et al., 1994].

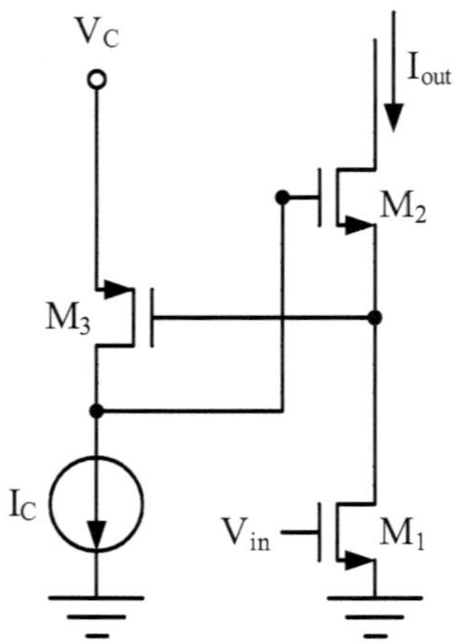

Fig. 3. RGC transconductor with PMOS gain stage

2.3 Transconductor using source degeneration

A simple differential transconductor is shown in Fig. 4(a). Assuming that M_1 and M_2 are in saturation and perfectly matched, the drain current is given by

$$I_D = \frac{\beta}{2}(V_{GS} - V_T)^2 \tag{11}$$

The transfer characteristic using (5) is given by

$$I_{out} = I_{out1} - I_{out2} = \sqrt{2\beta I_{SS}}\, V_i \sqrt{1 - \frac{\beta V_i^2}{8 I_{SS}}} = \sqrt{2\beta I_{SS}}\, V_i \sqrt{1 - \frac{V_i^2}{4(V_{GS} - V_T)}} \tag{12}$$

, where $V_i = (V_{in1} - V_{in2})$

If V_{GS} is large enough, the higher linearity can be achieved. Unfortunately, it can not be used in the low-voltage application and the linear input range is limited. Simplest techniques to linearize the transfer characteristic of MOS transconductor is the one with source degeneration using resistors as shows in Fig. 4(b). The circuit is described by

$$V_i - RI_{out} = V_{GS1} - V_{GS2} \tag{13}$$

A transfer characteristic derived from (13) is given by

$$I_{out} = \sqrt{2\beta I_{SS}}\left(V_i - RI_{out}\right)\sqrt{1 - \frac{\beta\left(V_i - RI_{out}\right)^2}{8 I_{SS}}} \tag{14}$$

The transconductance G_m is

$$G_m \approx \frac{g_m}{1 + g_m R} \tag{15}$$

where g_m is the transconductance of transistor M_1 and M_2.

We should notice that in (14), the nonlinear term depends on $V_i - RI_{out}$ rather than V_i. Higher linearity can be achieved when $R >> 1/g_m$. The disadvantage of this transconductor is that large resistor value is needed in order to maintain a wider linear input range. Owing to $G_m \approx 1/R$, the higher transconductance is limited by the smaller resistor. Hence, there is a tradeoff between wide linear input range and higher transconductance which is mainly determined by a resistor.

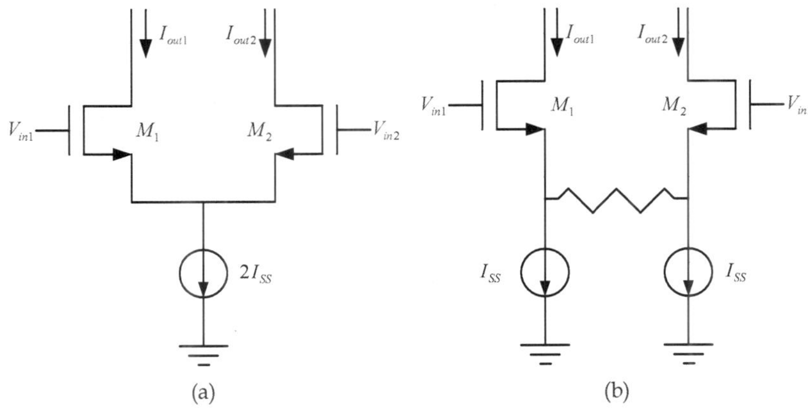

Fig. 4. (a) Simple differential MOS transconductor (b) MOS transconductor with resistive source degeneration

Another method to linearize the transfer characteristic of MOS transconductor is using source degeneration to replace the degeneration resistor with two MOS transistors operating in triode region. The circuit is shown in Fig. 5. Notice that the gates of transistor M_3 and M_4 connect to the differential input voltage rather than to a bias voltage. To see that M_3 and M_4 are generally in triode region, we look at the case of the equal input signals ($V_{in1}=V_{in2}$), resulting in

$$V_x = V_y = V_{in1} - V_{GS1} \tag{16}$$

Therefore, the drain-source voltages of M_3 and M_4 are zero. However, V_{DS} of M_3 and M_4 equal those of M_1 and M_2. Owing to (7), M_3 and M_4 are indeed in triode region. Assuming M_3, M_4 are operating in triode region, the small-signal drain-source resistance of M_3, M_4 is given by

$$r_{ds3} = r_{ds4} = \frac{1}{\beta_3 (V_{GS1} - V_T)} \tag{17}$$

It must be noted that in this circuit the effect of varying V_{DS} of M_1 and M_2 can not be ignored since the drain currents are not fixed to a constant value. The small-signal source resistance of M_1, M_2 is given by

$$r_{s1} = r_{s2} = \frac{1}{g_{m1}} = \frac{1}{\beta_1 (V_{GS1} - V_{T1})} \tag{18}$$

Using small-signal T model, the small-signal output current, i_{o1}, is equal to

$$\begin{aligned} i_{o1} &= \frac{V_{in1} - V_{in2}}{r_{s1} + r_{s2} + (r_{ds3} \,||\, r_{ds4})} \\ => i_{o1} &= \frac{2\beta_1 \beta_3}{\beta_1 + 4\beta_3} (V_{GS1} - V_{T1})(V_{in1} - V_{in2}) \end{aligned} \tag{19}$$

Assuming M_1 is in saturation region, the drain current of M_1 is given by

$$\begin{aligned} I_{SS} &= \frac{1}{2} \beta_1 (V_{GS1} - V_{T1})^2 \\ => (V_{GS1} - V_{T1}) &= \sqrt{\frac{2 I_{SS}}{\beta_1}} \end{aligned} \tag{20}$$

Using (20) substitutes for (19), that leads to

$$i_{o1} = \frac{2\beta_1 \beta_3}{\beta_1 + 4\beta_3} \sqrt{\frac{2 I_{SS}}{\beta_1}} (V_{in1} - V_{in2}) \tag{21}$$

The transconductance G_m is

$$G_m = \frac{2\beta_1 \beta_3}{\beta_1 + 4\beta_3} \sqrt{\frac{2 I_{SS}}{\beta_1}} \tag{22}$$

Linearity can be enhanced (assuming r_{ds3} >> r_{s1}) compared to that of a simple differential pair because transistors operated in triode region exhibits higher linearity than the source resistances of transistors operated in saturation region. When the input signal is increased, the small-signal resistance in one of two triode transistors in parallel, M_3 or M_4, is reduced. Meanwhile, the reduced resistance results in the lower linearity and the larger transconductance. As discussed in [Krummenacher & Joeh, 1988], if the proper size ratio of β_1/β_3 is chosen, the balance between higher linearity and stable transconductance can be achieved. How to choose the optimum size ratio of β_1/β_3 for the best linearity performance becomes slightly dependent on the quiescent overdrive voltage, $V_{GS}-V_T$. The size ratio of β_1/β_3=6.7 is used to achieve the best linearity performance.
According to (22), the transconductance can be tuned by changing I_{SS} and size ratio of β_1/β_3. Nevertheless, the nonlinearity error is up to 1% for I_{out}/I_{SS} < 80%. It is required to have a better linearity so as to achieve a THD of -60 dB or less in some filtering applications [Kuo & Leuciuc, 2001].

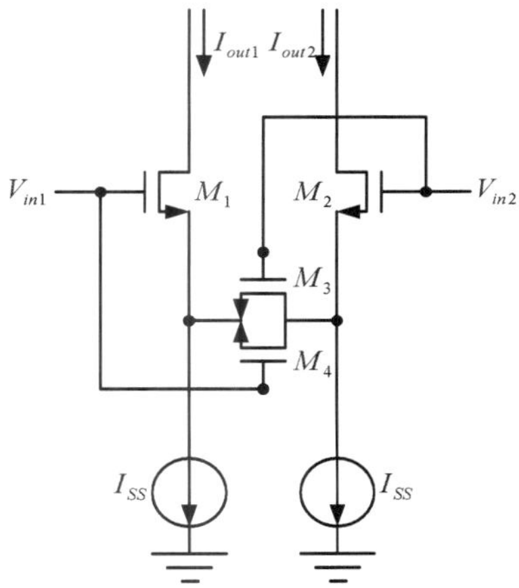

Fig. 5. Transconductor with source degeneration using MOS transistors

2.4 Transconductor using adaptive biasing

The transconductor using adaptive biasing is shown in Fig. 6. All transistors are assumed to be operated in saturation region, neglecting channel lengh modulation effect. First, transistor M_3 is absent, and output current as a function of two input voltages V_{in1} and V_{in2} is obtained as

$$I_1 = \frac{\beta}{2}(V_{GS1} - V_T)^2$$

$$I_2 = \frac{\beta}{2}(V_{GS2} - V_T)^2 \tag{23}$$

$$=> I_{out} = I_1 - I_2 = \sqrt{\beta I_{SS}}(V_{in1} - V_{in2})\sqrt{1 - \frac{\beta(V_{in1} - V_{in2})^2}{4I_{SS}}}$$

, where I_{SS} is a tail current and equals I_B.
An adaptive biasing technique is using a tail current containing an input dependent quadratic component to cancel the nonlinear term in (23). Consequently, the circuit in Fig. 6 changes the tail current by adding transistor M_3. The tail current will be changed by

$$I_{SS} = I_B + I_C \tag{24}$$

$$I_C = \frac{\beta}{4}(V_{in1} - V_{in2})^2 \tag{25}$$

, where I_B is tail current of differential pair and I_C is the compensating tail current that cancel nonlinear term.
Therefore, the transfer characteristic is changed by

$$I_{out} = \sqrt{\beta I_{SS}}(V_{in1} - V_{in2}) \tag{26}$$

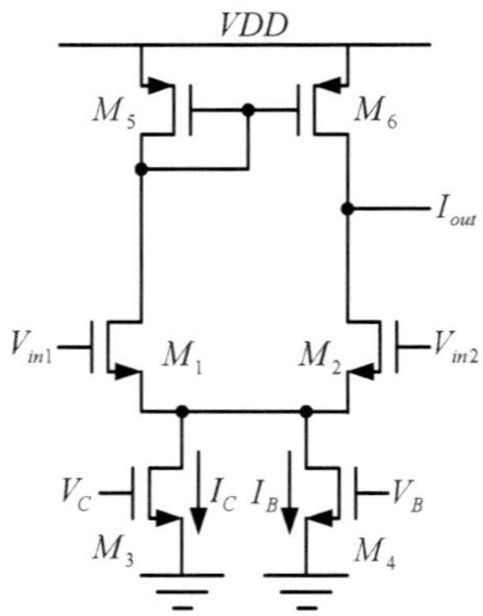

Fig. 6. Transconductor with adaptive biasing

3. New transconductor

The conventional structure which uses the constant drain source-voltage such as RGC with NMOS or PMOS can not operate at 1.8V or below. The main reason is that auxiliary amplifier under the low supply voltage can't provide enough gain to keep the constant drain-source voltage. Therefore, we propose a triode transconductor which uses new structure to replace the auxiliary amplifier. Fig. 7 shows the proposed triode transconductor structure.
MOS M_5, M_7, M_9 and M_{11} are made up a two-stage amplifier to replace the auxiliary amplifier. The two-stage amplifier is implemented using M_9 with the active loads M_{11} formed the first stage and M_5 with the active load M_7 formed the second stage. The first and second stages exhibit gains equal to

$$A_1 = gm_9(g_{m9}^{-1} || r_{O11}) \tag{27}$$

$$A_2 = gm_5(r_{O5} || r_{O7}) \tag{28}$$

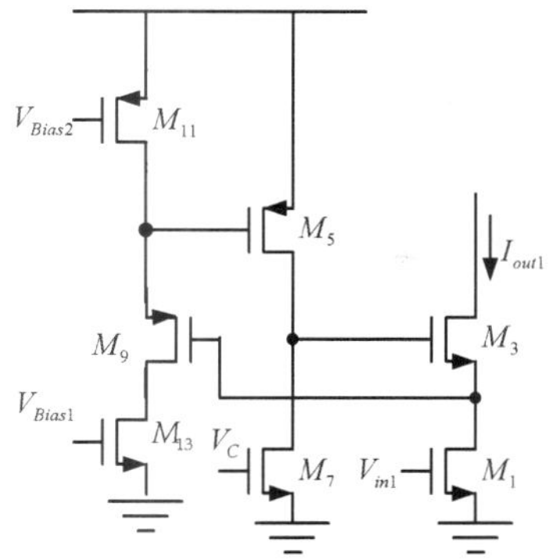

Fig. 7. Proposed triode transconductor

Therefore, the overall gain is

$$A_v = A_1 * A_2 = gm_9\left(g_{m9}^{-1} \,||\, r_{O11}\right)gm_5\left(r_{O5} \,||\, r_{O7}\right) \quad (29)$$

The proposed transconductor is shown in Fig. 8.

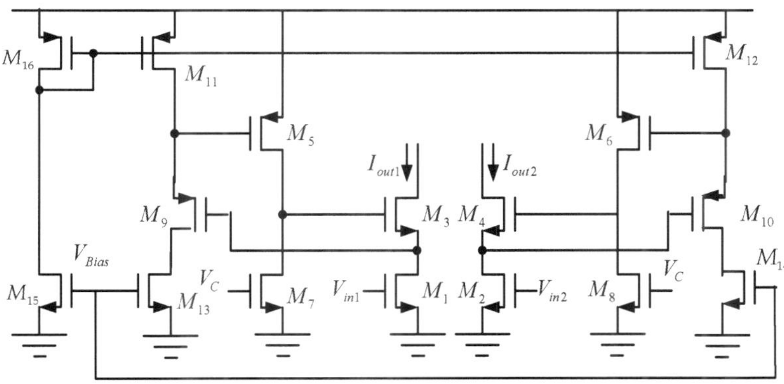

Fig. 8. The proposed transconductor

Considering that the large gain is achieved and is able to keep transistors M_1 and M_2 in triode region, the drain current of M_1 and M_2 is given by

$$_o \quad \beta_u \left[\left(_{G\,t} \quad _S \right) V_{D\,11} - \frac{V_{D1}^2}{2} \right] \quad S \qquad ($$

$$_o = \beta_u \left[\left(_{G\,t} \quad _S \right) V_{D\,22} - \frac{V_{D2}^2}{2} \right] \quad S \qquad ($$

T h e t r a n s f e

$$I_{out} = I_{out1} - I_{out2} = \beta_1 V_{DS1}(V_{in1} - V_{in2}) \tag{32}$$

, where $\beta_1 = \beta_2$, $V_{T1} = V_{T2}$, and $V_{DS1} = V_{DS2}$. Assuming that current I_9 flows from M_{11} through M_9 and MOS M_9 is in saturation region, V_{DS1} can be found in (33)

$$V_{GS3} + V_{DS1} = V_{DS7}$$

$$V_C - V_{T7} = V_{DS7}$$

$$=> V_{GS3} + V_{DS1} = V_C - V_{T7}$$

$$V_{DS1} = V_C - V_{T7} - V_{GS3} \tag{33}$$

According to (32)

$$I_{out} = \beta_1 V_{DS1}(V_{in1} - V_{in2}) = \beta_1 (V_C - V_{T7} - V_{GS3})(V_{in1} - V_{in2}) \tag{34}$$

The transconductance G_m is

$$G_m = \beta_1 (V_C - V_{T7} - V_{GS3}) \tag{35}$$

From (35), the transconductance can be tuned by control voltage V_C To keep M_1 and M_2 in triode region, the relation (36) needs to be satisfied.

$$V_{DS1} < V_{GS1} - V_{T1} \tag{36}$$

Using (33) to substitute (36)

$$V_C - V_{T7} - V_{GS3} \leq V_{GS1} - V_{T1} \;=>\; V_C \leq V_{GS1} + V_{GS3} - (V_{T1} - V_{T7}) \tag{37}$$

The proposed transconductor is suitable for low supply voltage and we choose 1.8V to achieve a wide linear range. Moreover, M_9 is needed to obtain a negative feedback to keep the drain-source voltage of M_1, V_{DS1}, constant. This new structure can provide enough gain to keep V_{DS1} constant at 1.8V supply voltage. It has a low control voltage V_C between 0.69V~0.72V and the large transconductance tuning range depending on applications. Besides, it has a simple structure so as to save area.

4. Simulation results

The circuits in Fig. 8 have been designed by using TSMC CMOS 0.18μm process with a single 1.8V supply voltage and simulated by Hspice. Fig. 9. shows the curve of input voltage transferring to the output current at V_C = 0.7V. The slope of the curve is linear when the input voltage varies from −1V to 1V. The slope in Fig. 9. is equal to the transconductance in Fig. 10. In order to verify the performance of the proposed transconductor, we define transconductance error (equation 39) as the linearity of the transconductance's output current. The transconductance error is less than 1% among ±0.9V input voltage, so the input linear range is up to 1.8V.

$$TE(\%) = \frac{G_m(V_{id}) - G_m(0)}{G_m(0)} * 100 \tag{39}$$

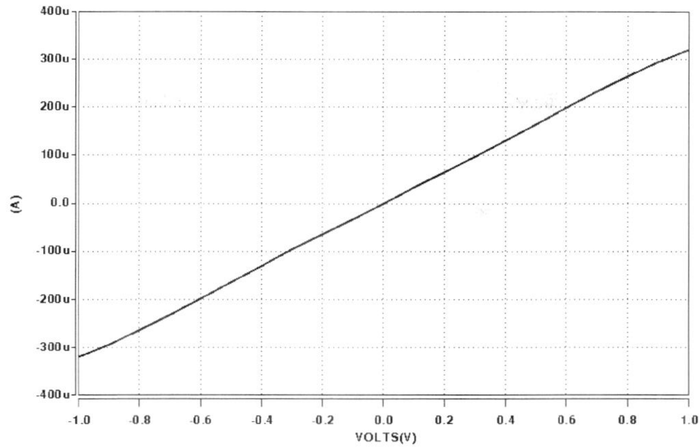

Fig. 9. V-I transfer characteristic

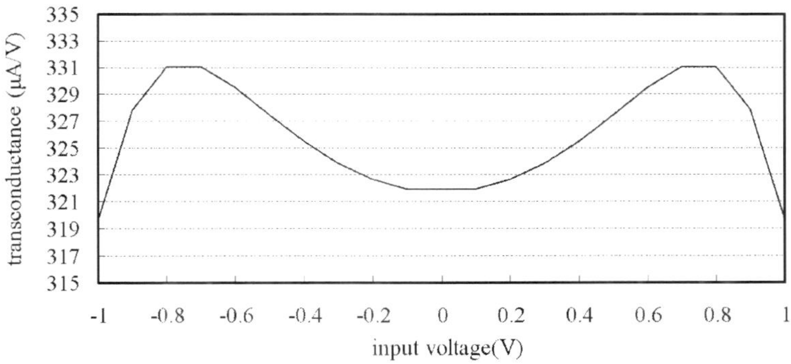

Fig. 10. The simulated transconductance at V_C=0.7V

In Fig. 11. it shows the drain-source voltage of the input transistors M_1 and M_2, V_{DS1} and V_{DS2}, changes with the input voltage. Within ±1V input voltage, V_{DS1} and V_{DS2} are very small. According to equation (40), V_{DS1} and V_{DS2} are too small such that transistors M_1 and M_2 can be set in triode region. Once the input voltage exceeds ±1V, V_{DS1} and V_{DS2} will increase rapidly. It results in that transistors M_1 and M_2 enter in saturation region. In other words, when M_1 and M_2 entering saturation region the proposed transconductor can not maintain the high linearity.

$$V_{DS} < V_{GS} - V_T \tag{40}$$

When V_C is set between 0.69V and 0.72V, the linear input range is up to 2.6V and the transconductance error is less than 1%. The smallest transconductance is 3.4μs and linear input range is 1.2V when V_C is 0.720V. The highest transconductance is 542μs and linear input range is 1.4V when V_C is 0.690V. Table 1 shows the linear input range and the transconductance tuned by different V_C. Therefore, the proposed transconductor achieve a large tuning range.

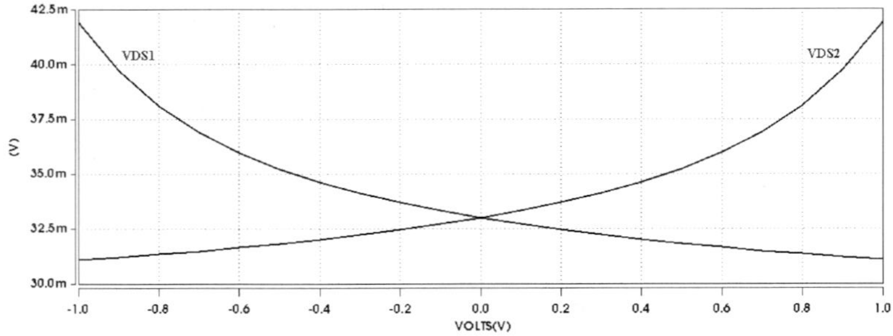

Fig. 11. The drain-source voltage of input transistor M_1 and M_2

V_C (V)	Linear input range (V)	Transconductance (μS)
0.690	1.4	542
0.695	1.8	434
0.700	1.8	326
0.705	2.2	219
0.710	2.4	122
0.715	2.6	42
0.720	1.2	3.4

Table 1. V_C versus Linear input range

In Fig. 12., the simulated THD as a function of the input frequency and input signal amplitude is plotted. The best THD is achieved at the low input voltage and the low frequency. When V_C is 0.7V, the linearity of the proposed transconductor is less than −60dB for 0.7Vpp at 100KHz.

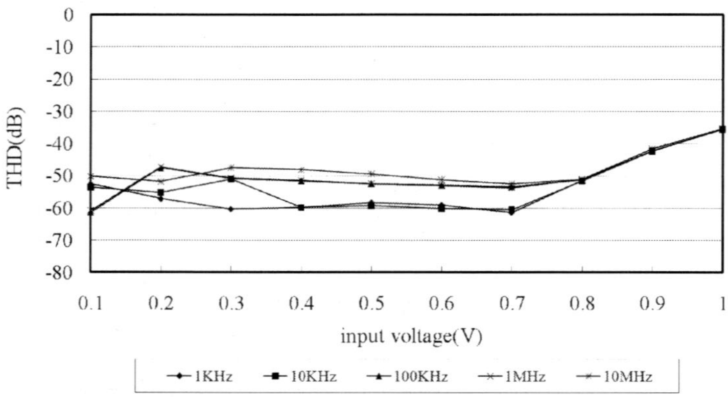

Fig. 12. Simulated THD for different input frequencies

Fig. 13. shows the linearity of transconductor in three linearization techniques. The transconductor using source degeneration with resistor is shown in Fig. 4(b), and the transconductance in Fig. 13(a) is tuned by different resistors. The transconductor using source degeneration with MOS transistors is shown in Fig. 5, and the transconductance in Fig. 13(b) is tuned by the different size ratio of β_1/β_3. The transconductor using adaptive biasing is shown in Fig. 6, and the transconductance in Fig. 13(c) is tuned by the different compensating tail current, I_C. Fig. 14. Shows the simulation result of the proposed technique and other techniques. Fig. 14(a) is the full plot of the different linearization techniques. From Fig. 14(b) it can be easily seen that the linearity achieved by the newly proposed technique is better than all other implementations.

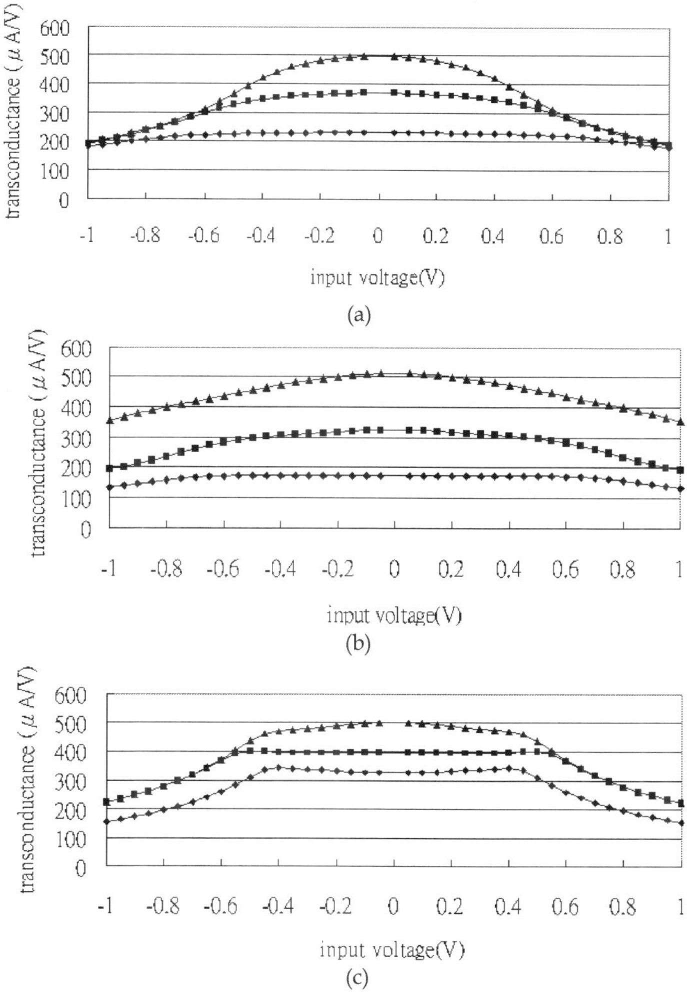

Fig. 13. Simulated transconductance of three linear transconductors (a) Source degeneration using resistor (b)Source degeneration using MOS transistors (c)Adaptive biasing

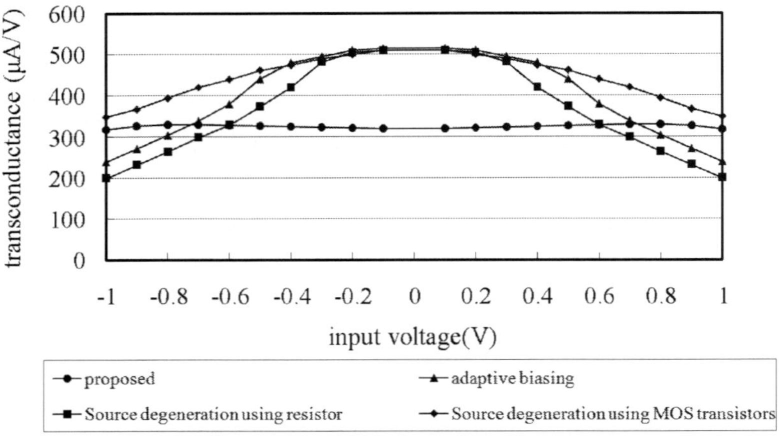

(a)

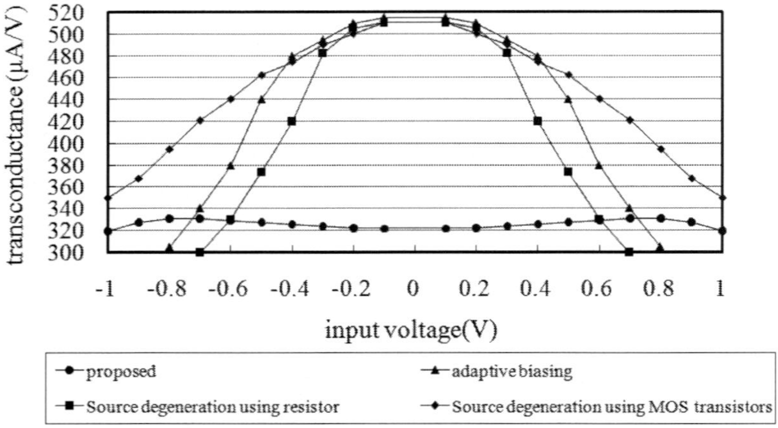

(b)

Fig. 14. Simulated transconductance for four linearization techniques (a) Full plot (b) Detail

The simulated THD of the output differential current versus the input signal amplitude for the four linearized transconductors is plotted in Fig. 15. The proposed transconductor achieves THD less than −61dB for the 0.7Vpp input voltage, 11dB better than the one using source degeneration using resistor, 24dB better than the one using source degeneration using MOS, and 31dB better than the one using adaptive biasing, at the same input range.

Table 2. shows the power consumption of the four linearized transconductors at the same transconductance. Power consumption changes with the different transconductances. Therefore, the same transconductance is chosen to be compared in each configuration. Table 3. shows different power consumption at the different transconductance of the proposed transconductor.

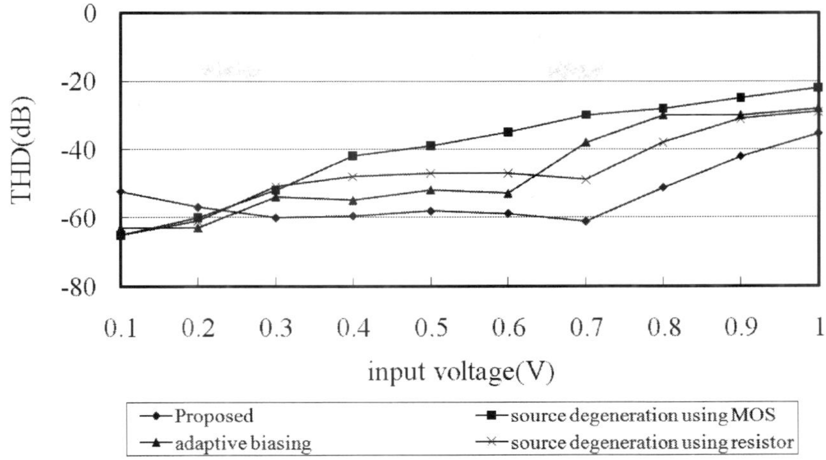

Fig. 15. Simulated THD at 1MHz for the four linearized transconductors

	Source degeneration using MOS	Source degeneration using resistor	Adaptive biasing	Proposed
Power (mW)	1.31	1.19	1.38	1.58

Table 2. The power consumption of four linearized transconductors

V_C (V)	Power (mW)	G_m (µA/V)
0.690	1.759	542
0.695	1.714	434
0.700	1.586	326
0.705	1.442	219
0.710	1.263	122
0.715	0.954	42
0.720	0.733	3.4

Table 3. The power consumption at different transconductances

Table 4. shows the comparison of performance with other transconductors at the low supply voltage (under 2V). The transconductor in [Fayed & Ismail 2005] also uses constant drain-source voltage. It modifies the basic structure of constant drain source voltage and uses the moderate amplifier. The proposed transconductor modifies the auxiliary amplifiers to obtain high gain under low supply voltage.

The layout including proposed transconductor, Common Mode Feedback, and bandgap is shown in Fig. 16. The proposed transconductor uses STC pure 1.8V linear I/O library in 0.18µm CMOS process. The chip area is 0.516mm^2.

	[Galan et. al 2002]	[Leuciuc & Chang 2002]	[Laguna et. al 2004]	[Sengupta 2005]	[Fayed & Ismail 2005]	Proposed
Process	0.8μm	0.25μm	0.8μm	0.18μm	0.18μm	0.18μm
Power supply	2V	1.8V	1.5V	1.8V	1.8V±10%	1.8V
THD	-40dB @10MHz	-80dB, 0.8Vpp, @2.5MHz	-33dB, 0.2Vpp, @5MHz	-65dB, 1Vpp, @1MHz	-50dB, 0.9Vpp, @50KHz	-60dB, 0.7Vpp, @100KHz
G_m (μA/V)	0.6~207	200~600	67~155	770	5~110	3.4~542
Linear input range	0.6Vpp	1.4Vpp	0.6Vpp	1Vpp	1.8Vpp	2.4Vpp
Year	2002	2002	2004	2005	2005	2009

Table 4. Comparison table

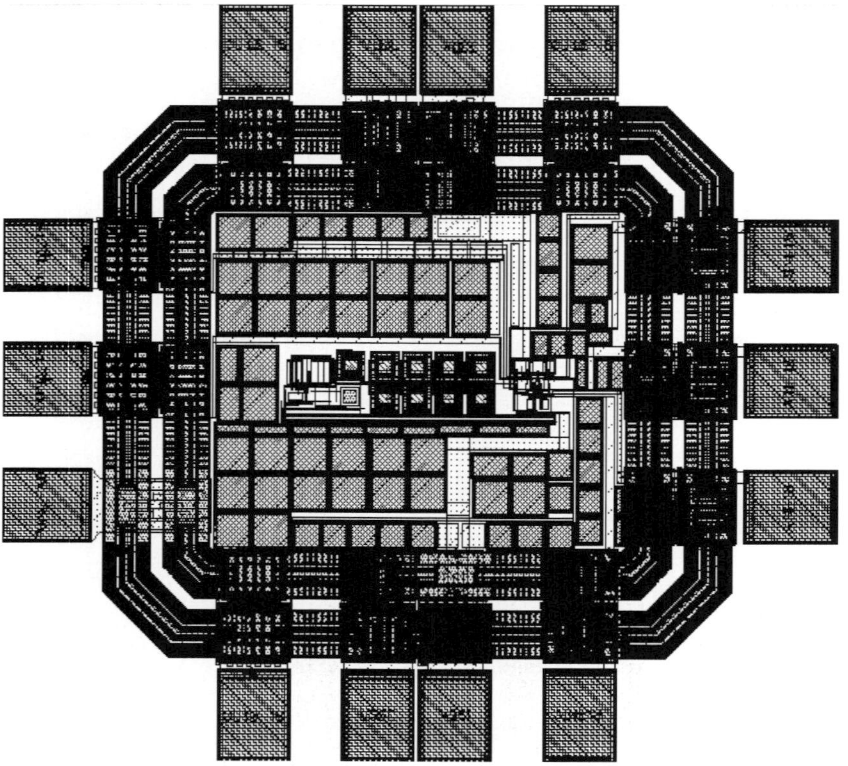

Fig. 16. The layout of proposed transconductor

5. Conclusion

The proposed low-voltage, highly linear, and tunable triode transconductor achieves the wide linear input range up to 2.4V. The total harmonic distortion is −60dB with a $0.7V_{pp}$ input voltage. The design uses TSMC 0.18μm CMOS technology and supply voltage is 1.8V. Moreover, it exhibits a large G_m tuning range from 3.4μS to 542μS and also keeps a wide linear input range. Finally, the performance comparison with other linear techniques shows that the proposed technique achieves better linearity, wider tuning range, and wider linear input range.

6. Acknowledgement

This work was supported in part by the National Science Council, Taiwan, ROC, under the grants: NSC 97-2221-E-110-078.

7. References

Degrauwe M. G.; Rijmenants J. & Vittoz E. A. (1982). Adaptive biasing CMOS amplifiers, *IEEE Journal of Solid-State Circuits*, Vol. 17, No. 3, (June 1982) pp. 522-528, ISSN 0018-9200

Elwan H.; Gao W.; Sadkowski R. & Ismail M. (2000). A low voltage CMOS class AB operational transconductance amplifier, *Electronics Letters*, Vol. 36, No.17, (Aug. 2000) pp. 1439-1440, ISSN 0013-5194

Fayed A. A. & Ismail M. (2005). A low-voltage, highly linear voltage-controlled transconductor, *IEEE Transactions on Circuits and Systems : I Express Briefs*, Vol.52, No. 12, (Dec. 2005) pp. 831-835, ISSN 1549-7747

Furth K. M. & Andreou A. G. (1995). Linearised differential transconcutors in subthres-hold CMOS, *Electronics Letters*, Vol. 31, No.7, (March 1995) pp. 1576-1581, ISSN 0013-5194

Galan A.; Carvajal R. G.; Munoz F.; Torralba A. & Ramirez-Angulo J. (2002). Design of linear CMOS transconduct- ance elements, *IEEE Proceedings of International Sympoisum of Circuits and Systems*, Vol. 2, pp. 9-12, ISBN 0-7803-7448-7, Phoenix-Scottsdale, Arizona, U.S.A, May 2002, Institute of Electrical and Electronics Engineers, Piscataway

Gharbiya A. & Syrzycki M. (2002). Highly linear, tunable, pseudo differential transconductor circuit for the design of Gm-C filters, *IEEE Proceeding of Canadian Conference on Electrical and Compiter Engineering*, Vol. 1, pp. 521-526, ISBN 0-7803-7448-7, Winnipeg, Manitoba, Canada, May 2002, Institute of Electrical and Electronics Engineers, Piscataway

Ismail A. M. & Soliman A. M. (2000). Novel CMOS wide-linear-range transconductance amplifier, *IEEE Transactions on Circuits and Systems*, Vol. 47, No. 8, (Aug. 2000) pp. 1248-1253, ISSN 0098-4094

Kim Y.; Park J.; Park M. & Yu H. (2004). A 1.8V triode-type transconductor and its application to a 10MHz 3rd-order chebyshev low pass filter, *IEEE Proceedings of the IEEE 2004 Custom Integrated Circuits Conference*, pp. 53-56, ISBN 0-7803-8495-4, Orlando, Florida, U.S.A, Oct. 2004, Institute of Electrical and Electronics Engineers, Piscataway

Kuo K. C. & Leuciuc A. (2001). A linear MOS transconductor using source degeneration and adaptive biasing, *IEEE Transactions on Circuits and Systems*, Vol.48, No. 10, (Oct. 2001) pp. 937-943, ISSN 0098-4094

Krummenacher F. & Joehl N. (2004). A 4-MHz CMOS continuous-time filter with on-chip automatic tuning, *IEEE Journal of Solid-State Circuits*, Vol. 23, No. 6, (Jun 2004) pp. 750-758, ISSN 0018-9200

Laguna M.; De la Cruz-Blas C.; Torralba A.; R. G. Carvajal R. G.; Lopez-Martin A. & Carlosena A. (2004). A novel low-voltage low-power class-AB linear transconductor, *IEEE Proceedings of International Sympoisum of Circuits and Systems*, Vol. 1, pp. 725-728, ISBN 0-7803-8251-X, Vancouver, British Columbia, Canada, May 2004, Institute of Electrical and Electronics Engineers, Piscataway

Lee S. O.; Park S. B. & Lee K. R. (1994). New CMOS triode transconductor, *Electronics Letters*, Vol. 30, No. 12, (June 1994) pp. 946-948, ISSN 0013-5194

Leuciuc A. & Zhang Y. (2002). A highly linear low-voltage MOS transconductor, *IEEE Proceedings of International Sympoisum of Circuits and Systems*, Vol. 3, pp. 735-738, ISBN 0-7803-7448-7, Phoenix-Scottsdale, Arizona, U.S.A, May 2002, Institute of Electrical and Electronics Engineers, Piscataway

Leuciuc A. (2003). A wide linear range low-voltage transconductor, *IEEE Proceedings of International Sympoisum of Circuits and Systems*, Vol. 1, pp. 161-164, ISBN 0-7803-7761-3, Bangkok, Thailand, May 2003, Institute of Electrical and Electronics Engineers, Piscataway

Likittanapong P.; Worapishet A. & Toumazou C. (1998). Tunable low-distortion BiICMOS transconductance amplifiers, *Electronics Letters*, Vol. 34, No. 12, (June 1998) pp. 1224-1225, ISSN 0013-5194

Mahattanakul J. & Toumazou C. (1998). Tunable low-distortion BiICMOS transconductance amplifiers, *Electronics Letters*, Vol. 34, No. 2, (Jan. 1998) pp. 175-176, ISSN 0013-5194

Nedungadi A. & Viswanathan T. R. (1984). Design of linear CMOS transconductance elements, *IEEE Transactions on Circuits and Systems*, Vol.31, No. 10, (Oct. 1984) pp. 891-894, ISSN 0098-4094

Razavi B. (2001). *Design of Analog CMOS Integrated Circuits*, McGraw-Hill, ISBN 0-07-118839-8, New York

Seevinck E. & Wassenaar R. F. (1987). A versatile CMOS linear transconductor/Square-Law function circuit, *IEEE Journal of Solid-State Circuits*, Vol. 22, No. 6, (June 1987) pp. 366- 377, ISSN 0018-9200

Sengupta S. (2005). Adaptive biased inear transconductor, *IEEE Transactions on Circuits and Systems : I Regular Papers*, Vol.52, No. 11, (Nov. 2005) pp. 2369-2375. ISSN 1549-8328

Torralba A.; Martinez-Heredia J. M.; Carvajal R. G. & Ramirez-Angulo J.(2002). Low-voltage transconductor with high linearity and large bandwidth, *Electronics Letters*, Vol. 38, No. 25, (Dec. 2002) pp. 1616-1617, ISSN 0013-5194

Wang A. & Guggenbuhl W. (1990). A voltage-controllable linear MOS transconductor using bias offset technique, *IEEE Journal of Solid-State Circuits*, Vol. 25, No. 2, (Feb. 1990) pp. 315-317, ISSN 0018-9200

Zeki A. (1999). Low-voltage CMOS triode transconductor with wide-range and linear tunability, *Electronics Letters*, Vol. 35, No. 20, (Sept. 1999) pp. 1685-1686, ISSN 0013-5194

3

A Dynamically Reconfigurable Device

Minoru Watanabe
Shizuoka University,
Japan

1. Introduction

To the present day, the performance of microprocessors has progressed dramatically. Recently, almost all computer systems use reduced instruction set computer (RISC) architectures. However, about 30 years ago, complex instruction set computer (CISC) architectures were widely used for almost all computer systems. The advantages and successes of RISC architectures are attributable to their simplified structures.

Conventional complex instruction set computer (CISC) architectures invariably included various and numerous instruction sets. Each instruction was able to execute a complicated multi-step operation. For that reason, the CISC architectures were useful in assembler-based programming environments and in systems with small amounts of memory. However, such complicated architectures prevent increases in clock frequency or a processor's processing power.

Therefore, RISC architectures—which use simple architectures based on single-step instruction sets—have been developed. The RISC architectures present advantages in terms of higher clock frequency, smaller implementation area, and lower power consumption than conventional complex instruction set computer (CISC) architectures. Observation of many examples reveals that, in circuit implementations, a simple structure is best to increase the overall performance. That principle is also applicable to programmable devices.

If clock-by-clock reconfigurable devices are used, even a single instruction set computer (SISC) can be implemented onto them. A single instruction set computer is one in which a processor has only a single instruction. During production, various single instruction set computers are prepared: a single instruction set computer with an AND logic function, a single instruction set computer with an adder function, and so on. These processor units are implemented at necessary times and at necessary places of a programmable device. In CISC and RISC architectures, the hardware is fixed. Its operations are switched using software commands, as portrayed in Fig. 1(a). In contrast, in a single instruction set computer, the operation changes are executed by hardware reconfigurations, as shown in Figs. 1(b) and 1(c). Therefore, in a single instruction set computer, a processor with a certain function itself can be reconfigured to another processor with another function.

The implementation of such single instruction set computers provides the following advantages under programmable device implementations. A single instruction set computer with the simplest architecture can operate at the highest clock frequency among all processor architectures. In RISC architectures, many selectors to change functions are implemented; such selectors have a certain delay. However, single instruction set computers

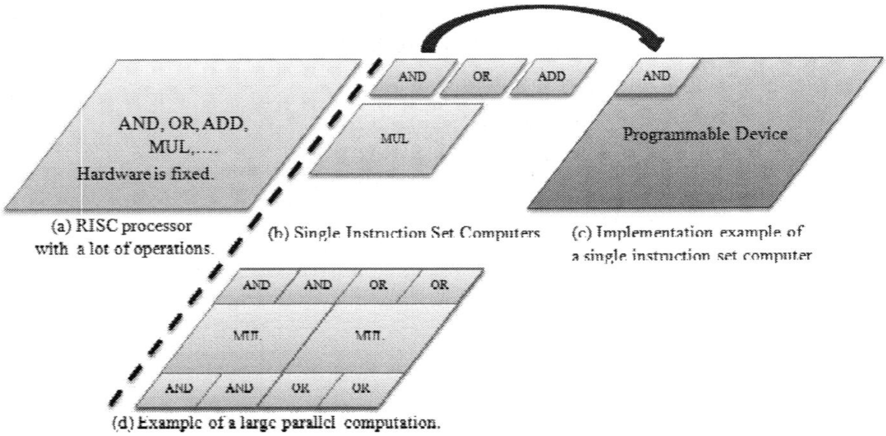

Fig. 1. RISC architecture and SISC architecture.

require no selector for use in function changes. Moreover, the inherent circuit complexity invariably increases the load capacitance and wiring capacitance at each circuit point. Large capacitance always decreases the maximum clock frequency. Therefore, the clock frequencies of simple architectures of single instruction set computers are higher than those of RISC and CISC architectures. As a result, the performance of single instruction set computers is superior to those of multi-instruction set computers.

Figure 1(d) shows that, since such a single instruction set computer can be implemented in a small area, large parallel computation can be achieved. Thereby, the total performance can be increased dramatically. However, to increase processing power using this concept, programmable devices must have a high-speed reconfiguration capability and a capability with numerous reconfiguration contexts to continue high-speed reconfigurations.

Currently, field programmable gate arrays (FPGAs) are widely used for many applications (1)-(3). Such FPGAs are always implemented with an external ROM. At power-on, a configuration context is downloaded from the external ROM to an internal configuration memory. However, such FPGAs have been shown to be unsuitable for dynamic reconfiguration applications because FPGAs require more than several milliseconds' reconfiguration time because of their serial transfer configuration mechanism.

On the other hand, high-speed reconfigurable devices have been developed, e.g. DRP chips (4). They include reconfiguration memories and a microprocessor array on a single chip. The internal reconfiguration memory stores the reconfiguration contexts of 16 banks, which can be substituted for one another during a clock cycle. Consequently, the arithmetic logic unit can be reconfigured on every clock cycle in a few nanoseconds. Unfortunately, increasing the internal reconfiguration memory while maintaining the number of processors is extremely difficult.

As with other rapidly reconfigurable devices, optically reconfigurable gate arrays (ORGAs) have been developed, which combine a holographic memory and an optically programmable gate array VLSI, as portrayed in Figs. 2 (5)-(9). Many configuration contexts can be stored in a holographic memory. Thereafter, they can be read out optically and programmed optically onto a gate array VLSI using photodiodes perfectly in parallel.

Therefore, high-speed configuration is possible in addition to numerous reconfiguration contexts. Such ORGA architectures present the possibility of opening the implementations of single instruction set computers.
This chapter introduces a VLSI design of an ORGA architecture: a dynamic ORGA architecture suitable for implementations of single instruction set computers.

2. ORGA architecture

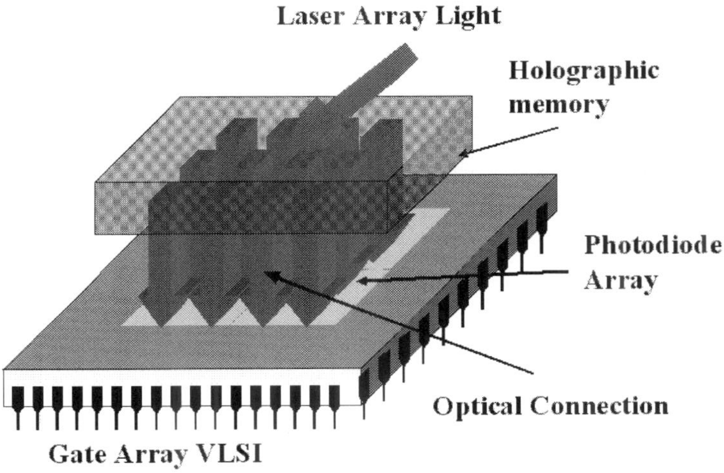

Fig. 2. Overall construction of an optically reconfigurable gate array (ORGA).

An overview of an Optically Reconfigurable Gate Array (ORGA) is portrayed in Fig. 2. An ORGA comprises a gate-array VLSI (ORGA-VLSI), a holographic memory, and a laser diode array. The holographic memory stores reconfiguration contexts. A laser array is mounted on the top of the holographicmemory for use in addressing the reconfiguration contexts in the holographic memory. One laser corresponds to a configuration context. Turning one laser on, the laser beam propagates into a certain corresponding area on the holographic memory at a certain angle so that the holographicmemory generates a certain diffraction pattern. A photodiode-array of a programmable gate array on an ORGA-VLSI can receive it as a reconfiguration context. Then, the ORGA-VLSI functions as the circuit of the configuration context. The reconfiguration time of such an ORGA architecture reaches nanosecond-order (5),(6). Therefore, very-high-speed context switching is possible. In addition to it, since the storage capacity of a holographicmemory is extremely high, numerous configuration contexts can be stored in a holographic memory. Therefore, the ORGA architecture can dynamically implement single instruction set computers.

3. Dynamic ORGA architecture

A configuration context is optically applied in ORGAs. In ORGA-VLSIs, a certain detection circuit must be used in addition to a programmable gate array. The detection circuit is called an optical reconfiguration circuit. Such an optical reconfiguration circuit is connected to each programming point of a programmable gate array. Therefore, the number of reconfiguration circuits can be as large as those of FPGAs. The resultant reduction of the implementation area of optical reconfiguration circuits is extremely important in ORGAs.

In major ORGAs (5),(6), each optical reconfiguration circuit consists of a photodiode, a refresh transistor, and a single-bit static configuration memory, as portrayed on the left side of Fig. 3. A reconfiguration procedure is initiated by charging the junction capacitance of the photodiode using refresh transistors. After charging, an optical configuration context is provided from a holographic memory and is received on the photodiodes. The electric charge in the junction capacitance of each light-received photodiode is discharged and the electric charge in the junction capacitance of each photodiode receiving no light is retained. The resultant difference is detectable by sensing the voltage between the anode and cathode of the photodiode. The sensed information is temporarily stored on a single-bit static configuration memory. Then, the context information is provided to each programming point of a gate array. Using this technique, a configuration context can be retained indefinitely in the ORGA-VLSI so that the state of the gate array can be maintained statically.

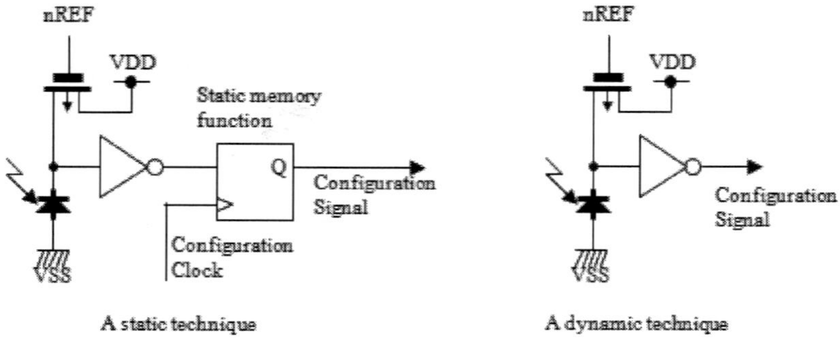

Fig. 3. Optical reconfiguration circuits of static and dynamic techniques.

However, the static configuration memory prevents realization of high gate count ORGA-VLSIs. The static configuration memory comes to occupy about 25% of the area of an entire VLSI chip. Moreover, using the memory function for storage during an indefinite period can be considered as over-capacity for implementation in single instruction set computers because a processor of a single instruction set computer is dynamically reconfigured. For that reason, its lifetime is very short. In addition, the configuration information is stored on a holographic memory; the information can therefore be read out anytime. Because of that feature, even when long-term functions are required, a certain refresh cycle enables such function implementations. Therefore, a Dynamic Optically Reconfigurable Gate Array (Dynamic ORGA) architecture without a long-term storable configuration memory was proposed (7). A photodiode invariably has junction capacitance. Therefore, the junction capacitance can maintain the state of a gate array for a certain time. The dynamic ORGA perfectly removes the static configuration memory to store a context and uses the junction capacitance of photodiodes as dynamic configuration memory, as shown on the right side of

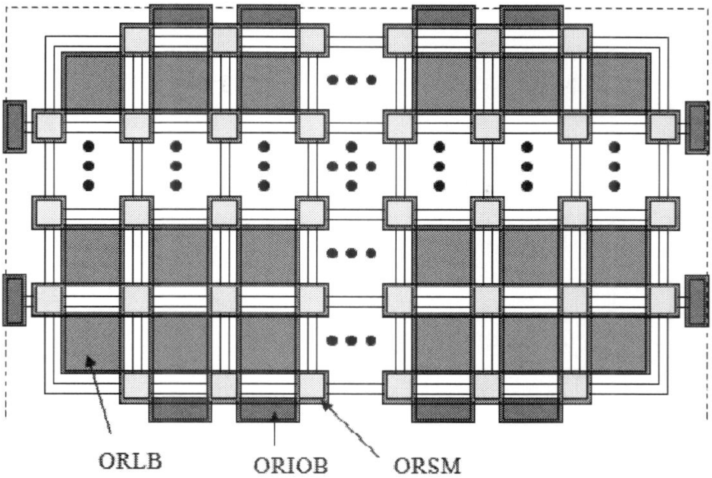

Fig. 4. An island-style gate array constructed by optically reconfigurable logic blocks (ORLBs), optically reconfigurable switching matrices (ORSMs), and optically reconfigurable I/O blocks (ORIOBs).

Fig. 3. Following such a concept of single instruction set computers, the junction capacitance of photodiodes is sufficient to retain the state of a gate array. This architecture is called a dynamic ORGA architecture. The dynamic ORGA architecture is a very advanced ORGA architecture in terms of gate density in ORGAs.

4. VLSI design with 51,272 gates

This section presents a description of the design of a 51,272 gate DORGA-VLSI. The 51,272-gate-count DORGA-VLSI chip was designed using a 0.35 μm standard complementarymetal oxide semiconductor (CMOS) process. The basic functionality of the DORGA-VLSI is fundamentally identical to that of currently available field programmable gate arrays (FPGAs). The DORGA-VLSI takes an island-style gate array or a fine-grain gate array.

4.1 Photodoide cell design

Always, the depletion depth of a photodiode between an N-well and a P-substrate is deeper than that of a photodiode between an N-diffusion and a P-substrate. However, the minimum size of a photodiode between an N-well and a P-substrate is always larger than that of a photodiode between an N-diffusion and a P-substrate. Since an ORGA requires many photodiodes, the implementation area reduction is very important. For that reason, photodiodes were constructed between the N-diffusion and the P-substrate. The acceptance surface size of the photodiode is $8.8 \times 9.5\ \mu m^2$. In addition, the photodiode cell size is $21.0 \times 16.5\ \mu m^2$. Such a cell was designed as a full custom design. The fourth metal layer is used for guarding transistors from light irradiation; the other three layers were used for wiring.

Technology	0.35 μm double-poly four-metal CMOS process
Chip size [mm^2]	14.2 ×14.2
Supply Voltage [V]	Core 3.3 , I/O 3.3
Photodiode size [μm^2]	9.5 ×8.8
Horizontal distance between photodiodes [μm]	28.5-42
Vertical distance between photodiodes [μm]	12-21
Number of photodiodes	170,165
Number of logic blocks	1,508
Number of switching matrices	1,589
Number of I/O bits	272
Gate count	51,272

Table 1. Specifications of a high-density DORGA.

4.2 Optically reconfigurable logic block

A block diagram of an optically reconfigurable logic block of the DORGA-VLSI chip is presented in Fig. 5. Each optically reconfigurable logic block consists of 2 four-input one-output look-up tables (LUTs), 10 multiplexers, 8 tri-state buffers, and 2 delay-type flip-flops

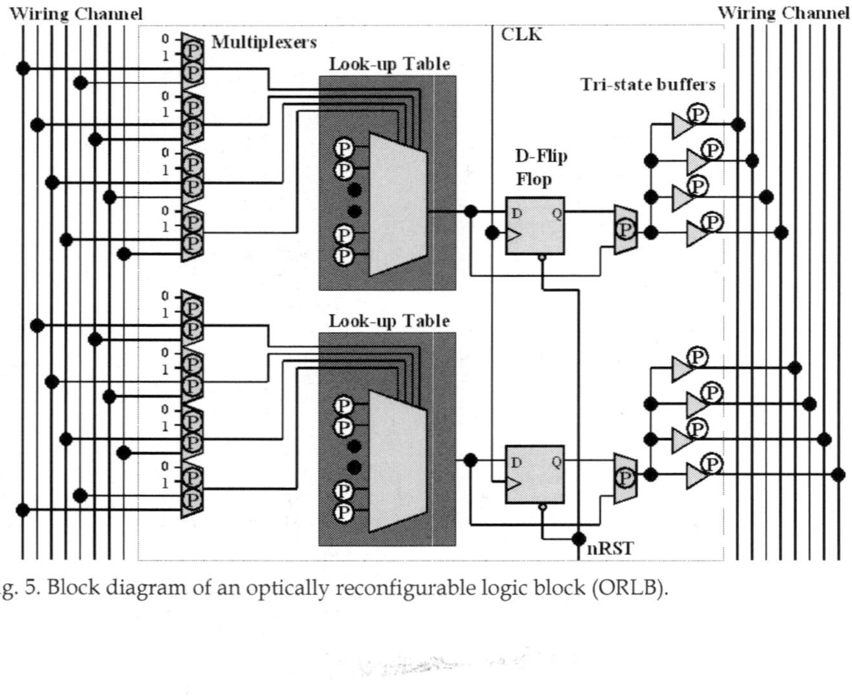

Fig. 5. Block diagram of an optically reconfigurable logic block (ORLB).

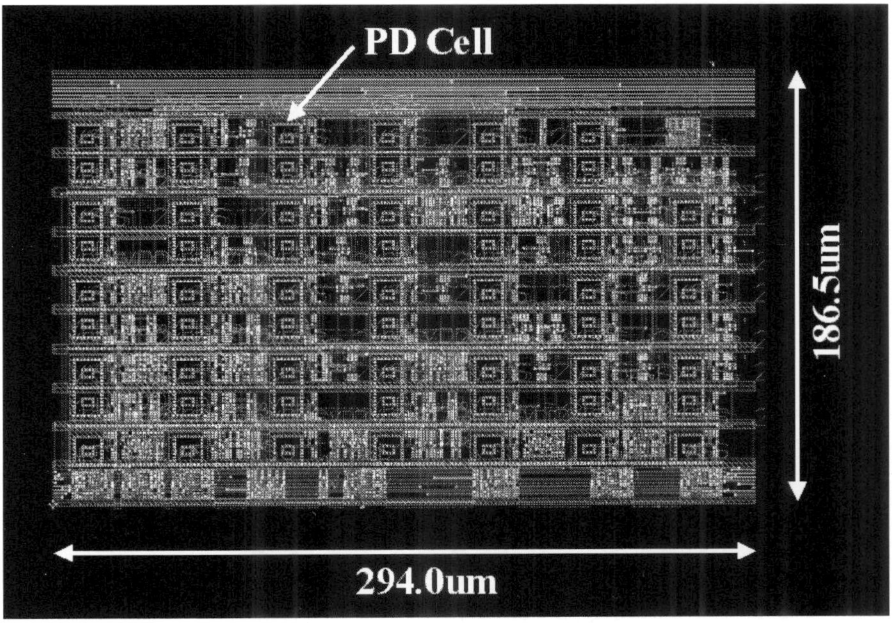

Fig. 6. CAD layout of the optically reconfigurable logic block (ORLB).

with a reset function. The input signals from the wiring channel, which are applied through some switching matrices and wiring channels from optically reconfigurable I/O blocks, are transferred to LUTs through eight multiplexers. The LUTs are used for implementing Boolean functions. The outputs of an LUT and of a delay-type flip-flop connected to the LUT are connected to a multiplexer. A combinational circuit and sequential circuit can be chosen by changing the multiplexer, as in FPGAs. Finally, outputs of the multiplexers are connected to the wiring channel again through eight tri-state buffers. As a result, each four-input one-output LUT, multiplexer, and tri-state buffer has 16 photodiodes, 2 photodiodes, and 1 photodiode, respectively. In all, 58 photodiodes are used for programming an optically reconfigurable logic block. The optically reconfigurable logic block can be reconfigured perfectly in parallel. The CAD layout is depicted in Fig. 6. This is a standard-cell based design. The cell size is $294.0 \times 186.5\ \mu m^2$. Wiring between cells was executed using the first to the third metal layers while avoiding the aperture area of the photodiode cell. Such optically reconfigurable logic block design is based on a standard cell design, except for custom designs of transmission gate cells and photodiode cells. Each photodiode is arranged at $42.0\ \mu m$ horizontal intervals and at 12.0-$21.0\ \mu m$ vertical intervals.

4.3 Optically reconfigurable switching matrix

Similarly, optically reconfigurable switching matrices are optically reconfigurable. The block diagram of the optically reconfigurable switching matrix is portrayed in Fig. 7. Its basic construction is the same as that used by Xilinx Inc. Four-directional switching matrices with 48 transmission gates were implemented in the gate array. Each transmission gate can be

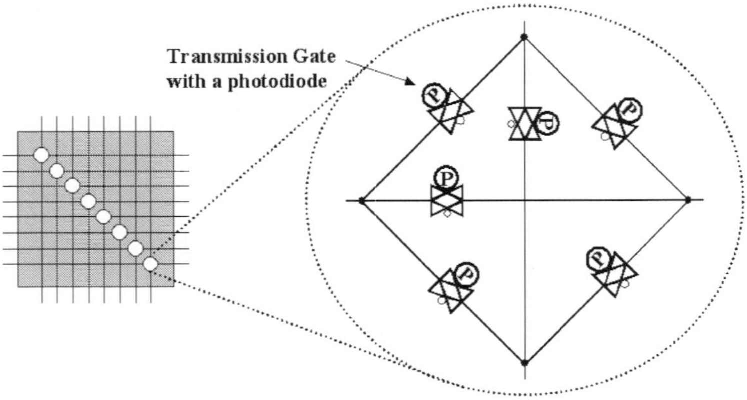

Fig. 7. Block diagram of an ORSM.

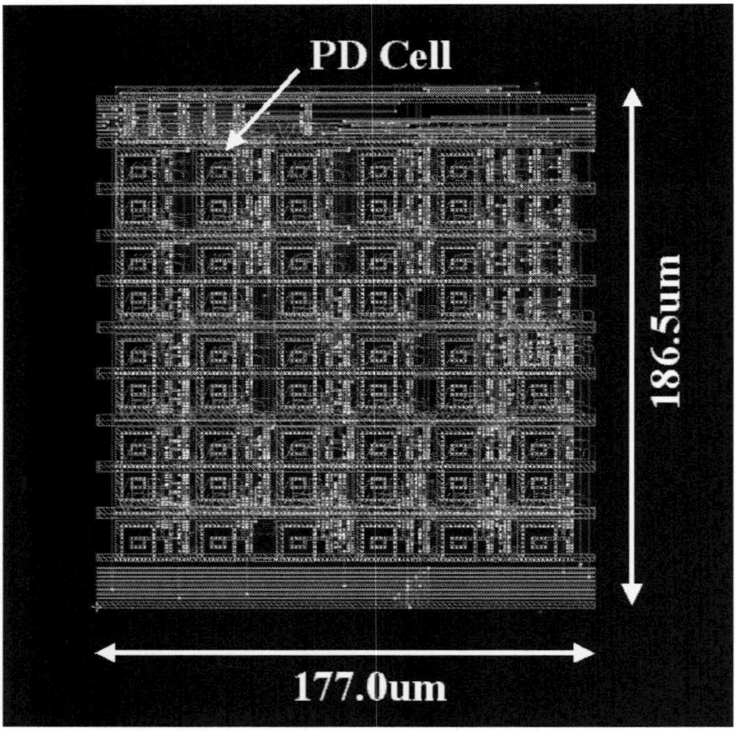

Fig. 8. CAD layout of an ORSM.

considered as a bi-directional switch. A photodiode is connected to each transmission gate; it controls whether the transmission gate is closed or not. Based on that capability, four-direction switching matrices can be programmed as 48 optical connections. The CAD layout is portrayed in Fig. 8. The cell size is 177.0 × 186.5 μm^2. As with the ORLBs, wiring was executed using the first to the third metal layers, thereby avoiding the aperture area of the photodiode cell. Such an optically reconfigurable switching matrix was designed using custom cells of photodiode cells and transmission gate cells, except for some buffers. Each photodiode is arranged at 28.5 μm horizontal intervals and at 12.0-21.0 μm vertical intervals.

4.4 Gate array

Figure 4 depicts the gate array structure. Table 1 presents its specifications. The gate array was designed using the Design Compiler logic synthesis tool and the Apollo place and route tool (Synopsys Inc.). The ORGA-VLSI chip consists of 1,508 optically reconfigurable logic blocks (ORLB), 1,589 optically reconfigurable switching matrices (ORSM), and 272 optically reconfigurable I/O bits (ORIOB). Each optically reconfigurable logic block is surrounded by wiring channels. In this chip, one wiring channel has eight connections. Switching matrices are located on the corners of optically reconfigurable logic blocks. Each connection of the switching matrices is connected to a wiring channel.

The accepted surface size of the photodiode and photodiode-cell size, including an optical reconfiguration circuit are, respectively, 8.8 × 9.5 μm^2 and 21.0 × 16.5 μm^2. The photodiode cells were arranged at 28.5-42.0 μm horizontal intervals and at 12.0–21.0 μm vertical intervals: in all, 170,165 photodiodes were used. The fourth metal layer is used for guarding transistors from light irradiation; the other three layers were used for wiring.

4.5 Reconfiguration performance

The retention time and configuration time of photodiode memory architecture in a DORGA-VLSI were estimated experimentally using another DORGA-VLSI chip. That other VLSI chip was fabricated using the same CMOS process. In addition, the VLSI chip has identical photodiode construction and characteristics. Therefore, although a 51,272 DORGA-VLSI chip has never been fabricated, its characteristics were measured using the other DORGA-VLSI chip. As a result, the retention time of photodiode was measured as longer than 45 s. That retention time is much longer than that of current DRAMs. Consequently, the storage time is sufficient for the implementation of single instruction set computers. Additionally, the product of the photodiode response time and laser power for each photodiode was measured as $T_{reconfiguration} \cdot P_{laser} = 12.7\ pJ$. That measurement demonstrates that nanosecond-order configuration is possible.

5. Conclusion

This chapter has introduced and explained important concepts related to single instruction set computers. Such single-instruction set computers constitute an acceleration method used with microprocessor operations. To implement them, clock-by-clock dynamically reconfigurable devices are desired. However, using current VLSI technologies, simultaneous realization of fast reconfiguration and numerous reconfiguration contexts is impossible. To realize such clock-by-clock dynamically reconfigurable devices, another technology must be developed. As one possibility, this chapter has introduced and described an optically reconfigurable gate array VLSI. Currently, the gate count and performance of such ORGA-VLSIs are insufficient. Nevertheless, such architecture presents the possibility of overcoming current VLSI limitations. Realizing a device to overcome those limitations remains as a subject for future works.

6. References

[1] Altera Corporation, "Altera Devices," http://www. altera.com.

[2] Xilinx Inc., "Xilinx Product Data Sheets," http://www. xilinx.com.

[3] Lattice Semiconductor Corporation, "LatticeECP and EC Family Data Sheet," http://www. latticesemi.co.jp/products, 2005.

[4] H. Nakano, T. Shindo, T. Kazami, M. Motomura, "Development of dynamically reconfigurable processor LSI," NEC Tech. J. (Japan), vol. 56, no. 4, pp. 99–102, 2003.

[5] M. Miyano, M. Watanabe, F. Kobayashi, "Optically Differential Reconfigurable Gate Array," Electronics and Computers in Japan, Part II, Issue 11, vol. 90, pp. 132-139, 2007.

[6] M. Nakajima, M. Watanabe, "A four-context optically differential reconfigurable gate array," IEEE/OSA Journal of Lightwave Technology, Vol. 27, No. 24, 2009.

[7] M. Watanabe, F. Kobayashi, "Dynamic Optically Reconfigurable Gate Array," Japanese Journal of Applied Physics, Vol. 45, No. 4B, pp. 3510-3515, 2006.

[8] D. Seto, M. Watanabe, "A dynamic optically reconfigurable gate array - perfect emulation," IEEE Journal of Quantum Electronics, Vol. 44, Issue 5, pp. 493-500, 2008.

[9] D. Seto, M. Watanabe, "An 11,424 Gate-Count Dynamic Optically Reconfigurable Gate Array with a Photodiode Memory Architecture," Asia and South Pacific Design Automation Conference, pp. 117-118, 2009.

4

Evolutionary Memory: Unified Random Access Memory (URAM)

Yang-Kyu Choi and Jin-Woo Han
Korea Advanced Institute of Science and Technology (KAIST)
Republic of Korea

1. Introduction

Over the course of the past three decades, we have witnessed dramatic changes in our lifestyles. This is attributed to an unprecedented revolution of information technology (IT). The key element of the IT revolution is the continuing advancement of semiconductor technology. A major driving force of semiconductor technology lies in silicon. The silicon semiconductor has been applied to logic chips as well as memory chips for various applications. Meanwhile, the silicon memory has been at the center of an ongoing battle to manufacture the smallest, highest density, and most innovative product. Since their invention in the early 1970s, silicon memory devices have advanced at a remarkable pace. Silicon based memories such as dynamic random access memory (DRAM), static random access memory (SRAM), and Flash memory have been crucial elements for the semiconductor chip industry in the areas of density, speed, and nonvolatility, respectively. An important growth engine is scaling, which has enabled multiple devices to be integrated within a given area, resulting in an exponential increase in density and a decrease in bit-cost (Moore, 1965). The traditional scaling approach, however, is now confronting physical and technical challenges toward the end-point of the international technology roadmap for semiconductors (ITRS), indicating that the revenue from downscaling will diminish as scaling slows. Thus, an entirely new concept is required to ensure that silicon memory technology remains competitive. To meet this stringent requirement, this chapter will exploit a new paradigm of memory technology.

An ideal memory device should satisfy three requirements: high speed, high density, and nonvolatility. Unfortunately, a memory satisfying all requirements has yet to be developed. Memory devices have consequently been advanced by pursuing just one of these virtues, and appear in many different forms. SRAM dominates high speed on-chip caches for advanced logic and DRAM occupies applications for high-density and high-speed computation; but DRAM's data is volatile, and Flash memory is widely used for high density and non-volatile data storage. Therefore, if a single memory transistor can process different memory functions, a paradigm shift from 'scaling' to 'multifunction' can continue the evolution of silicon technology. In this chapter, the prototype of the fusion memory, named unified-random access memory (URAM), is introduced that can simplify device architecture, reduce power consumption, increase performance, and cut bit-cost.

2. Operational principle of URAM

URAM is composed of a single memory transistor, which must be of the smallest cell size. It can perform nonvolatile functions or high-speed operations according to the set of operational biases. In other words, circuit designers can specify URAM to be Flash memory or DRAM in order to comply with their specifications. Before discussing the details of URAM, each underlying operation principle is briefly introduced.

2.1 Flash memory operation

Advancements in high quality and ultra-thin oxides have paved the way for nonvolatile memory for silicon-oxide-nitride-oxide-silicon (SONOS) devices, which have replaced conventional floating-gate memory (Brown & Brewer, 1998). Fig. 2-1 shows the SONOS device structure and the program/erase operations. The device has a multiple gate dielectric stack consisting of tunnel oxide/nitride/control oxide (O/N/O), and the charges are stored in discrete traps in the nitride layer sandwiched between the upper/lower oxide barriers. The stored charges are positive or negative depending on whether negative or positive voltage is applied to the gate electrode. If positive programming voltage is applied to the gate, the electrons quantum-mechanically tunnel from the inverted channel through the tunnel oxide, and these electrons are stored in the deep-level traps in the nitride layer. During erasing, the holes are injected into the traps in the nitride in a manner similar to the program operation. The data is identified by the difference in the drain current. Once the charges are stored, the information is retained for up to 10 years with 10^6 to 10^7 program/erase cycles. Due to the superior ability of data retention, SONOS memory is called nonvolatile memory. From the perspective of speed, however, the writing requires few to few tens of microseconds, which might be too long to transfer high density data. Thus, the SONOS has been mainly utilized for portable applications, such as MP3 players, digital cameras, and memory stick solution.

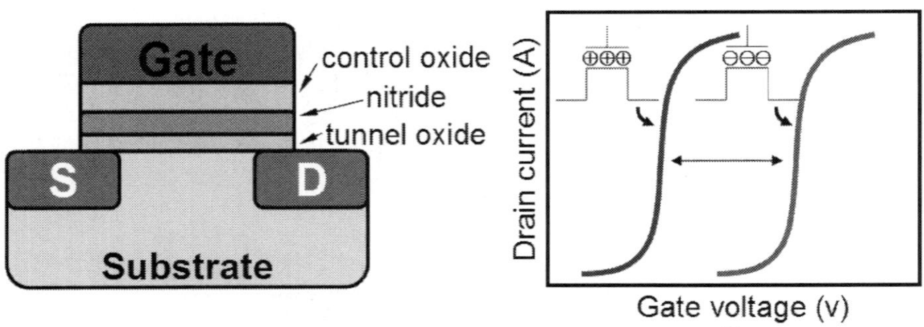

(a) (b)

Fig. 2-1. Operational principle of a SONOS Flash memory. (a) schematic of SONOS structure and (b) drain current versus gate voltage characteristics for two data states. The information is stored as a form of nitride trapped charges. The polarity and amount of charges stored in nitride layer determine the threshold voltages. The data is distinguished by measuring drain current flow at a given voltage. Once the charges are stored, the data is retained for over ten years so that the Flash memory is referred as nonvolatile memory.

2.2 Capacitorless 1T-DRAM operation

In conventional one transistor/one capacitor DRAM (1T/1C DRAM), *Moore's Law* tends to be invalid as the device scaling advances. While the cell transistors continue scaling, the cell capacitors cannot shrink much because they should store a detectable amount of charge, which is equivalent to the minimum cell capacitance, 30fF/cell. Therefore, the size mismatch between the transistors and capacitors leads to complexity in the fabrication process. In 2001 (Okhonin et al., 2001), the densest and cheapest DRAM, which is called capacitorless 1T-DRAM or zero-capacitorless RAM (ZRAM), was developed. The capacitorless 1T-DRAM replaces the large and complicated capacitor to be fabricated with a floating-body capacitor. The capacitorless 1T-DRAM exploits inherent properties, known as the floating body effects or history effects, of transistors made on silicon-on-insulator (SOI) substrates. The floating body effects are generally considered as parasitic by circuit designers because they cause the current overshoot, and obstruct to model and implement into circuit simulator (Gautier, 1997). While the majority of efforts are made to suppress these effects, Okhonin *et al.* found out that they can be a method to temporarily store the information. Fig. 2-2 illustrates the principle of the capacitorless 1T-DRAM. In the program, the impact ionization process generates pairs of electron and holes. While the electrons exit the channel through the drain, the holes are repelled by the drain, charging the body. Since the body is isolated vertically by the energy band offset of the buried-oxide and gate oxide, and laterally by the built-in potential energy of the n^+ source and n^+ drain with a p-type body, the confined holes are stored inside the floating body, as shown in Fig. 2-2. During erase, the negative drain voltage pulls the holes out of the floating body. The information is identified by turning on the transistor and measuring the amount of current flow. More current flows at programmed state as the positive body charges contribute to lowering the channel potential. Since the holes can disappear by recombination at the programmed state and the holes can be generated by band-to-band tunneling or thermal generation at the erased state, the data is volatile. However, as the generation and removal of holes only takes a few nanoseconds, the capacitorless 1T-DRAM can be embedded for high-speed applications such as the caches of microprocessor, digital signal processor (DSP), system-on-chip (SOC), *etc*.

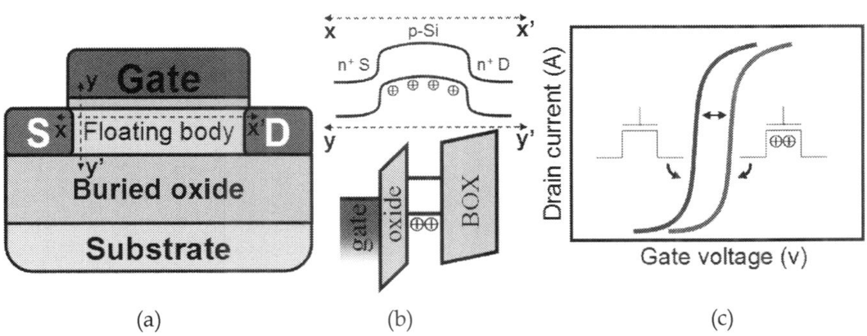

Fig. 2-2. Operational principle of a capacitorless 1T-DRAM. (a) Schematic of floating body structure, (b) energy band diagram of capacitorless 1T-DRAM, and (c) drain current versus gate voltage characteristics for two data states. The information is stored as a form of floating body charges. The excess holes inside the floating body increase the drain current. Since the stored charges disappear in a second, it is referred as DRAM.

2.3 Operation principle of URAM

The basis of URAM lies in the difference of the inherent operational biases for Flash and capacitorless 1T-DRAM (Han et al., 2007). Fig. 2-3 shows an operational bias domain for two memory modes. For erase operation, the two erase bias regions are distinctive. For program operation, even though the two regions partially overlap, the Flash memory utilizes relatively higher biases than the capacitorless 1T-DRAM. The overlapping region might cause them to disturb each other, a problem that will be solved in Section 5. If the proper biases are selected, two functions can work without disturbance from each other. In order to realize two functions in a single transistor, O/N/O gate dielectric is embodied onto a floating body transistor. When the Flash memory mode is activated, relatively higher voltages are used. On the other hand, relatively lower voltages are utilized to activate the capacitorless 1T-DRAM mode. Once the mode of URAM is determined, the operational biases are accordingly selected.

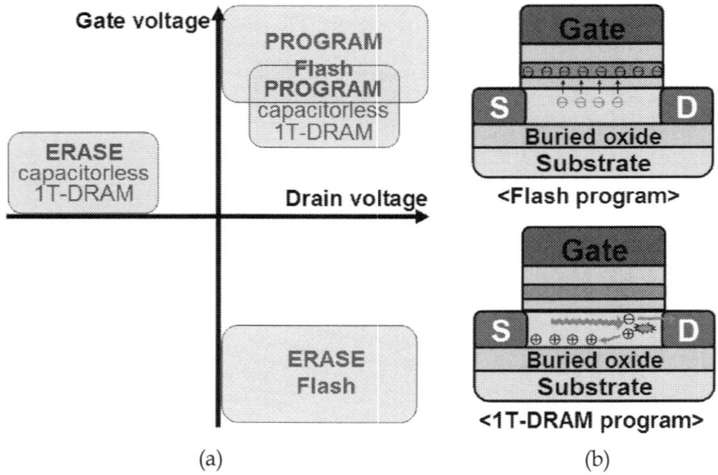

Fig. 2-3. (a) Operational bias domain of URAM and (b) schematics device structure and the program mechanism of two functions. The inherent difference stems from the distinctive operational domain, which allows independent functions in a single memory transistor.

The operational sequence is presented in Fig. 2-4. The memory block is firstly selected, and the operation mode is then decided. If the nonvolatile mode is chosen, the Flash operation is activated. Similarly, the capacitorless 1T-DRAM is activated if the high speed mode is needed. When the mode transits from Flash to capacitorless 1T-DRAM, the cell transistors in the selected block should be initialized to have a threshold voltage of 0.2V. If the threshold voltages are not initialized and high value remains, the greater gate voltage would be required to bias the fixed gate overdrive voltage. The high gate voltage can gradually impose stress on the gate oxide, which gradually increases the threshold voltage. On the other hand, if the initialized threshold voltage is small or even negative, excess holes can be generated, even in the zero gate voltage (off-state) since the carrier supplement is sufficient to trigger impact ionization, which can cause drain disturbance. It should be noted that the impact ionization process for the program operation of the capacitorless 1T-DRAM can adversely affect the charge trapping into the O/N/O layer. Here, the undesirable threshold voltage shift caused by capacitorless 1T-DRAM program is referred to as a soft-program. Since the soft-program causes unstable operation, the threshold voltage should be periodically monitored to find out whether the cells have suffered from soft-programming. The memory block would be re-initialized if the cells failed the verification test. This verification and re-initialization loop is an essential but time consuming process. The method to minimize and, furthermore, eliminate this redundant loop will be discussed in Section 5.

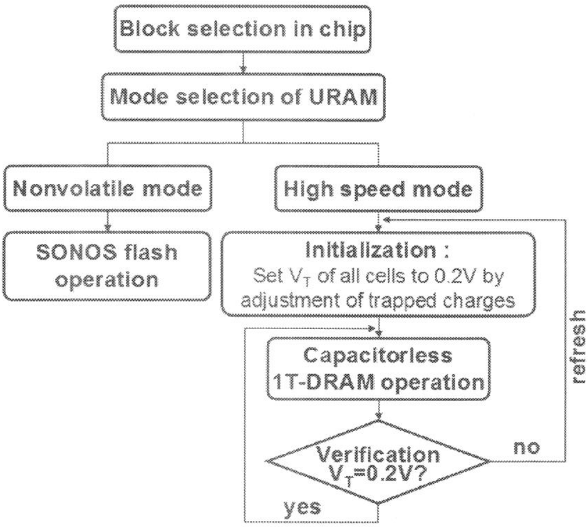

Fig. 2-4. Operational sequence diagram for URAM. The mode is selected according to the designer's demand. Since the program mechanism between two modes partially overlaps, the verification and re-initialization loop is inserted in high-speed mode.

3. Device fabrication and various quantum substrates

3.1 Various quantum substrates for URAM

To date, silicon-on-insulator (SOI) substrate has been utilized for the capacitorless 1T-DRAM. As embedded DRAM (eDRAM) now occupies more than 50% of the total chip area, and advanced processors have started to pick up SOI, the capacitorless 1T-DRAM made on SOI substrate is highly attractive for embedded memory. However, since a bulk substrate still occupies a significant portion of the market share, if the floating body effect is found in the bulk substrate, a chip built on the bulk substrate will be fully blessed with the benefits from the bulk substrate technology. It is true that major memory industries are conservative to adopt SOI substrate for their stand-alone memory products mainly due to the cost issue. Therefore, the capacitorless 1T-DRAM fabricated on the bulk substrates will be explored in terms of not only the embedded memory, but also stand-alone memory applications. In this section, the various quantum substrates, in particular the bulk substrates embodied with the quantum energy band structure, are introduced and the device fabrication process is illustrated.

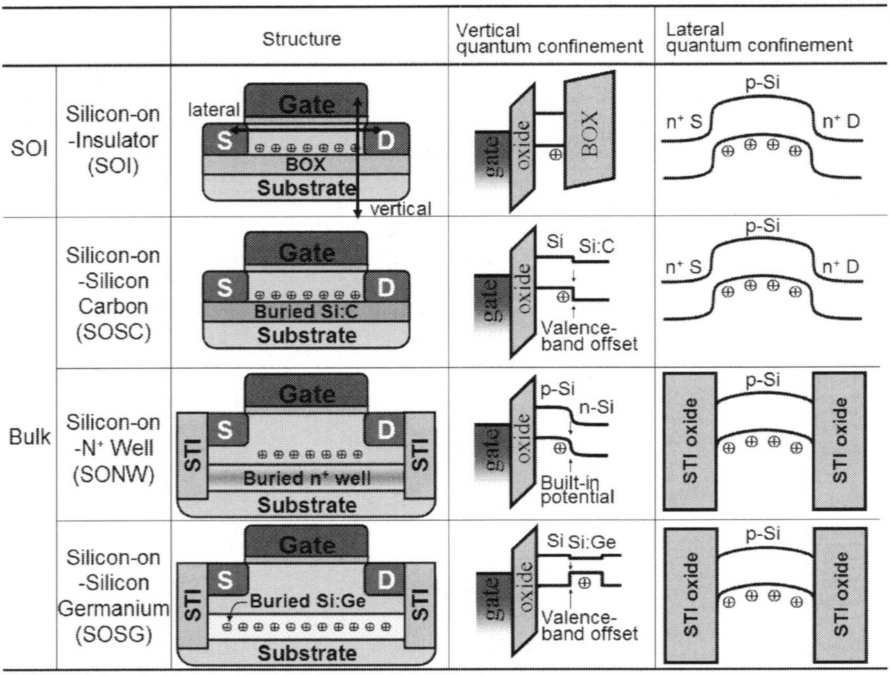

Fig. 3-1. Various templates and their energy band diagrams for excess hole storage. The SOI, SOSC, and SONW are potential barrier types, and the SOSC is potential well types. Unlike the SOI and SOSC, the SONW and SOSG confine holes by shallow trench isolation (STI) oxide in the lateral direction.

The quantum substrates used for the device fabrication and their corresponding energy band diagram for hole storage are comparatively shown in Fig. 3-1. In SOI substrate, the excess holes are vertically confined between the tunnel oxide and the buried oxide and are horizontally isolated by the built-in potential barrier of the n^+ source/drain and p-type body. Next, three methods for the floating body in bulk substrates are introduced. The hetero-epitaxial growth of semiconductor can imitate the energy band lineup of SOI. The introduction of carbon (C) into the silicon substrate enlarges the energy band-gap (Kim & Osten, 1997). Thus, the sequential growth of $Si_{1-y}C_y$ and Si on the bulk wafer can mimic SOI substrate. Here, $Si_{1-y}C_y$ serves as the role of the buried oxide. This substrate is named SOSC after the abbreviation of silicon-on-silicon carbon. Similar to the SOI substrate, the tunnel oxide and the valence band barrier at $Si/Si_{1-y}C_y$ confine holes in the vertical direction, and the built-in potential at the junction boundary confines in the horizontal direction. The n^+ ion deep implantation onto the p-type bulk substrates forms the buried n-type well structure (Ranica et al., 2005). The n-type well and p-type body forms a built-in potential barrier that prevents the holes from flowing out to the substrate terminal. This template is named SONW after the abbreviation of silicon-on-n^+ well. Whereas the holes are horizontally isolated by the junction barrier at SOI and SOSC, the SONW confines hole by shallow trench isolation (STI) oxide. In order to an avoid electrical short between the n^+ source/drain and n^+ well, the n^+ well should be buried much deeper than the junction depth of the source/drain. This requirement inevitably imposes a minimum space between the two junctions. Therefore, that opened space should be filled by the STI oxide. The aforementioned three substrates: SOI, SOSC, and SONW, vertically confine holes with the potential barrier. Similarly, a potential well can also store the excess holes as the potential barrier did. Similar to the SOSC preparation, the introduction of germanium (Ge) into the silicon substrate reduces the energy bandgap. Thus, the sequential epitaxial growth of $Si_{1-x}Ge_x$ and Si forms the potential well (Ni & Hansson, 1990). In order to avoid the loss of the excess holes via recombination at the source/drain junction, buried $Si_{1-x}Ge_x$ is placed under the source/drain junction boundary. As a result, the STI oxide blocks the evacuation of stored holes along the lateral direction. The silicon-on-silicon germanium is referred to SOSG. In addition to the fundamental interest in the well-type storage media, $Si_{1-x}Ge_x$ is more frequently studied in the literature than $Si_{1-y}C_y$, and $Si_{1-x}Ge_x$ has been already adopted for the strained technology in the mass production so that SOSG technology might be more practical.

3.2 Device fabrication

There are two common types of Flash memory array architectures: NAND and NOR which follow to the logical form of the cell configuration. The cell layout of URAM is the same as that of NOR type Flash because the drain voltage should be applied to each memory cell to trigger the impact ionization. The cell layout of URAM is shown in Fig. 3-2. The gates of each cell are coupled by a row line, and their drains are coupled with column lines. Since the individual memory cells are connected in parallel, random access is allowed. NOR architecture generally has one contact per two neighboring cells by sharing the source contact, thereby reducing the chip area. Some types of URAM, however, cannot use the shared source contact. While the shared source is possible for SOI and SOSC, SONW and SOSG require each source contact for all cells because each cell should be isolated by the STI oxide. The cross-sectional schematic along the bit-line direction in Fig. 3-3 shows that the source can be shared in SOI and SOSC. If the source is shared at SONW and SOSG, however, the lateral migration of excess holes can disturb the body charges of the neighboured cell. In other words, every cell should be isolated by the STI oxide and have their own source line. As a result, the layout efficiency of SOI and SOSC is better than that of SONW and SOSG.

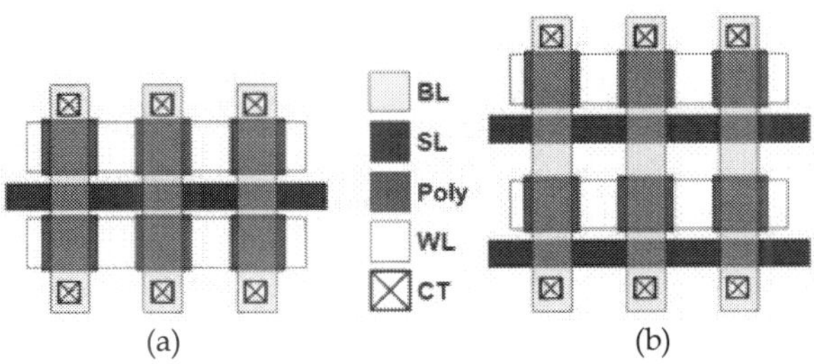

Fig. 3-2. Two types of URAM configuration. (a) Shared source line uses one contact for two cells and (b) divided source lines require individual contact for each cell.

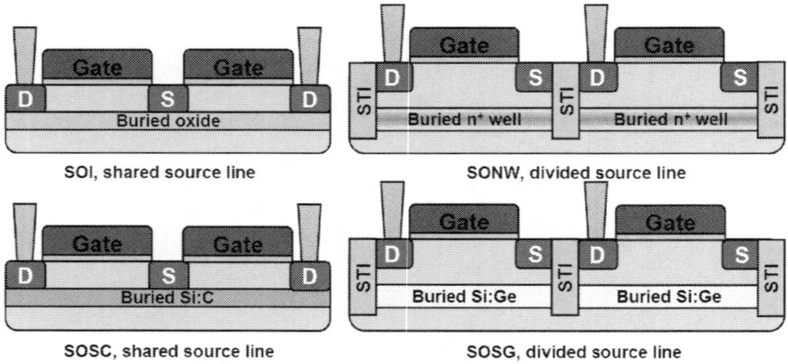

Fig. 3-3. Schematics of the cells along the word line direction. Whereas the SOI and SOSC use the shored bit line, the SONW and SOSG should utilize the divided source line.

The schematic of the process flow is shown in Fig. 3-4 (Han et al., 2009). Except the SOI substrate, SOSC, SONW, and SOSG utilize the bulk silicon wafer. Whereas SOI itself provides the intrinsic floating body, bulk substrates require the energy band engineering to form the extrinsic floating body. The n+ deep ion implantation is carried out for the SONW, $Si_{1-y}C_y/Si$ is epitaxially grown for SOSC, and $Si_{1-x}Ge_x/Si$ is epitaxially grown for SOSG. After the various types of the templates are prepared, the subsequent processes are similar. A photolithography process with a 0.18μm design rule is applied for channel definition. The photoresist is then trimmed down to a line width of 30nm by plasma ashing. The silicon is etched by reactive ion etching (RIE), resulting in the a 30nm width fin shaped channel. High density plasma (HDP) oxide is deposited and planarized by chemical mechanical polishing (CMP) and partially recessed by diluted HF until the upper part of the fin is exposed. The remaining lower part of the fin is covered by the isolation STI oxide, and the exposed upper part of the fin becomes the active area. The gate dielectric stack, tunnelling oxide/nitride/control oxide, is formed, and *in-situ* doped n+ polysilicon for the gate is sequentially deposited. After the gate patterning, source/drain implantation and activation are carried out followed by forming gas annealing.

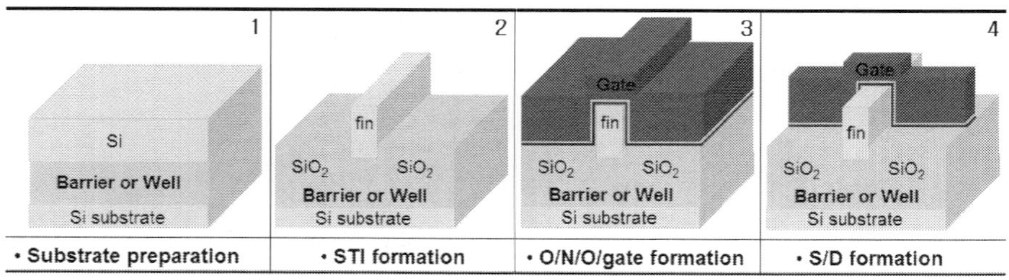

Fig. 3-4. Process flow of the URAM. After the quantum substrates for the floating body are prepared, the subsequent process flow is identical. Whereas SOI itself provides the intrinsic floating body, bulk substrates are hindered by the energy band engineering to form the extrinsic floating body. The n+ deep ion implantation is carried out for the SONW, $Si_{1-y}C_y/Si$ is epitaxially grown for SOSC, and $Si_{1-x}Ge_x/Si$ is epitaxially grown for SOSG.

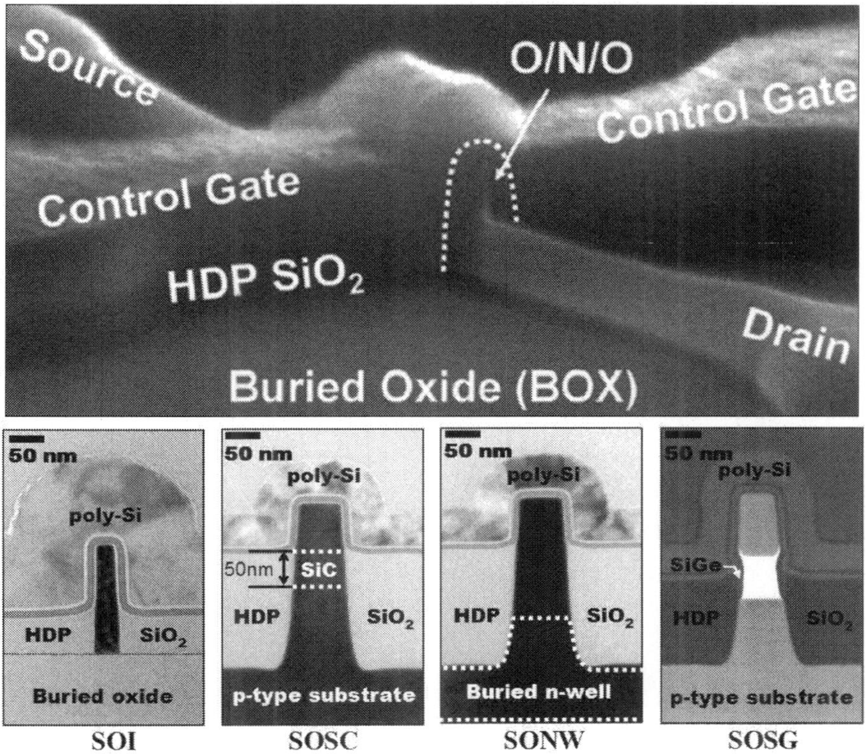

Fig. 3-5. Tilted view of the SOI URAM (upper), and cross-sectional view of four types of URAM (lower).

Tilted scanning electron microscopy (SEM) image and cross-sectional transmission electron microscopy (TEM) images of the fabricated device on various quantum templates are shown in Fig. 3-5. Table 3-1 summarizes the geometric dimensions.

	SOI	SOSC	SONW	SOSG
Gate length	180nm	180nm	180nm	180nm
Fin width	30nm	50nm	50nm	50nm
Fin height	110nm	50nm	100nm	85nm
O/N/O thickness	3/6/3nm	4/6/4nm	4/6/4nm	3/6/4nm
Doping concentration	$5 \times 10^{17}/cm^3$	undoped	$5 \times 10^{18}/cm^3$	undoped

Table 3-1. Summary of geometric dimensions for four types of URAM.

4. Device performance

4.1 Direct Current (DC) characteristics of URAM

Once the devices are fabricated, the fundamental properties should be investigated to find out whether the current-voltage characteristics are acceptable. The drain current (I_D) versus gate voltage (V_G), *i.e.* transfer characteristics, is commonly monitored, providing important parameters such as threshold voltage (V_T), on-current (I_{on}), off-current (I_{off}), subthreshold slope (SS), drain induced barrier lowering (DIBL), *etc.* Fig. 4-1 shows the transfer plot for URAM. The SOI exhibits the steepest SS due to the well known fact that the depletion capacitance is the smallest at SOI. Whereas V_T of the SOI, SOSG, and SOSG are similar, that of the SON is larger than others because of the high body doping concentration. In order to avoid an electrical short between n^+ source/drain and n^+ well, body doping concentration should, reluctantly, be high. Thus, driving current degrades due to the mobility degradation stemming from impurity scattering. It is found that the other parameters are superior to the counter devices (planar single-gate structure), which is attributed to the three-dimensional device structure. The device parameters are summarized in Table 4-1.

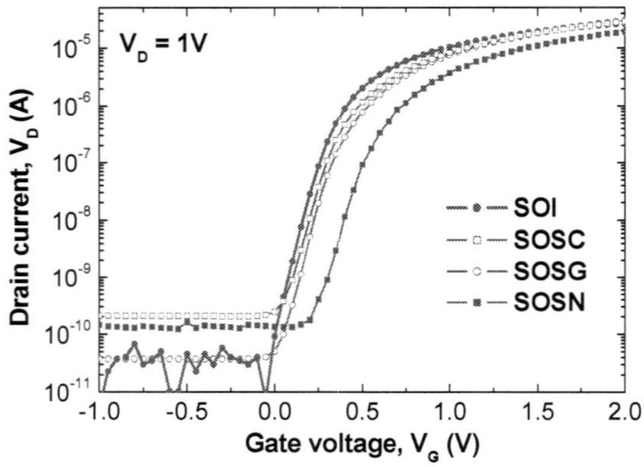

Fig. 4-1. Drain current versus gate voltage characteristics for various types of URAM. The superior device properties are attributed to the three dimensional device structure.

	SOI	SOSC	SOSG	SONW
Threshold voltage	0.21V	0.29V	0.33V	0.5V
Subthreshold slope	85mV/dec	93mV/dec	95mV/dec	101mV/dec
DIBL	32mV/V	110mV/dec	115mV/dec	151mV/dec
On-current	$9.3x10^{-11}$A	$2.4x10^{-10}$A	$1.3x10^{-10}$A	$5.1x10^{-11}$A
Off-current	$1.0x10^{-5}$A	$8.2x10^{-6}$A	$3.7x10^{-6}$A	$7.6x10^{-6}$A

Table 4-1. Summary of the device performances. The high threshold voltage in SONW is attributed to the high body doping concentration.

The simplest method to verify whether the impact ionization generates excess holes that will be stored inside the body is to examine the kink point in the drain current (I_D) versus drain voltage (V_D), *i.e.* output characteristics. As the drain voltage increases, the impact ionization process begins to occur beyond a certain drain voltage, generating pairs of electrons and holes. While the generated electrons flow out toward the drain terminal, the generated holes are repulsed to the body by positive drain voltage. In bulk substrates, generally, these holes are collected by a grounded substrate terminal, appearing as a form of substrate current. If the body is electrically floated, however, the holes are accumulated, contributing as an extra quasi-gate. Therefore, the accumulation of excess holes causes current increase at certain drain voltage, and anomalous output characteristics can be found. Fig. 4-2 shows the output characteristics for URAM. The kink points assure that the excess holes are effectively accumulated, even at the bulk substrates, which are quantum mechanically engineered.

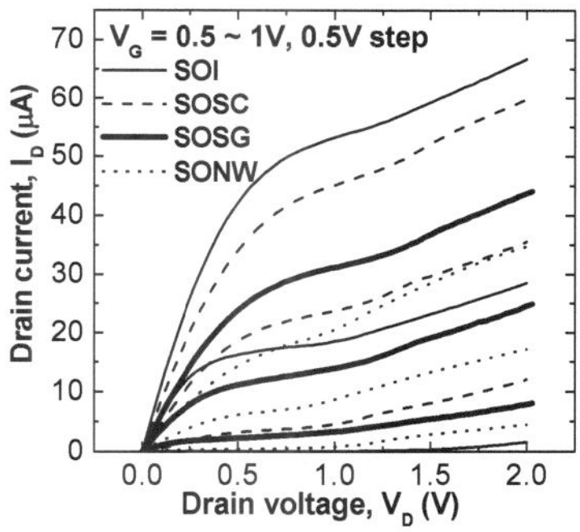

Fig. 4-2. Drain current versus drain voltage characteristics for various types of URAM. As the drain voltage increases, the excess holes generated by the impact ionization are stored in the floating body, resulting in a kink in the saturation region.

4.2 Flash memory characteristics

Flash memory performance is normally evaluated in terms of four aspects: program speed, erase speed, data retention time, and endurance cycles. For nonvolatile memory application, the cells should satisfy the 10-years data retention requirement with 10^7 program/erase endurance cycles. The ability to store and recover data after ten years is called 'retention', and the ability to withstand repeated program/erase cycles is called 'endurance'. The program/erase can be carried out by Fowler-Nordheim (FN) tunneling or hot-carrier injection (HCI). In this study, the program/erase is enabled by FN tunneling. Fig. 4-3 shows representative program/erase transient characteristics. The characteristics are obtained from the SOI, representatively. As the program/erase voltages are increased, a higher threshold voltage shift is achieved. In addition, as the program/erase time is increased, the threshold voltage is at first shifted and then saturated after a certain time. Normally, the erase speed is slower than the program speed because the tunneling efficiency of holes is lower than that of electrons due to the high effective mass and energy barrier height in the valence band side. Thus, in a memory array, erase operation is normally carried out by block erasing to improve the erasing throughput. Here, a V_T window of 3.3V is achieved at the program of 11V with 80μsec and the erase of -11V with 10msec.

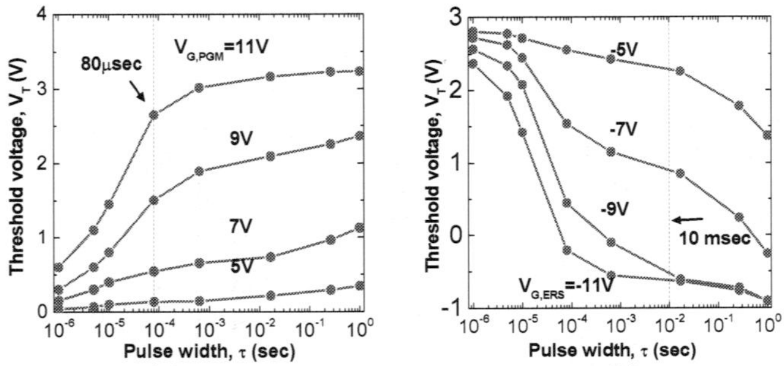

(a) (b)

Fig. 4-3. Program/erase transient characteristics of Flash memory mode at SOI URAM. (a) program transient and (b) erase transient characteristics. (Han et al., 2007)

The retention and endurance are crucial factors that determine the reliability of the Flash. Fig. 4-4 shows that the 10-years retention and 10^7 cycles are guaranteed with a 1.9V detection window. Table 4-2 summarizes the reliability factors for various templates. No memory retention and endurance failure are obtained as long as the detectable threshold voltage window is greater than 1V. The endurance failure for SOSG is speculated to be caused not by a structure related failure, but by a process induced failure of O/N/O.

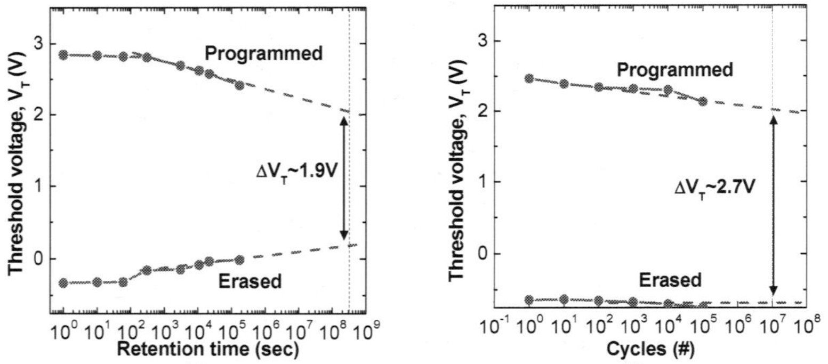

(a) (b)

Fig. 4-4. (a) Data retention and (b) endurance characteristics of Flash mode at SOI URAM. (Han et al., 2007)

	SOI	SOSC	SOSG	SON
ΔV_T (11V / 80μsec, -11V / 10msec)	3.3V	3.6V	3.0V	3.6V
ΔV_T (after 10 years)	1.9V	3.5V	2.6V	1.6V
ΔV_T (after 10^7 cycles)	2.7V	3.2V	Fail	2.7V

Table 4-2. Summary of program/erase efficiency and reliability for various types of URAM.

4.3 Capacitorless 1T-DRAM characteristics

A capacitorless 1T-DRAM mode is characterized by the customized system. Fig. 4-5 shows the measurement system. The computer controls the pulse generator (Agilent 81110A), oscilloscope (Agilent 54542C), and current amplifier (Keithley 428). The pulse generator applies voltage patterns to the device. The source current is amplified by the current amplifier, changed into a form of voltage, and monitored by the oscilloscope. For low noise measurement, a low noise cable with length of 50cm is used. The device is tested under the probe station (Cascade R4840). All operations utilize the gate voltage of 1V, which is not an indispensable condition, but for monotone waveform to simplify the sensing circuit circuitry.

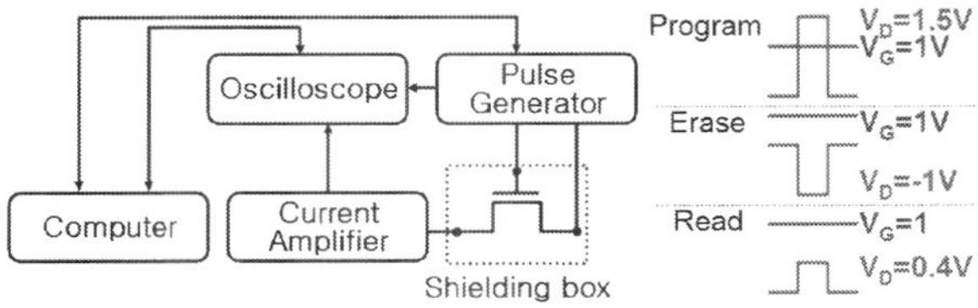

Fig. 4-5. Customized measurement system and the operational pulse waveform.

In order to minimize the leakage paths in the three-dimensional FinFET, the fin width should be as narrow as possible. This means that a fully depleted body is desirable in terms of scalability. The capacitorless 1T-DRAM, however, requires a partially depleted body, *i.e.* wider fin width, to store the detectable amount of holes. To compromise the scalability and the performance functionality, the fin is divided into two regions (Han et al., 2009). Fig. 4-6 compares the conventional FinFET and the proposed one as counterpart structures. Whereas the fin is fully surrounded by the gate at the conventional FinFET, the fin of the proposed one is partially covered. The essence of the proposed one is that the hole accumulation region is spatially separated from the inverted channel. The upper part covered by the gate, which is fully depleted, provides a conduction path. The lower part covered by STI oxide, which is partially depleted, serves for a hole storage. Therefore, scalability and performance functionality (the floating body effect) are attained at the same time.

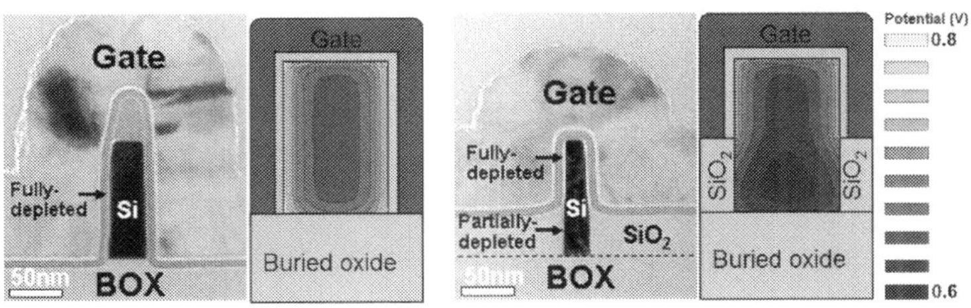

(a) (b)

Fig. 4-6. Comparative images of (a) the conventional fully-depleted FinFET SONOS and (b) the proposed half fully-depleted and half partially-depleted FinFET SONOS. The contours of the body potential supported by simulation assure that the existence of a partially depleted region to accommodate more holes is attractive for proper 1T-DRAM operation. Consequently, the proposed FinFET is superior to the conventional FinFET. (Han et al., 2007)

Fig. 4-7 shows the program/erase characteristics of the capacitorless 1T-DRAM. As mentioned in Section 2.3, the program/erase voltage should be optimized in order to avoid undesired charge trapping in the O/N/O layer. The program uses $V_{G,PGM}$=1V and $V_{D,PGM}$=1.5V, the erase uses $V_{G,ERS}$=1V and $V_{D,ERS}$=−1V, and the read voltages are $V_{G,READ}$=1V and $V_{D,READ}$=0.4V. Before utilizing the capacitorless 1T-DRAM mode, the initial V_T is set to 0.2 V. The data states are clearly distinguished with a 7μA sensing window after 80msec data retention, whereas the conventional device exhibits a smaller sensing window. This is attributed to the presence of increased excess hole accumulation as shown in Fig. 4-6.

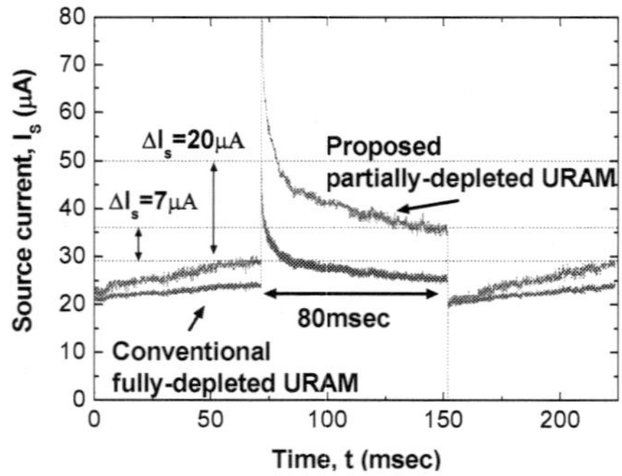

Fig. 4-7. Source current for the capacitorless 1T-DRAM mode. The two data states are clearly identified because more holes are accumulated in the partially-depleted URAM. However, the source current difference is relatively small in the conventional fully-depleted URAM. (Han et al., 2007)

In the SONW substrate, the buried n-type well is embedded inside a p-type bulk substrate. The junction of the p-type body and the n-type well forms the pn built-in potential barrier, thus the excess holes can be retained inside the p-type body region. In order to prove that the excess holes can really be confined, the simulated contours of the hole concentration after the program are shown in Fig. 4-8. In conventional bulk substrates, excess holes are generally collected by the grounded substrate. At the SONW substrate, the holes confront the n-well junction barrier, and the holes are thus accumulated inside the body.

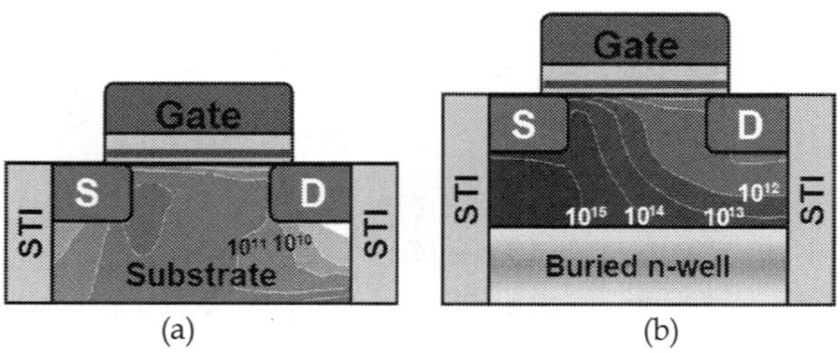

Fig. 4-8. Simulated contours of the hole concentration biased at hold condition after impact ionization, (a) convention bulk FinFET, and (b) SONW URAM. In contrast to the conventional case, SONW URAM stores the excess holes in the body region. (Han et al., 2008a)

Fig. 4-9 shows the program/erase characteristics. The important feature in the bulk substrates is that the barrier height can be modulated by the substrate voltage. In other words, the ability to retain holes can be improved by proper substrate voltage. An applying a weak positive voltage and enlarging the hole barrier height can enhance the sensing window. The sensing window with retention time is increased from 4μA with 8msec to 7μA with 30msec as the substrate voltage is increased from 0V to 0.3V. In the case of strong positive voltage, however, the capacitorless 1T-DRAM cannot work because the forward biased source/drain to the body junction diode is inevitably turned on.

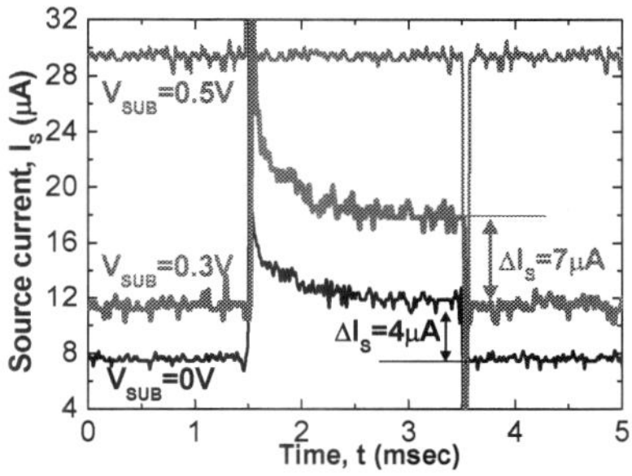

Fig. 4-9. Source current for capacitorless 1T-DRAM of SONW. The sensing window is widened at a small positive substrate voltage. (Han et al., 2008a)

Since the SONW substrate needs the deep implantation process, it is hard to define an accurate and abrupt quantum engineered junction profile. The SOSC would be preferred as its energy band is determined by epitaxial growth and the mole fraction of C in $Si_{1-y}C_y$. In addition, whereas the buried n-well should be located far from the source/drain junction in order to avoid the electrical short, the band offset interface of $Si/Si_{1-y}C_y$ can be at closer to the source/drain so that the influence of stored holes on the inverted channel becomes stronger. Therefore, SOSC gives a rise to improvements in performance. Fig. 4-10 shows the program/erase characteristics. The sensing window of 11μA with a retention time of 50msec at a substrate voltage of 0.3V is wider compared to that of SONW (Han et al., 2008). Also, the sensing window is wider at V_{SUB}=0.3V than at V_{SUB}=0V as predicted.

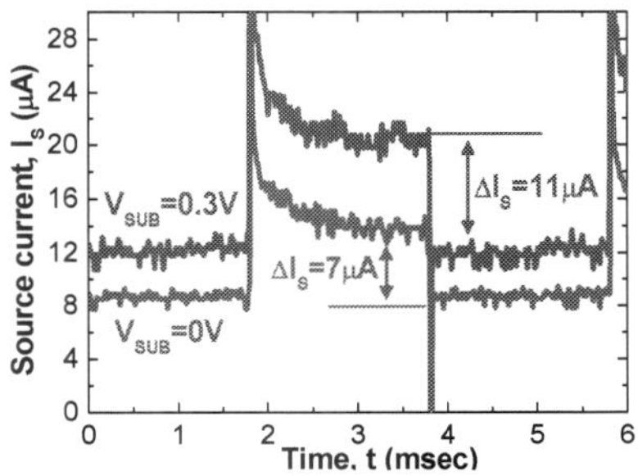

Fig. 4-10. Source current for the capacitorless 1T-DRAM of SOSC. The small positive substrate voltage raises the sensing current window. (Han et al., 2008a)

The above three substrates, SOI, SONW, and SOSC, showed a quantum barrier, *i.e.* the energy band of the floating body is above that of the quantum engineered substrate. In contrast, SOSG is the quantum well structure because the energy band of the floating body is below that of the quantum engineered substrate. The quantum barrier type substrates use their bodies for the conduction path as well as the storage region, simultaneously. This condition can cause the excess holes to easily disappear by recombination with the inverted electrons, leading to degradation in the data retention time. However, the quantum well can separate the excess holes and conduction electrons; thus, the stored charge loss via the recombination process with an inverted electron is expected to be minimized, and improved performance is predicted. The Si/$Si_{1-x}Ge_x$/Si, SOSG, forms the potential well structure because the valence band energy of $Si_{1-x}Ge_x$ is higher than that of Si, as shown in Fig. 4-7. In SOSG, the top Si serves as the conduction channel, and the centered $Si_{1-x}Ge_x$ is devoted to the hole storage region. The major advantage compared to SOSC is that, whereas the solid solubility of carbon in silicon is limited to 5%, the germanium content can be adjusted from 0% to 100%, which allows wide band offset modulation by changing stoichiometry of $Si_{1-x}Ge_x$. Therefore, the SOSG can provide more degrees of freedom in the energy band design because the depth of the potential well is favorably determined by the germanium content. The impact of germanium on the band offset has been theoretically reported, and it turns out that the valence band offset between Si and $Si_{1-x}Ge_x$ is linearly increased with content x. The simulated distribution of excess holes after programming and maximum hole concentration for various content x are shown in Fig. 4-11. The holes are found to be preferentially accumulated in the $Si_{1-x}Ge_x$ layer. The hole concentration is exponentially increased as the valence band offset is increased. However, the hole concentration starts to saturate after a band offset of 0.24eV, which corresponds to germanium content of x=0.4. This means that the usage of Ge higher than x=0.4 will be ineffective in terms of the ability to store holes. In other word, a very deep potential well is not always necessary for higher performance (Han et al., 2008b).

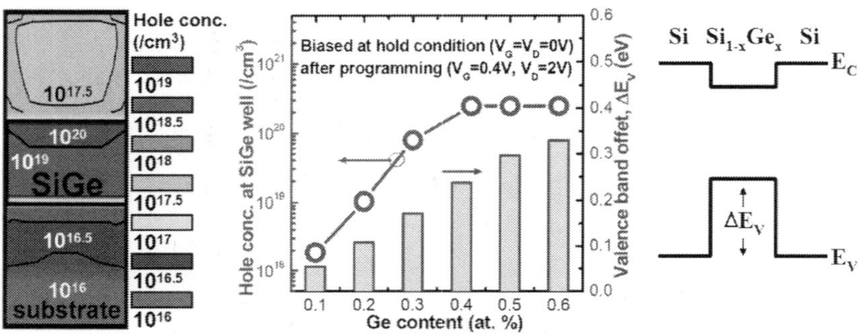

Fig. 4-11. (a) Simulation results of hole concentration biased at hold condition after impact ionization. The stored excess holes are found in a $Si_{1-x}Ge_x$ potential well, (b) the valence band offset as a function of germanium content and resultant hole concentration, and (c) the energy band offset in the SOSG structure. The amount of stored holes is exponentially increased as the valence band offset is increased, but the hole concentration starts to saturate after a band offset of 0.24eV. (Han et al., 2008b)

Fig. 4-12 shows the program/erase characteristics for two germanium contents, x=0.3 and x=0.5. Despite of the larger sensing window at x=0.5, the retention is found to be inferior to that at x=0.3 due to the fact that the higher defect density caused by an atomic lattice mismatch at x=0.5 induces faster recombination during read operation. In addition to the retention degradation at the programmed state, a deeper potential well also is found to degrade the retention at the erased state. The reason is speculated to be that the holes are easily diffused into the potential well from the neighbored p-type silicon layers, as illustrated in Fig. 4-13. As a result of the trade off between sensing current window and retention time, the optimized stoichiometry of $Si_{1-x}Ge_x$ is x=0.3. The retention time (the order of microseconds) appears to be insufficient to practical application, however, refinement of the epitaxial process and geometric optimization of the 3-D structure will enhance the performance.

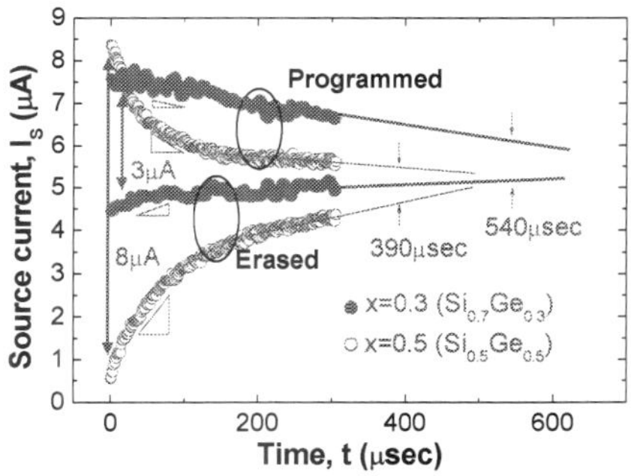

Fig. 4-12. Capacitorless 1T-DRAM characteristics for different germanium content in $Si_{1-x}Ge_x$ of SOSG. The higher x exhibits a wider sensing window at the beginning of the sensing, but also faster charge loss. (Han et al., 2008b)

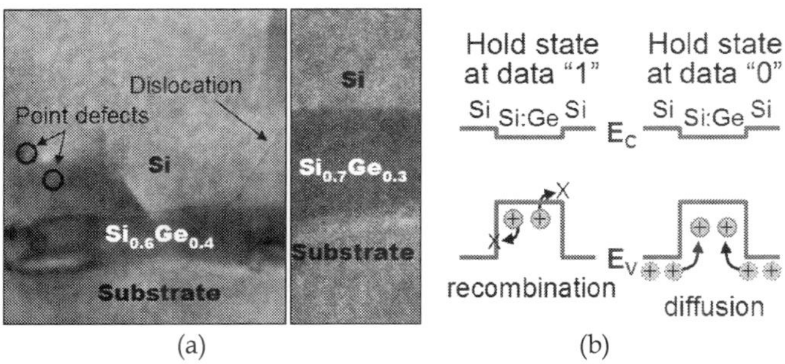

Fig. 4-13. (a) Transmission electron microscopy images of x=0.4 and 0.3 and (b) schematics for retention degradation mechanisms. The high germanium content induces a high lattice mismatch because the lattice constant of germanium is larger than that of silicon. The defects originating from the lattice mismatch reduce the data retention at the programmed state via charge recombination, and a deeper potential well degrades the data retention at the erased state due to hole-to-hole repulsion and its repellent diffusion mechanism. (Han et al., 2008b)

Table 4-3 summarizes the features and performances of four types of URAM. The SOI substrate exhibits the fastest write speed, the widest sensing window, and the longest retention among the four substrates. Among the bulk types, the SOSC substrate displays superior performance.

	SOI	SONW	SOSC	SOSG
substrate	SOI	bulk	bulk	bulk
energy band type	valence band barrier	built-in potential	valence band barrier	valence band well
energy band abruptness	abrupt	gradual	abrupt	abrupt
band offset / built-in potential	4.8 eV	0.9 eV	0.18 eV	0.05~0.32eV
channel and storage region	share	separate	share	separate
program speed	6 nsec	20 nsec	20 nsec	50 nsec
sensing window	~20 μA	~7 μA	~11 μA	~8 μA
retention time	~ 80 msec	~ 30 msec	~ 50 msec	~ 600 μsec

Table 4-3. Summary of the features and performance of various URAMs. All data were measured at 300K.

5. Soft-programming issue and solutions

5.1 Soft-programming issue

URAM can be realized by combining the O/N/O gate dielectric to store electrons and the floating body to capture holes in a single transistor. Unfortunately, impact ionization for the program of the capacitorless 1T-DRAM can adversely affect the stored charges in the O/N/O layer. This gives a rise to an undesired threshold voltage shift, which is called 'soft-programming'. The strong impact ionization condition provides faster program speed, a wider sensing window, and longer retention time, but this simultaneously increases the hot-electron injection into O/N/O, leading to instability as a result of the disturbance between the Flash and capacitorless 1T-DRAM modes. Thus, the program condition of capacitorless 1T-DRAM has reluctantly been bounded in order not to disturb the Flash memory states. This is becoming an increasingly important concern.

In order to clarify the soft-programming, Fig. 5-1 shows the capacitorless 1T-DRAM performance after 10^5 cyclic operations. When the drain voltage for a program is 1.8V, a sensing current window of 6μA is sustained, which means the interference is negligible. To improve the performance, when the drain voltage is increased to 2.2V, hot-electron injection is unpropitiously caused. This causes a gradual charge trapping into O/N/O, and the resultant sensing window is decreased. Therefore, a soft-program poses a constraint on the maximum program voltage. In order to overcome this issue, soft-programming immune device structures and operational methods are suggested in the following subsections.

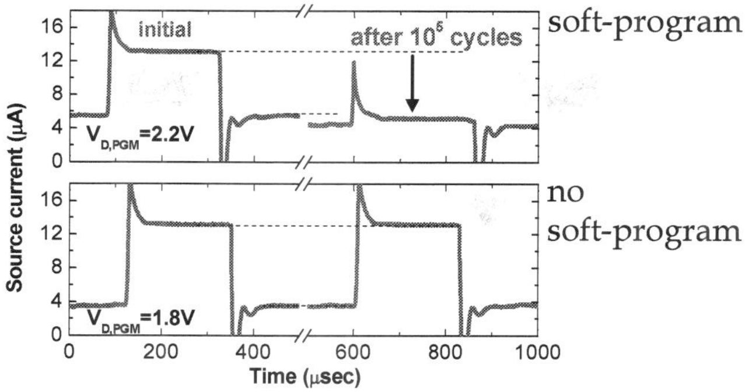

Fig. 5-1. Program/erase characteristics by the impact ionization method for different program voltages. The sensing window is reduced at a high drain voltage as the stress cycles increase due to hot electron injection into the nitride layer. (Han et al., 2009a)

5.2 Soft-programming immune structure: gate-to-S/D nonoverlap structure

The soft-program tends to occur as the impact ionization process is triggered under the gate. Thus, if the impact ionization region is steered out of the O/N/O layer, hot electron injection can be mitigated. The impact ionization process occurs at the region with the highest electric field, *i.e.*, drain end. Thus, a gate-to-source/drain nonoverlap creates an impact ionization region located outside of the gate. Even though the impact ionization triggers a hot-electron injection, the charge trapping is alleviated since there are no trap sites. Thus, the constraint of program bias is relieved. For this purpose, junction nonoverlap structure is fabricated and compared to the conventional overlap structure. Fig. 5-2 shows the fabricated device images. The nonoverlap length is 20nm.

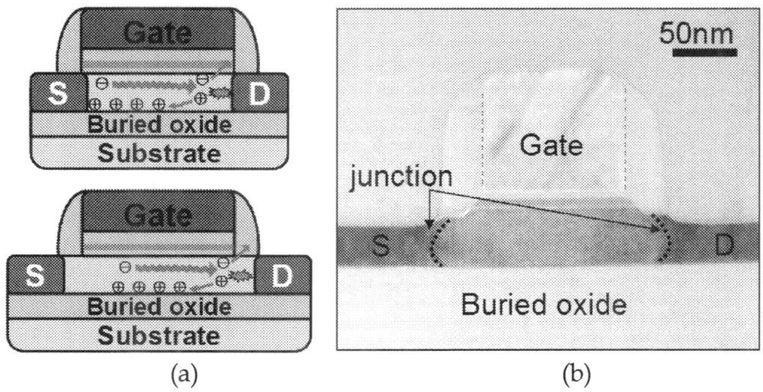

Fig. 5-2. (a) Schematics of gate-to-source/drain overlap and nonoverlap structure and (b) transmission electron microscopy image of the gate-to-source/drain nonoverlap devices. The body thickness is 50nm, the gate length is 110nm, and the nonoverlap length is 20nm. (Han et al., 2009a)

Fig. 5-3 shows the memory characteristics for both structures. Even though the nonoverlap device may suffer from degradation in the impact ionization efficiency, the sensing current window of the nonoverlap device is found to be wider than that of an overlap one (Fig. 5-3a) because the nonoverlap reduces the junction leakage and recombination rate. In addition, the effective volume of the floating body is extended by the amount of nonoverlap. As a result, reduced impact ionization efficiency can be compromised (Song et. al., 2008). In the Flash memory characteristics shown in Fig. 5-3b, a threshold voltage window of 4.3V is achieved. The threshold voltage for a fresh device is higher in a nonoverlap than at an overlap device, and the threshold voltage window of a nonoverlap structure is narrower than that of an overlap structure. Despite of the degradation in the threshold voltage window for flash memory, the window of 4.3V is acceptable to identify the data states (Han et al., 2009).

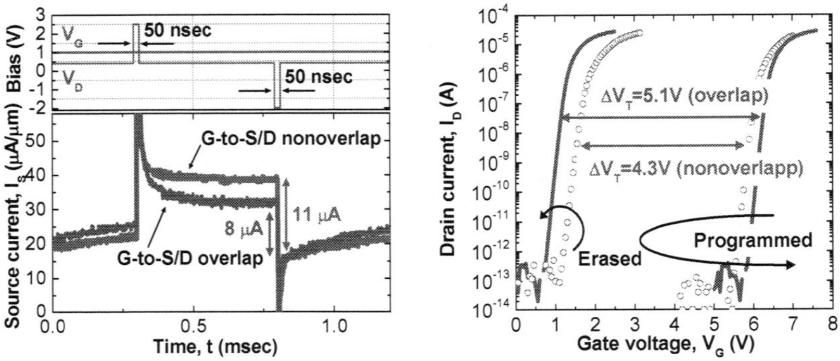

(a) (b)

Fig. 5-3. (a) Capacitorless 1T-DRAM and (b) Flash memory characteristics. The nonoverlap structure shows a wider sensing current window in capacitorless 1T-DRAM mode, but narrow threshold voltage window in Flash mode. (Han et al., 2009a)

In order to evaluate the soft-programming immunity, a stress test is carried out. While the program voltage is applied, the threshold voltage shift is periodically monitored during the operation cycles. As shown in Fig. 5-4, whereas the overlap structure shows a threshold voltage shift of 0.2V, the nonoverlap device exhibits distinctively superior immunity against the soft-program to the overlap device.

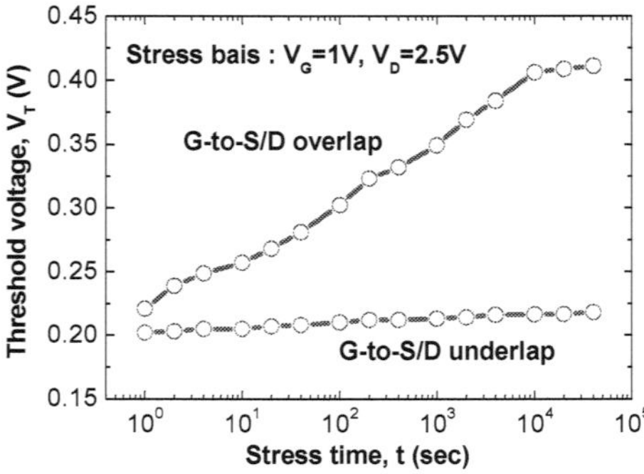

Fig. 5-4. The threshold voltage shift monitored during the cyclic operations. The nonoverlap structure shows superior immunity against the soft-program. (Han et al., 2009a)

It is worthwhile to note that the major weakness of the nonoverlap junction structure is that the parasitic voltage drops via series resistance at the nonoverlap region reduces the impact ionization efficiency. This drawback can be countervailed by increasing the dielectric constant of the gate offset spacer. The high dielectric constant of the spacer can increase the effect of the gate fringing field to the nonoverlap region, which is expected to boost the impact ionization rate (Ma et. al., 2007). Therefore, the abundance of excess holes can further improve the current drivability and recover performance of the capacitorless 1T-DRAM against the sacrificed impact ionization efficiency.

5.3 Soft-programming immune operation: gate-induced-drain-leakage program

To date, impact ionization was commonly used to create excess holes in the body. However, in place of the impact ionization, there is another method to generate excess holes; the gate-induced-drain-leakage (GIDL). A device biased on the GIDL condition, *i.e.*, negative gate and positive drain voltage, creates excess holes in the body by band-to-band tunneling. The impact ionization program significantly wastes power since it is triggered by high drain current. However, GIDL current does not require such drain current; thus, low power operation is feasible. If the O/N/O is in the erase saturation state prior to activating the capacitorless 1T-DRAM mode, the hole injection into O/N/O is effectively restricted. In addition, the hole injection is even suppressed because the effective mass and energy band barrier of the hole in the valence band side are high. Fig. 5-5a shows the program/erase pulse waveform of the GIDL program method and resultant sensing current. The current window of 12μA with 50msec data retention facilitates the data sensing. In order to verify the immunity against the soft-program, the stress test is carried out by the impact ionization and GDIL program methods. The amount of trapped charges is evaluated by monitoring the shift of the threshold voltage. Fig. 5-5b shows the impact of cyclic capacitorless 1T-DRAM on the threshold voltage shift. Whereas the impact ionization condition induces the charge trapping and results in a threshold voltage shift, the GIDL method does not. Thus, the GIDL method is the effective tool to achieve a soft-program immune operation (Han et al, 2009).

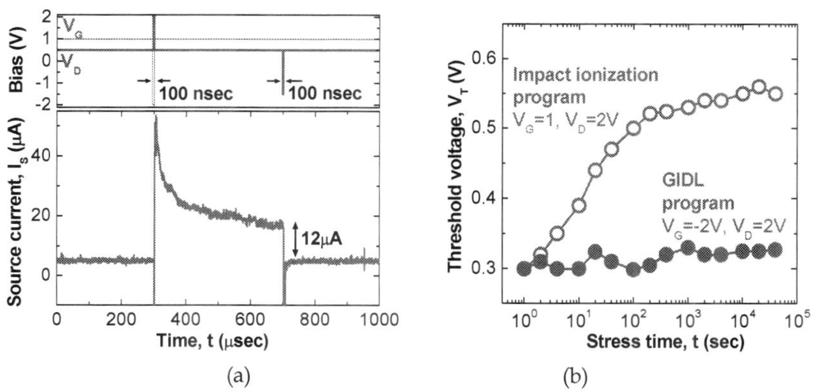

Fig. 5-5. (a) Program/erase characteristic of GIDL program method and (b) threshold voltage shift versus cyclic stress time. The GIDL program method does not shift the threshold voltage, while the impact ionization program does. (Han et al., 2009b)

It is worthwhile to note that the GIDL program method in URAM may be inefficient in terms of the generation efficiency of holes. As the Flash memory utilizes O/N/O gate dielectric, such a thick gate dielectric hampers to achievement of sufficient band bending for band-to-band tunneling. Thus, a programming time of 100nsec was used, which would be too long to apply for the embedded system. Since the thickness of O/N/O is no longer scalable to sustain acceptable nonvolatility, a higher gate voltage would be required, but it also poses power issues. In order to overcome this drawback, a p^+ polysilicon gate on the p-type body can be used. Since the flat band voltage difference between p^+ polysilicon and n^+ drain is higher than that of n^+ polysilicon and n^+ drain, a higher GIDL current is induced at a given gate voltage (Lindert et. al, 1996). Therefore, the implementation of the p^+ polysilicon gate can yield improved memory characteristics in the GIDL method.

5.4 High performance and soft-programming immune operation: parasitic BJT read

In the first prototype of URAM, the impact ionization program condition caused a soft-program issue. Next, despite the suppression of the soft-program, the GIDL program tends to sacrifice program efficiency. In summary, both methods have their distinctive strengths as well as weaknesses, simultaneously. In this section, a third method is introduced for improved performance with soft-program free operation. It is important to note that the floating body MOSFET contains a parasitic lateral bipolar junction transistor (BJT) composed of n^+ source, p-type body, and n^+ drain, which correspond to an emitter, base, and collector, respectively. As the p-type body is floated, the BJT with the floating base cannot be activated in the normal MOSFET operational conditions. However, if the high voltage is applied to the drain, the hole injection to the floating base can turn on the parasitic BJT, and the drain current is maintained even though the MOSFET is supposed to be turned off (Chen et. al, 1988). Fig. 5-6 shows the double-sweep transfer characteristics. At a low drain voltage, the normal MOSFET transfer curved is shown, and there is no hysteresis. At high drain voltage, however, the subthreshold slope approaches 0mV/dec at the time of the parasitic BJT activation, and a hysteresis loop is generated. Thus, even at the given read voltage, bistable current-voltage characteristics can be utilized as a single memory transistor, as indicated in the Fig. 5-6 (Okhonin et. al, 2007).

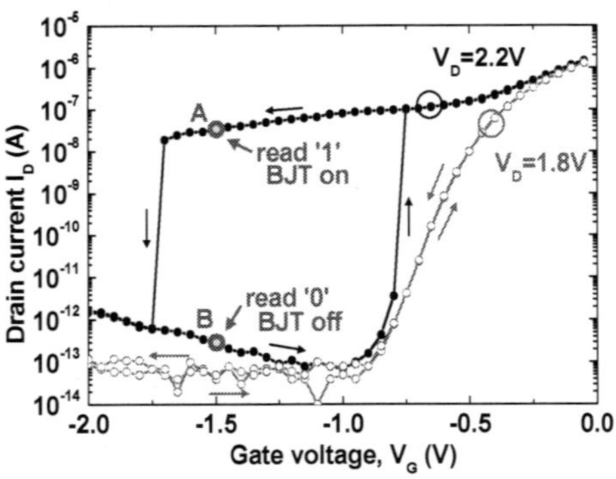

Fig. 5-6. Double sweep drain current versus the gate voltage characteristics at the SOI URAM. At V_D=1.8V, the device shows normal MOSFET transfer characteristics. At V_D=2.2V, the parasitic BJT alternatively begins to work, and the hysteresis loop is created.

Fig. 5-7 comparatively shows the capacitorless 1T-DRAM characteristics with the previous two program methods and the BJT read method. A pulse width of 5nsec is applied for program and erase in the BJT mode. The programming is carried by impact ionization or GIDL, and the erase is fulfilled by forward junction current. The difference lies in the read condition. The previous methods used low drain voltage, typically V_D<0.6V, in order not to disturb the body charged state, resulting in a small sensing current. In addition the sensing current window was gradually narrowed by the generation and recombination processes. In contrast, the parasitic BJT read uses a high drain voltage at least V_D>2V to activate bipolar action. While the negative gate voltage turns off the MOSFET, the parasitic BJT can either be activated or deactivated according to an excessive number of holes or lack of holes. When excess holes exist in the body, a parasitic BJT is activated in which the current corresponds to the point A in Fig. 5-6. On the other hand, when excess holes are eliminated, the parasitic BJT is deactivated, thereby causing the current not to flow, which corresponds to point B in Fig. 5-6. In particular, once the parasitic BJT is activated, the high BJT current is latched despite the MOSFET being in off state because the hole is continuously supplied as long as the read voltage is applied. Therefore, the BJT read method is considered to be completely non-destructive, and the sensing current window is high enough that a sense amplifier may not be necessary to identify the data.

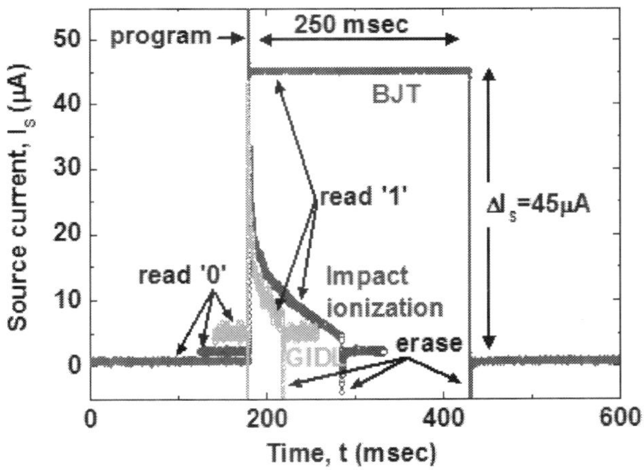

Fig. 5-7. Comparison of the capacitorless 1T-DRAM characteristics with various methods. In the conventional read method, the sensing current window is gradually narrowed with the read time. In the BJT read method, the source current remains constant because the stored data at BJT read condition is latched.

It is important to note that the read and program operations correspond to the hot-hole injection conditions, which can cause a threshold voltage shift during cyclic capacitorless 1T-DRAM operations. This situation seems similar to the soft-program, but it turns out that the soft-program in BJT is negligible, as shown in Fig. 5-8. If the nitride traps are saturated with holes that can be carried out by an initialization step before the capacitorless 1T-DRAM mode, there are no additional threshold voltage shifts because there are no extra available trap sites in the nitride. According to the stress test data, the threshold voltage shift is found to be negligible, resulting in stable operation. The soft-program free scheme can exclude the operation loop of the verification of the soft-programming and re-initialization that is supposed to be required in the conventional methods, as shown in Fig. 5-9. The elimination of the redundant loop can greatly conserve the integrity of the gate oxide, which can otherwise be degraded by repeated initialization processes.

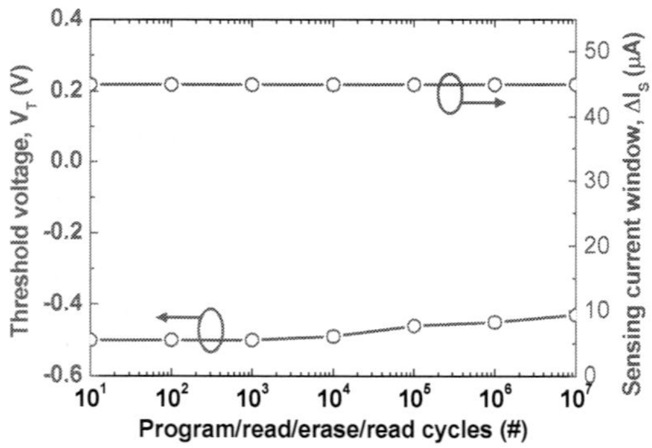

Fig. 5-8. Threshold voltage and sensing current window versus the operation cycles. The threshold voltage shift and sensing current window degradation are found to be negligible, which guarantees very stable URAM operation without soft-programming.

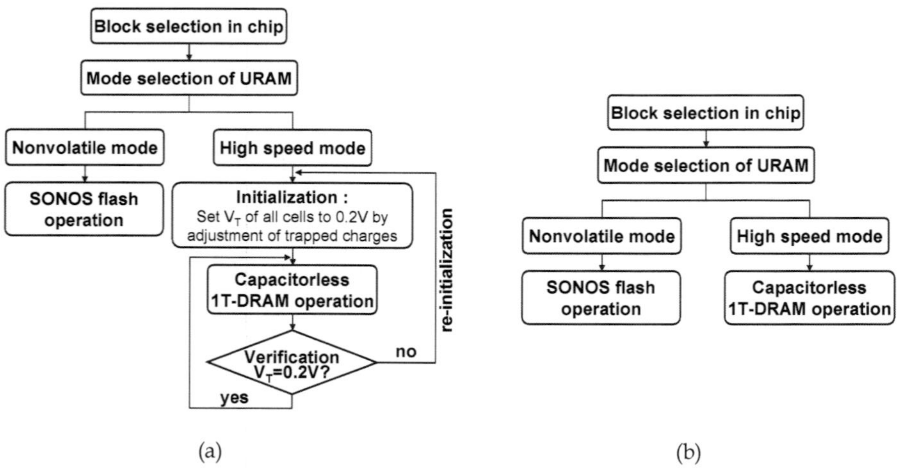

Fig. 5-9. Operational sequence for URAM. (a) The conventional read method and (b) the parasitic BJT read method. In the conventional read method, the verification and re-initialization loop is necessary due to the soft-program issue. In contrast, the parasitic read method excludes the redundant loop because the interference between the two modes is eliminated.

6. Conclusions

In this chapter, as we confront challenges of current memory technology and as the design rule deviates from the historical scaling paradigm, a novel memory scheme is proposed to continue the roadmap beyond the end point of silicon based memory. Over the scaling to multi-bit era, the multi-functional paradigm is proposed. Whereas the conventional fusion memory pursues high-cost multi-chip-package technology, multi-function is realized in a single memory transistor. The functions of nonvolatile Flash memory and high-speed DRAM are co-integrated, and this memory is named Unified-RAM or URAM. The combination of oxide/nitride/oxide gate dielectric and the floating body structure provide two functions in a single memory cell. In addition, the inherent operational bias domain for two functions allows independent function depending on the end user's demand. The various floating body substrates designed with consideration of the quantum mechanics were proposed in order to confine the excess hole to operate the capacitorless 1T-DRAM. In addition to the conventional silicon on insulator (SOI) substrate, three bulk type floating body substrates were developed. The silicon on n-well (SONW) formed by the deep ion implantation and the silicon on $Si_{1-y}C_y$ (SOSC) formed by epitaxial growth were presented for the potential barrier type approach. Furthermore, the silicon on $Si_{1-x}Ge_x$ (SOSG) was developed for the potential well type substrate. Even though the performance of bulk might be inferior to that of SOI, the bulk can be still useful in terms of cost-effective manufacturing and heat dissipation with a moderate sensing window. After the soft-program issue was certificated, the parasitic bipolar junction transistor (BJT) read method was newly proposed for powerful performance with soft-programming immunity.

As URAM is implemented by using standard semiconductor design and fabrication facility, new products can be manufactured quickly, reducing development time and investment cost. The beauty of URAM lies in the fact that it does not require exotic semiconductor materials, oddly structured parts, exploratory insulator, or an extra photolithography step. URAM is considered to be the next generation for advanced memory technology, which will open a new paradigm shift, and it will be a viable successor to the future embedded memory.

7. References

Brown, W. D. & Brewer, J. E. (1998). *Nonvolatile semiconductor memory technology*, IEEE Press, ISBN-13:978-0780311732, Piscataway, NJ

Chen, C.-E. D. et. al. (1988). Single-transistor latch in SOI MOSFET's, *IEEE Electron Device Lett.*, Vol. 9, No. 12, pp. 636-638, IEEE, ISSN:0741-3106

Gautier, J. et. al. (1997). SOI Floating-Body, Device And Circuit Issues, *IEEE IEDM Tech. Dig.* pp. 407-410, IEEE, ISBN:978-1-4244-237-4

Han, J.-W. et. al. (2007). A Unified-RAM (URAM) Cell for Multi-Functioning Capacitorless DRAM and NVM, *IEEE IEDM Tech. Dig.* pp. 929-932, IEEE, ISBN:978-1-4244-237-4

Han, J.-W. et. al. (2008a). Band offset FinFET-based URAM (Unified-RAM) built on SiC for multi-functioning NVM and capacitorless 1T-DRAM, *IEEE VLSI Symp. Tech. Dig.*, pp. 200-201, IEEE, ISBN:4-900784-00-1

Han, J.-W. et. al. (2008b). Energy Band Engineered Unified-RAM (URAM) for Multi-Functioning 1T-DRAM and NVM, *IEEE IEDM Tech. Dig.* pp. 227-230, IEEE, ISBN:978-1-4244-237-4

Han, J.-W. et. al. (2009a). Gate-to-Source/Drain Nonoverlap Device for Soft-Program Immune Unified RAM (URAM), *IEEE Electron Device Lett.*, Vol. 30, No. 5, pp. 544-546, IEEE, ISSN:0741-3106

Han, J.-W. et. al. (2009b). Gate-Induced Drain-Leakage (GIDL) Programming Method for Soft-Programming-Free Operation in Unified RAM (URAM), *IEEE Electron Device Lett.*, Vol. 30, No. 2, pp. 189-191, IEEE, ISSN:0741-3106

Kim, M. & Osten, H. J. (1997). X-ray photoelectron spectroscopic evaluation of valence band offsets for strained $Si_{1-x}Ge_x$, $Si_{1-y}C_y$, and $Si_{1-x-y}Ge_xC_y$ on Si(001), *Appl. Phys. Lett.*, Vol. 70, No. 20, pp. 2702-2704, AIP, ISSN:1077-3118

Lindert, N. et. al. (1996). Comparison of GIDL in P+-poly PMOS and n+-poly PMOS devices, *IEEE Electron Device Lett.*, Vol. 9, No. 12, pp. 636-638, IEEE, ISSN:0741-3106

Moore, G. E. (1965). Cramming more components onto integrated circuits, *Electronics*, Vol. 38, No. 8, pp. 114-117

Ma, M.-W. et. al. (2007). Impact of High-*k* Offset Spacer in 65-nm Node SOI Devices, *IEEE Electron Device Lett.*, Vol. 28, No. 3, pp. 238-241, IEEE, ISSN:0741-3106

Ni, W.-X. & Hansson, G. V. (1990). Band offset in pseudomorphically grown $Si/Si_{1-x}Ge_x$ heterostructure studied with core-level x-ray photoelectron spectroscopy, *Phys. Rev. B*, Vol. 42, No. 5, pp. 42-50, APS, ISSN:0163-1829

Okhonin, S. et. al. (2007). New Generation of Z-RAM, *IEEE IEDM Tech. Dig.* pp. 929-932, IEEE, ISBN:978-1-4244-237-4

Okhonin, S. et. al. (2001). A capacitor-less SOI 1T-DRAM concept, *IEEE Int. SOI Conf.* pp. 153-154, IEEE, ISBN:0-7803-7439-8

Ranica, R. et. al. (2005). Scaled 1T-Bulk devices built with CMOS 90 nm technology for low-cost eDRAM applications, *IEEE VLSI Symp. Tech. Dig.*, pp. 38–39, IEEE, ISBN:4-900784-00-1

Song, K.-W. et. al. (2008). 55 nm Capacitor-less 1T DRAM Cell Transistor with Non-Overlap Structure, *IEEE IEDM Tech. Dig.* pp. 727-780, IEEE, ISBN:978-1-4244-237-4

5

Low-Voltage Fully Differential CMOS Switched-Capacitor Amplifiers

Tsung-Sum Lee
National Yunlin University of Science and Technology
Taiwan (R.O.C.)

1. Introduction

Analog signal amplification in discrete-time system can be performed by switched-capacitor amplifiers (Martin et al., 1987). Switched-capacitor amplifier has been used in the design of digital-to-analog converter (Yang & Martin, 1989). The schematic for the switched-capacitor amplifier is shown in Figure 1.

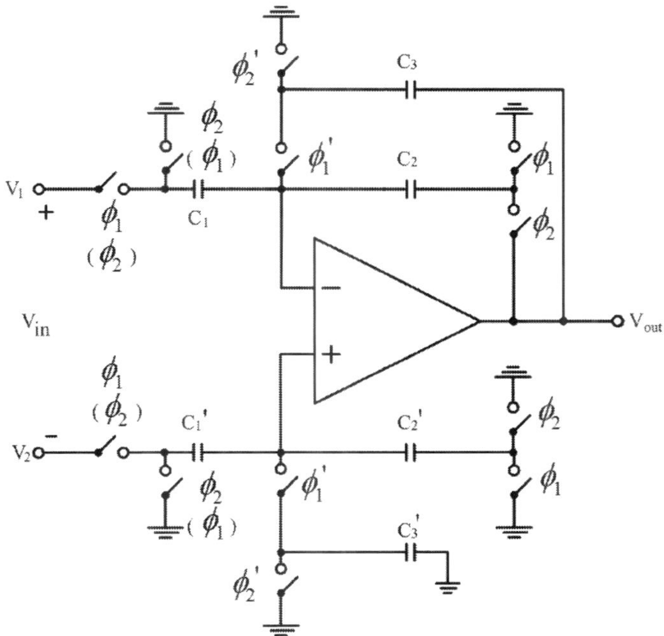

Fig. 1. A differential-to-single-ended CMOS switched-capacitor amplifier. Depending on the input-stage clock signals, the amplifier can be either noninverting (as shown) or inverting (input-stage clocks shown in parentheses).

Assuming an infinite op amp gain, the output voltage at end of ϕ_2 is given by

$$V_{out}(nT) = \frac{C_1}{C_2} V_{in}(nT - \frac{T}{2}), \tag{1}$$

irrespective of the op amp offset voltage. If the clock waveforms shown in parentheses are used, then an inverting function is realized, and

$$V_{out}(nT) = -\frac{C_1}{C_2} V_{in}(nT), \tag{2}$$

again independent of the op amp input offset voltage. During the reset phase (ϕ_1), C_3 is connected in feedback around the op amp which causes the output change only by the op amp input offset voltage. The switches are realized as CMOS transmission gate. For low supply voltages, a conductance gap begins to appear around the middle of the supply range (Crols & Steyaert, 1994). This means that under low-voltage operation, this configuration no longer works. Existing solutions of low-voltage operation of switched-capacitor circuits include using low threshold voltage process (Matsuya & Yamada, 1994), switched-opamp technique (Baschirotto & Castello, 1997; Cheung et al., 2001; Cheung et al., 2002; Cheung et al., 2003; Crols & Steyaert, 1994; Peluso et al., 1997; Peluso et al., 1998; Sauerbrey et al., 2002; Waltari & Halonen, 2001; Wu et al., 2007), opamp-reset switching technique (Chang, & Moon, 2003; Keskin et al., 2002; Wang &. Embabi, 2003), voltage multiplier (charge pump) technique (Nicollini et al., 1996; Rombouts et al., 2001), clock multiplier (clock booster) technique (Au & Leung, 1997; Rabii & Wooley, 1997), and bootstrapping switch technique (Abo & Gray, 1999; Dessouky & Kaiser, 2001; Park et al., 2004). First, the use of low-threshold transistors involves special and high-cost technology (Matsuya & Yamada, 1994). The switched-opamp technique (Baschirotto & Castello, 1997; Cheung et al., 2001; Cheung et al., 2002; Cheung et al., 2003; Crols & Steyaert, 1994; Peluso et al., 1997; Peluso et al., 1998; Sauerbrey et al., 2002; Waltari & Halonen, 2001; Wu et al., 2007) and opamp-reset switching technique (Chang, & Moon, 2003; Keskin et al., 2002; Wang &. Embabi, 2003) can only be applicable to filters, delta-sigma modulators, and pipelined analog-to-digital converters. The main limitations of voltage multiplier (charge pump) technique (Nicollini et al., 1996; Rombouts et al., 2001) regards: the gate-oxide breakdown reliability, the need to supply a dc current to the op amps from the multiplied supply (this necessitates the use of an external capacitor, with additional cost), and the conversion efficiency of the charge pump (which is lower than 100%). The clock multiplier (clock booster) technique (Au & Leung, 1997; Rabii & Wooley, 1997) suffers from the technology limitation associated with the gate oxide breakdown. Device reliability can be assured in the bootstrapped switch technique (Abo & Gray, 1999; Dessouky & Kaiser, 2001; Park et al., 2004), owing to keeping the terminal-to-terminal voltages of the MOSFET devices within the rated operating supply voltage of the technology. The bootstrapped switch provides a small, nearly constant input resistance. The switch linearity is also improved, and signal-dependent charge injections is reduced.

To improve the overall linearity, minimize the effect of common-mode interference and noise, the fully differential approach has obtained wider acceptance for accurate and/or high-speed signal processing. The switched-capacitor amplifier in (Martin et al., 1987) is a differential-to-single-ended design. A fully differential switched-capacitor amplifier using series compensation MOSFET capacitors has been presented in (Yoshizawa et al., 1999).

However its operating voltage is ±2.5-V. Consequently there is an increasing demand to extend these improvements to this circuit.
This chapter describes the design of two 1V fully differential CMOS switched-capacitor amplifiers in a standard CMOS technology using improved bootstrapped switches. In section 2, the circuit realization of these two switched-capacitor amplifiers is addressed. In section 3 the circuit design of low-voltage building blocks is described. Experimental results are presented in section 4 to support the ideas put forth in paper. Finally conclusion is given.

2. Circuit Description

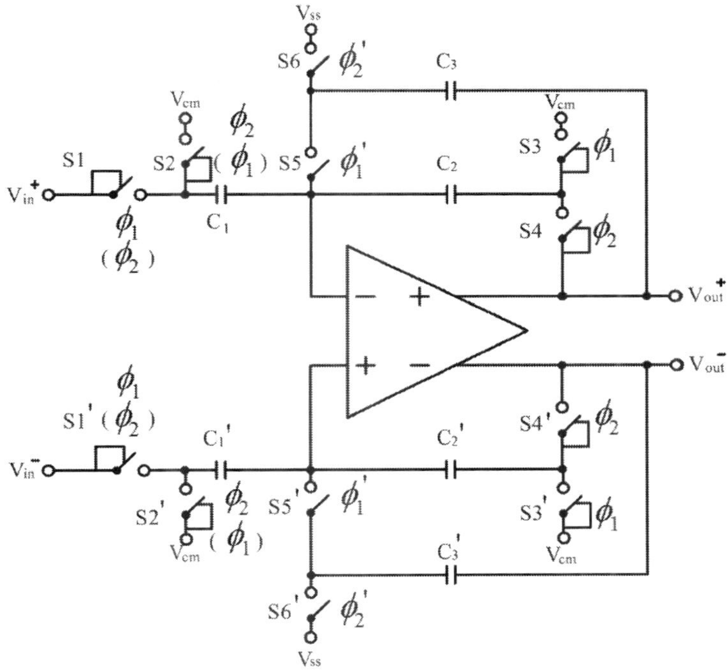

Fig. 2. First low-voltage fully differential CMOS switched-capacitor amplifier. Depending on the input-stage clock signals, the amplifier can be either noninverting (as shown) or inverting (input-stage clocks shown in parentheses).

Figure 2 shows the first low-voltage fully differential CMOS switched-capacitor amplifier based on improved bootstrapped switches described in section 3.2, where switches S1-S4 and S1′-S4′ are matched improved bootstrapped switch pairs and switches S5-S6 and S5′-S6′ are NMOS matched switch pairs. In order to minimize the number of improved bootstrapped switches, two analog reference voltages are used: V_{SS} at the op amp input where a normal NMOS switch can be used to switch the lowest supply voltage, and a $\frac{V_{DD}+V_{SS}}{2}$ common-mode voltage at the op amp output and the circuit input to maximize the signal swing. The improved bootstrapped switch is used to switch signals at this voltage level. Figure 3 is the single-ended version of Figure 2.

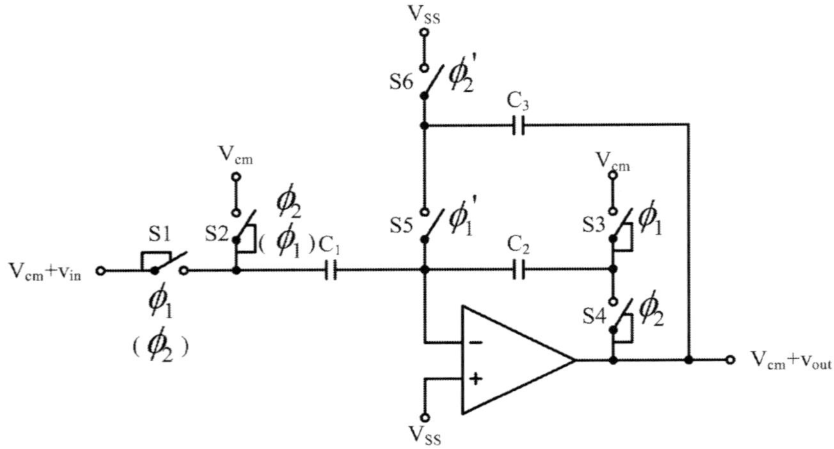

Fig. 3. Single-ended version of Fig. 2.

To see how this circuit operates, consider the inverting circuit during the reset phase (ϕ_1) and during valid output phase (ϕ_2), as shown in Figure 4. Then based on charge conservation principle we can write:

$$C_1(V_{SS}+V_{off}-V_{cm})+C_2(V_{SS}+V_{off}-V_{cm})$$

$$=C_1[V_{SS}+V_{off}-V_{cm}-v_{in}(nT)]+C_2[V_{SS}+V_{off}-V_{cm}-v_{out}(nT)],$$

$$\text{or } v_{out}(nT)=-\frac{C_1}{C_2}v_{in}(nT)\,. \tag{3}$$

It should be noted that the clock waveforms with the primed superscripts change before the nonprimed waveforms in order to reduce nonlinearities due to charge injection.

Another technique to further reduce the number of improved bootstrapped switches is shown in Figure 5, where switches S1 and S4 and S1′ and S4′ are matched improved bootstrapped switch pairs. Those switches connected to V_{SS} are realized with NMOS transistors, while those switches connected to V_{DD} are realized with PMOS transistors. In Figure 5 a single reference voltage at V_{SS} is used. However, the signal still varies around $\frac{V_{DD}+V_{SS}}{2}$ at the circuit input as well as at the op amp output to preserve the maximum swing. The difference between the two reference voltages is compensated by injecting a fixed amount of charge at the op amp input using extra capacitor pairs $C_{M1}=\frac{C_1}{2}$ and $C_{M2}=\frac{C_2}{2}$ ($C'_{M1}=\frac{C'_1}{2}$ and $C'_{M2}=\frac{C'_2}{2}$) switching between V_{DD} and V_{SS} (Baschirotto & Castello, 1997). Figure 6 is the single-ended version of Figure 5.

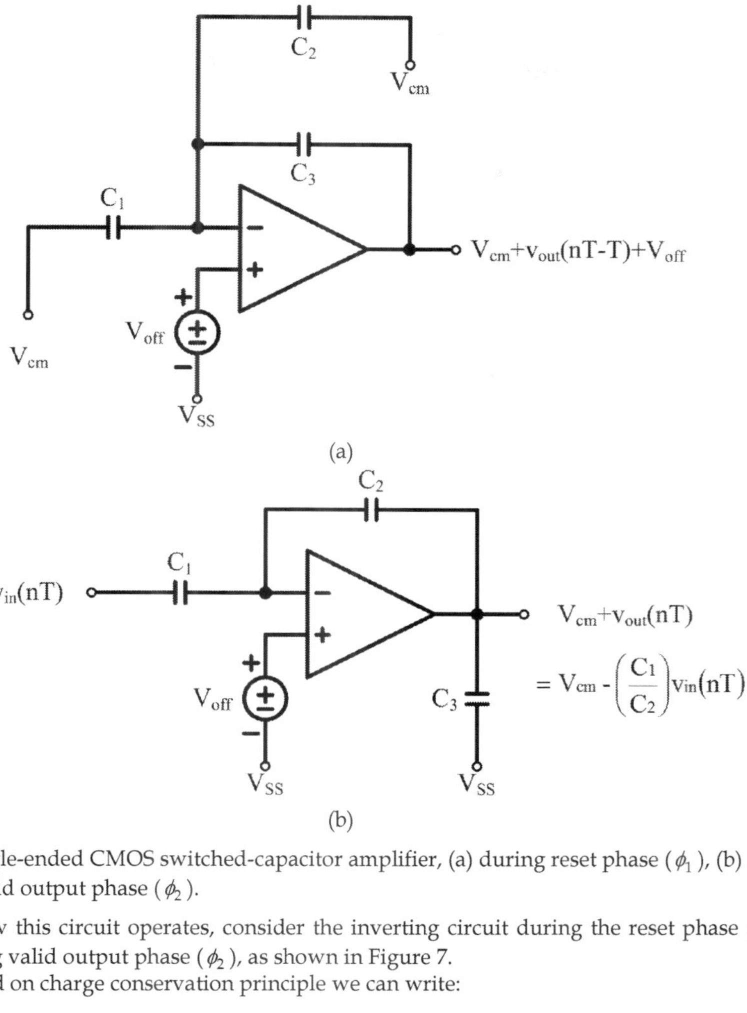

Fig. 4. Single-ended CMOS switched-capacitor amplifier, (a) during reset phase (ϕ_1), (b) during valid output phase (ϕ_2).

To see how this circuit operates, consider the inverting circuit during the reset phase (ϕ_1) and during valid output phase (ϕ_2), as shown in Figure 7.
Then based on charge conservation principle we can write:

$$C_1(V_{SS}+V_{off}-V_{SS})+C_2(V_{SS}+V_{off}-V_{SS})+(C_{M1}+C_{M2})(V_{SS}+V_{off}-V_{DD})$$

$$=C_1[V_{SS}+V_{off}-V_{cm}-v_{in}(nT)]+C_2[V_{SS}+V_{off}-V_{cm}-v_{out}(nT)] +(C_{M1}+C_{M2})(V_{SS}+V_{off}-V_{SS}),$$

$$\text{or } v_{out}(nT)=-\frac{C_1}{C_2}v_{in}(nT)\,. \tag{4}$$

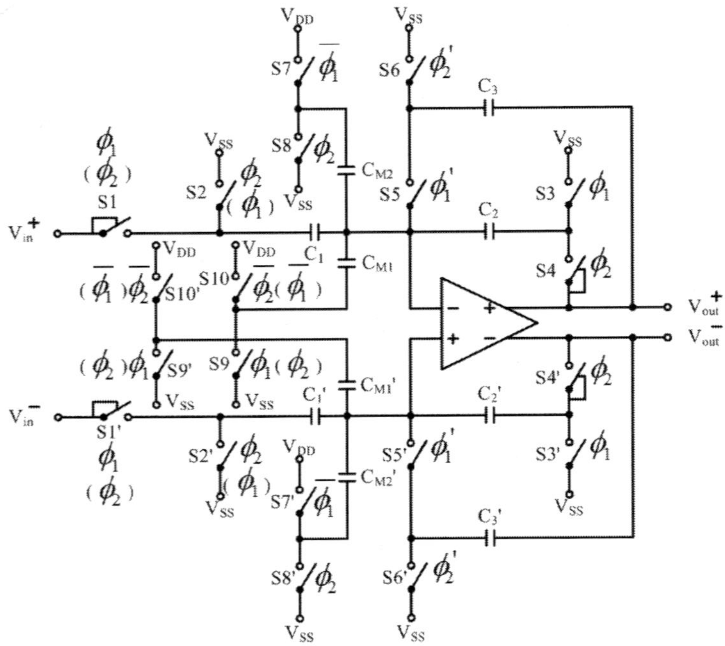

Fig. 5. Second low-voltage fully differential CMOS switched-capacitor amplifier. Depending on the input-stage clock signals, the amplifier can be either noninverting (as shown) or inverting (input-stage clocks shown in parentheses).

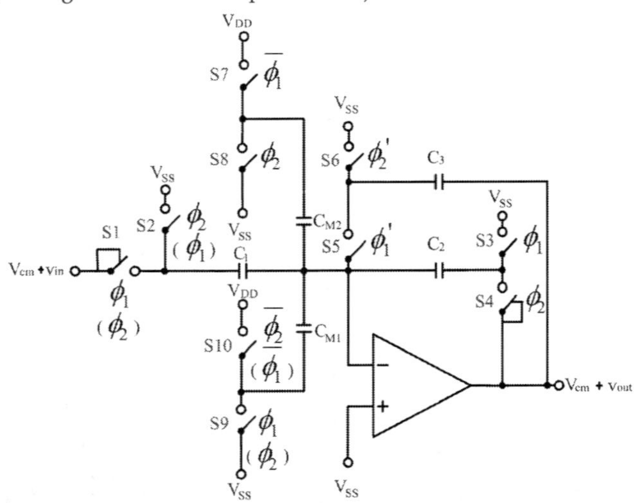

Fig. 6. Single-ended version of Fig. 5.

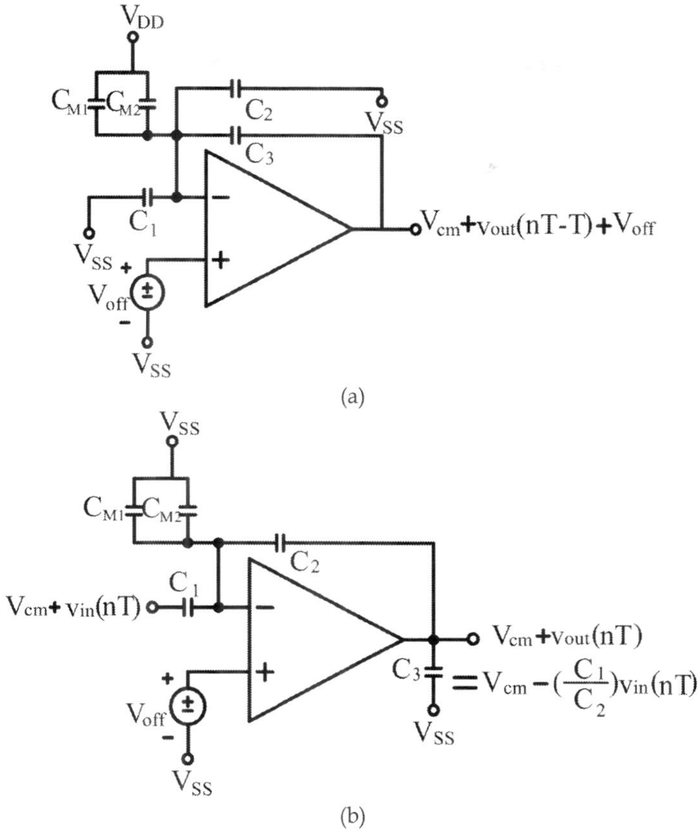

Fig. 7. Single-ended CMOS switched-capacitor amplifier, (a) during reset phase (ϕ_1), (b) during valid output phase (ϕ_2).

3. Low-voltage building blocks

In this section, the low-voltage circuit building blocks used in the two fully differential CMOS switched-capacitor amplifiers are discussed

3.1 Op Amp

Figure 8 shows the used op amp. It is based on a fully differential folded-cascode p-type two-stage Miller-compensated configuration. The second stage is a common-source amplifier with active load which also allows a large output swing. In order to avoid the common-mode feedback (CMFB) circuit for the first stage, transistors *M51, M52, M61,* and *M62* are used, which is similar to (Waltari & Halonen, 1998). For the second stage, a simple passive switched-capacitor CMFB circuit, shown in Figure 9, is used. The improved bootstrapped switches are used to connect and disconnect the common-mode sensing capacitor.

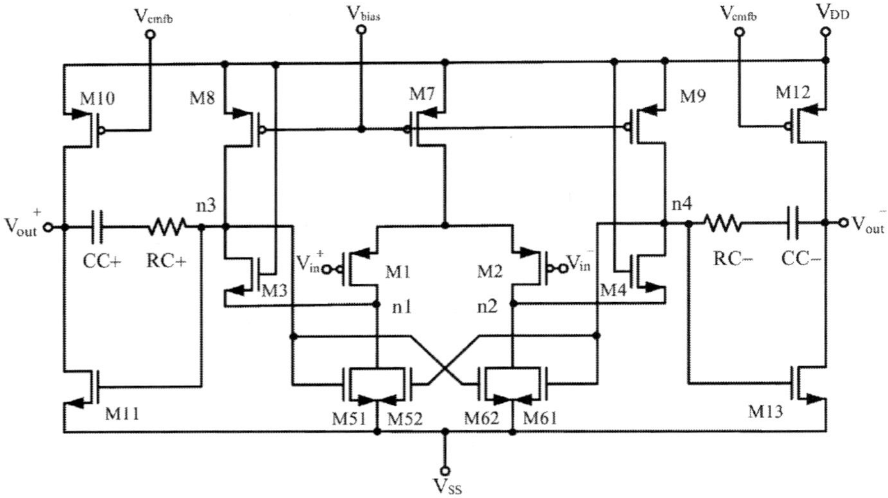

Fig. 8. Low-voltage op amp.

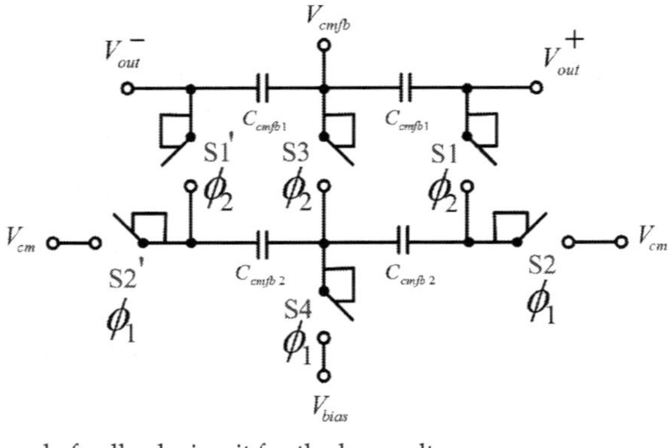

Fig. 9. Common-mode feedback circuit for the low-voltage op amp.

3.2 Improved bootstrapped switch

The improved bootstrapped switch shown in Figure 10 is utilized in the proposed circuit. The circuitry is improved version of that presented in (Abo & Gray, 1999). In the circuit presented in (Abo & Gray, 1999), the voltage at the drain side of the main switch *M11* must be always higher than that at the source side at the switching moment to prevent the gate-drain voltage from exceeding V_{DD} during the turn-on transient. In order to overcome this limitation, an additional transistor *M14* has been added on the drain side, such that the switch *M11* becomes completely symmetrical. This bootstrapping circuit thus allows switch operation (transistor *M11*) from rail-to-rail while limiting all gate-source/drain voltages to V_{DD} avoiding any oxide overstress.

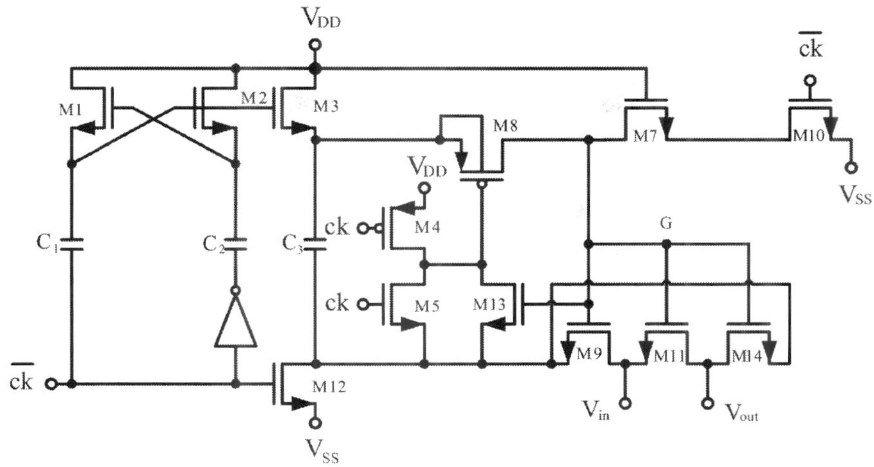

Fig. 10. Improved bootstrapped switch.

4. Experimental results

Based on the principles presented earlier, we have designed two 1-V fully differential CMOS switched-capacitor amplifiers. These two switched-capacitor amplifiers were operated with ±0.5-V. The capacitor sizes used were C_1 =1.25-pF, C_2 =0.25-pF, and C_3 =0.25-pF, for a nominal gain of -5. The circuits of Figure 2 and Figure 5 were fabricated using a TSMC 0.35-μm double-poly four-metal CMOS technology. Figure 11 and Figure 12 show the photomicrographs of Figure 2 and Figure 5, respectively. The chip areas of Figure 2 and Figure 5 excluding bonding pads are 414×278-μm^2 and 460×330-μm^2, respectively.

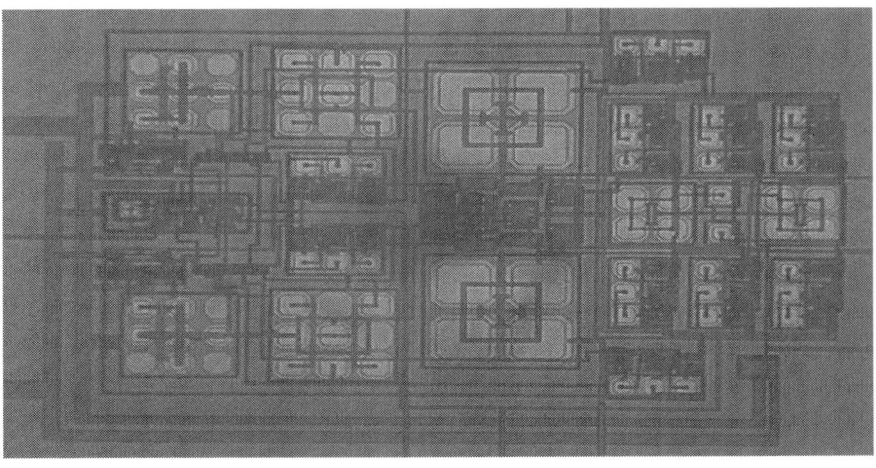

Fig. 11. Photomicrograph of Fig. 2.

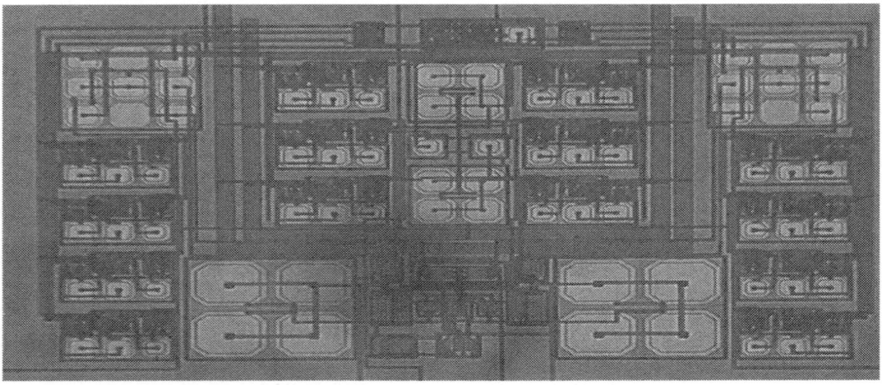

Fig. 12. Photomicrograph of Fig. 5.

Two figures of the measured input/output waveforms for 0.2V peak-to-peak sinusoidal differential input signal are shown in Fig. 13 and Fig. 14, respectively. The input signal was at 10kHz whereas the clock signal was at 1MHz. It can be seen that the gain is very close to the nominal value of -5.

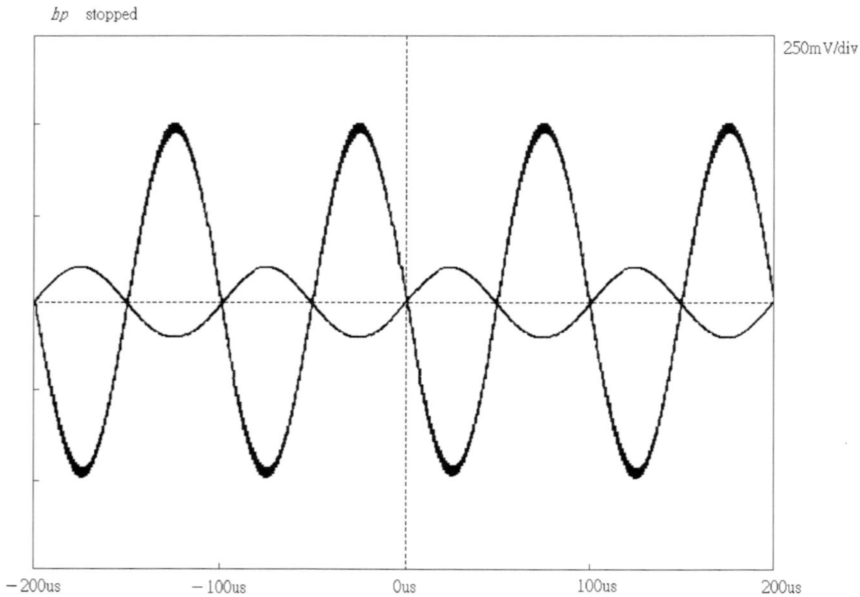

Fig. 13. Measured differential input and output waveforms of Fig. 2 (f_{clk}=1-MHz, f_{in}=10-kHz, sinusoidal differential input voltage=0.2-V_{pp}).

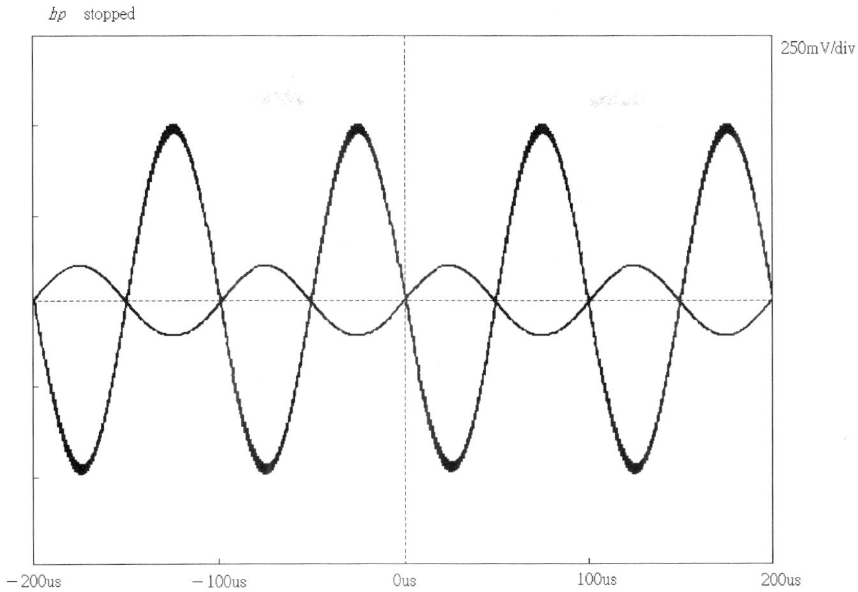

Fig. 14. Measured differential input and output waveforms of Fig. 5 (f_{clk}=1-MHz, f_{in}=10-kHz, sinusoidal differential input voltage=0.2-V_{pp})

Fig. 15 and Fig. 16 show the resulting output spectrum. As shown in Fig. 15 and Fig. 16, the even-order harmonics have been largely attenuated by the fully differential topology and 59dB and 52dB spurious-free dynamic range (SFDR) are exhibited, respectively. The circuits of Fig. 2 and Fig. 5 dissipate 206.5μW and 206.6μW, respectively with a 1V power supply.

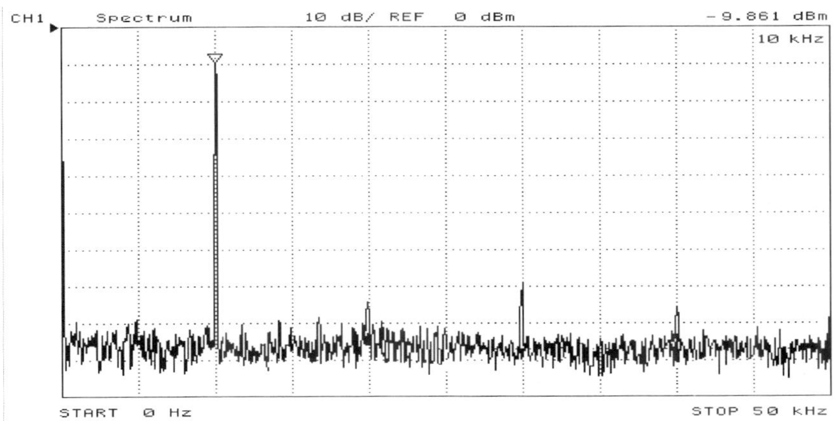

Fig. 15. Measured output spectrum of Fig. 2.

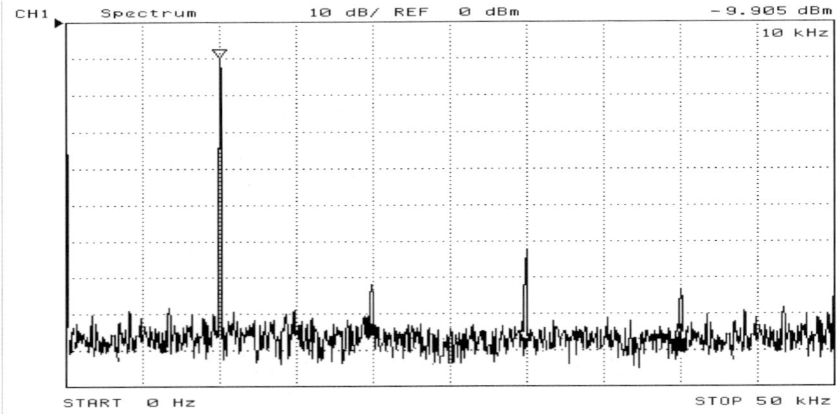

Fig. 16. Measured output spectrum of Fig. 5.

5. Conclusion

Two fully differential CMOS 1-V switched-capacitor amplifiers have been described. Rail-to-rail operation of improved bootstrapped switches allows very low voltage robust switched-capacitor designs in standard CMOS technologies while avoiding transistor gate oxide overstress. The circuits have been fabricated and all aspects of their performance have been confirmed.

6. References

Abo , A. M. & Gray , P. R. (1999). A 1.5-V, 10-bit, 14.3-MS/s CMOS pipeline analog-to-digital converter, *IEEE J. Solid-State Circuits*, May , vol. 34, pp. 599-606 ,ISSN: 0018-9200.

Au, S. & Leung, B. H., (1997). A 1.95-V, 0.34-mW, 12-b sigma-delta modulator stabilized by local feedback loops, *IEEE J. Solid-State Circuits*, March, vol. 32, pp. 321-328, ISSN: 0018-9200.

Baschirotto, A. & Castello, R. (1997). A 1-V 1.8-MHz CMOS switched-opamp SC filter with rail-to-rail output swing, *IEEE J. Solid-State Circuits*, December, vol. 32, pp. 1979-1986, ISSN: 0018-9200.

Chang, D. Y. & Moon, U.-K. (2003). A 1.4-V 10-bit 25-MS/s pipelined ADC using opamp-reset switching technique, *IEEE J. Solid-State Circuits*, August, vol. 38, pp. 1401-1404, ISSN: 0018-9200.

Cheung,V. S.-L. et al. (2001). A 1-V CMOS switched-opamp switched-capacitor pseudo-2-path filter, *IEEE J. Solid-State Circuits*, Jan.2001, vol. 36, pp. 14-22, ISSN: 0018-9200.

Cheung,V. S. L. et al. (2002). A 1-V 10.7-MHz switched-opamp bandpass ΣΔ modulator using double-sampling finite-gain-compensation technique, *IEEE J. Solid-State Circuits*, October, vol. 37, pp. 1215-1225, ISSN: 0018-9200.

Cheung V. S.-L et al. (2003). A 1-V 3.5-mW CMOS switched-opamp quadrature IF circuitry for Bluetooth receivers, *IEEE J. Solid-State Circuits*, May., vol. 38, pp. 805-816, ISSN: 0018-9200.

Crols, J. & Steyaert, M., (1994). Switched-opamp: an approach to realize full CMOS switched-capacitor circuits at very low power supply voltage, *IEEE J. Solid-State Circuits*, August, vol. 29, pp. 936-942, ISSN: 0018-9200.

Dessouky, M. & Kaiser, A. (2001). Very low-voltage digital-audio ΣΔ modulator with 88-dB dynamic range using local switch bootstrapping, *IEEE J. Solid-State Circuits*, March, vol. 36, pp. 349-355, ISSN: 0018-9200.

Keskin, M. et al. (2002). A 1-V 10-MHz Clock-Rate 13-Bit CMOS ΣΔ modulator using unity-gain-reset opamps, *IEEE J. Solid-State Circuits*, July, vol. 37, pp. 817-824, ISSN: 0018-9200.

Martin, K. et al. (1987). A differential switched-capacitor amplifier, *IEEE J. Solid-State Circuits*, February, vol. 22, pp. 104-106, ISSN: 0018-9200.

Matsuya, Y. & Yamada, J. (1994). 1-V power supply, low-power consumption A/D conversion technique with swing-suppression noise shaping, *IEEE J. Solid-State Circuits*, December, vol. 29, pp. 1524-1530, ISSN: 0018-9200.

Nicollini,G. A. et al. (1996). A -80dB THD, 4-*Vpp* switched capacitor filter for 1.5-V battery-operated systems, *IEEE J. Solid-State Circuits*, August, vol. 31, pp. 1214-1219, ISSN: 0018-9200.

Park, J.-B. et al. (2004). A 10-b 150-MSample/s 1.8-V 123-mW CMOS A/D converter with 400-MHz input bandwidth, *IEEE J. Solid-State Circuits*, August, vol. 39, pp. 1335-1337, ISSN: 0018-9200.

Peluso, V. et al. (1997), A 1.5-V 100-μW ΣΔ modulator with 12-b dynamic range using the switched-opamp technique, *IEEE J. Solid-State Circuits*, July, vol. 32, pp. 943-952, ISSN: 0018-9200.

Peluso,V. et al. (1998). A 900-mV low-power ΣΔ A/D converter with 77-dB dynamic range," *IEEE J. Solid-State Circuits*, December, vol. 33, pp. 1887-1897, ISSN: 0018-9200.

Rabii, S. & Wooley, B. A. (1997). A 1.8-V digital-audio sigma-delta modulator in 0.8-μm CMOS, *IEEE J. Solid-State Circuits*, June, vol. 32, pp. 783-796, ISSN: 0018-9200.

Rombouts, P. et al. (2001). A 13.5-b 1.2-V micropower extended counting A/D converter, "*IEEE J. Solid-State Circuits*, February, vol. 36, pp. 176-183, ISSN: 0018-9200.

Sauerbrey, J. et al. (2002). A 0.7-V MOSFET-only switched-opamp ΣΔ modulators in standard digital CMOS technology, *IEEE J. Solid-State Circuits*, December, vol. 37, pp. 1662-1669, ISSN: 0018-9200.

Waltari, M. & Halonen, K. A. I. (2001). 1-V 9-Bit pipelined switched-opamp ADC," *IEEE J. Solid-State Circuits*, January, vol. 36, pp. 129-134, ISSN: 0018-9200.

Waltari, M. & Halonen, K. (1998). Fully differential switched opamp with enhanced common-mode feedback, *Electron. Lett.* , November, vol. 34, no. 23, pp. 2181-2182, ISSN:0013-5194..

Wang, L. &. Embabi S. H. K. (2003). Low-voltage high-speed switched-capacitor circuits without voltage bootstrapper, *IEEE J. Solid-State Circuits*, August, vol. 38, pp. 1411-1415, ISSN: 0013-5194.

Wu, P. Y. et al. (2007). A 1-V 100-MHS/s 8-bit CMOS Switched-Opamp Pipelined ADC Using Loading-Free Architecture, *IEEE J. Solid-State Circuits,* April, vol. 42, pp. 730-738, ISSN:0013-5194.

Yang, J. W. & Martin, K. W. (1989). High-resolution low-power D/A converter, *IEEE J. Solid-State Circuits*, October, vol. 24, pp. 1458-1461, ISSN: 0013-5194.

Yoshizawa, H. et al. (1999). MOSFET-only switched-capacitor circuits in digital CMOS technology," *IEEE J. Solid-State Circuits*, June, vol. 34, pp. 734-747, ISSN: 0013-5194.

6

Multi-Mode, Multi-Band Active-RC Filter and Tuning Circuits for SDR Applications

Kang-Yoon Lee
Konkuk University
Republic of Korea

1. Introduction

The prevalence of wireless standards and the introduction of dynamic standards/ applications, such as software-defined radio, necessitate filters with wide ranges of adjustable bandwidth/power, and with selectable degrees and shapes. The baseband filters of transceivers often utilize a significant portion of the power budget, especially when high linearity is required. Likewise, a widely tunable filter designed for its highest achievable frequency consumes more power than necessary when adjusted to its lowest frequency. Because power consumption is proportional to the dynamic range and frequency of operation, power-adjustable filters have recently gained popularity, as they can adapt their power consumption dynamically to meet the needs of the system.

Dynamic variation in filter attributes (e.g. frequency, order, type) coupled with companies' desire to reuse IP has popularized highly programmable filters. The lowpass-filter cutoff frequencies of several wireless/wireline standards are within the 1–20 MHz frequency range. Many of these standards are irregularly spaced in frequency and do not lend themselves well to standard binary-weighted resistor arrays. In previous designs frequency is solely controlled digitally, and hence, digital circuitry or an ADC is used to tune the frequency of the filters. Gm–C filters offer continuous frequency tunability and can operate at higher frequencies than their active-RC counterparts. MOSFET-C filters can also provide continuous frequency tuning, but both Gm–C and MOSFET-C filters lack good linearity. MOSFET-C filters additionally suffer from reduced tuning range at lower supply voltages; also, in MOSFET-C most of the voltage drop occurs over nonlinear MOSFET triode resistors, which appreciably degrades its linearity. Along these lines, filters with good linearity have been developed that tune on the basis of duty-cycle control in switched-R-MOSFET-C filters; however, duty-cycle control by nature necessitates a discrete-time filter. Active-RC filters, known to have good linearity, have been used for continuous-time programmable filters. Some such filters have narrow frequency ranges, with a few others providing wider but solely discrete ranges. Purely discrete switched resistor tuning limits filter frequency tuning to discrete frequencies determined by the overall frequency range and number of bits used. Tuning to different frequency bands precisely would require very fine resistor stripes to meet precision requirements for the low-frequency end.

Newer technologies offer increased integration with smaller feature sizes, allowing the filter to be on the same chip with other transceiver blocks. This integration especially promotes reconfigurable architectures, as DSPs can be integrated with the transceiver and can control

the mode of operation. However, as minimum feature size shrinks, supply voltages also reduce, which complicates classical linearization techniques of Gm–C filters. In this respect, active-RC filters possess an intrinsic linearity advantage.
The third generation standards will create a demand for cellular phones capable of operating both in the new wideband and in the existing narrower band systems. Second generation system has a channel bandwidth in range of tens of kilohertz, whereas channel bandwidths of wideband systems are in megahertz range. So the corner frequency of an analog channel select filter must be tunable over at least a decade of frequency. This will increase the power consumption and the area. The proposed multi-mode baseband filter can minimize the area and optimize the power consumption by sharing the capacitors and resistors. And new tuning method can reduce the number of switches in programmable capacitor arrays which can be dominant noise sources.
This chapter is organized as follows. In Section 2, the multi-mode, multi-band active-RC filter architecture is described. Section 3 describes filter tuning circuits. Section IV shows experimental results from a 0.35 μm CMOS implementation and Section V concludes the paper.

2. Multi-mode, multi-band active-RC filter architecture

2.1 Multi-mode, multi-band active-RC low-pass filter

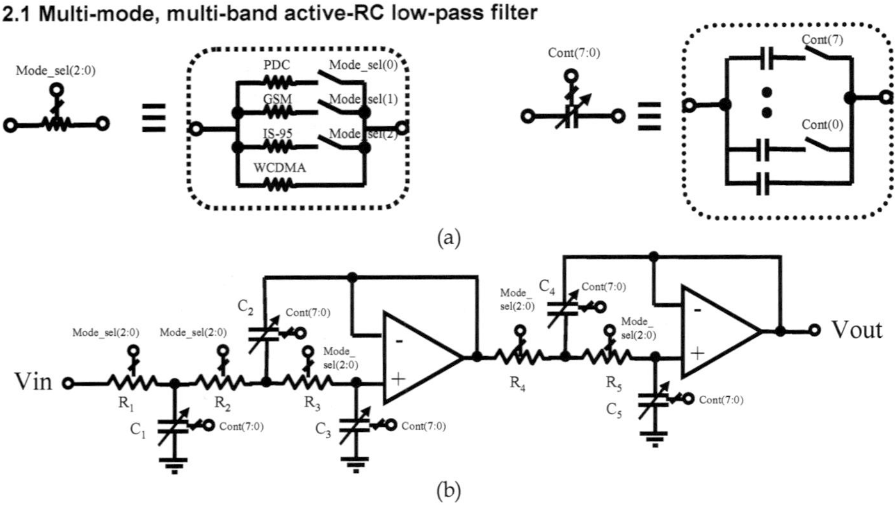

Fig. 1. (a) Resistor matrices and capacitor matrices (b) schematic of the baseband filter.

Fig. 1 shows the designed active-RC 5th-order Chebyshev filter. Resistor matrices and capacitor matrices are shown in Fig. 1(a). Resistor matrics are composed of resistors and switches. Switches are controlled by Mode_sel(3:0), as defined in Table 1. Resistor for WCDMA is not connected to any switch.
The bandwidths of PDC, GSM, IS-95, and WCDMA are 13 kHz, 100 kHz, 630 kHz, and 2.1 MHz, respectively. Mode_sel(3:0) bits are set through the serial interface and represented in thermometer code. The corner frequency was made tunable by using programmable

capacitor matrices. Capacitor matrices are composed of capacitors and switches. The control bits required for each mode are 2-bits. Thus, total number of 8-bits are required. The tuning bits, cont(7:0) are determined from the on-chip tuning block based on the mode. In PDC mode, the low corner frequency leads to large passive components occupying a lot of die area [2]. Because the capacitor matrices dominate the area, capacitors were shared between modes. There are trade-offs between resistor values and capacitor values. When resistor values are reduced to make thermal noise small, capacitor values become large. That leads to a large area. On the other hand, as capacitor values become smaller to reduce the area, the noise level rises. So, capacitor values and resistor values were optimized.

Mode_sel(3:0)	Standard	Bandwidth
"0001"	PDC	13 kHz
"0010"	GSM	100 kHz
"0100"	IS-95	630 kHz
"1000"	WCDMA	2.1 MHz

Table 1. Mode definition and corresponding bandwidths

2.2 Tunable active-RC complex band-pass filter

Fig. 2 shows the Adjacent Channel Interference (ACI) of the PHS system. The nearest interferer is located at 600 kHz, and its magnitude is 50 dB larger than the wanted signal.

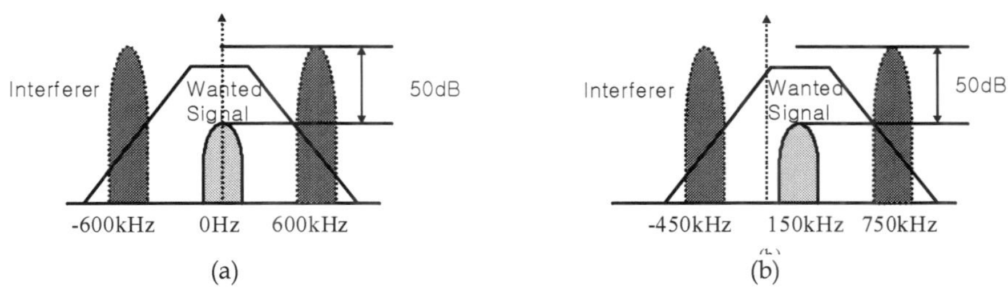

Fig. 2. Characteristic of (a) Lowpass Filter in Direct-Conversion Receiver (b) Complex Bandpass Filter in Low-IF Receiver

As shown in Fig. 2(a), if the direct conversion receiver architecture is used, the interferences are located at ±600 kHz, which can be attenuated by the lowpass filter. However, if the IF frequency is 150 kHz, the interferences are shifted to -450 kHz, +750 kHz, respectively as shown in Fig. 2(b). If the lowpass filter is used, the attenuation characteristic is tighter because the worst case interferer is seemed to be located at 450 kHz. Therefore, the complex bandpass filter whose center frequency is located at 150 kHz is designed for the ACS performance.

The transfer function of the complex bandpass filter is found by frequency translating a low-pass filter.

$$H_{bp}(j\omega) = H_{lp}(j\omega - j\omega_c) \tag{1}$$

The translation of a single pole is given in Eq. (2) and (3)

$$H_{lp}(j\omega) = \frac{1}{1 + j\omega/\omega_o} \tag{2}$$

$$\begin{aligned} H_{bp}(j\omega) &= \frac{1}{1 - j\omega_c/\omega_o + j\omega / \omega_o} \\ &= \frac{1}{1 - 2jQ + j\omega / \omega_o} \end{aligned} \tag{3}$$

A single complex pole cannot be realized with a real filter. Only complex pole pairs can be realized. The result of Eq. (2) is a single complex pole. The translated version of a single complex pole is also given with Eq. (3). The complex part must just be added to or subtracted from the complex term $2jQ$.

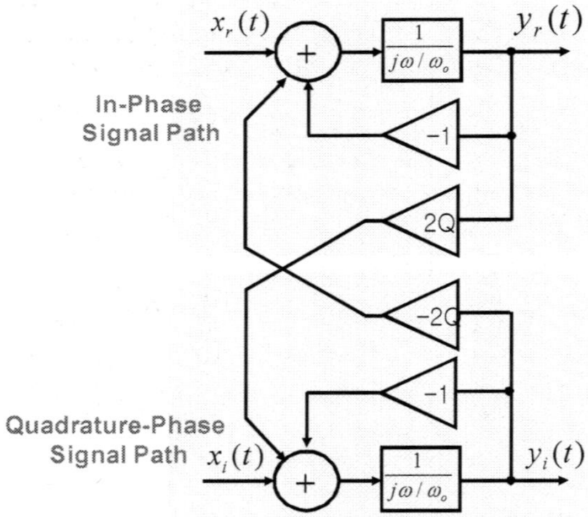

Fig. 3. Block scheme for the realization of a single complex pole.

The realization of for a single pole is given in Fig. 3. It is nothing more than the direct synthesis of the transfer function. Fig. 3 is the full block schematic with building blocks for real signals.

Fig. 4(a) shows the designed 3rd-order Chebyshev complex bandpass filter. The wanted signal is composed of the in-phase signal and quadrature signal, which are separated by the 90° phase. Complex bandpass filter uses both signals to perform the complex operations. As shown in Fig. 4(a), the complex bandpass filter has the in-phase signal path and the quadrature-phase signal path. Internal nodes of each paths are inter-connected to other paths. Therefore, I/Q mismatches is one of the most critical design issues in the complex filter. In this design, because I/Q mismatch compensation scheme is applied, I/Q mismatch is drastically reduced.

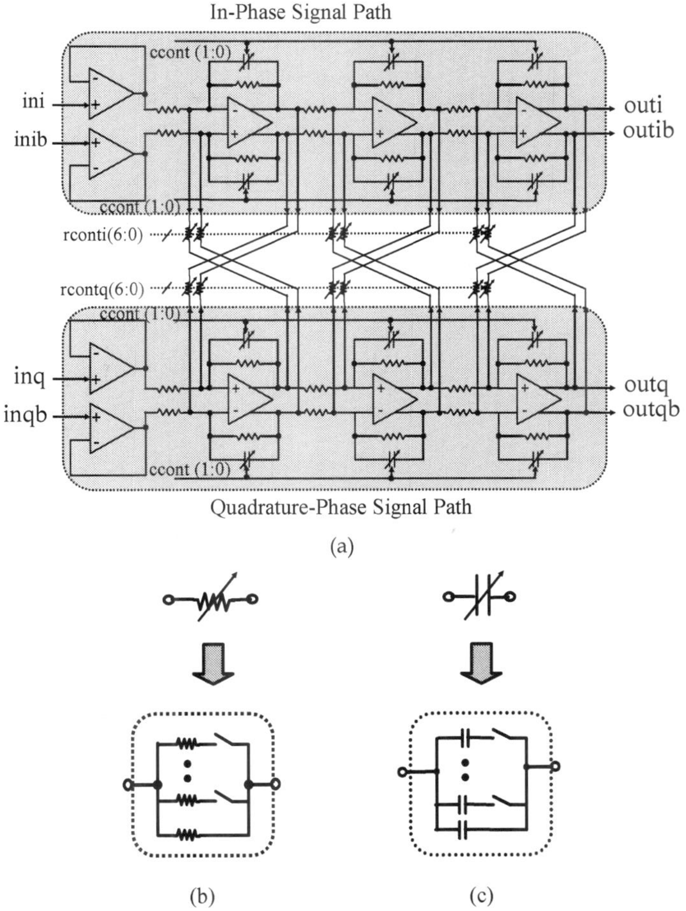

Fig. 4. (a) Schematic (b) resistor arrays (c) capacitor arrays of the complex bandpass filter

Resistor arrays and capacitor arrays are shown in Fig. 4(a). Resistor arrays are composed of resistors and switches. Resistor arrays control the center frequency of the bandpass filter and its control signals, rconti(6:0), rcontq(6:0), are set through the serial interface and represented in thermometer code.

The corner frequency was made tunable by using programmable capacitor arrays. Capacitor arrays are composed of capacitors and switches. The tuning bits, ccont(1:0), are determined from the on-chip tuning block. There are trade-offs between resistor values and capacitor values. When resistor values are reduced to make thermal noise small, capacitor values become large. That leads to a large area. On the other hand, as capacitor values become smaller to reduce the area, the noise level rises. So, capacitor values and resistor values were optimized.

3. Tuning circuit of active-RC filter

Resistors and capacitors are usually varied about ±15% due to the process variation. In continuous time filters, this leads to a large variation of the corner frequency, which in most case, must be compensated by adjusting the component values.

Conventional full analog tuning circuit based on VCO is shown in Fig. 5(a). However, this tuning circuit is not suitable to tune an active-RC filter with programmable capacitor matrices. The output of the loop filter in the PLL is analog voltage, which cannot be interfaced directly with the capacitor matrices.

On the other hand, too many digital bits are required for fine resolution in the conventional full digital circuit shown in Fig. 5(b). Thus, the area and noise level are too high due to many number of switches and capacitor matrices.

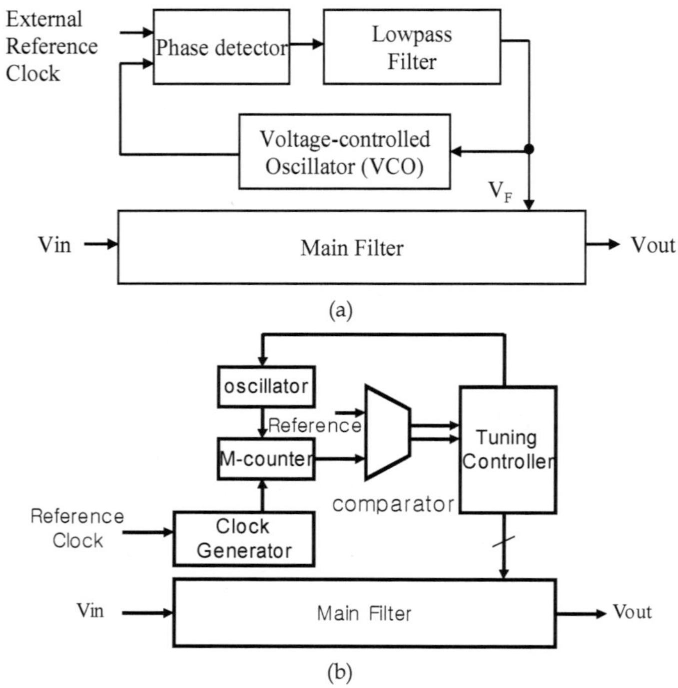

Fig. 5. (a) Full analog tuning method (b) Full digital tuning method.

The concepts of the full analog, full digital and proposed two-step tuning method are shown in Fig. 6(a), (b) and (c), respectively.

The block diagram of the proposed two-step tuning scheme is shown in Fig. 7. The clock generator provides the clocks, clk0, clk1 to coarse and fine tuning controllers. C_{tu} is charged during clk0 is high, and VCOMP is sampled by clk1. Reference voltages for comparators, Vref, Refl, RefH, RefM are generated in the reference voltage generator block. The operation is as follows. Before main capacitor tuning steps, the reference tuning loop is enabled to

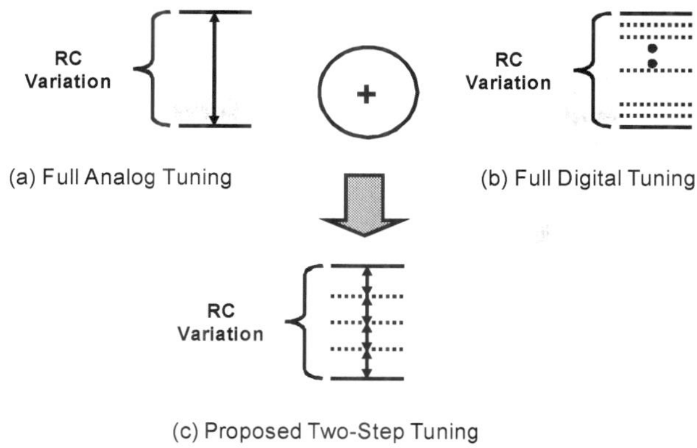

Fig. 6. Concept of (a) full analog tuning method (b) full digital tuning method (c) proposed two-step tuning method.

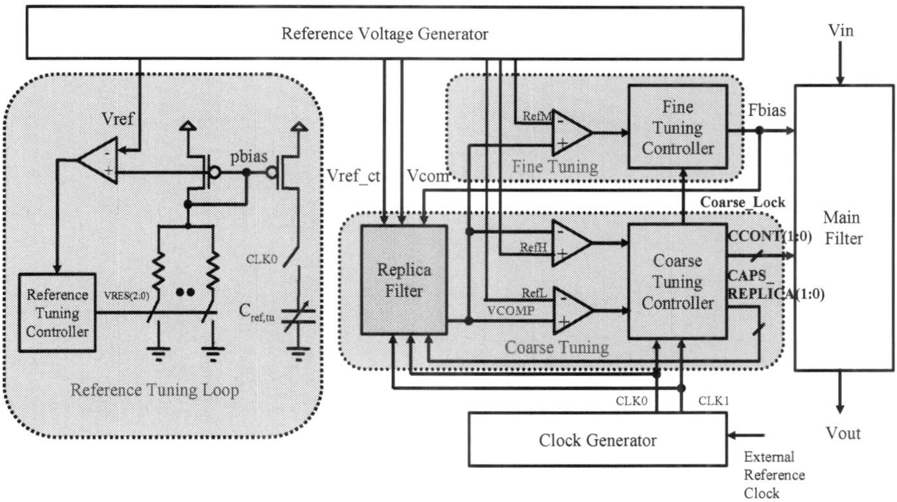

Fig. 7. Block diagram of the proposed two-step tuning method.

compensate the resistor variation. Pbias is compared with Vref, and vres(2:0) is controlled according to the result. When pbias is larger than Vref, resistor load should be smaller, so vres(2:0) is increased. On the other hand, if pbias is smaller than Vref, resistor load should be larger, so vres(2:0) is decreased. Reference tuning is completed when pbias crosses the Vref.

After the reference tuning, main capacitor tuning is done in two-steps, that is, the coarse tuning and the fine tuning.

Fig. 8 shows the timing diagram of the proposed two-step tuning method.

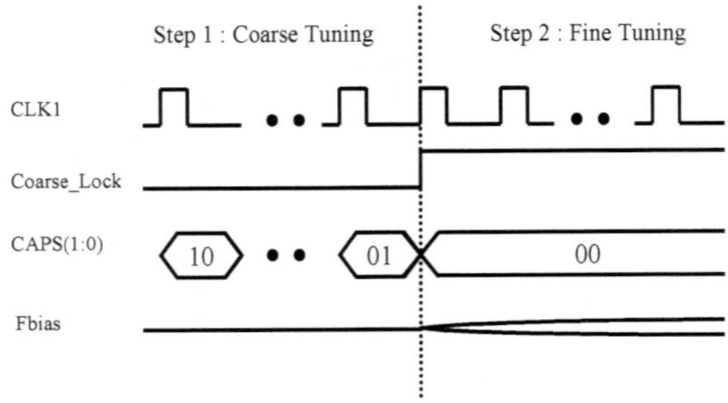

Fig. 8. Timing diagram of the proposed tuning method.

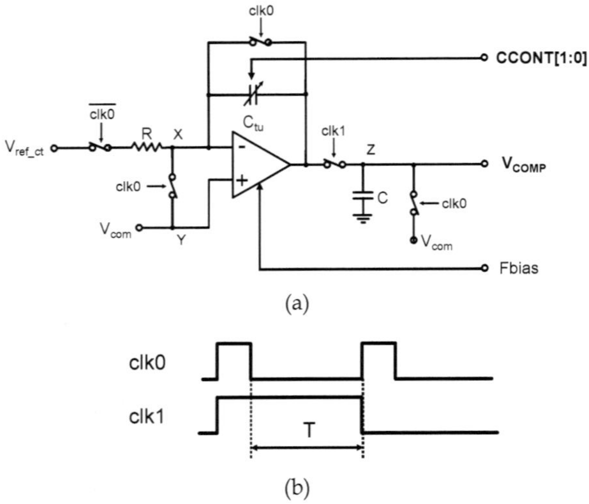

Fig. 9. (a) Block diagram (b) timing diagram of the replica filter in coarse tuning block.

Fig. 9 shows the block diagram and the timing diagram of the replica filter in the coarse tuning block. When VCOMP voltage is precharged when clk0 and clk1 are high. When clk0 goes from high to low, the VCOMP voltage is determined as Eq. 4.

$$VCOMP = V_{com} - \frac{1}{RC_{tu}}\left(V_{ref\_ct} - V_{com}\right)T \tag{4}$$

First, CCONT(1:0) are tuned until VCOMP is located between pre-determined ranges. VCOMP is compared with refL and refH. When VCOMP is higher than refH, CCONT(1:0) are increased. Whereas, if VCOMP is lower than refL, CCONT(1:0) are decreased. When VCOMP is located between refL and refH, coarse_Lock signal goes from low to high. Usually, many tuning capacitance levels are required for fine resolution. But, only two bits are sufficient in this design with the two-step tuning method. After the coarse_lock signal is asserted, the corner frequency is tuned by the fine tuning control block.

Fbias controls the tail current of the op-amp. Thus, the DC-gain of the op-amp is changed according to the Fbias voltage. If the DC-gain of the op-amp is infinite, the cut-off frequency does not change. However, because the DC-gain of the op-amp is finite, the cut-off frequency of the filter is changed as the DC-gain of the op-amp changes. Fbias is compared with refM. When VCOMP is larger than refM, Fbias should be increased. On the other hand, Fbias should be decreased when VCOMP is smaller than refM. The range of Fbias is 0.8 V to 1.2 V. The bandwidth of the op-amp is adjustable according to the mode to save the power. And the transistor sizes in the op-amp are designed to be very large to reduce the 1/f noise.

4. Experimental results

4.1 Multi-mode, multi-band active-RC low-pass filter

The multi-mode, multi-band active-RC low-pass filter was fabricated using a 0.35 μm CMOS process. The chip area is 3.8 mm^2 and the supply voltage is 3 V.

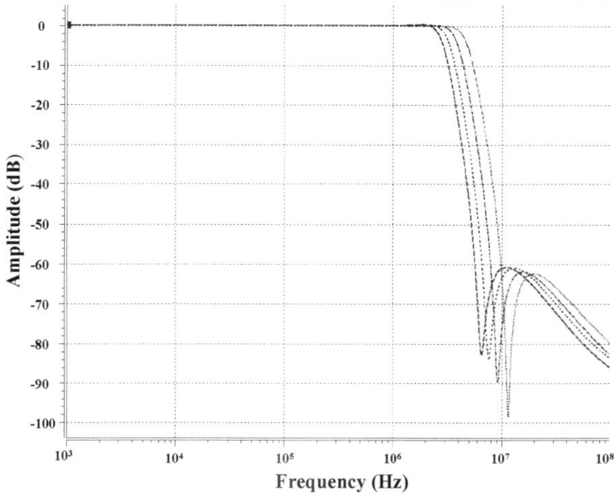

Fig. 10. Amplitude response of the filter in WCDMA mode.

Fig. 10 shows the amplitude response of the filter in WCDMA mode at every code. The cut-off frequency is 2.1 MHz in WCDMA mode. The current consumptions are 5.2 mA, 6.1 mA, 7.1 mA, and 8.4 mA, respectively. The current in PDC mode is less than that of WCDMA mode because the gain and the bandwidth of the op-amp are smaller in PDC mode. The frequency tuning range is from 10 kHz to 3MHz. Out-of-band IIP3 was determined by performing IM3 test. In PDC mode, when two tones of +15 dBm at 20 kHz and 30 kHz are applied, IM3 is -77 dBm. In WCDMA mode, out-of-band IM3 is -68 dBm, when two tones of +14 dBm at 1.8 MHz and 3.0 kHz are applied. Input-referred average passband noise densities of the filter are 250, 130, 85, and 54 $nV/\sqrt{Hz}$ for PDC, GSM, IS-95, and WCDMA, respectively. The passband ripple is less than 0.5 dB in all modes. The stopband rejections are 79, 79, 75, and 75 dB for PDC, GSM, IS-95, and WCDMA, respectively. Table 2 summarizes the performance of the filter.

Technology	0.35 μm CMOS			
Chip area	3.8 mm^2			
Supply voltage	3 V			
Tuning range	10 kHz ~ 3 MHz			
	PDC	GSM	IS-95	WCDMA
Current (mA)	5.2	6.1	7.1	8.4
IIP3 (dBm)	28	25	23	21
Noise ($nV/\sqrt{Hz}$)	250	130	85	54
Passband ripple	0.5	0.5	0.5	0.5
Stopband rejection	79	79	75	75

Table 2. Performance summary

4.2 Tunable active-RC complex band-pass filter

The complex bandpass filter was fabricated using a 0.35 μm CMOS process. The chip area is 3.8 mm^2. The supply voltage is 3 V. Fig. 11 shows the microphotograph.

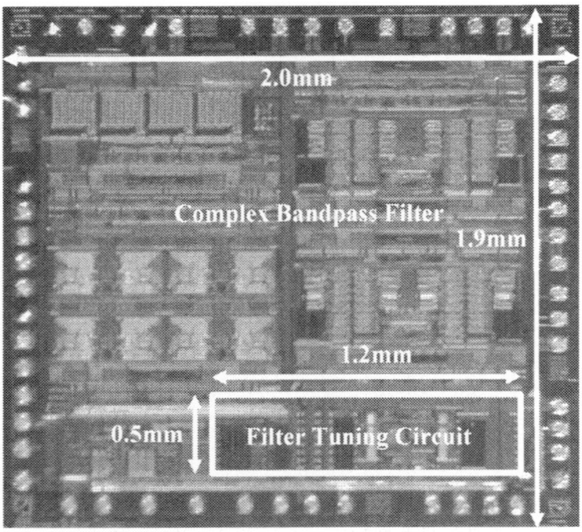

Fig. 11. Chip microphotograph

Fig. 12 shows measured amplitude response of the complex baseband filter, when the temperature is changed from -10°C to 60°C. The cut-off frequency is 150 kHz ± 110 kHz. The cut-off frequency is almost constant as the temperature is changed due to the proposed filter tuning method.

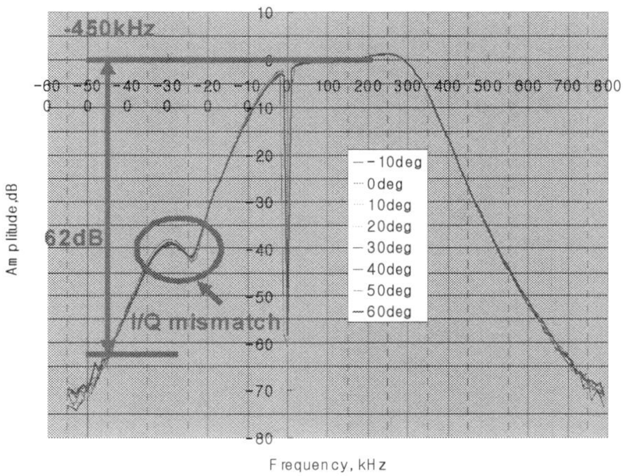

Fig. 12. Measured amplitude response of the complex bandpass filter

The power consumption is 13 mW. The frequency tuning range is 100 kHz, which is the 50% of the signal bandwidth. That is, the cut-off frequency can be adjusted by 100 kHz although it is shifted due to the temperature, the supply voltage, and the process variations.

Out-of-band IIP3 was determined by performing IM3 test. When two tones of -44 dBm at 600 kHz and 1.2 MHz are applied, IIP3 is +25 dBm.

Input-referred average passband noise density of the filter is $85\,nV/\sqrt{Hz}$. The passband ripple is less than 0.8 dB, and the stopband rejection at -450 kHz is 66 dB. Table 3 summarizes the performance of the filter.

Technology	0.35 μm CMOS
Chip area	3.8 mm^2
Supply voltage	3 V
Tuning range	100 kHz
Power (mW)	13
IIP3 (dBm)	25
Noise ($nV/\sqrt{Hz}$)	85
Passband ripple	0.8

Table 3. Performance summary

5. Conclusion

The CMOS multi-mode, multi-band low-pass filter and complex baseband filter are presented. Capacitors and resistors were shared to minimize the area. Proposed two-step tuning method can reduce the number of switches and thus, can reduce the noise and the area.

6. References

A. M. Durham, W. Redman-White, and J.B. Hughes, "High linearity continuous-time filters in 5-V VLSI CMOS," *IEEE J. Solid-State Circuits*, vol. 27, pp 1270-1276, Sept. 1992.

H. A. Alzaher, H. O. Elwan, and M. Ismail "A CMOS Highly Linear Channel-Select Filter for 3G Multistandard Integrated Wireless Receivers," *IEEE J. Solid-State Circuits*, vol. 37, pp 27-37, Jan. 2002.

H. A.-Aslanzadeh, E. J. Pankratz, E. S.-Sinencio, "A 1-V +31 dBm IIP3, Reconfigurable, Continuously Tunable, Power-Adjustable Active-RC LPF", *IEEE J. Solid-State Circuits*, vol. 44, no. 2, February 2009.

J. Crols, M. S. J. Steyaert, "Low-IF Topologies for High-Performance Analog Front Ends of Fully Integrated Receivers", *IEEE Transactions on Circuits and Systems – II*, vol. 45, no. 3, March 1998.

J. Hughes, N. Bird, and R. Soin, "Self-tuned RC-active filters for VLSI," *IEE Electronics Letters*, vol. 34, pp. 1479-1480, July 1998.

R. Gastello, F. Montecchi, F.Rezzi, and A. Baschirotto, "Low-voltage analog filters," *IEEE Trans. Circuits Syst. I*, vol. 42, pp 827-840, Nov. 1995.

T. Hollman, S. Lindfors, M. Lansirinne, J. Jussila, and K. A.I. Halonen, "A 2.7-V CMOS Dual-Mode Baseband Filter for PDC and WCDMA," *IEEE J. Solid-State Circuits*, vol. 36, pp 1148-1153, July 2001.

T. Salo, S. Lindfors, T. Hollman, "Programmnable Direct Digital Tuning Circuit for a Continuous-Time Filter," *Proc. ESSCIRC'00*, pp. 168-171, Sept. 2000.

Y. G. Pu, S. K. Jung, D. J. Park, J. K. Kim, J. H. Jung, C. Nam, and K. Y. Lee, "A CMOS baseband complex bandpass filter with a new Automatic tuning method for PHS applications" *Proc. ESSCIRC'07*, pp. 500 - 503, Sept. 2007.

7

A Novel Multiclad Single Mode Optical Fibers for Broadband Optical Networks

Rostami[1,2] and S. Makouei[1]
[1]*Photonics and Nanocrystal Research Lab. (PNRL), Faculty of Electrical and Computer Engineering, University of Tabriz, Tabriz 51666,*
[2]*School of Engineering-Emerging Technologies, University of Tabriz, Tabriz 51666, Iran*

1. Introduction

The demand for data communication is growing rapidly due to the increasing popularity of the Internet and other factors [1]. The unprecedented success of information technology in recent years ushers an explosive growth in demand for the Internet access in the 21st century. Ultra-wideband transmission media are needed in order to provide high-speed communications for a much larger number of users. A promising solution to the capacity crunch can come from wavelength-division-multiplexed optical fiber communication systems that are shown to provide enormous capacities on the order of terabit per second over long distances. These systems utilize single-mode fibers, in conjunction with erbium-doped fiber amplifiers, as the transmission medium [2]. The optical silica fiber might be the only proper choice to realize this task. Optical fiber based communication is the excellent alternative for these purposes which needs low dispersion as well as dispersion slope and large bandwidth supported by optical physical medium[3]. Optical transmission began to be used in trunk cables about 1990; the capacity of those systems was several hundred Mbit/s per fiber. The capacity jumped to 2.5 (5.0) Gbit/s per fiber with the introduction of optical repeaters using erbium-doped fiber amplifiers in 1995. It jumped again in 1998, to 10 (20) Gbit/s per fiber, with the introduction of wavelength division multiplexing (WDM) [5]. Overall transmission capacity now exceeds 100 Gbit/s per fiber due to improvements in WDM techniques [6]. Usually, those techniques are called dense WDM (DWDM). In recent years, the increasing demands for transmission capacity have led to intense research activities on high capacity DWDM communication system [7].

Nowadays, applications such as optical time division multiplexing (OTDM) and dense wavelength division multiplexing (DWDM) are usual tasks in industry [1]. Therefore by considering these applications, providing a large bandwidth and high-speed communication possibility using optical fibers is highly interesting [3].

In the following, we review requirements for DWDM, Dispersion properties, optical nonlinearity, loss properties, and design of optical fiber for DWDM.

2. Requirements for DWDM

The number of wavelengths (channels) in the fibers of DWDM systems is increasing. As discussed earlier, a WDM signal typically occupies a bandwidth of 30 nm or more, although it is bunched in spectral packets of bandwidth ~ 0.1 nm (depending on the bit rate of individual channels) [1]. To use such new optical transmission systems, the DWDM fiber should overcome three transmission related limitations: dispersion, optical fiber nonlinearity and loss properties. For 10-Gb/s channels, the third-order dispersion does not play an important role as relatively wide (> 10 ps) optical pulses are used for individual channels. However, because of the wavelength dependence of β□2, or the dispersion parameter *D*, the accumulated dispersion will be different for each channel [4]. Wavelength division multiplexing (WDM) systems have been widely introduced for large capacity transmission. In order to further increase the transmission capacity, several techniques have been investigated, such as higher bit-rate transmission [8], enhancement of the spectral efficiency [9], and use of new transmission bands [10, 11]. In those systems, optical fibers are required more strongly to reduce the nonlinearity and the dispersion slope [5].

3. Dispersion properties

Silica fibers suffer from some disadvantages especially dispersion and dispersion slope. Meanwhile, these two factors cause severe restrictions for high-speed pulse propagation [1]. The dispersion value becomes larger by the wavelength increasing in the conventional optical fibers. So owing to the dissimilar broadening for different channels, the multi-channel application realization would be hard. A suitable optical fiber should meet the small dispersion as well as the small dispersion slope in the predefined wavelength interval. The dispersion properties are the dispersion itself and the dispersion slope of the optical fiber. The dispersion value cannot be in the zero-value region because FWM causes interaction between signals (optical channels) in DWDM systems when there is phase matching between the optical channels due to zero dispersion. Therefore, the dispersion value in the signal wavelength region must have the proper non-zero value. The sign of the dispersion value should be positive for short-distance transmission and negative for ultra-long-distance transmission [12] because of modulation instability in the positive dispersion in a long link. When the signal wavelength band becomes wider, the difference in the dispersion values at the edges of the wavelength band becomes larger. Dispersion compensation thus becomes difficult for long distance DWDM transmission. To achieve both long distance and high speed transmission with easy dispersion compensation for a wide wavelength band, the dispersion slope should be reduced. For single wavelength communication, dispersion shifted fiber is enough. But for applications such as DWDM this method cannot provide high speed possibility. In these applications, the physical media should provide the flat, minimum, and uniform dispersion as well as dispersion slope ideally. An important limitation induced by chromatic dispersion and its slope is broadening factor which restricts the bit rate parameter.
To minimize pulse broadening in an optical fiber, the chromatic dispersion should be low over the wavelength range used. A fiber in which the chromatic dispersion is low over a broad wavelength range is called a dispersion-flattened fiber.

4. Optical nonlinearity

The response of any dielectric to light becomes nonlinear for intense electromagnetic fields, and optical fibers are no exception. Even though silica is intrinsically not a highly nonlinear material, the waveguide geometry that confines light to a small cross section over long fiber lengths makes nonlinear effects quite important in the design of modern light wave systems [1]. The much higher power level due to simultaneous transmission of multiplexed channels and propagation over much longer distances made possible with the utilization of fiber amplifiers, cause the otherwise weak and negligible fiber nonlinearities to affect the signal transmission significantly [2]. When the number of signal wavelengths carried in an optical fiber increases, the average transmission power density becomes larger than that in conventional systems. Consequently, optical-fiber nonlinearities have emerged as a main issue. This nonlinearity seriously limits transmission capacity with various nonlinear interactions, which are generally categorized as scattering effects and optical signal interactive modulation. Because the signal power density is stronger due to the greater number of channels in DWDM systems, optical fiber nonlinearity limits the number/spacing of the channels and the length/speed of the transmission. In general, the refractive index of optical fiber has a weak dependence on optical intensity (equal to signal power (p) per effective area (A_{eff}) in the fiber. Optical fiber nonlinearity arises from modulation of the refractive index caused by changes in the optical intensity of the signal. This cause four wave mixing (FWM), self-phase modulation (SPM) and cross phase modulation (XPM) can be observed in the fiber. The XPM and SPM distort the signals. Therefore, optical fiber nonlinearity must be reduced. The most practical way to do this is to enlarge A_{eff} [13]. The relationship between A_{eff} and mode field diameter (MFD) is direct and proportional. As a result, enlarging MFD is a practical solution for low nonlinearity. The choice of dispersion shifted fibers (DSFs) along with erbium-doped fiber amplifiers (EDFAs), for operation at 1550nm window, would be an ideal one to achieve greater transmission distance and utilize full capacity of transmission system [5,7,15]. However, when the system is operated at the zero dispersion wavelengths, the nonlinear interaction between the channels and noise components is increased. The system working slightly away from the zero dispersion wavelengths can reduce these unwanted interactions. The WDM system reduces the nonlinear effects and enables multi-wavelength transmission through non-zero dispersion shifted fibers having very small dispersion in duration 1530-1610 nm. In order to increase the information carrying capacity, latest high speed communication system is based on the dense wavelength division multiplexing/demultiplexing (DWDM) [16, 17]. In such systems, nonlinear effects like four wave mixing (FWM), which arise due to simultaneous transmission at many closely spaced wavelengths and high optical gain from EDFA, imposes serious limitations on the use of a DSF with zero dispersion wavelength at 1550 nm [18,19]. To overcome this difficulty, the nonzero dispersion shifted fibers having small dispersion in the range ~ 2−4 ps/km/nm over the entire gain window of EDFA have been proposed [20, 21]. In such fibers, the phase matching condition is not satisfied and hence the effect of FWM becomes negligible due to small dispersion [15].

5. Loss properties

Progress in optical fiber fabrication technologies has resulted in a routine production of low loss single mode fibers. This enables us to apply the single mode fibers promisingly in high bit rate and long haul optical transmission systems. Structural optimization must be established so as to provide desirable transmission characteristics for given operating conditions. A basic design consideration has been made by taking into account transmission characteristics such as fiber intrinsic loss, bending loss, splice loss, and launching efficiency [22, 23]. Use of commercially available erbium doped fiber amplifiers (EDFA), which forces optical communication systems to be operated in the 1550 nm window, has significantly reduced the link length limitation imposed by attenuation in the optical fiber [15]. The fiber loss is one of the significant restrictions in the optical fiber communication links. It is one of some reasons limit the maximum distance that information can be sent without presence of the repeaters. Meanwhile, due to the loss, the pulse amplitude reduces so that the initial information cannot be restored in the noisy conditions. Seeing that, in the fiber design one likes to shift the zero dispersion wavelength to the region that the fiber has the lowest level attenuation. The combination of natural attenuations has a global minimum around 1.55 μm and that is why most optical communication systems are operated at this wavelength [4, 18]. A kind of loss which must be taken into account in fiber design is the bending loss. Every time an optical fiber is bent, radiation occurs. When a bent occurs, a portion of the power propagating in the cladding is lost through radiation.

6. Design of DWDM fiber

There are three methods to increase the capacity of a DWDM transmission system, using a broad wavelength range, narrowing channel spacing and increasing a bit rate per channel. However, one of disadvantages for the last two methods is the degradation of the transmission performance due to optical nonlinear effects. In this area, there are three categories which cover all designs. There are based on using zero dispersion shifted fibers (ZDSFs), non-zero dispersion shifted fiber (NZDSFs) and dispersion flattened fibers (DFFs).

7. Zero Dispersion Shifted Fibers (ZDSFs)

Use of commercially available erbium doped fiber amplifiers (EDFA), which forces optical communication systems to be operated in the 1550 nm window, has significantly reduced the link length limitation imposed by attenuation in the optical fiber. However, high bit rate (~10 Gb/s) data transmission can be limited by the large inherent dispersion of the fiber. Dispersion shifted fibers (DSF), which has zero dispersion around 1550 nm, have been proposed and developed to overcome this problem. Dispersion shifted fiber for single wavelength optical communication is a proper choice. The much higher power level due to simultaneous transmission of multiplexed channels and propagation over much longer distances made possible with the utilization of fiber amplifiers, cause the otherwise weak and negligible fiber nonlinearities to affect the signal transmission significantly. The effects of fiber nonlinearities on pulse propagation and on the capacity of fiber optic communication systems have been studied extensively by many researchers. To mitigate the nonlinear effects in long fiber optic communication systems by zero dispersion shifted fiber, a new generation of optical fibers, referred to as large effective area fibers, has been introduced. As said earlier, in order to reduce nonlinear effects, it is preferred to increase effective area. Gathering zero dispersion and large effective area together will be an appropriate solution in this task. The large effective area fibers allow a much smaller light intensity inside the guiding region, thus resulting in less refractive index nonlinearity than the conventional single mode fibers. In addition to reduced nonlinearities, large effective area fibers must also provide low attenuation, low bending and micro-bending losses, low chromatic dispersion, and low polarization mode dispersion. In recent years, a variety of large effective area fiber designs have been reported in the literature. These designs may be broadly classified into two groups based on their refractive index profiles; R-type and M-type. Each of two types is divided to two other categories too named type I and II. A small pulse broadening factor (small dispersion and dispersion slope), as well as small nonlinearity (large effective area) and low bending loss (small mode field diameter) are required as the design parameters in Zero dispersion shifted fibers [24]. The performance of a design may be assessed in terms of the quality factor. This dimensionless factor determines the trade-off between mode field diameter, which is an indicator of bending loss and effective area, which provides a measure of signal distortion owing to nonlinearity [25]. It is also difficult to realize a dispersion shifted fiber while achieving small dispersion slope. Here, we attempted to present an optimized MII triple-clad optical fiber to obtain exciting performance in terms of dispersion and its slope [24]. The index refraction profile of the MII fiber structure is shown in Fig. 1. According to the LP approximation [26] to calculate the electrical field distribution, there are two regions of operation and the guided modes and propagating wave vectors can be obtained by using two determinants which are constructed by boundary conditions [27].

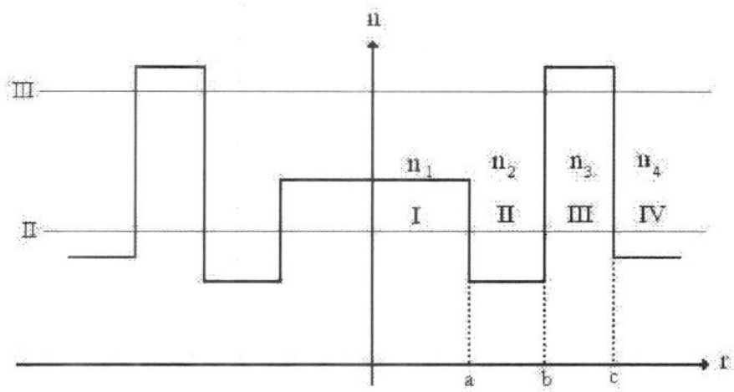

Fig. 1. Refractive index Profile for MII Structure.

For calculation of dispersion and dispersion slope the following parameters are used.

$$P = \frac{b}{c}, \tag{1}$$

$$Q = \frac{a}{c}, \tag{2}$$

where P and Q are geometrical parameters. Also, the optical parameters for the structure are defined as follows.

$$R_1 = \frac{n_3 - n_1}{n_3 - n_2}, \tag{3}$$

$$R_2 = \frac{n_2 - n_4}{n_3 - n_2}. \tag{4}$$

For evaluating of the index of refraction difference between core and cladding the following definition is done.

$$\Delta = \frac{n_3^2 - n_4^2}{2n_4^2} \approx \frac{n_3 - n_4}{n_4} \tag{5}$$

Here, we propose a novel methodology to make design procedure systematic. It is done by the aim of optimization technique and based on the Genetic Algorithm. A GA belongs to a class of evolutionary computation techniques [28] based on models of biological evolution. This method has been proved useful in the domains that are not understood well; search spaces that are too large to be searched efficiently through standard methods. Here, we concentrate on dispersion and dispersion slope simultaneously to achieve to the small dispersion and its slope in the predefined wavelength duration. Our goal is to propose a special fitness function that optimizes the pulse broadening factor. To achieve this, we have defined a weighted fitness function. In fact, the weighting function is necessary to describe the relative importance of each subset in the fitness function [24]; in other words, we let the pulse broadening factor have different coefficient in each wavelength. To weight the mentioned factor in the predefined wavelength interval, we have used the Gaussian weighting function. The central wavelength (λ_0) and the Gaussian parameter (σ) are used for the manipulation of the proposed fitness function and their effects on system dispersion and dispersion slope. To express the fiber optic structure, we considered three optical and geometrical parameters. According to the GA technique, the problem will have six genes, which explain those parameters. It should be mentioned that the initial range of parameters are chosen after some conceptual examinations. The initial population has 50 chromosomes, which cover the search space approximately. By using the initial population, the dispersion (β_2) and dispersion slope (β_3), which are the important parameters in the proposed fitness function, can be calculated. Consequently elites are selected to survive in the next generation. Gradually the fitness function leads to the minimum point of the search zone with an appropriate dispersion and slope. Equation (6) shows our proposal for the weighted fitness function of the pulse broadening factor.

$$F = \sum_{\lambda} e^{-\frac{(\lambda-\lambda_0)^2}{2\sigma^2}} \sum_{Z} [1 + (\frac{\beta_2(\lambda)Z}{t_i^2})^2 + (\frac{\beta_3(\lambda)Z}{2t_i^3})^2]^{\frac{1}{2}}, \tag{6}$$

where $\lambda_0, \sigma, t_i, Z, \beta_2$ and β_3 are central wavelength, Gaussian parameter, full width at half maximum, distance, second and third order derivatives of the wave vector respectively. In the defined fitness function in Eq. (6), internal summation is proposed to include optimum broadening factor for each length up to 200 km. By applying the fitness function and running the GA, the fitness function is minimized. So, the small dispersion and its slope are achieved. This condition corresponds to the maximum value for the dispersion length and higher-order dispersion length as well. By using this proposal, the zero dispersion wavelengths can be shifted to the central wavelength (λ_0). Since, the weight of the pulse broadening factor at λ_0 is greater than others in the weighted fitness function; it is more likely to find the zero dispersion wavelength at λ_0 compared to the other wavelengths. In the meantime, the flattening of the dispersion curve is controlled by Gaussian parameter (σ). To put it other ways, the weighting Gaussian function becomes broader in the predefined wavelength interval by increasing the Gaussian parameter (σ). As a result, the effect of the pulse broadening factor with greater value is regarded in different wavelengths, which causes a considerable decrease in the dispersion slope in the interval. Consequently, the zero dispersion wavelength and dispersion slope can be tuned by λ_0 and σ respectively. The advantage of this method is introducing two parameters (λ_0 and σ) instead of multi-designing parameters (optical and geometrical), which makes system design easy.

The flowchart given in Fig. 2 explains the foregoing design strategy clearly.

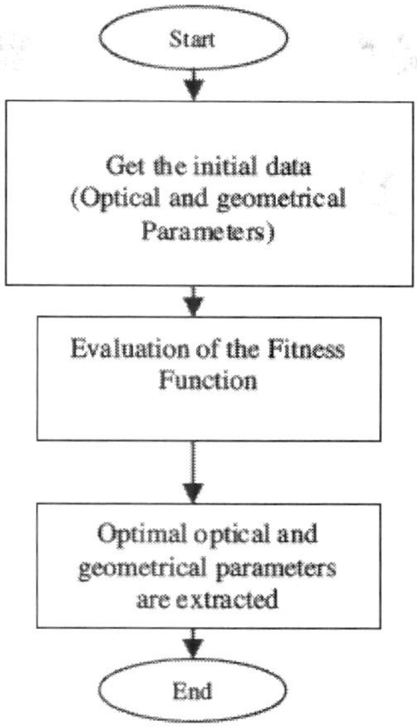

Fig. 2. The scheme of the design procedure

To illustrate capability of the suggested technique and weighted fitness function, the MII triple-clad optical fiber is studied, and the simulated results are demonstrated below. In the presented figures, we consider four simulation categories including dispersion related quantities, nonlinear behavior of the proposed fibers, electrical field distribution in the structures, and fiber losses.

For all the simulations, we consider λ0=1500, 1550 nm and σ = 0, 0.027869 and 0.036935 μm as design constants. To apply the GA for optimization, we consider the search space illustrated in Table 1 for each parameter as a gene. The choice of these intervals is done according to two items. The designed structure must be practical in terms of manufacturing and have high probability of supporting only one propagating mode [24].

Parameter	a (μm)	p	Q	R_1	R_2	Δ
duration	[2-2.6]	[0.4-0.9]	[0.1-0.7]	[0.05-0.99]	[(-0.99)- (-0.05)]	[2×10^{-3} - 1×10^{-2}]

Table 1. Optimization Search Space of Optical and Geometrical Parameters

The wavelength and distance durations for optimization are selected as follows. For λ_0=1550nm: 1500 nm<λ< 1600 nm, for λ_0=1500 nm: 1450 nm <λ< 1550 nm, and 0 < Z < 200 km. In this design method Z is variable. In the simulations an un-chirped initial pulse with 5 ps as full width at half maximum is used. Considering the information in Table 1 and GA method, optimal parameters are extracted and demonstrated in Table 2.

	λ_0 (μm)	a (μm)	Δ	R_1	R_2	p	Q
σ=0	1.55	2.0883	8.042e-3	0.5761	-0.4212	0.7116	0.3070
	1.5	2.1109	7.036e-3	0.6758	-0.2785	0.8356	0.2389
$\sigma = 2.7869\times10^{-8}$	1.55	2.0592	9.899e-3	0.7320	-0.2670	0.7552	0.2599
	1.5	2.5822	9.111e-3	0.5457	-0.4237	0.7425	0.2880
$\sigma = 3.6935\times10^{-8}$	1.55	2.2753	9.933e-3	0.5779	-0.4218	0.6666	0.3428
	1.5	2.5203	9.965e-3	0.4867	-0.3841	0.6819	0.3324

Table 2. Optimized Optical and Geometrical Parameters at λ_0=1500, 1550 nm and three given Gaussian parameters

It is found that optimization method for precise tuning of the zero dispersion wavelengths as well as the small dispersion slope requires large value for the index of refraction difference (Δ). That is to say that providing large index of refraction is excellent for the simultaneous optimization of zero dispersion wavelength and dispersion slope. First, we consider the dispersion behavior of the structures. To demonstrate the capability of the proposed algorithm for the assumed data, the obtained dispersion characteristics of the structures are illustrated in Fig. 3. It shows that the zero dispersion wavelengths can be controlled precisely by controlling the central wavelength. Meanwhile, the Gaussian parameters are used to manipulate the dispersion slope of the profile. Considering Fig. 3 and Table 3, it is found that the zero value for the Gaussian parameter can tune the zero dispersion wavelengths accurately (~100 times better than other cases).

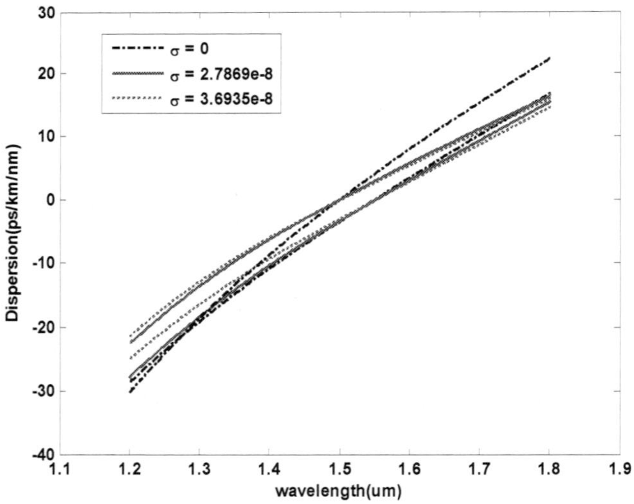

Fig. 3. Dispersion vs. Wavelength at λ0=1500nm, 1550nm with σ as parameter.

Second, the dispersion slope is examined. The presented curves say that by increasing the Gaussian parameter the dispersion slope becomes smaller, and it is going to be smooth in large wavelengths. Furthermore it is clear that there is a trade-off between tuning the zero dispersion wavelengths and decreasing the dispersion slope as shown in Figs. 3, 4, and Table 3.

type	$\lambda_0(\mu m)$	Dispersion $(ps/km/nm)$	Dispersion Slope $(ps/km/nm^2)$	Effective Area (μm^2)	Mode Field Diameter (μm)	Quality Factor
$\sigma = 0$	1.55	-2.57e-4	0.0695	191.92	7.95	3.04
	1.5	2.55e-5	0.0828	344.15	9.76	3.61
$\sigma = 2.7869 \times 10^{-8}$	1.55	-0.013	0.0647	194.79	7.12	3.85
	1.5	0.008	0.0597	209.95	6.70	4.68
$\sigma = 3.6935 \times 10^{-8}$	1.55	-0.085	0.0592	150.05	6.82	3.22
	1.5	-0.089	0.0564	164.21	6.55	3.82

Table 3. Dispersion, Dispersion Slope, Effective Area, Mode Field Diameter and Quality Factor at λ0=1500nm, 1550nm and three given Gaussian parameters

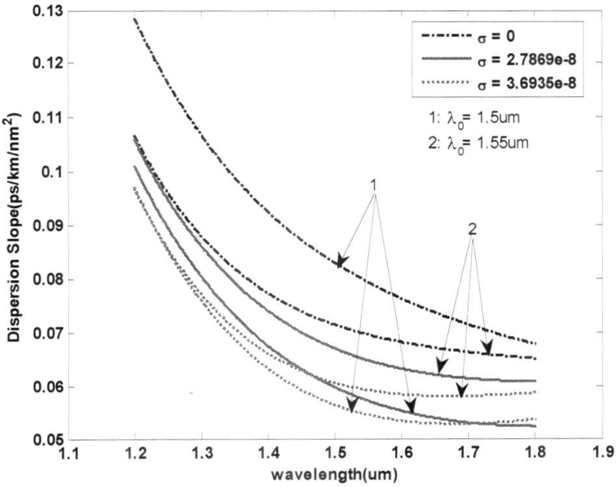

Fig. 4. Dispersion slope Vs. Wavelength at λ0=1500nm, 1550nm with σ as parameter.

The normalized field distribution of the MII based designed structures is illustrated in Figs. 5 and 6. Because of the special structure, the field distribution peak has fallen in region III. As such most of the field distribution displaces to the cladding region. In addition it is observed that the field distribution peak is shifted toward the core, and its tail is depressed in the cladding region by increasing the Gaussian parameter (except σ=0). On the other hand the field distribution slope increases inside the cladding region by increasing of the Gaussian parameter.

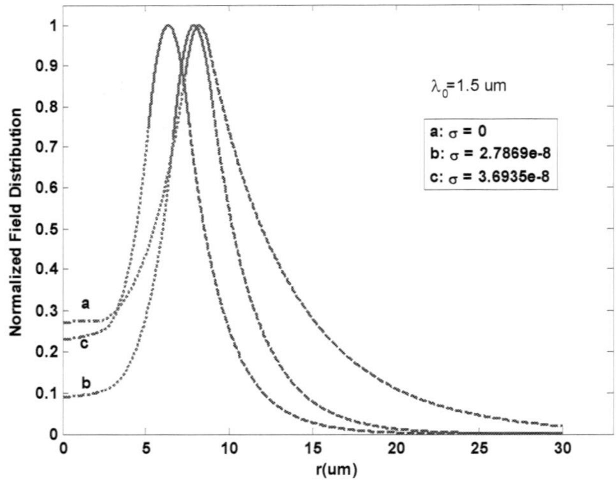

Fig. 5. Normalized Field distribution versus the radius of the fiber at λ_0=1500nm with σ as parameter (dashed-dotted, dotted, solid line, and dashed curves represent regions I, II, III and IV respectively).

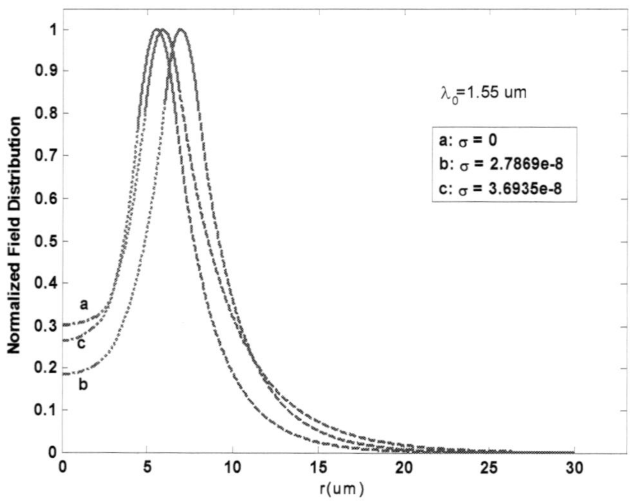

Fig. 6. Normalized Field distribution versus the radius of the fiber at λ_0=1550nm with σ as parameter (dashed-dotted, dotted, solid line, and dashed curves represent regions I, II, III and IV respectively).

The effective area or nonlinear behavior of the suggested structures is illustrated in Fig. 7. It is observed that the effective area becomes smaller by increasing the Gaussian parameter. Figs. 5-7, and Table 3 indicate a trade-off between the large effective area and the small dispersion slope. The results illustrated in Fig. 4 show that the dispersion slope reduces by increasing the Gaussian parameter. However the field distribution shifts toward the core, which concludes the small effective area in this case. Foregoing points show that there is an inherent trade-off between these two important quantities.

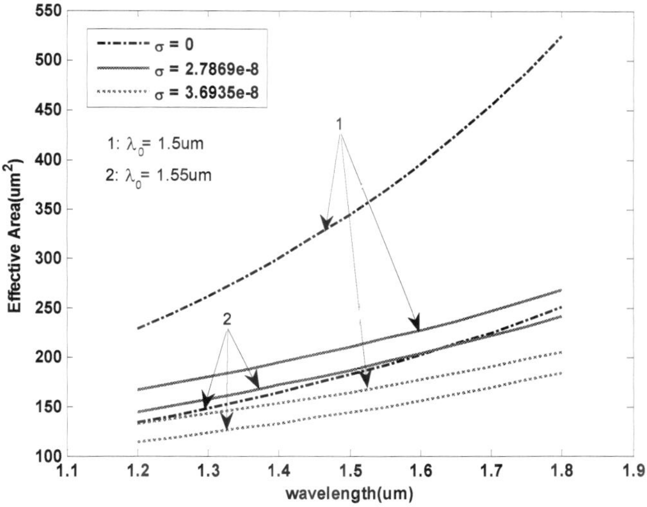

Fig. 7. Effective area versus wavelength at λ_0=1550nm, 1500nm with σ as the parameter.

The mode field diameter that corresponds to the bend loss is illustrated in Figs. 8 and 9 for both central wavelengths. It is clearly observed that the mode field diameter decreases by increasing the Gaussian parameter. In other words, the Gaussian parameter is suitable for the bend loss manipulation in these structures. Furthermore, Table 3 shows that the mode field diameter is ~7μm in the designed structure.

As another concept to consider, Table 3 says that the mode field diameter is not affected noticeably by increasing the effective area. This is the origin of raising the quality factor in these structures. This is a key point why the average amount of the quality factor in the proposed structures is increased in Fig. 9. The quality factor of the designed fibers is illustrated in Fig. 10. The calculations show that the quality factor is generally larger than 3. It is mentionable that the quality factor is smaller than unity in the inner depressed clad fibers (*W* structures) and around unity in the depressed core fibers (*R* structures). This feature shows the high quality of the putting forward methodology. It is observed that the quality factor decreases by increasing the Gaussian parameter. It is strongly related to the effective area reduction.

As another result the dispersion length is illustrated in Fig. 11 for the given Gaussian parameter and two central wavelengths. The narrow peaks at λ=1500nm and 1550nm imply the precise tuning of the zero dispersion wavelengths. The higher-order dispersion length of the designed fibers is demonstrated in Fig. 12. It is clear that the higher-order dispersion length increases by raising the Gaussian parameter.

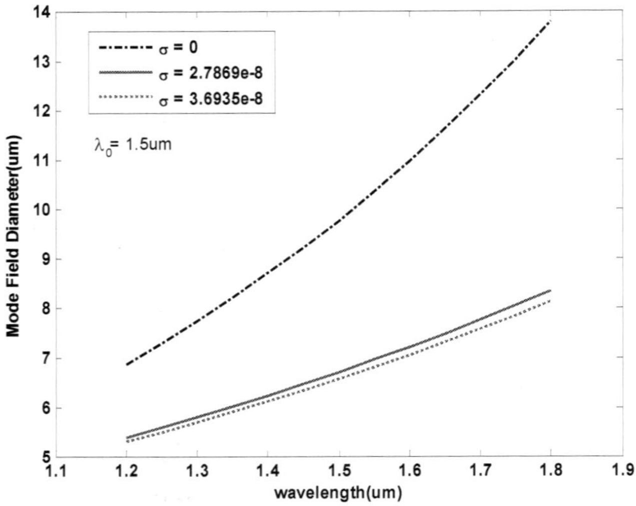

Fig. 8. Mode Field Diameter versus wavelength at λ_0=1500nm with σ as parameter.

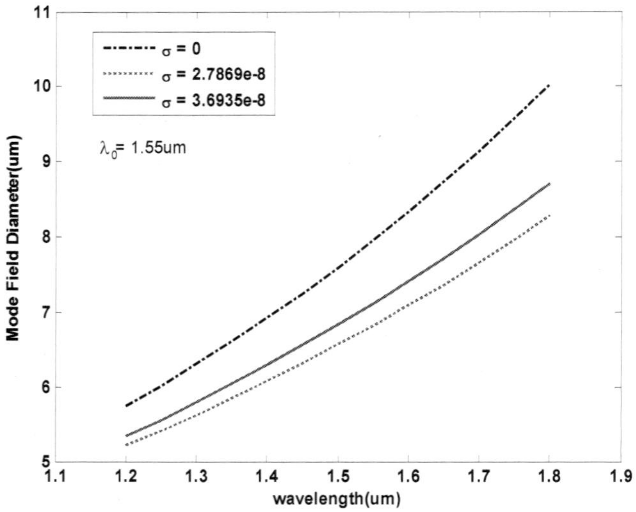

Fig. 9. Mode Field Diameter versus wavelength at λ_0=1550nm with σ as parameter.

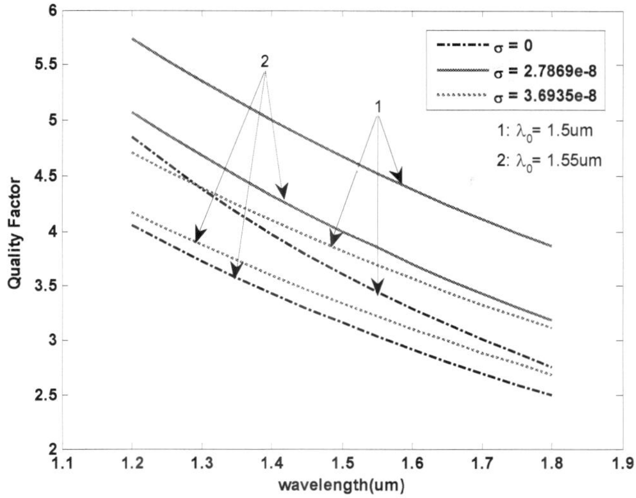

Fig. 10. Quality Factor versus wavelength at λ_0=1500nm, 1550nm with σ as parameter.

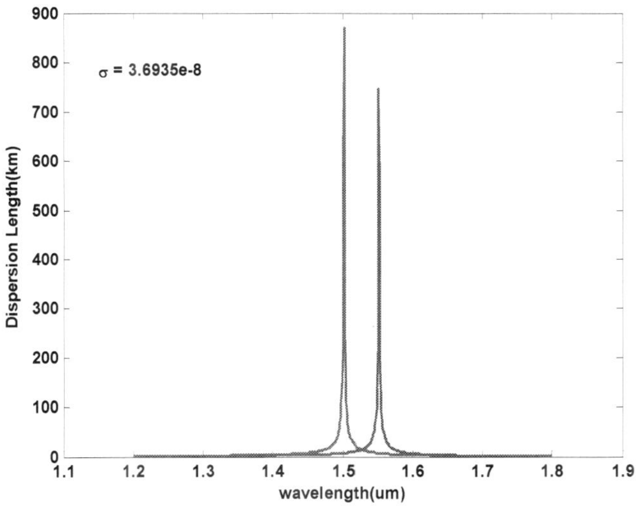

Fig. 11. Dispersion Length vs. Wavelength at $\lambda_0 = 1.5, 1.55\,\mu m$.

In the following, the nonlinear effect length for 1 mW input power is illustrated in Fig. 13. First, it can be extracted that the suggested structures have the high nonlinear effect length. For the general distances, these simulations show that the fiber input power can become some hundred times greater to have the nonlinear effect length comparable with the fiber dispersion length. Second, the nonlinear effect length decreases and increases, respectively, by raising the Gaussian parameter and wavelength.

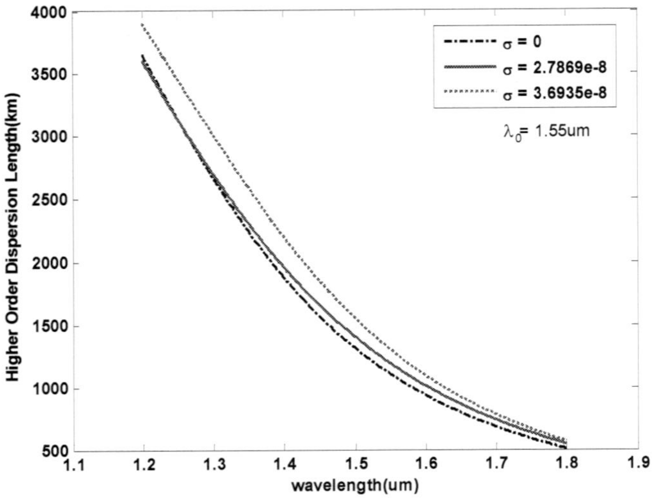

Fig. 12. Higher Order Dispersion Length vs. Wavelength at $\lambda_0 = 1.55\,\mu m$ and Variance of the weight function as parameter.

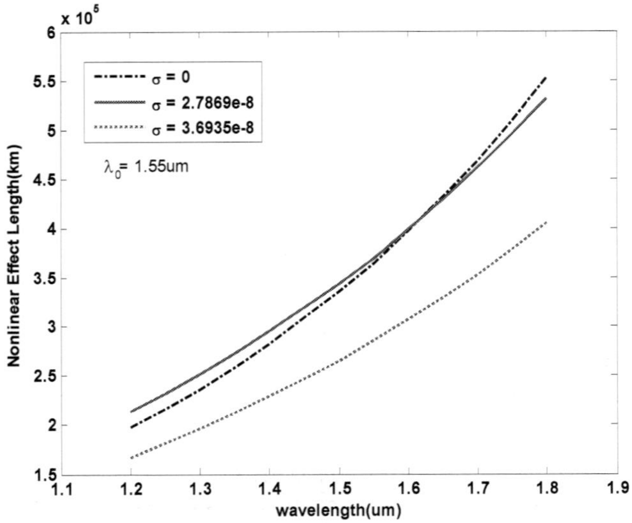

Fig. 13. Nonlinear Effective Length versus wavelength at λ_0=1550nm with σ as parameter.

The amount of the fiber bending loss strongly depends on the bend radius and the mode field diameter. Figures 14 and 15, respectively, illustrate the bending loss (dB/m) versus the bending radius (mm) at λ_0= 1550 nm and 1500nm with variance of the weighting function (σ) as a parameter. According to Figs. 8, 9, 14, and 15, it is clear that smaller mode field diameter yields to the greater tolerance to the bending loss.

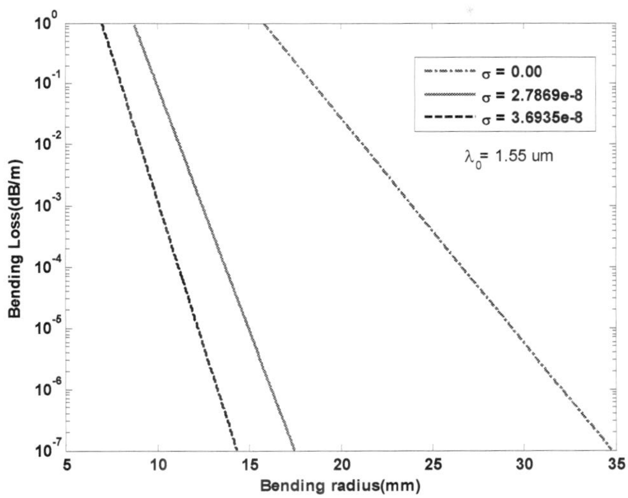

Fig. 14. Bending loss (dB/m) Vs. Bending radius at λ_0=1550nm with σ as parameter.

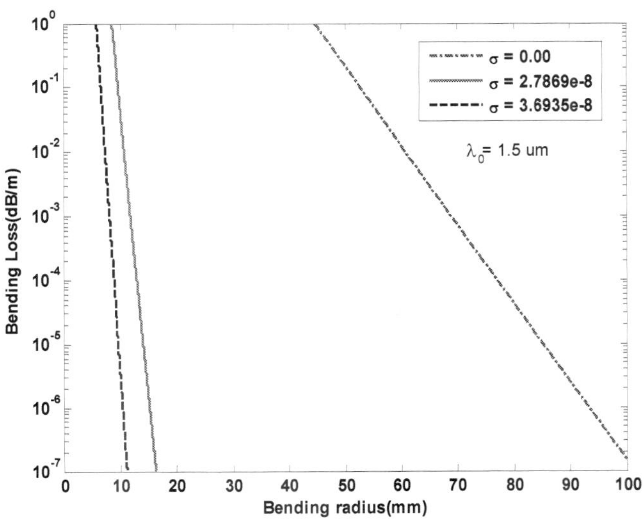

Fig. 15. Bending loss (dB/m) Vs. Bending radius at λ_0=1500nm with σ as parameter.

All of the presented outcomes show that the suggested idea has capability to introduce a fiber including higher performance. We have presented a novel method that includes the small dispersion, its slope, high effective area, and small mode field diameter simultaneously [24]. So all options required for the zero dispersion shifted communication system are achieved successfully. This advantage is obtained owing to the selection of the basic fiber structure as well as the method of optimization. Our selected fiber structure is the MII, and we use the weighted fitness function applied in the GA for optimization. By combining the suitable structure and the novel optimization method, all of the stated advantages can be gathered simultaneously. The features of the proposed method are capable of being extended to all of fiber structures, introduce two parameters instead of multi-designing parameters, and tune the zero dispersion wavelengths precisely.
The ring index profiles fibers have been closely paid attentions because it has the larger effective-areas that can minimize the harmful effects of fiber nonlinearity [29]. For the proposed MII fiber structures, the small dispersion and its slope have been obtained thanks to a design method based on genetic algorithm. But there is not any concentration on the bending loss characteristic at the design process. Here we want to enter bending loss effect on the fitness function directly and attempt to present an optimized RII triple-clad optical fiber to obtain the wondering performance from dispersion, its slope, and bending loss points of view. The index refraction profile of the RII fiber structure is shown in Fig. 16.

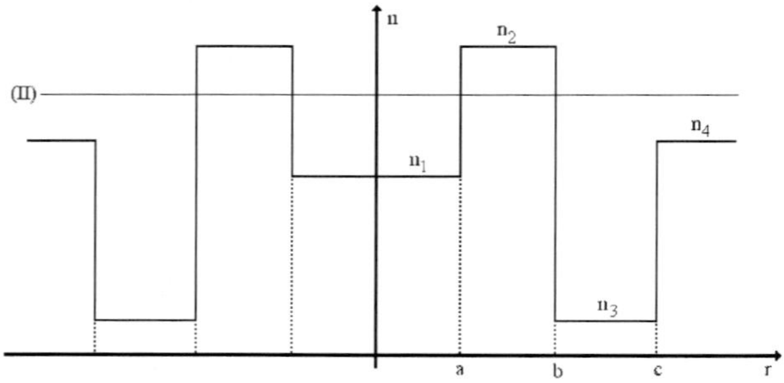

Fig. 16. Index of Refraction Profile for RII Structure

To calculate the dispersion, its slope and bending loss characteristics of the structure, the geometrical and optical parameters are defined as follows.

$$P = \frac{b}{c}, \; Q = \frac{a}{c} \tag{7}$$

$$R_1 = \frac{n_2 - n_3}{n_2 - n_1}, \; R_2 = \frac{n_1 - n_4}{n_2 - n_1}, \; \Delta = \frac{n_2^2 - n_4^2}{2n_4^2} \approx \frac{n_2 - n_4}{n_4}. \tag{8}$$

The design method is based on the combination of the Genetic Algorithm (GA) and Coordinate Descent (CD) approaches. It is well known that the GA is the scatter-shot and the CD is the single-shot searching technique. The single-shot search is very quick compared to the scatter-shot type, but depends critically on the guessed initial parameter values. This description indicates that for the CD search, there is a considerable emphasis on the initial search position. In this method, it is possible to define a fitness function and evaluate every individuals of the population with it. So we have combined the CD and GA methods to improve the initial point selection with the help of generation elite and inherit the quick convergence of coordinate descent [30]. In other words, we cover and evaluate the answer zone by initial population and deriving few generations and use the elite of the latest generation as an initial search position in the CD (Fig. 17).

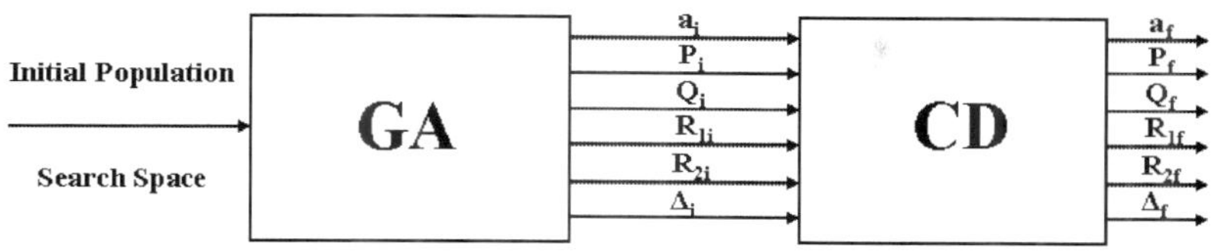

Fig. 17. The Block Diagram of The Proposed Method

To derive the suggested design methodology, the following weighted cost function is introduced. We have normalized the pulse broadening factor in the manner to be comparable with bending loss. This normalization is essential to optimize the pulse broadening factor and bending loss simultaneously. If not, the bending loss impact will be imperceptible and be lost in the broadening factor term.

$$F=\sum_{\lambda} e^{-\frac{(\lambda-\lambda_0)^2}{2\sigma^2}}\left(\frac{1}{z}\sum_{Z}[(1+\frac{\beta_2(\lambda)Z}{t_i^2})^2+(\frac{\beta_2(\lambda)Z}{t_i^2})^2+(\frac{\beta_3(\lambda)Z}{2t_i^3})^2]^{\frac{1}{2}}\right)+BL(\lambda), \tag{9}$$

The bending radius is set on *1 cm* and kept still. The fitness function includes dispersion (β_2), dispersion slope (β_3), and bending loss (*BL*) impacts. In the defined weighted fitness function, internal summation is proposed to include optimum broadening factor for each length up to 200 km. as said at the beginning of this section, one can adjust the zero dispersion wavelength at λ_0 and dominate the dispersion slope by Gaussian parameter (σ). The obtained dispersion behaviors of the structures are illustrated in Fig. 18 which obviously demonstrates the λ_0 and σ parameters influences. It is clear that the zero-dispersion wavelength is successfully set on λ_0 and the dispersion curve is become flatter in the higher σ cases.

To show the capability of the proposed algorithm, Table 4 is presented to clarify the different characteristics of these three structures. By considering on Fig. 18 and Table 4, it is clear that there is a trade-off between the zero dispersion wavelength tuning and the dispersion slope decreasing. In other words, it is found out that the zero value for the σ parameter can tune the zero dispersion wavelength accurately (~100 times better than other cases).

The effective area or nonlinear behavior of the suggested structures is listed in Table 4. These values are high enough for the optical transmission applications. Owing to the special structure of the RII type fiber, the field distribution peak has fallen in the first cladding layer. As such most of the field distribution displaces to the cladding region. This is the origin of large effective area in the designed structures. The normalized field distribution of the RII based designed structures is illustrated in Fig. 19.

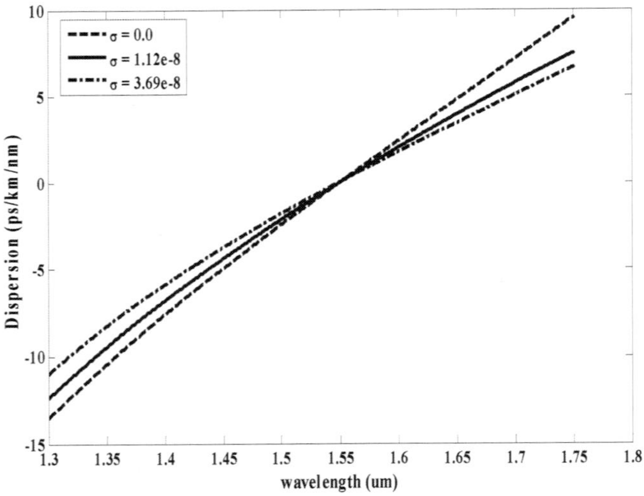

Fig. 18. Dispersion vs. Wavelength at λ_0=1.55 μm.

type	$D(\lambda=1.55\ \mu m)$ (ps/km/nm)	$S(\lambda=1.55\ \mu m)$ (ps/km/nm^2)	$BL(\lambda=1.55\ \mu m)$ (dB/m)	$A_{eff}(\lambda=1.55\ \mu m)$ (μm^2)
σ = 0.0	1.38e-4	0.048	1.90e-2	86.84
σ =1.12e-8	-6.15e-4	0.041	1.67e-1	82.53
σ =3.69e-8	4.50e-2	0.035	4.66e-2	86.01

Table 4. Dispersion, Dispersion Slope, Bending Loss, and Affective Area at λ_0=1.55 μm and Three Given Gaussian Parameters

Due to the refractive index thermo-optic coefficient and the thermal expansion coefficient, the optical and geometrical parameters are altered. Consequently, the optical transmission characteristics of the optical fiber such as dispersion, its slope and bending loss are confronted to change. In order to evaluate the thermal stability of the designed structures, the following results are extracted and presented in Table 5. The dD/dT, dS/dT, $d\lambda_0/dT$, and dBL/dT expressions are respectively the chromatic dispersion, its slope, zero dispersion wavelength, and bending loss thermal coefficients at 1.55μm. It is found out that this environmental factor must be considered in the desired optical fiber design. For example, in the worst case, the zero dispersion wavelengths can be shifted more than 3 nm with 100°C.

In the least design we have focused on RII depressed core triple clad single mode optical fiber and presented a combined optimization approach to obtain desirable design goals. Furthermore, we have used the special fitness function including dispersion, its slope and bending loss impacts simultaneously. With application of this fitness function in the case of higher σ, we could obtain the dispersion and dispersion slope in [1.5 - 1.6] μm interval to be

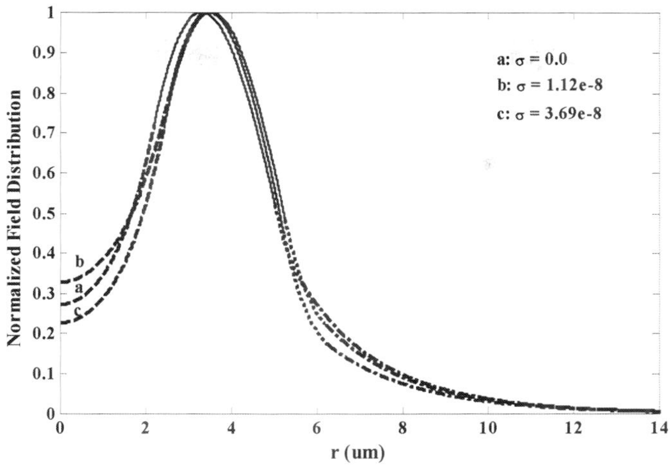

Fig. 19. Normalized field distribution versus the radius of the fiber at λ=1.55 μm with σ as parameter (dashed, solid line, dotted, and dashed-dotted curve represent the core and three cladding layers, respectively).

Type	dD/dT (ps/km/nm/°C)	dS/dT (ps/km/nm^2/°C)	$d\lambda_0/dT$ (nm/°C)	dBL/dT (dBL/m/°C)
$\sigma = 0.0$	-1.22×10^{-3}	$+2.83\times10^{-6}$	$+2.5\times10^{-2}$	$+3.97\times10^{-6}$
σ =1.12e-8	-1.21×10^{-3}	$+2.93\times10^{-6}$	$+3.33\times10^{-2}$	$+2.70\times10^{-5}$
σ =3.69e-8	-1.21×10^{-3}	$+2.93\times10^{-6}$	$+2.5\times10^{-2}$	$+8.79\times10^{-6}$

Table 5. Dispersion, Dispersion Slope, and Bending Loss Thermal Coefficients at λ_0=1.55 μm and Three Given Gaussian Parameters

[(-1.77) - (+1.77)] ps/km/nm and [(0.037) – (0.033)] ps/km/nm^2. Also the amount of bending loss at 1.55 μm with 1cm radius of curvature and effective area are 4.66e-2 dB/m and 86.01 μm^2 respectively. In the meantime, the thermal stabilities of the designed structures are evaluated. It is possible to design zero dispersion shifted by using graded index structure. The main options are dispersion value and the effective area at 1550nm to minimize pulse broadening and nonlinearity effects. Excess investigation of large mode area fibers show that there is not serious focusing on design of zero dispersion shifted fiber based on the graded index refractive structures. The index refraction profile of the triangular-core graded index optical fiber structure, which is suggested by us for the first time, is shown in Fig. 20. It is clear that the proposed graded index fiber has a linear variation in core region. According to the TMM approach, it is assumed that the refractive index of the fiber with an arbitrary but axially symmetric profile is approximately expressed by a staircase function. So the field distribution and guided modes are calculated [26].

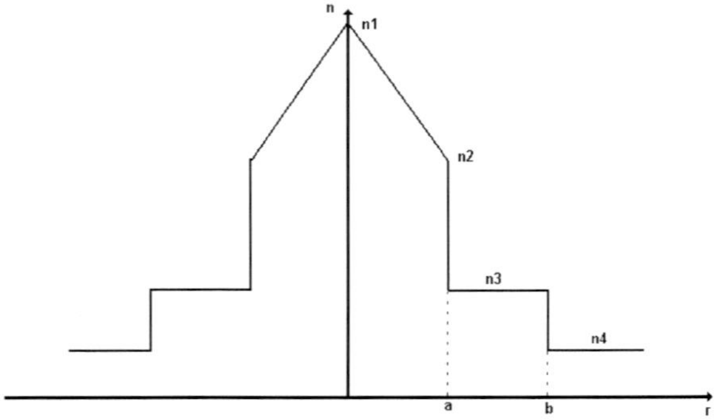

Fig. 20. Refractive Index Profile for Triangular Core Graded Index Fiber Structure

Also for easy handling of the problem and calculating the dispersion and its slope of the proposed fiber, following optical and geometrical parameters are defined.

$$\Delta = \frac{n_1^2 - n_4^2}{2n_4^2} \approx \frac{n_1 - n_4}{n_4}, \quad P = \frac{a}{b}, \; n_3 = n_4 + \eta\left(n_2 - n_4\right), n_2 = n_3 + \mu\left(n_1 - n_3\right). \tag{9}$$

In order to explain layers' refractive index, η and μ coefficients are introduced and set between 0 and 1. It must be declared that P parameter relates the cladding layer thickness with the core radius. The design method is based on the limited coordinate descent (CD) approach [30]. Based on the extensive investigation, it is found out that the smaller core radius and larger refractive index difference lead the zero dispersion wavelength to around 1.55μm. Therefore in order to design the dispersion shifted optical fiber, Δ and core radius are set to 8×10^{-3} and 1.8μm respectively. Also it is assumed that the core and first cladding layer have same thickness. Then direct search is done for η and μ parameters in the [0,1] interval. To derive the suggested design methodology, the following fitness function is introduced which includes the pulse broadening factor.

$$F = \sum_{Z} [(1 + \frac{\beta_2(\lambda_0)Z}{t_i^2})^2 + (\frac{\beta_2(\lambda_0)Z}{t_i^2})^2 + (\frac{\beta_3(\lambda_0)Z}{2t_i^3})^2]^{\frac{1}{2}} \tag{10}$$

In the defined fitness function, the summation is proposed to include optimum broadening factor for each length up to 200 km. The short glance on eq.(10) shows that the above fitness function is a limited version of the past one [24], which the wavelength duration optimizing is abbreviated. Also, it is not weighted because the fitness function is evaluated at single wavelength (λ_0). One can adjust the zero dispersion wavelength at λ_0. It should be kept in mind that in the fiber design, one likes to shift the zero dispersion wavelength to the region that the fiber has the lowest level attenuation. The optical attenuation has a global minimum around 1.55μm wavelength and that is why the most optical communication systems are operated at this wavelength. Seeing that, the λ_0 parameter of the applied method is set to 1.55μm to achieve the desired zero dispersion shifted structure. Considering the parameter presented above and CD method, the design procedure is driven and optimal parameters are extracted and demonstrated in Table 6.

Core radius	Δ	p	η	μ
*1.8μm	8×10^{-3}	0.5	0.25	0.44

*b=2a

Table 6. The Optical and Geometrical Parameters for the Designed Structure.

The value of the dispersion is -0.04549ps/km/nm at 1.55μm, which is properly small. The nonlinear effects in a single mode fiber are the ultimate restricting factors for bit rate and distance in long haul optical fiber communication system. Therefore, the large effective area single mode fibers have been the subject of considerable studies recently. The effective area of the suggested structure is 65.3 μm^2 at 1.55μm, which is acceptable for this application. The associated mode field diameter at aforesaid wavelength is 9.03μm.

8. Non-Zero Dispersion Shifted Fibers (NZDSFs)

Use of commercially available erbium doped fiber amplifiers (EDFA), which forces optical communication systems to be operated in the 1550 nm window, has significantly reduced the link length limitation imposed by attenuation in the optical fiber. However, high bit rate (~ 10 Gb/s) data transmission can be limited by the large inherent dispersion of the fiber. Dispersion shifted fibers (DSF), which has zero dispersion around 1550 nm, have been proposed and developed to overcome this problem. However in order to increase the information carrying capacity, latest high speed communication system is based on the dense wavelength division multiplexing/demultiplexing (DWDM). In such systems, nonlinear effects like four wave mixing (FWM), which arise due to simultaneous transmission at many closely spaced wavelengths and high optical gain from EDFA, imposes serious limitations on the use of a DSF with zero dispersion wavelength at 1550 nm. To overcome this difficulty, the nonzero dispersion shifted fibers having small dispersion in the range ~ 2–4 ps/km/nm over the entire gain window of EDFA have been proposed. In such fibers, the phase matching condition is not satisfied and hence the effect of FWM becomes negligible due to small dispersion. Nonlinear effects like cross phase modulation (XPM), which limits the numbers of different wavelength signals, can be reduced by increasing the mode field diameter (MFD) and hence effective area of the fiber. Therefore large effective area nonzero dispersion shifted fibers have been developed [31]. To achieve large effective area and low bending and splice loss with conventional fiber, a refractive index profile, as shown in Fig. 21, is designed, which is mathematically described by Eq. (11).

$$
\begin{cases}
n_1 & |r| \le R_1 \\
n_1\sqrt{1-2\Delta_1\left(\frac{r-R_1}{R_2-R_1}\right)^{\alpha}} & R_1 < |r| \le R_2 \\
n_2 & R_2 < |r| \le R_3 \\
n_3\sqrt{1-2\Delta_2\left(\frac{R_{34}-r}{R_{34}-R_3}\right)^{\alpha}} & R_3 < |r| \le R_{34} \\
n_3\sqrt{1-2\Delta_2\left(\frac{r-R_{34}}{R_4-R_{34}}\right)^{\alpha}} & R_{34} < |r| \le R_4 \\
n_2 & |r| > R_4
\end{cases}
\tag{11}
$$

where R_1, R_2, R_3, R_4 are radius parameters, $R_{34}=(R_3+R_4)/2$ is the center of the side core, n_1, n_2, n_3 are the highest refractive index of central core, refractive index of cladding, and highest refractive index of side core, respectively, $\Delta_1=(n_1-n_2)/n_2$ and $\Delta_2=(n_3-n_2)/n_2$ are the relative profile heights of central core and side core, respectively, and α is the curve parameter. This fiber consists of three parts, including a central core, side core and a cladding layer. The side core inside the core region (i.e., the region $r<R_4$) is designed to allow more signal energy to flow into the cladding region so that a large effective area can be obtained [31].

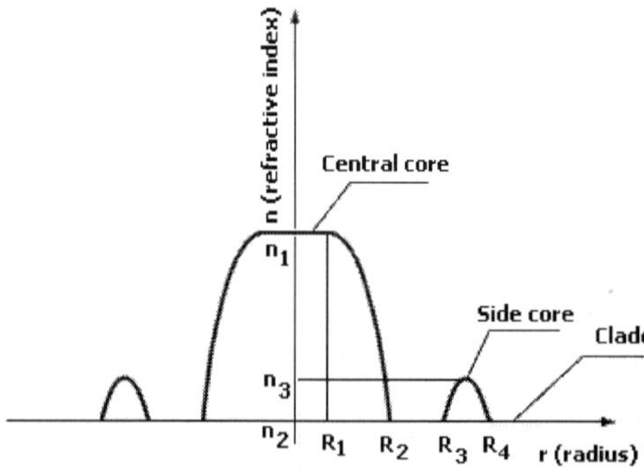

Fig. 21. Refractive index profile of newly designed large effective area, low bending and splice loss NZ-DSF.

From Eq. (11) it is found that the exact refractive index profile are controlled by seven parameters (i.e., R_1, R_2, R_3, R_4, Δ_1, Δ_2, and α). To ensure the single mode operation, $R4$ must be less than certain number. Thus, there are six parameters need to be optimized for achieving the required dispersion slope, large effective area, low splice loss, low bending loss, and low Rayleigh scattering loss simultaneously. By using the random searching method, it is found that the fiber has the optimum performance under the following conditions:

$$\begin{cases} a = 3.2\mu m\ (radius\ of\ fiber\ core) \\ R_1 = 0.218309a = 0.6986\mu m \\ R_2 = 0.568242a = 1.8184\mu m \\ R_3 = 0.7325779a = 2.3442\mu m \\ R_4 = 1.0a = 3.2\mu m \\ \Delta_1 = 0.6425\% \\ \Delta_2 = 0.2378\% \\ \alpha = 1.6 \end{cases} \tag{12}$$

The Gaussian approximation method is used for calculating electrical field distribution for the designed refractive index profile. The designed fiber has a dispersion about 4 ps/km/nm and dispersion slope about 0.06 ps/km/nm^2 at 1.55 μm operating wavelength, which can be used to avoid four-wave mixing (FWM). Also the zero dispersion wavelength is adjusted near 1480nm. In addition, calculation also shown that the designed fiber not only has a large effective area over 100 μm^2 but also has low bending loss (<1.3×10^{-3} dB with 30 mm bending radius and 100 turns) and low splice loss (<6.38×10^{-3} dB) with conventional

fiber. In order to broaden the wavelength range - not only the conventional C-band (1530-1565 nm) - but also the L-band (1565-1620 nm) and S-band (1460-1530 nm), the dispersion slope should be as low as possible, which can decrease the cost of dispersion compensation and suppress the self-phase modulation, especially for the 40 Gb/s communication system. Now we present a new depressed core index dual ring profile, which can provide a large A_{eff} and a low dispersion slope simultaneously, and the zero dispersion wavelength is less than 1430 nm [32]. The fibers with this refractive index profile have the low bending loss and low intrinsic loss. The splice loss also can reach the accepted value as it is spliced with conventional single mode fiber. The principal design requirements for the fibers are large A_{eff}, small dispersion slope, low bending and intrinsic loss, low polarization-mode dispersion, and zero dispersion wavelength that should be lower than 1430 nm.
Relative refractive index Δn_i is defined by the equation: $\Delta n_i=(n_i-n_c)/n_c$, where n_c is the outer cladding layer's refractive index. The dopant in the glass can decrease the glass viscosity, i.e, more dopant concentration means less glass viscosity. High difference value between Δn_1 and Δn_2 may cause high difference value in viscosity property of the depressed core layer and the first raised ring. Therefore, more mechanical stress will be built in the optical fibers during the drawing process [33]. On the other hand, high difference value between layers' refractive index can increase the compositional variation in the optical fiber. Therefore, more thermal stress can be also caused by the radial variation of thermal expansion coefficient due to the big compositional variation in optical fiber [34]. The residual stress in optical fibers can not only weaken the strength of optical fibers, but also increase the fiber's attenuation and polarization mode dispersion values. According to the above description, the refractive index profile shown in figure 22 is proposed to overcome problems. Every parameters of the fiber profile are set out in Table 6, where Δn_i are the relative refractive index of different layer from the depressed core layer to the cladding, respectively.

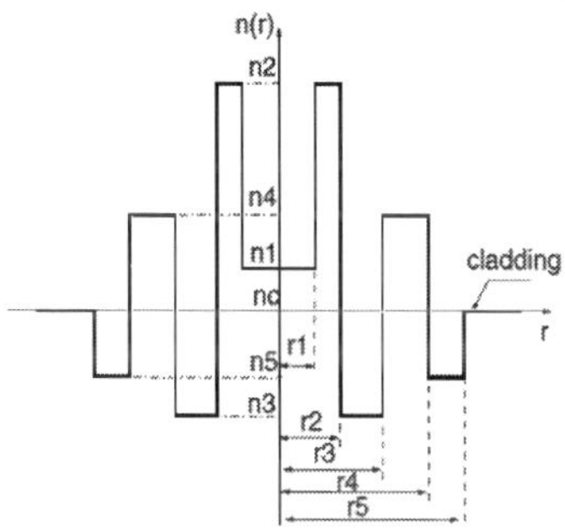

Fig. 22. Improved refractive index profile with dual ring and depressed outer ring based on the depressed core-index.

Δn_1 (%)	Δn_2 (%)	Δn_3 (%)	Δn_4 (%)	Δn_5 (%)	r_1 (μm)	r_2 (μm)	r_3 (μm)	r_4 (μm)	r_5 (μm)
0.14	0.57	-0.27	0.30	-0.18	2.50	4.10	6.88	9.98	12.41

Table 6. Parameters of refractive index profile shown in Fig. 22

Table 7 shows the optical characteristic of fabricated fiber designed according to the refractive index profile parameters as Table 6, where MFD, RDS are the mode field diameter and relative dispersion slope, respectively. It is noted that the fiber has a large A_{eff} of 105 μm² and a small dispersion slope of about 0.065 ps/ km /nm² simultaneously. Macro bending loss at 1550 nm is less than 0.05 dB/km (100 turns on the 60 mm diameter mandrel).

Parameters item	Wavelength (nm)	Typical value
Dispersion (ps/km/nm)	1460	3.465
	1550	9.569
	1625	14.324
Dispersion Slope(ps/km/nm²)	1550	0.0648
Zero dispersion wavelength (nm)	-	1413
Attenuation (dB/km)	1550	0.210
MFD (μm)	1550	10.2
A_{eff} (μm²)	1550	105.6
A_{eff} ×dispersion	1550	1010.5
RDS (nm^{-1})	1550	0.0068
PMD($ps/km^{0.5}$)	1550	0.04
Macro bending loss for 100 turns on the 60mm diameter (dB/km)	1550	0.006
	1625	0.015

Table 7. Optical characteristics of the fabricated fiber

From the table7 we can see that the zero dispersion wavelength is below 1430 nm, the dispersion at 1460, 1550 and 1625 nm are 3.465, 9.569 and 12.324 ps/km/nm, respectively. Therefore, this fiber not only can be used at the conventional C band for transmission link, but also can be suited for S-band and L-band. With the progress being made in the practical application of the Raman amplifier, this fiber also can be applicable in transmission links using distributed Raman amplifier in the future. Furthermore, the value of A_{eff} × dispersion is also large enough to suppress the dispersion-related non-linear effects in the transmission system [35]. It is obvious that a large effective area fiber with non-zero dispersion about 4 ps/km/nm at 1.55 μm wavelength band is a good approach to avoid four-wave mixing (FWM) effect, which in turn enhances the performance of wavelength division multiplexing (WDM) system [31]. However, such large effective area fibers have relatively large dispersion slope which also restricts the numbers of different wavelength signals [21, 36-38]. Therefore, considerable efforts are being made to reduce the dispersion slope of such fibers to deal with a rapid progress of DWDM system [39,40]. The refractive index profile of the fiber is shown in Fig. 23. This profile has been named RI type in the literatures.

In order to obtain the flat modal field over the entire central dip region, the effective index (n_{eff}) of the mode should be equal to the refractive index of the central dip (i.e., $n_{eff} = n_1$) [15]. The values of various parameters used in design of the dispersion characteristics are tabulated in Table 8. Here, the relative index difference Δi is given by $\Delta i = (n_i^2 - n_4^2)/2n_i^2$, i=1, 2, 3.

The values of the MFD and A_{eff} associated with the modal field of the proposed design are 8.3 μm and 56.1 μm², respectively. The total dispersion coefficient (D), which includes both waveguide and material dispersion of the fiber, is calculated in the wavelength range of 1530 to 1610 nm, which covers the entire C- and L-bands of erbium doped fiber amplifiers. In order to study the tolerance of the various characteristics of the proposed fiber design, we have randomly changed the values of thickness and Δ of each region by 1%. The actual and the corresponding perturbed refractive index profiles are shown schematically in Fig. 23 by the solid and dashed curves, respectively. Figure 24 shows the variation of the total dispersion (*D*) with wavelength. The solid and dashed curves correspond to the actual and the perturbed refractive index profiles, respectively. This figure indicates that over the entire wavelength range of 1530 to 1610 nm, the dispersion value, which is within 2.6-3.4 ps/km/nm, for both profiles are within the appropriate range (2-4 ps/km/nm) needed to avoid four wave mixing (FWM) [41].

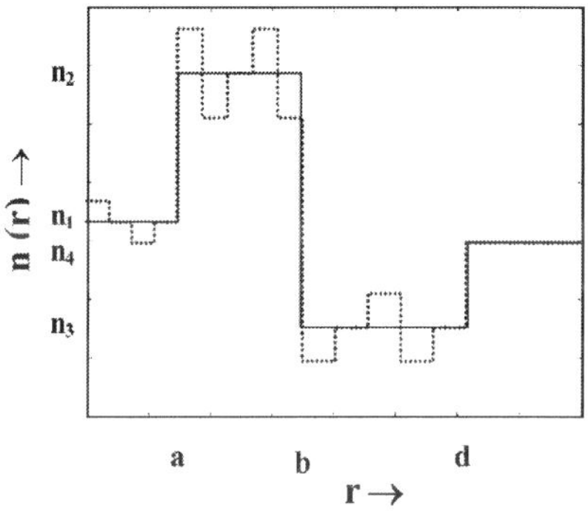

Fig. 23. Schematic of the refractive index profiles of the proposed fiber (solid curve). The dashed curve corresponds to the perturbed refractive index profile.

a(μm)	b(μm)	d(μm)	Δ_1(%)	Δ_2(%)	Δ_3(%)
1.0	3.1	6.5	0.03	0.48	-0.20

Table 8. The values of various parameters used in design

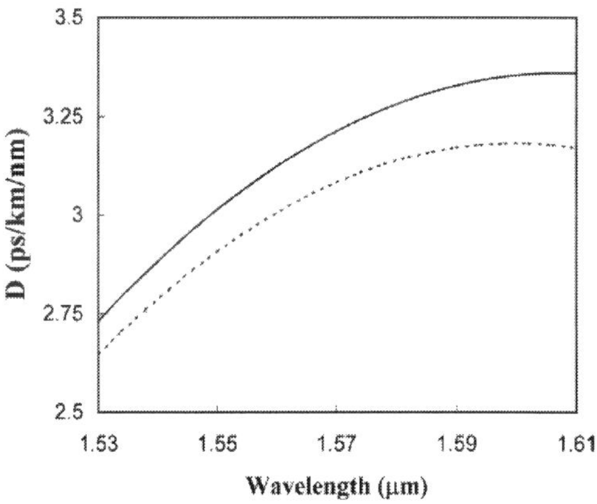

Fig. 24. Variation of the total dispersion (*D*) as a function wavelength. The solid and dashed curves correspond to the proposed and the perturbed refractive index profiles (shown schematically in Fig. 23), respectively.

The dispersion and dispersion values of the designed fiber at λ_0 = 1550 nm is 3.0 ps/km/nm and 0.01 ps/km/nm^2, respectively. The maximum value of dispersion slope is 0.015 and 0.014 ps/km/nm^2 for the actual and perturbed refractive index profiles over the entire wavelength range of 1530 to 1610 nm. This shows that the perturbation in the refractive index profile does not make much change in the dispersion slope.

9. Dispersion Flattened Fibers (DFFs)

The dispersion value becomes larger by the wavelength increasing in the conventional optical fibers. So owing to the dissimilar broadening for different channels, the multi-channel application realization would be hard. A suitable optical fiber should meet the small dispersion as well as the small dispersion slope in the predefined wavelength interval [42]. The concept of providing the attractive option of low dispersion over a range of wavelengths was first suggested by Kawakami and Nishida in 1974 [43, 44]. They proposed the original "W" fiber structure and explained the importance of a relatively narrow depressed cladding region in modifying the waveguide dispersion to give a curve which turned over to give two wavelengths for zero dispersion [45]. To minimize pulse broadening in an optical fiber, the chromatic dispersion should be low over the wavelength range used. A fiber in which the chromatic dispersion is low over a broad wavelength range is called a dispersion-flattened fiber. The rms value, or the function f, to be minimized is:

$$f = \left(\frac{1}{\lambda_2-\lambda_1}\int_{\lambda_1}^{\lambda_2} C^2(\lambda_0)d\lambda_0\right)^{1/2} \tag{13}$$

where C is the chromatic dispersion. A normalized W profile is given by

$$N(r) = \begin{cases} N_1 & 0 \le r < b \\ N_2 & b \le r < a \\ 1 & r \ge a \end{cases} \tag{14}$$

where $N_1 > 1$ and $N_2 < 1$. The constraint that the first higher-order mode should appear exactly at 1.25 µm is imposed. Thus there are four variables, namely, (N_1, N_2, b, a), and one constraint. Assume that N_1 and N_2 are given certain fixed values. The values $N_1 = 1.02$ and $N_2 = 0.99$ will prove to be interesting. If $b = a$, then the W profile has degenerated into a step-index profile. The core radius, a, of this step-index fiber is easily calculated with the exact cutoff condition V = 2.405, where V is the normalized frequency. The value $N_1 = 1.02$ yields $b = a = 1.64$µm. Direct numerical calculation shows that if the outer radius, a, is increased then the inner radius, b, must also be increased to keep the cutoff wavelength at 1.25 µm. Hence the constraint λ_c = 1.25 µm corresponds to a curved line in the a-b plane. The rms value of the chromatic dispersion along this line is given in Fig. 25. The point of minimum dispersion is easily located.
This procedure is repeated for different combinations of N_1 and N_2, and the result is given in Table 9. The first column of this table, i.e., $N_2 = 1$, corresponds to step-index profiles.
According to Table 9, the global minimum is 0.9 ps/km /nm, and the corresponding optimal W fiber is (N_1, N_2, b, a) = (1.02, 0.99, 1.91, and 2.85 µm); see Fig. 26. It should be observed that the global minimum is flat; i.e., there is a valley in Table 9 giving roughly the same rms dispersion. Another observation is that the dependence of N_2 in Table 9 is weak if N_1 is less than 1.01. On the other hand, the dependence of N_2 is strong if N_1 is greater than 1.01 and N_2 is close to unity.

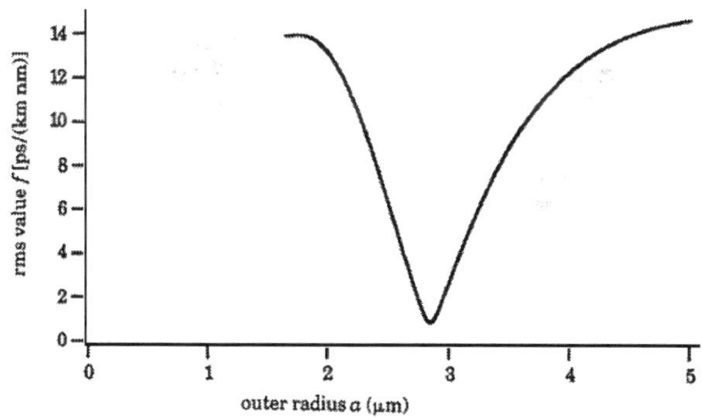

Fig. 25. The rms value of the chromatic dispersion over the vacuum wavelength range (1.25 μm, 1.60 μm) as a function of the outer radius a in a W fiber. The cutoff vacuum wavelength is 1.25 μm. The relative refractive-index increases in the core and in the inner cladding are 1.02 and 0.99, respectively.

	N_2				
N_1	1.000	0.995	0.990	0.985	0.980
1.002	12	12	12	12	12
1.005	8.6	8.6	8.5	8.5	8.5
1.010	4.8	4.3	4.2	4.1	4.1
1.015	7.9	2.3	1.8	1.6	1.5
1.020	14	1.3	0.9	1.0	1.2
1.025	21	1.1	1.7	2.4	2.9
1.030	27	4.0	2.4	3.7	4.5

Table 9. Minimum rms chromatic dispersion (ps/km/nm) for different doping level in the core & cladding

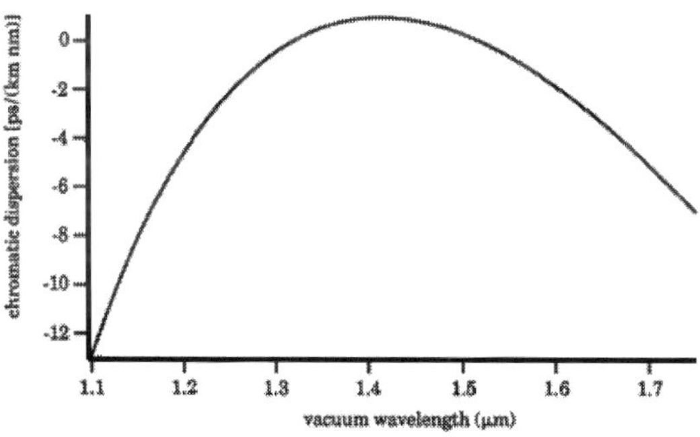

Fig. 26. Chromatic dispersion for the optimal W fiber $(N_1, N_2, b,$ a) = (1.02, 0.99, 1.91 μm, 2.85 μm). The rms value of the chromatic dispersion over the vacuum wavelength range (1.25μm, 1.60 μm) is equal to 0.9 ps/km/ nm. The cutoff vacuum wavelength is 1.25 μm.

An exhaustive method for calculating the minimum rms chromatic dispersion in W fibers has been presented [45]. The procedure is to generate all W fibers with a certain cutoff wavelength and then find the minimum rms dispersion by one-dimensional minimization followed by direct inspection. It was found, in the case investigated, that the W fiber is capable of dispersion flattening only if high doping levels are used. Dispersion and its slope are responsible on the pulse broadening [4]. So we believe that mixing and gathering these parameters on the design procedure would lead us to a fiber with exciting performances. In other words, we concentrate on dispersion and slope simultaneously to achieve the small dispersion and its slope in the predefined wavelength interval, the band which we want to have flat dispersion behavior [42]. We use WII-type optical fiber structure which its refractive index profile is shown in Fig. 27.

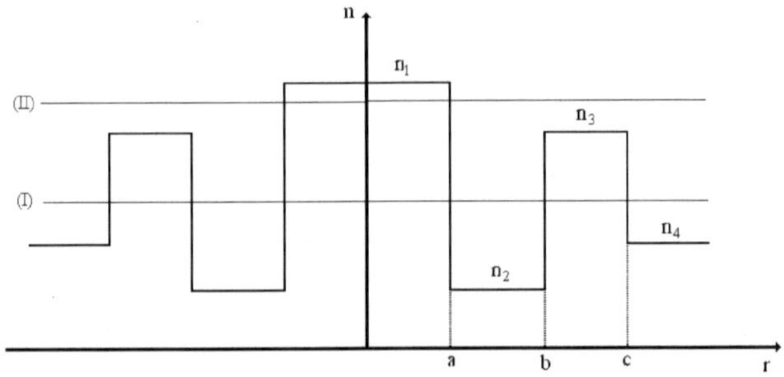

Fig. 27. The index of refraction profile for the proposed structures (WII)

Also, for easy handling of the problem the following optical parameters are defined as follows.

$$R_1 = \frac{n_1 - n_3}{n_3 - n_2},\ R_2 = \frac{n_2 - n_4}{n_3 - n_2},\ \Delta = \frac{n_1^2 - n_4^2}{2n_4^2} \approx \frac{n_1 - n_4}{n_4}. \tag{15}$$

For this structure the geometrical parameters are introduced in the following.

$$P = \frac{b}{c}, Q = \frac{a}{c}. \tag{16}$$

To achieve the flattening purpose, we have defined a weighted fitness function. Equation (17) shows our proposal for the weighted fitness function of the un-chirped pulse broadening factor.

$$F = \sum_{\lambda} e^{-\frac{(\lambda-\lambda_0)^2}{2\sigma^2}} \sum_{Z} [1 + (\frac{\beta_2(\lambda)Z}{t_i^2})^2 + (\frac{\beta_3(\lambda)Z}{2t_i^3})^2]^{\frac{1}{2}}, \tag{17}$$

It is useful to say that at first we applied this fitness function to design dispersion shifted fiber. But outcomes presentation will show that this function is appropriate in Flattening application too [42]. By using this fitness function, the zero dispersion wavelength can be shifted to the central wavelength (λ_0). Since, the weight of the pulse broadening factor at λ0 is greater than others in the weighted fitness function; it is more likely to find the zero dispersion wavelength at λ_0 compared to the other wavelengths. In the meantime, the flattening of the dispersion curve is controlled by Gaussian parameter (σ). To put it in another way, the weighting Gaussian function becomes broader in the predefined wavelength interval by increasing the Gaussian parameter (σ). As a result, the effect of the pulse broadening factor with greater value is regarded in different wavelengths, which causes a considerable decrease in the dispersion slope in the interval. Consequently, the zero dispersion wavelength and dispersion slope can be tuned by λ_0 and σ, respectively. The wavelength and the distance durations for the design are defined as follows: 1.50 μm < λ < 1.60 μm; 0 < z < 200 km. In the simulations un-chirped initial pulse with 5 ps as full-width at half-maximum is used. From Figure 28, it is clear that the zero dispersion wavelength is successfully set at λ_0 which is equal to 1.55 μm. Furthermore, the dispersion curve becomes so flat by adding the Gaussian weighting term to the fitness function. In other words, in the absence of weighting function, the optimized dispersion has higher slope compared to its presence.

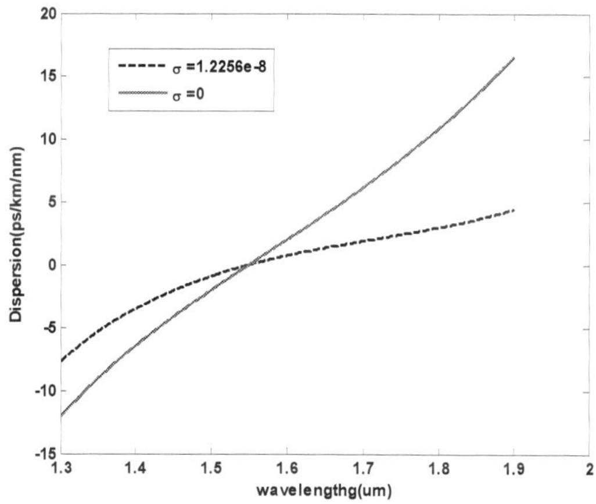

Fig. 28. Dispersion versus Wavelength with and without weighting function

The impact of sigma parameter on the dispersion and its slope is illustrated in Figures 29 and 30. It is obvious that the dispersion value reduces by the sigma parameter increase in the predefined wavelength interval and the curve is the smoothest in the highest sigma case. This event can be described based on the fact that the weighting function (Gaussian function) has large values around the central wavelength by the increase in the Gaussian parameter. So, a large band of the wavelength around the central wavelength has almost the same chance for optimization and thus the dispersion will be small and uniform for this band. In other words, the duration of this band can be controlled by the sigma parameter.

The dispersion slope is strongly affected by the presence of σ in such a manner that its increase has the considerable influence on the dispersion slope and decreases it obviously. This result is easily visible in Figure 30 which shows the dispersion slope versus wavelength with variance of the σ as a parameter. According to the presented weighted function based GA optimization, the following optical and geometrical parameters are obtained. We find out that the optimal value of R_2 for all Gaussian parameters are near to -1.

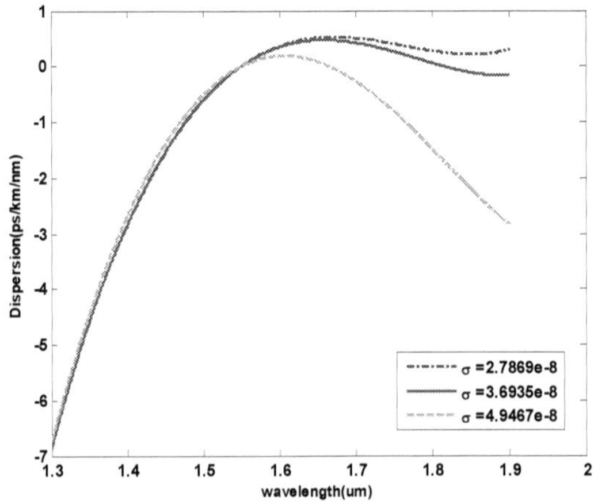

Fig. 29. Dispersion versus Wavelength for different Gaussian parameter

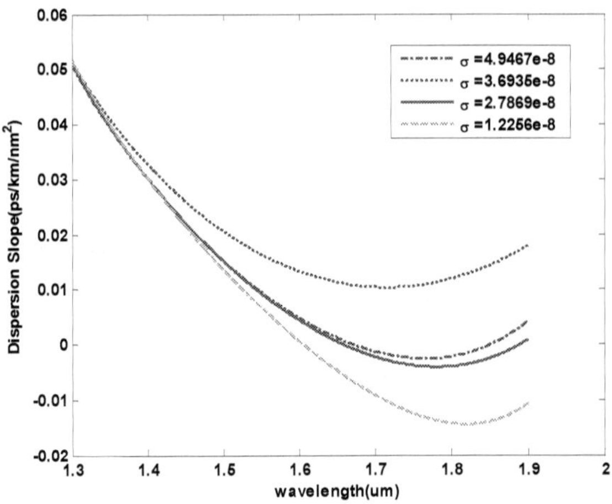

Fig. 30. Dispersion Slope versus Wavelength for different Gaussian parameter

type	$a(\mu m)$	P	Q	R_1	R_2	Δ
$\sigma = 0.00$	2.5462	0.6942	0.3046	7.0897	-0.9517	4.886e-3
$\sigma = 1.2256e-8$	2.4450	0.7189	0.4049	3.0497	-0.9854	8.159e-3
$\sigma = 2.7869e-8$	2.5763	0.8478	0.4771	2.2137	-0.9877	7.178e-3
$\sigma = 3.6935e-8$	2.4374	0.8461	0.4098	1.7966	-0.7076	8.064e-3
$\sigma = 4.9467e-8$	2.5914	0.8942	0.4728	2.2583	-0.9780	6.812e-3

Table 10. Optimal Values for the Optical and Geometrical Parameters

In tables 11 and 12 the simulated numerical values of dispersion and dispersion slope based on the presented algorithm for wavelength duration in $[1.5-1.6]\mu m$ are given. Also, dispersion and dispersion slope difference for this band are presented for each case. For these cases, we show that there is about 6 times difference between traditional optimization and weighted function based optimization in dispersion case.

type	$D(\lambda=1.5\mu m)$ $ps/km/nm$	$D(\lambda=1.55\mu m)$ $ps/km/nm$	$D(\lambda=1.6\mu m)$ $ps/km/nm$	$\Delta D=D(1.6)-D(1.5)$ $ps/km/nm$
$\sigma=0.00$	-2.039	-0.032e-3	2.002	4.041
$\sigma=1.2256e-8$	-0.899	0.0238	0.765	1.664
$\sigma=2.7869e-8$	-0.618	0.0032	0.363	0.981
$\sigma=3.6935e-8$	-0.606	0.0051	0.346	0.952
$\sigma=4.9467e-8$	-0.510	0.0034	0.191	0.701

Table 11. Simulated Numerical Results for Dispersion for wavelength duration $[1.5-1.6]\mu m$

type	$S(\lambda=1.5\mu m)$ $ps/km/nm^2$	$S(\lambda=1.55\mu m)$ $ps/km/nm^2$	$S(\lambda=1.6\mu m)$ $ps/km/nm^2$
$\sigma=0.00$	0.0416	0.0402	0.0401
$\sigma=1.2256e-8$	0.0206	0.0164	0.0133
$\sigma=2.7869e-8$	0.0153	0.0096	0.0488
$\sigma=3.6935e-8$	0.0152	0.0093	0.0044
$\sigma=4.9467e-8$	0.0137	0.0068	0.0007

Table 12. Simulated Numerical Results for Dispersion Slope for wavelength duration $[1.5-1.6]\mu m$

As a final result of these simulations, we should point out that for zero value of the Gaussian parameter zero-dispersion wavelength has high accuracy compared to nonzero-values of the Gaussian parameters cases. By comparing presented results and the ones demonstrated earlier as a dispersion flattened optical fiber, it is clear that the least design has considerable band width, the band between to zero dispersion wavelength. Moreover, the dispersion value tolerance in this interval is so small which is a direct result of its small slope.

10. Conclusion

In this chapter some special fiber structures for covering broadband optical fiber communications were reviewed. For these three cases R, W and M two different types for each of them were considered and discussed in detail. We have been shown that using the proposed design method in this chapter systematic approach for broadband applications can be found and considering the fibers in this chapter ultra broadband communications are available.

11. References

[1] kazumasa Ohsono, Tomoyuki Nishio, Takahiro Yamazaki, Tomomi Onose, and Kotaro Tan, "Low Non-linear Dispersion-shifted Fiber for DWDM Transmission", HITACHI CABLE REVIW, Vol. 19, 2000.

[2] J. A. Baghdadi, A. Safaai-Jazi, and H.T. Hattori, "Optical fibers with low nonlinearity and low polarization-mode dispersion for terabit communications," Optics & Laser Technology, Vol. 33, pp. 285-291, 2001.

[3] S. Makouei, M. savadi-Oskouei, A. Rostami, and Z.D.K. Kanani, "DISPERSION FLATTENED OPTICAL FIBER DESIGN FOR LARGE BANDWIDTH AND HIGH-SPEED OPTICAL COMMUNICATIONS USING OPTIMIZATION TECHNIQUE", Progress In Electromagnetics Research B, Vol. 13, pp. 21-40, 2009.

[4] Govind P. Agrawal, "Fiber-Optic Communication Systems",(John Wiley & Sons, Third Edition), 2002.

[5] T. Kato, M. Hirano, A. Tada, K. Fukuada, T. Fujii, T. Ooishi, Y. Yokoyama, M. Yoshida, and M. Onishi, "Dispersion flattened transmission line consisting of wide-band non-zero dispersion shifted fiber and dispersion compensating fiber module", Optical Fiber Technology, Vol. 8, pp. 231-239, 2002.

[6] R. Tewari, B.P. Pal, and U.K. Das, "Dispersion shifted dual shape core fibers: Optimization based on spot size definitions" , Ligthwave Technology, vol. 10, pp. 1-5, 1992.

[7]Dipankar Ghosh, Debashri Ghosh, and Mousumi Basu, "Designing a graded index depressed clad non-zero dispersion shifted optical fiber for wide band transmission system", Optik Optics, vol. 119, pp. 63-68, 2008.

[8] B. Mikkelsen, G. Raybon, B. Zhu, R.J. Essiambre, P.G. Bernasconi, K. Dreyer, L.W. Stulz, S.N. Knudsen, " High spectral efficiency (0.53 bit/s/Hz) WDM transmission of 160 Gb/s per wavelength over 400 km of Fiber", Technical Digest of OFC 2001, 2001, Paper ThF2.

[9] T. Ito, K. Fukuchi, K. Sekiya, D. Ogasawara, R. Ohhira, T. Ono, " 6.4 Tb/s (160 × 40 Gb/s) WDM transmission experiment with 0.8 bit/s/Hz spectral efficiency" , ECOC 2000, 2000, Postdeadline paper PD1.1.

[10] J. Kani, K. Hattori, M. Jinno, T. Kanamori, K. Oguchi, " Tripple-wavelength-band WDM transmission over cascaded dispersion-shifted fibers", Technical Digest of OAA'99, 1999, Paper WC2.

[11] K. Fukuchi, T. Kasamatsu, M. Morie, R. Ohhira, T. Ito, K. Sekiya, D. Ogasawara, T. Ono, "10.92-Tb/s (273 × 40-Gb/s) triple-band/ultra-dense WDM optical-repeatered transmission experiment", OFC 2001, Postdeadline paper PD24, 2001.

[12] A. Naka et al., Lightwave Technol., vol.12, no.2, February (1994) pp. 280-287

[13] Y. Liu et al., OFC'96, WK15, 1996.

[15] R.K. Varshney, A.K. Ghatak, I.C. Goyal, and C. siny Antony, "Design of a flat field fiber with very small dispersion slope," Optical Fiber Technology, vol. 9, pp. 189-198, 2003.

[16] J. Sakamoto, J. Kani, M. Jinno, S. Aisawa, M. Fukui, M. Yamada, K. Oguchi, "Wide wavelength band (1535-1560 nm and 1574-1600 nm), 28 × 10 Gbit/s WDM transmission over 320 km dispersion shifted fiber", Electron. Letter, vol. 34, pp. 392-394, 1998.

[17] S. Yashida, S. Kuwano, K. Iwashita, "10 Gbit/s ×10 channel WDM transmission experiment over 1200 km with repeater spacing of 100 km without gain equalization or pre-emphasis", Optical Fiber Communication (OFC') 96, San Jose, CA, TuD6, 1996.
[18] A.R. Chraplyvy, "Limitation on lightwave communications imposed by optical fiber nonlinearities", Lightwave Technology, vol. 8, pp. 1548-1557, 1990.
[19] R.W. Tkach, A.R. Chraplyvy, F. Fabrizio, A.H. Gnauck, R.M. Derosier, "Four photon mixing and high speed WDM systems", Lightwave Technology, vol. 13, pp. 841-849, 1995.
[20] Y. Akasaka, "New optical fibers for high bit rate and high capacity transmission", SPIE Proc., Vol. 3666, pp. 23-29, 1999.
[21] Y. Liu, A.J. Antos, V.A. Bhagavatula, M.A. Newhouse," Single mode dispersion shifted fiber with effective area larger than 80 μm^2 and good bending performance", Proc. of ECOC'95, TuL2.4, 1995.
[22] M. Tateda, Y. Kato, S. Seikai, and N. Uchida, "Design consideration on single mode fiber parameters," Pans. IECE Japan, (in Japanese) vol. J56-B, pp. 324-331, 1982.
[23] K.I. Kitayama, Y. Kato, M. Ohashi, Y. Ishida, and N. Uchida, "Design considerations for the structural optimization of a single mode fiber," Ligthwave Technology, vol. 1, pp. 363-369, 1983.
[24] M. Savadi-Oskouei, S. Makouei, A. Rostami, and Z.D. Koozeh Kanani, "Proposal for optical fiber designs with ultrahigh effective area and small bending loss applicable to long haul communications", applied Optics, vol. 46, pp. 6330-6339, 2007.
[25] Y. Namihira, "Relationship between nonlinear effective area and mode-diameter for dispersion shifted fiber," Electron. Letter, vol. 30, pp. 262-264, 1994.
[26] M. Hautakorpi and M. Kaivola, "Modal analysis of M-type dielectric-profile optical fibers in the weakly guiding approximation," Optical Society of America A., vol. 22, pp. 1163-1169, 2005.
[27] F. D. Nunes and C. A. de Souza Melo, "Theoretical study of coaxial fibers," Applied Optics, vol. 35, pp. 388-398, 1996.
[28] D. E. Goldberg, *Genetic Algorithms in Search, Optimization and Machine Learning* (Addison-Wesley, 1989).
[29] X. Tian and X. Zhang, " Dispersion flattened design of large effective area single mode fibers with ring index profiles", Optics Communications, vol. 230, pp. 105-113, 2004.
[30] Z. O. Tseng, "On the Convergence of the Coordinate Descent Method for Convex Differentiable Minimization", Optimization Theory and Applications, vol. 72, pp. 7-35, 1992.
[31] S. Yin, K.w. Chung, H. Liu, P. Kurtz, and K. Reichard, " A new design for non-zero dispersion shifted fiber (NZ-DSF) with a large effective area over 100μm^2 and low bending and splice loss", Optics Communications, vol. 177, pp. 225-232, 2000.
[32] X. Jiang and R. Wang, "Non-zero dispersion shifted optical fiber with ultra large effective area and low dispersion slope for terabit communication systems", optics Communications, vol. 236, pp. 69-74, 2004.
[33] B.H. Kim, Y. Park, D.Y. Kim, U.C. Paek, W.-T. Han, in: OFC_2002, Technical Digest, 2002, pp. 173-174.
[34] U.C. Pack, C.R. Kurkjian, J. Am. Ceram. Soc. 58 (1975)330.

[35] S. Matsuo, S. Tanigawa, K. Himeno, K. Harada, in: OFC_2002, Technical Digest, 2002, pp. 329–330.

[36] A. Safaai-Jazi, "Large effective area fibers: propagation properties and optimum designs", SPIE Proc., Vol. 3666, pp. 30–39, 1999.

[37] P. Nouchi, P. Sansonetti, J. Von Wirth, and C. Le Sergent," New dispersion shifted fiber with effective area larger than 90 μm^2", Proc. of ECOC'96, MoB3.2, 1996.

[38] V. Silva, "A new design for dispersion shifted fiber with an effective area larger than 100 µm2 and good bending characteristics," Proc. of OFC'98, ThK1, p. 301, 1998.

[39] N. Kumano, K. Mukasa, S. Matsushita, T. Yagi, "Zero dispersion slope NZ-DSF with ultra wide bandwidth over 300 nm", Proc. of ECOC'2002, 2002.

[40] B. Zhu, L.E. Nelson, L. Leng, S. Stulz, S. Knudsen, D. Peckham, "1.6 Tbits/s (40 × 42.7 Gbit/s) WDM transmission over 2400 Km of fiber with 100 Km dispersion managed spans", Electron. Letters, vol. 38, pp. 647–648, 2002.

[41] P. Nouchi, P. Sansonetti, J. Von Wirth, C. Le Sergent," New dispersion shifted fiber with effective area larger than 90 µm2", Proc. of ECOC'96, MoB3.2, 1996.

[42] M. Savadi-Oskouei, A. Rostami, and S. Makouei," A novel fiber design strategy for simultaneously introducing ultra small dispersion and dispersion slope using genetic algorithm", European Transaction on Telecommunications, vol. 20, pp. 37-47, 2009.

[43] S. Kawakami and S. Nishida, "Anomalous dispersion of new doubly clad optical fibers," Electron Letters, vol. 10, pp. 38-40, 1974.

[44] S. Kawakami, S. Nishida, and M. Sumi, "Transmission characteristics of W-type optical fibers", Proc. Inst. Elec. Eng., vol. 123, pp. 586-590, 1976.

[45] B. James and C.R. Day, "A review of single mode fibers with modified dispersion characteristics", Lightwave Technology, vol. 4, pp. 967-979, 1986.

8

Continuous-Time Analog Filtering: Design Strategies and Programmability in CMOS Technologies for VHF Applications

Aránzazu Otín, Santiago Celma and Concepción Aldea
Group of Electronic Design, Aragón Institute for Engineering Research (I3A). Zaragoza University, Zaragoza. Spain

1. Introduction

The evolution of wireless applications (the performance as well as the number of users) has undergone explosive growth in the last years, resulting in an increasing demand for smaller, low-cost wireless transceivers with low power consumption. In order to meet this demand, continuous development must take place both in CMOS technology and in RF electronics, the goal of which should be to achieve a fully-integrated single-chip receiver in a low-cost CMOS process. This demand for complex read channel and multi-standard receiver ICs calls for the design and implementation of one category of analog interface chips as continuous-time (CT) filters, suitable for high speed with variable bandwidths over a wide frequency range, preferably using the G_m-C approach rather than other existing solutions.
Filters based on the G_m-C technique were used quite early on with bipolar technology and they have now become the dominant option to implement monolithic filters for very high frequency. The basic building block of a G_m-C filter is the integrator, which involves the use of transconductors and capacitors only and whose structure is therefore simpler than others, such as operational amplifiers. The simplicity of the transconductor coupled with the open-loop operation, which does not involve any complex frequency compensation schemes, point to this cell as the basic active element to be considered and the best option to operate in a VHF range with low supply voltages.

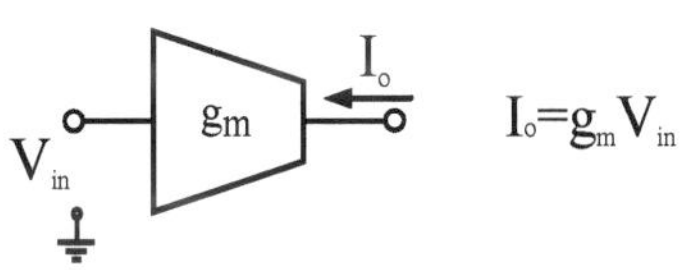

Fig. 1. Ideal transconductor V_{in} to I_o converter of transconductance g_m (conversion factor).

All the benefits of the G_m-C approach lie in the ideal behaviour of the transconductor. Nevertheless, its use as the basic element in the VHF active filter implementation forces one to consider some drawbacks related with the non-idealities of this fundamental cell: finite output resistance, finite bandwidth, noise, non-linearity, etc. The main disadvantages inherent to this technique are its high sensitivity to parasitic capacitors and the non-linear behaviour of the transconductor, to the extent of appearing a distortion brought about mainly by the non-linearities generated in the V-I conversion. Certain specific strategies require to be used to minimize these effects.

By taking differential or balanced transconductor structures into account, distortion is reduced (even non-linear components are cancelled) and better immunity to common-mode noise is obtained. Furthermore, the use of tuning techniques compensates parameter deviations due to process and temperature variations. These ideas, together with a careful layout, a detailed study of the technology and a deep analysis of the device, lead to an improvement in the transconductor behaviour, and consequently, in the filter performance. Thus, while developing the design of an active G_m-C filter, the effects of transconductor non-idealities must be analysed in depth to achieve optimum filter performance. The implementation of the transconductor should show a trade-off between dc-gain, linearity and low phase-error at the cut-off frequency.

Any pole or zero frequency in filters based on the G_m-C technique is of the G_m/C type. This means that there are two fundamental ways of programming the frequency response of the filter: keeping G_m constant and varying C, or vice versa. The choice of filter approach will affect noise and power dissipation (Pavan et al., 2000). The constant-C approach has the advantage of maintaining the noise specifications constant over the entire programming range while decreasing the power consumption for lower frequencies. Due to the above considerations, the constant-C scaling technique is the preferred approach for implementing filters operating in a very high frequency range, focusing on the design of tunable CMOS transconductors. On the other hand, discrete tuning is currently being more widely used than continuous tuning, both to preserve the dynamic range and take advantage of the digital system in mixed design to determine the control signal that calibrates and reconfigures the filter. A possible discrete tuning technique is based on a parallel connection of transconductors, where the desired time-constant can be digitally programmed (Pavan et al., 2000a). This approach succeeds in keeping the Q-factor constant and maintains an adequate dynamic range over the entire bandwidth setting.

The target is to implement a transconductor that is compatible with the latest low-cost pure digital CMOS technologies and programmable over a very high frequency range while maintaining an adequate dynamic range (DR). The concrete values of these specifications depend on each particular application. This work does not focus on a concrete application but on carrying out an overall analysis to seek the structure that provides the best trade-off between operation frequency, programmability, dynamic range and power consumption.

Considering all these points, an optimal solution for digitally programmable analog filters in the VHF/UHF range is to take advantage of current-mode pseudo-differential topologies and endow them with digital programmability. The design strategy is therefore as follows: after analysing the transconductor parameters that limit its ideal behaviour, a very well-known current-mode topology (Smith et al., 1996; Zele et al., 1996) will be characterized; starting from the Zele-Smith architecture, two different transconductors will be presented and in-depth analysis will be carried out, following which all the characteristic parameters of each active cell will be obtained; programmability will then be added to the VHF transconductors and the experimental results of a low-cost 0.35 μm CMOS implementation will be presented. As the active cell is based on a classical structure, a broad diversity of digitally programmable and continuously tunable CT filters can be obtained, where the programmability exhibited by the filter is achieved due to the design of a generic programmable transconductor. Due to the lack of special capacitor structures in standard digital technologies, the use of the MOS structure as

an intended passive device is probably as old as the MOS transistor concept itself. An alternative to implementing linear capacitors is to use the *gate-to-channel capacitance* of MOSFET devices as capacitors, where the gate-oxide thickness is a well-controlled variable in the process. This option will be considered in this work.
Therefore, in this chapter we will show the best way to implement key analog building blocks of a high-speed system in a CMOS technology with a wide programmable frequency range; considering new design techniques and uncovering potential problems associated with the design of high-speed analog circuits using short-channel and low-voltage devices. These are the challenges of CMOS filter design at very high frequencies and this study addresses the theoretical and practical problems encountered in the design of robust, programmable continuous-time filters with very high bandwidths implemented in low-cost digital CMOS technologies.

2. The Integrator: building-block in the Gm-C technique

The majority of continuous-time (CT) integrated filters, circuits where high frequency at low cost of silicon and power is required, present a frequency response controlled by time-constants, and one of the simplest implementations for these factors is taking advantage of the integrator structure. Therefore, the integrator is the dominant building block for many high-frequency active circuit design techniques, and its frequency response and linearity directly determine the filter performance.
Accordingly, systems based on the G_m-C technique are the first option for implementing CT filters, thanks to their acceptable performance over the VHF range. The active building element used by the G_m-C filter approach, based on an open-loop integrator, is the transconductor cell (Fig. 1), which ideally delivers an output current proportional to the input signal voltage:

$$I_o = g_m V_{in} \tag{1}$$

where g_m is the transconductance of the element. When a grounded capacitor is connected to the output node of the transconductor in order to take this current out, an integrator is obtained leading to V_{in}-V_o conversion, as shown in Fig. 2(a). It turns out that an ideal voltage-mode integrator has been obtained with a simple transconductance-capacitor combination. Nevertheless, a second structure can be considered taking into account the current-mode signal processing, whereby two different, yet completely equivalent, topologies are obtained. In this case, the input current is taken across the integration capacitance in order to obtain the transconductor input voltage and then, after the active cell, the output current. Thus, Fig. 2(b) shows the I_{in}-I_o conversion.

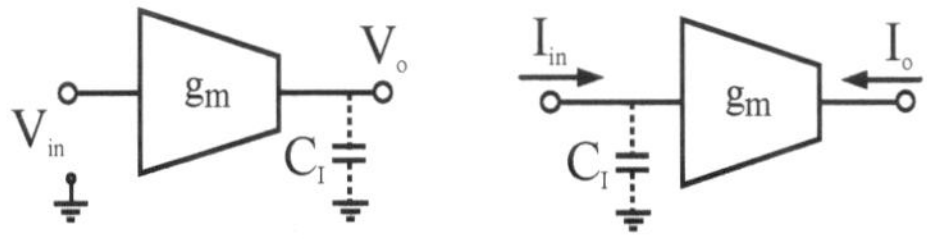

(a) (b)

Fig. 2. Ideal integrator; (a) voltage-mode: $\frac{V_o}{V_{in}} = -\frac{g_m}{sC_I}$, (b) current-mode: $\frac{I_o}{I_{in}} = \frac{g_m}{sC_I}$.

Due to the grounded location of most parasitic capacitors of the active cell (the total output/input node effective parasitic capacitance, depending on the configuration), they must be considered by constituting a percentage of the total integration capacitance C_I, which is particularly significant at high frequencies. An extreme situation can be reached when considering the proposed transconductor as an integrator where total integration capacitance C_I is constituted only by these parasitic capacitances, with no need for any external capacitor. Nevertheless, these capacitances are not linear and, depending on their contribution, the total linearity of the system will be affected. As technological process variations will also affect the value of these parasitic capacitances, sensitivity to these capacitors requires a detailed study of the device models and integration technology together with a careful system layout.

The ideal integrator has an infinite dc-gain and no parasitic effects, thus obtaining a phase of $-\pi/2$ for all the frequencies. The unity-gain frequency is $\omega_t=g_m/C_I$. Nevertheless, a real integrator presents a non-zero transconductor output conductance g_{out} and parasitic poles and zeros, which distort the transfer function:

$$H(s) = A_{DC}\frac{\left(1-{s}/{\omega_2}\right)}{\left(1+{s}/{\omega_1}\right)} \tag{2}$$

where $A_{DC}=g_m/g_{out}$ is the dc-gain and $\omega_1=\omega_t/A_{DC}=g_{out}/C_I$ is the frequency of the dominant pole. The effects of parasitic poles and zeros at frequencies much higher than the frequency range of the transconductor can be modelled with a single effective zero ω_2: positive ω_2 results in an effective parasitic RHP-zero and negative ω_2 in an LHP-zero.

Non-zero transconductor output conductance g_{out} causes finite dc-gain in real integrators in the filter. In addition, parasitic poles and zeros in the integrator transfer function, together with finite A_{DC}, generate deviations of the inverter integrator phase response from $-\pi/2$, and it is well-known that phase error is the main source of misfunctions in filters. In particular, phase deviations around ω_t can cause significant errors in the filter transfer, depending on filter quality factors. The accuracy of the overall frequency response of the filter depends on how closely the individual integrators in the filter follow the ideal response. The filter remains very close to the ideal one if the integrator phase at its unity-gain frequency ω_t is equal to its ideal value $-\pi/2$; the amount by which the phase at ω_t deviates from this quantity will be called $\Delta\varphi(\omega_t)$.

$$\varphi(\omega_t) = -\frac{\pi}{2} + \tan^{-1}\left(\frac{\omega_1}{\omega_t}\right) - \tan^{-1}\left(\frac{\omega_t}{\omega_2}\right) \Rightarrow \Delta\varphi(\omega_t) \approx \tan^{-1}\left(\frac{g_{out}}{g_m}\right) - \tan^{-1}\left(\frac{\omega_t}{\omega_2}\right) \tag{3}$$

Low dc-gain causes a leading phase error, and parasitic high-frequency poles and zeros in the integrator create lagging ($\omega_2>0$, RHP-zero) or leading ($\omega_2<0$, LHP-zero) phase errors. The acceptable worst case value of $\Delta\varphi(\omega_t)$ depends on the specifications for the high-frequency response of the overall filter and the poles and quality factor of the transconductor transfer function. The integrator phase error can be modelled with a frequency-dependent integrator quality factor Q_{int} (Nauta, 1993), concluding that a high and accurate filter quality factor puts strong constraints on the integrators phase error, i.e. on Q_{int}.

$$\frac{1}{Q_{int}(\omega)} \approx \Delta\varphi(\omega) = \frac{\omega_1}{\omega} - \frac{\omega}{\omega_2} \tag{4}$$

The filter performance is dominated by the performance of the transconductors, since the filter specifications (dynamic range, dissipation and chip area) depend not only on filter properties (Q, cut-off frequency, impedance level) but also on transconductor properties (A_{DC}, ω_t, ω_1, ω_2, noise behaviour, linearity, area and power consumption). It is therefore useful to put effort into the study of a high-performance transconductor that would improve all its specifications, in order to obtain a proper design for these VHF filter building blocks.

3. Fully-balanced pseudo-differential transconductor cell

In this section, the development of a fully-balanced current-mode integrator based on a classical structure is described, which is characterized by low-power, high rejection of supply noise and VHF potential application. Fig. 3 shows the conceptual scheme of the Zele-Smith pseudo-differential integrator (Smith et al., 1996; Zele et al., 1996), a complete fully-balanced transconductance cell arranged for using a current-mode integrator.

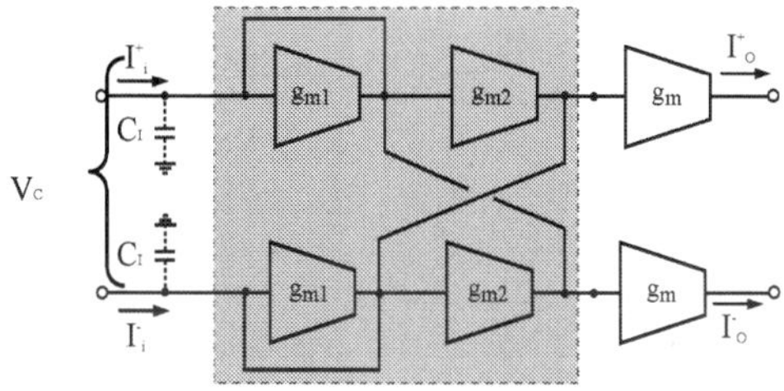

Fig. 3. Conceptual scheme of the complete fully-balanced current-mode transconductor.

To understand the basic operation we analyse the simple first-order model of the proposed transconductor, considering each unit cell as a simple transistor, i.e., single common-source stages as shown in Fig. 4. Under these conditions, the small-signal analysis gives the expression for the differential gain of the integrator (Eq. 5), where gmi is the i-cell transconductance and g_o' is the sum of output conductances g_{dsi} at the input node.

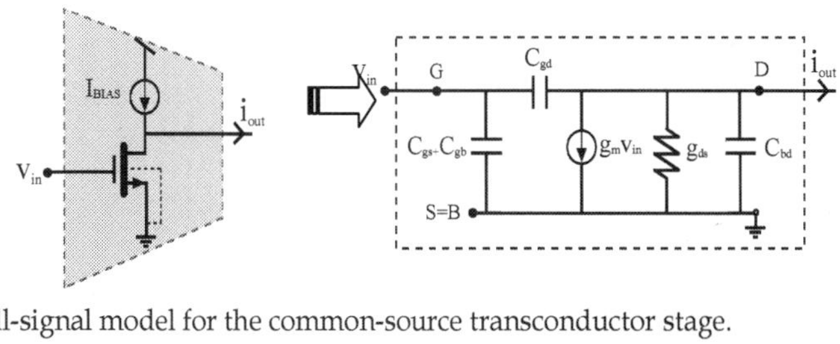

Fig. 4. Small-signal model for the common-source transconductor stage.

By analysing this expression and considering a first-order approximation, i.e., neglecting the g_{ds} effects of each transistor, an infinite dc-gain is achieved if perfect matching is obtained between g_{m1} and g_{m2}, so that $\delta g_m = g_{m1} - g_{m2} = 0$. Nevertheless, the effect of the output conductances is not avoidable and the implementation of a negative resistance ($\delta g_m<0$), inherent to this topology, provides the possibility of achieving dc-gain enhancement. Note that by making $\delta g_m+g_o'\rightarrow 0$, then $|A_{DC}|\rightarrow\infty$. In practice, mismatching between transistors limits the differential gain by up to 55 dB at most. Another equivalent way for analysing this improvement is to consider the differential-mode input resistance of the transconductor cell.

$$A_{DC}=\left|\frac{I_O^+-I_O^-}{I_i^+-I_i^-}\right|\approx\frac{g_m}{g_{m1}-g_{m2}+g_o'}\cong\frac{g_m}{\delta g_m+g_o'};\quad R_D(in)\approx\frac{2}{g_{m1}-g_{m2}+g_o'}=\frac{2A_{DC}}{g_m} \tag{5}$$

As a result, this scheme shows the basic pseudo-differential structure obtained by considering two dual transconductor cells (g_m), leading to current integration through input capacitance C_I. Thanks to the additional negative resistance shown in grey in the same figure, dc-gain is increased by providing positive feedback compensation for the signal current and boosting the input resistance of the transconductor.

The approximate common-mode gain, which must be less than unity to guarantee stability in closed-loop configurations, is constrained by device ratios to a stable value over all frequencies (Eq. 6). Common-mode stability is assured by designing $(g_{m1}+g_{m2})/g_m>1$. Common-mode behaviour analysis can be also carried out by calculating common-mode input resistance.

$$|A_{CM}|=\left|\frac{I_O^++I_O^-}{I_i^++I_i^-}\right|\approx\frac{g_m}{g_{m1}+g_{m2}+g_o'};\quad R_{CM}(in)\approx\frac{1}{2(g_{m1}+g_{m2}+g_o')}=\frac{|A_{CM}|}{2g_m} \tag{6}$$

The common-mode feedback resulting from the interconnection of the negative resistance provides both a naturally high differential gain and low common-mode gain for the integrator, improving these limits attached to a real integrator structure. Consequently, the basic operation of the transconductor will be best understood by explaining, first, that the common-mode control and dc-enhancement circuitry is connected at the input of the circuit and then, that the linear V-I conversion mechanism occurs in the output stage.

The gain of the basic current integrator is independent of the supply voltage to the first-order approximation. When fully-differential current topologies are used, the small remaining supply noise feedthrough is common to both sides of the signal and thus has no direct effect, except through random device mismatch. Therefore, the integrator has good immunity to supply noise. Device mismatch can be minimized with careful layout and specific design techniques to around 0.1-1 %, in many applications (Croon et al., 2002; Otin et al., 2004; Otin et al., 2005).

The use of an integrator based on transconductance cells implemented by using single transistors (no internal nodes), results in a proper frequency response because the only nodes are at the inputs and at the outputs. To a first-order approximation, no parasitic poles or zeros exist in the differential ac-response of the basic integrator circuit. Both differential and common-mode gains can be independently set by the different values of g_{m1} and g_{m2}. The ideal integrator function is a result of setting $\delta g_m+g_o'=0$ and the phase error at the unity-gain frequency, $\omega_t=g_m/C_I$, can be calculated by:

$$\Delta\varphi(\omega_t)=\tan^{-1}\left(\frac{\omega_1}{\omega_t}\right)\approx\tan^{-1}\left(\frac{\delta g_m+g_o'}{g_m}\right)=\frac{\pi}{2}-\tan^{-1}\left(\frac{g_m}{\delta g_m+g_o'}\right) \tag{7}$$

To summarize, infinite differential input impedance can be obtained if $\delta g_m + g_o' \rightarrow 0$ while maximizing the differential dc-gain and minimizing the phase error at ω_t, and common-mode input impedance can be reduced by maximizing the sum of the transconductances $(g_{m1}+g_{m2})$. Consequently, the common-mode rejection ratio (CMRR) is improved. Nevertheless, an important concept should be borne in mind: as the dc-gain depends on the difference $(g_o' - |\delta g_m|)$, the structure can lead to instability if this quantity becomes negative (total negative input conductance) due to overcompensation.

By analysing the small signal model of each common-source stage forming the complete transconductor topology (Fig. 4), the need to solve a frequency problem arises: the feedforward ac-current path from the gate (input) to the drain (output), through the overlap parasitic capacitance C_{gd}. When considering the stages forming the negative resistance, the repercussion of this effect is not important because the contribution to the total behaviour of the cell decreases as these capacitances are short-circuited with their respective g_{mi}. However, the feedforward current through C_{gd} in the fully-balanced output-stage g_m of the transconductor structure generates a transmission zero (g_m/C_{gd}) in the right complex half-plane. This parasitic RHP-zero modifies the integrator frequency response and creates a phase-lag at the unity-gain frequency. Furthermore, the Miller effect, also associated with this parasitic capacitor, introduces larger equivalent input capacitance ($C_{in}=C_{gs}+C_{gb} \Rightarrow C_{in}=C_{gs}+C_{gb}+(1+A)C_{gd}$) and an additional component to the equivalent output capacitance ($C_{out}=C_{bd} \Rightarrow C_{out}=C_{bd}+(1+A^{-1})C_{gd}$), where $A \approx g_m/g_{ds}$. Therefore, the neutralization of this effect will involve bandwidth enhancement.

Many methods have been proposed to minimize the Miller effect: the cascode technique (Gray et al., 2001), the inductor shunt peaking technique (Mohan et al., 2000), the capacitive compensation technique (Wakimoto et al., 1990; Vadipour, 1993), the distributed amplification technique (Ahn et al., 2002) and the active inductor technique (Säckinger et al., 2000). They all have the advantages of low-voltage compatibility and low area; however, the solutions considered in this work will be the use of cascode structures together with the capacitive compensation technique.

Differential systems allow the C_c-cancellation technique, using positive feedback to generate negative capacitances, which can cancel the positive ones to yield bandwidth increases. These C_c capacitors are the overlap C_{gd} parasitic capacitances of dummy compensation transistors used in a cross-coupled way to neutralize the feedback action of these Miller capacitors. The connection is between the output and the opposite sign input, available in a balanced configuration. Under these conditions, the RHP-zero is moved to infinity, i.e., the cause of phase lag is removed, thus expanding the bandwidth of the transconductor. At the same time, compensation capacitor C_c will cancel the Miller effect and a lower input node effective capacitance is obtained due to the reduction of the feedforward effect. This technique depends on feeding back a current that is precisely the same as the one flowing through the Miller capacitance C_{gd} and, in consequence, the neutralization capacitor must match precisely. However, it is remarkable that C_{gd} is voltage-dependent and compensation can only work with small signals. In the case of mismatch between C_{gd} and C_c, parasitic zero is not at infinity and can cause a small phase lag or lead.

However, this is not the full story of the high frequency behaviour of the transconductor cell and there are more frequency limits. Mismatch in common-mode feedback circuits can result in unexpected parasitic poles and zeros. In addition, high frequency models of the MOS transistor show that g_m is not independent of frequency, but has a finite delay $g_m(s)$

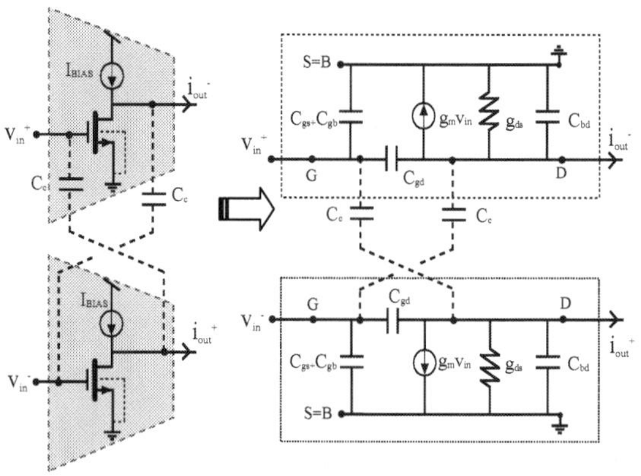

Fig. 5. Cancellation of transmission zero g_m/C_{gd} and neutralization of the Miller effect: C_c-cancellation technique.

and begins to roll off at very high frequencies. Although the frequency where this roll-off begins can be in the GHz range, the phase shift from this effect can become significant at much lower frequencies. Since most active filters are very sensitive to small phase changes in the integrator response, it is thus important to take this effect into account.

The first way to minimize these effects is to eliminate the internal nodes or, if this is not possible, to design them as low-impedance nodes. This procedure can be carried out by considering cascode topologies. Moreover, their use further prevents bandwidth reduction generated by transmission zero because one side of C_c is connected to the internal low-impedance node, i.e., to the low-gain point of the cascode transistor source.

Therefore, an enhancement of the integrator dc-gain has been obtained with this topology by means of the differential negative resistance, increasing the differential input resistance of the transconductor and keeping the common-mode gain lower than unity. With regard to frequency limit-related problems, transmission zero has been reduced by using C_c-capacitor compensation, taking advantage of the pseudo-differential topology. This solution can also be improved by considering cascode topologies, giving higher dc-gains in a natural way, which will also reduce the frequency drawbacks associated to the internal nodes of other topologies by avoiding internal high-impedance nodes in the signal path.

As a result, a low-voltage transconductor with high linearity, very high operation frequency and high power efficiency has been designed where cascode structures should be considered to obtain an improvement in the high-frequency behaviour of the basic topology. The main advantage of using cascode stages instead of single common-source stages is the higher dc-gain while maintaining a good frequency response. Hence, a higher quality factor of the integrator is expected due to the higher differential dc-gain (Abidi, 1988). Basic cascode circuits require high supply voltages to operate due to the large overhead bias from threshold voltages. However, variations of the cascode technique exist which can be used with lower voltage supplies. Two options are considered in this work, the so-called **high-swing cascode (HS)** stage and the **folded cascode (FC)** stage (Baker et al., 1998; Sansen et al. 1999; Sedra et al., 2004). The unit cells replacing the common-source stages previously used are shown in Fig. 6. The complete fully-balanced current-mode transconductance cells implemented by using these cascode stages are described in the following sections.

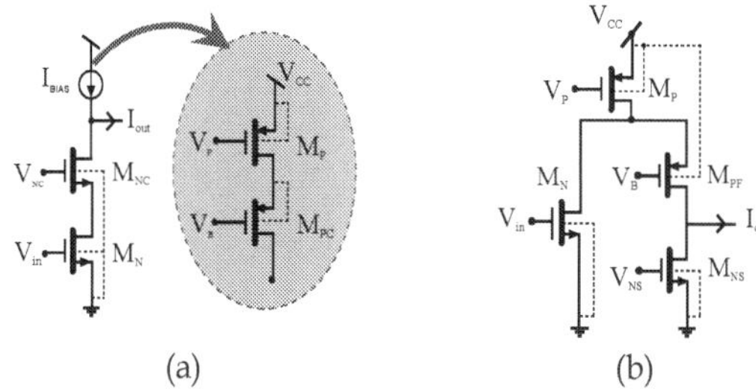

Fig. 6. Unit transconductance cell: (a) high-swing (HS) and (b) folded cascode (FC) topology.

3.1 High-swing cascode section: HS topology

Fig. 7(a) shows the transconductor arranged for using the current-mode integrator described in Fig. 3, where the unit cells have been implemented by using high-swing cascode stages (Baker et al., 1998; Sansen et al. 1999; Sedra et al., 2004). As illustrated in the corresponding HS unit cell (Fig. 6(a)), current sources are also implemented by using high-swing cascode elements. The substrate terminals of NMOS transistors are connected to the reference voltage as usual, and those of the PMOS transistors are connected to the corresponding source node of each transistor.

The use of high-swing cascode elements offers as high accuracy as using basic cascode stages to implement each unit cell of the transconductor but, because of the slightly different connection between transistors, needs lower supply voltage and has fewer internal parasitic poles, generating nodes between the input and the output, giving a better frequency response of the integrator. The main disadvantage of the improved cascode topology is that due to biasing constraints, the gate-source voltages must be kept small, resulting in larger devices for a bias current level.

3.2 Folded cascode section: FC topology

In order to obtain an improvement in biasing flexibility and further reduction of the supply voltage in the design of the transconductor cell, we can also take advantage of folded-cascode sections (Sansen et al. 1999; Sedra et al., 2004). The schematic used to describe the complete integrator based on the proposed current-mode pseudo-differential transconductor is shown in Fig. 7(b), where the unit cells have been implemented by using FC stages illustrated in Fig. 6(b). In this case, current sources are implemented by using single elements, both bias sources I_{BIAS} and cascode sources M_{NSi}. The substrate terminals of NMOS transistors are connected to the reference voltage as usual, and those of PMOS transistors, both those used to implement current sources I_{BIAS} and those implementing folded transistors M_{PFi}, are connected to the V_{CC} node.

The use of folded cascode elements exhibits a substantial improvement in biasing flexibility, because of the increased drain-voltage of the transistors, at the cost of additional current sources and bias voltages. Another significant benefit of using these stages is that by avoiding the biasing constraints associated to the high-swing cascode structure, we obviate the need to keep gate-source voltages low, which results in smaller and simpler devices for a given bias current level, lower voltage supply and larger unity-gain frequencies.

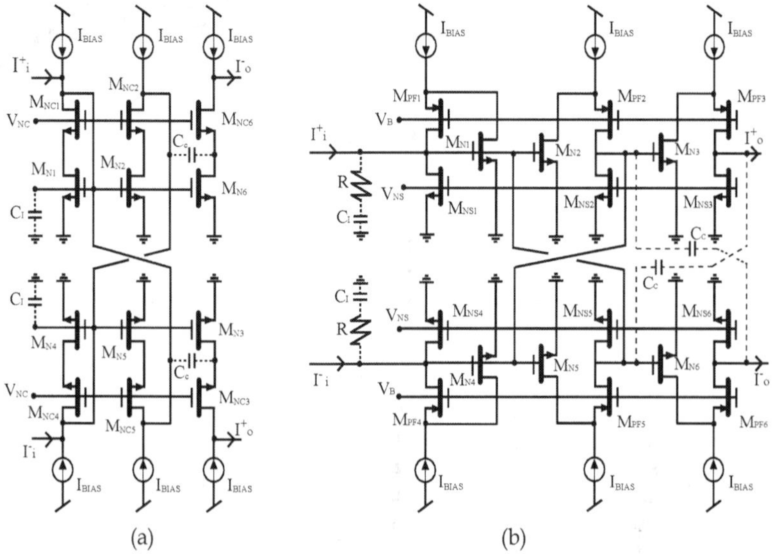

Fig. 7. Fully-balanced pseudo-differential current-mode cell, based on (a) high-swing cascode unit stages: **HS transconductor**; (b) folded-cascode unit stages: **FC transconductor**.

3.3 General considerations: basic principle

A similar notation and index-linking have been adopted in both transconductor implementations in order to simplify the description of the basic principle and to unify the topology analysis. Assuming ideal behaviour for the integrator, the balanced input current flows entirely into integration capacitance C_I. Diode-connected stages $M_{1,4}$ adequately bias cascode output $M_{3,6}$, whilst transistors $M_{2,5}$ provide positive feedback compensation for the signal current flowing into $M_{1,4}$ and boost the input resistance of the integrator.

Regarding the gain enhancement by means of the negative resistance formed by transistors $M_{2,5}$, this technique is absolutely necessary for Zele-Smith topologies and other voltage-mode transconductors (Nauta, 1993), in order to obtain reasonably high dc-gain values. Theoretically, the dc-gain could be infinity by adjusting the equivalent negative resistance, but in practice mismatching limits the dc-gain by about 40 dB in single transistor stages. However, the use of cascode topologies leads to a natural enhancement of this parameter and differential dc-gain values of up to 55 dB can be reached with identical transistors under identical bias conditions by means of a lower mismatching sensitive design. Nevertheless, there is a key difference between the HS and the FC cascode structure:

- The high output resistance is directly guaranteed thanks to the true cascode output stage exhibited by the HS transconductor. Therefore, positive feedback compensation is not necessary to boost differential resistance or enhance the dc-gain.
- On the other hand, the output-node for the FC transconductor is not a very high impedance node, and the negative resistance proves necessary to obtain real input resistance enhancement. In this approach, positive feedback compensation for the signal current flowing into $M_{3,6}$ is essential, boosting the input resistance of the integrator and increasing dc-gain.

The reason for this difference is that the FC unit cell considered and shown in Fig. 6(b) is a pseudo-cascode topology. To obtain similar output impedance to that of the HS approach, the M_{NS} current source would have to be implemented by using a cascode current source. Nevertheless, even if the HS implementation undergoes an immediate dc-gain/input resistance improvement, both topologies require positive feedback compensation (negative resistance) to reduce common-mode gain (common-mode input resistance) and stabilize common-mode voltages. On the other hand, the use of pseudo-differential structures requires careful and efficient control over the common-mode behaviour of the circuit. It is worth noting that this structure not only stabilises the common-mode voltage, but also rejects common-mode signals by means of partial positive feedback. This idea has already been used for high-frequency transconductors in (Nauta, 1993) and for a class of current-mode filters (Smith et al., 1996; Zele et al., 1996).

Thus, considering this topology implemented by using cascode stages, dc-gains of about 55 dB and CMRR of 60 dB can be obtained with inherent stability of common-mode voltages. Note that the propagation of common-mode (CM) signals in balanced circuits can cause instability and distortion. Further, current consumption, linearity and transconductance value are strongly dependent on the CM input signals. Additional techniques can be used in the proposed topology if a greater CMRR is needed, such as feedforward cancellation of the input CM signal. Balanced transconductors with high-input common-mode rejection that are capable of operating with low-voltage supplies are obtained by using an additional transconductor that is only sensitive to CM signals (Baschirotto et al., 1994; Wyszynski et al., 1994). Considering this technique, CMRR values up to 70 dB could be obtained.

4. High frequency response

In this section, the bandwidth of the transconductor will be analysed. Note that if single transistor stages and unrealistic simplified models are used in the proposed topology (Fig.3), the bandwidth could be infinite owing to the absence of internal nodes influencing the transfer function of the integrator. A more complete model of the MOS transistor does predict a finite bandwidth due to the second-order frequency effects such as the transmission zero associated to overlap parasitic capacitance C_{gd}, frequency dependence of the transconductance $g_m(s)$ and mismatch in common-mode feedback circuits. A closer explanation of MOS behaviour at high-frequencies (splitting it into an **intrinsic** and an **extrinsic** part) is required before starting the study of the complete integrator. Taking into account a non-quasistatic model (Tsividis, 1996), the high-frequency behaviour of the current mode integrator will be calculated.

When analysing transconductor bandwidth, several general factors must be considered:

- The output of the complete transconductor may be assumed to be short-circuited for ac-signals when calculating the frequency response of the integrator.
- In all the equations: g_m is the transconductance, g_{ds} the output conductance, g_{mb} the bulk-transconductance and C_{gd}, C_{gs}, C_{ds}, C_{bs} and C_{bd} the parasitic capacitances.
- All the unit cells are designed to seek perfect matching between them. Therefore, all similar transistors have the same properties except for transconductance g_m of the N-transistor processing the signal (M_N transistor in both unit cells shown in Fig. 6). In this way, considering the notation and index-linking previously used in Fig. 3: $g_m(N_1)=A_N g_m(N)$; $g_m(N_2)=A_P g_m(N)$; $g_m(N_3)=g_m(N)$. Consequently, $\delta g_m=(A_N-A_P)g_m$

represents the difference between M_1 and M_2, or, M_4 and M_5 due to the difference in dimensions and bias currents of N-transistors, which gives rise to the negative resistance that enhances the dc-gain of the system.

- Total integration capacitance C_I comprises not only the external capacitance, but also the contribution of parasitic capacitors ($C_I=C_{ext}+C_p$). Intrinsic capacitance C_{gs} is the main contribution of these parasitic capacitances C_p and consideration of it as a great percentage of total integration capacitance acquires great significance.
- External capacitor C_{ext} can be implemented by using double-poly, metal-metal or MOS capacitors, depending on the technological process.
- Current source is modelled with a Norton equivalent circuit, where G_s and C_s are the admittance and capacitance components. The external capacitance C_{ext} connected to the transconductor input is in parallel with C_s. For purposes of simplicity, from now on, C_s will include the equivalent capacitance of the current source and the external capacitance ($C_s \equiv C_s+C_{ext}$). Therefore, total integration capacitance can be expressed as: $C_I=C_s+C_p$, including parasitic effects, the external capacitance and the equivalent-model of the non-ideal current source.

Firstly, a model for the V-I conversion of the unit cell is derived, in both implementations to show the calculation process of the bandwidth of the complete transconductance cell.

4.1 High-frequency model of the HS unit cell

The following circuit represents the high frequency model for the HS unit cell previously shown in Fig. 6(a), where X(N) denote the parameters associated to the M_N-transistor, X(NC) those associated to the cascode transistor, X(P) those associated to current sources I_{BIAS} and X(PC) those associated to the cascode current sources as shown in Fig. 6(a). Table 1 summarizes the parameters associated to the impedances shown in the small-signal equivalent circuit. The rest of the elements: $g_m(N)$, $g_{ds}(N)$, $g_{ds}(NC)$, $g_{ds}(PC)$, $g_m(PC)$, $C_{gd}(N)$ and $C_{ds}(NC)$ directly represent the parameters of the respective transistor.

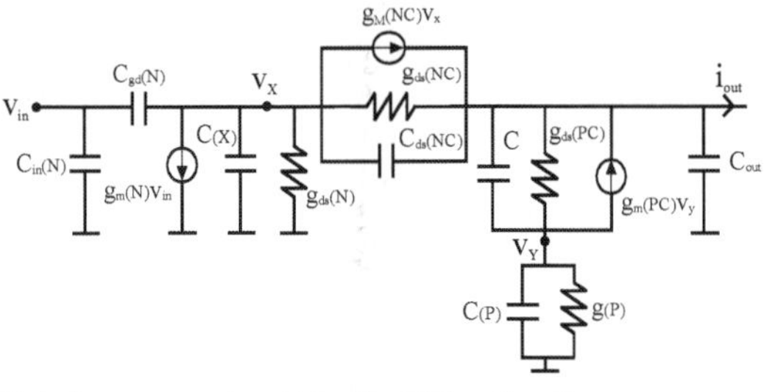

Fig. 8. Equivalent high-frequency circuit for the HS unit cell.

$g_M(NC)=g_m(NC)+g_{mb}(NC)$	$g(P)=g_{ds}(P)$
$C_{in}(N)=C_{gs}(N)+C_{gb}(N)$	$C(x)=C_{ds}(N)+C_{bd}(N)+C_{gs}(NC)+C_{bs}(NC)$
$C=C_{ds}(PC)+C_{bd}(PC)$	$C_{out}=C_{gb}(NC)+C_{bd}(NC)+C_{gd}(PC)$
$C(P)=C_{gd}(P)+C_{ds}(P)+C_{bd}(P)+C_{gb}(PC)+C_{gs}(PC)$	

Table 1. Small-signal parameters for the HS unit cell.

4.2 High-frequency model of the FC unit cell

The following circuit represents the high frequency model for the FC unit cell previously shown in Fig. 6(b), where X(N) are the parameters associated to the M_N-transistor, X(PF) those associated to the folded transistor, X(P) those associated to the current sources I_{BIAS} and X(NS) those associated to the current source of the folded transistor, which is implemented with a single NMOS transistor as previously illustrated.

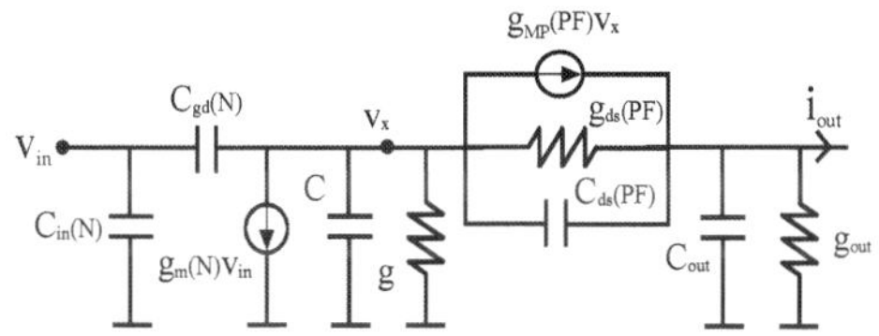

Fig. 9. Equivalent high-frequency circuit for the FC unit cell.

Table 2 summarizes the parameters associated to the impedances shown in the small-signal equivalent circuit. The rest of the elements: $g_m(N)$, $g_{ds}(PF)$, $C_{gd}(N)$ and $C_{ds}(PF)$ directly represent the parameters of the respective transistor.

$g = g_{ds}(N) + 2g_{ds}(P)$	$C_{in}(N) = C_{gs}(N) + C_{gb}(N)$
$g_{out} = g_{ds}(NS)$	$C = C_{ds}(N) + C_{bd}(N) + 2C_{gd}(P) + 2C_{ds}(P) + 2C_{bd}(P) + C_{gs}(PF) + C_{bs}(PF)$
$g_{MP}(PF) = g_m(PF) + g_{mb}(PF)$	$C_{out} = C_{gd}(PF) + C_{bd}(PF) + C_{gd}(NS) + C_{ds}(NS) + C_{bd}(NS)$

Table 2. Small-signal parameters for the FC unit cell.

Great similarity is obtained in the description of both unit cells. Even the FC capacitive parameter C will be equivalent to C+C(x) in the HS description. This parallelism will also appear in the complete transconductor analysis.

4.3 High-frequency model of the complete transconductance cell

Under these conditions, the differential gain of the proposed transconductor cell (in both implementations) can be calculated by:

$$H(s) = k\frac{(s - s_0)(s + s_1)}{s^2 + x_1 s + x_2} \approx K\frac{(s - s_0)}{(s + \delta_1)(s + \delta_2)} \tag{8}$$

where:

$$K = \frac{-A_{DC}\,\delta_1\,\delta_2}{s_0}; \quad x_1 = \frac{\beta}{\gamma}; x_2 = \frac{\alpha}{\gamma}; \quad \delta_1 = \frac{\alpha}{\beta} = \frac{x_2}{x_1}; \quad \delta_2 = \frac{\beta}{\gamma} - \delta_1 \approx \frac{\beta}{\gamma} = x_1 \tag{9}$$

Denominator factorization of Eq.(8) leads to obtain two parasitic poles, $-\delta_1$ and $-\delta_2$, but only if the approximation $(\alpha \cdot \gamma / \beta^2) \ll 1$ is verified. Furthermore, parasitic zero $-s_1$ must be negligible in the frequency range of interest. Both considerations will be demonstrated, either in the HS or in the FC approach ($s_1 \gg s$ and $s_1 \gg s_0$: $s_1(HS)=1600$ $s_0=1100$ GHz, $s_1(FC)=30$ $s_0=1900$ GHz).

Due to the use of a pseudo-differential structure, a careful study of the common-mode behaviour is mandatory. Thanks to the topology proposed, the common-mode voltage is stabilized by means of partial positive feedback as previously explained. Under these assumptions:

$$A_{CM}(s) = k_{CM}\frac{(s-s_0)(s+s_1)}{s^2+y_1 s+y_2} \approx K_{CM}\frac{(s-s_0)}{s^2+y_1 s+y_2} \quad (10)$$

$$K_{CM} = \frac{-A_{CM}\, y_2}{s_0}; \quad y_1 = \frac{\beta_{CM}}{\gamma_{CM}}; y_2 = \frac{\alpha_{CM}}{\gamma_{CM}} \quad (11)$$

Eq.(10) once more shows the need to neglect parasitic zero $-s_1$, leading to the same consideration as in the differential gain equation. In this analysis, denominator factorization is more difficult to accomplish. However, the approximate equations are easily derived and, making use of the figures obtained in each particular design, the possibility of using the approximate common-mode transfer function will be analysed. Therefore, if $(\alpha_{CM}\cdot\gamma_{CM}/\beta_{CM}^2)<<1$ is verified, the common-mode gain can be expressed as:

$$A_{CM}(s) = K_{CM}\frac{(s-s_0)}{(s+\xi_1)(s+\xi_2)} \quad (12)$$

$$\xi_1 = \frac{\alpha_{CM}}{\beta_{CM}} = \frac{y_2}{y_1}; \quad \xi_2 = \frac{\beta_{CM}}{\gamma_{CM}} - \xi_1 \approx \frac{\beta_{CM}}{\gamma_{CM}} = y_1 \quad (13)$$

Both transfer functions are characterized by two parasitic poles and one RHP-zero; the differential-mode by $-\delta_1$, $-\delta_2$, and s_0; and the common-mode by $-\xi_1$, $-\xi_2$, and s_0 (zero s_0 is the same in both transfer functions)[1]. Consequently, in order to obtain a transconductor design that is compatible with all the requirements of the active G_m-C filter implementation, a proper analysis and characterization of these parasitic elements becomes a top-priority challenge. From this study, their origin and frequency location may lead to some design considerations to improve the integrator frequency response.

Differential dc-gain, common-mode dc-gain, s_0 and s_1 are summarized in table 3 for both implementations. The parasitic poles can be calculated by using α, β, γ, α_{CM}, β_{CM}, γ_{CM}, as shown in Eqs.(9) and (13). As the resulting relations are very complicated, it is necessary to look for the dominant terms and obtain approximate expressions to draw conclusions. They can be simplified to analyse and understand the behaviour of the transconductance cell and its frequency limits that are associated to second order effects, which differentiate between the frequency behaviour of the proposed topology and the expected ideal response.

4.4 HS transconductance cell frequency response

Considering the previous study of the HS unit cell, a detailed analysis of the frequency response of the complete HS integrator is carried out. The dominant terms of these expressions are subsequently obtained in this section. In order to simplify the notation, the small-signal parameters are redefined in table 4.

---

[1] According to stable systems, negative poles $-\delta_1$, $-\delta_2$, $-\xi_1$ and $-\xi_2$ have been obtained in the transfer functions for both implementations. For purposes of simplicity, when referring to these poles, their associated frequency (δ_1, δ_2, ξ_1, ξ_2) will be the considered magnitude.

	HS transconductor	FC transconductor
S_0	$s_0 = \frac{g_m(N)}{C_{gd}(N)}$	$s_0 = \frac{g_m(N)}{C_{gd}(N)}$
S_1	$s_1 = \frac{g_M(NC)+g_{ds}(NC)}{C_{ds}(NC)}$	$s_1 = \frac{g_{MP}(PF)+g_{ds}(PF)}{C_{ds}(PF)}$
A_{DC}	$A_{DC} = \frac{g_m(N)(g_M(NC)+g_{ds}(NC))}{\alpha}$	$A_{DC} = \frac{g_m(N)(g_{MP}(PF)+g_{ds}(PF))}{\alpha}$
A_{CM}	$A_{CM} = \frac{-g_m(N)(g_M(NC)+g_{ds}(NC))}{\alpha_{CM}}$	$A_{CM} = \frac{-g_m(N)(g_{MP}(PF)+g_{ds}(PF))}{\alpha_{CM}}$

Table 3. Summary of the high-frequency parameters.

$G = g_{ds}(N)+g_M(NC)+g_{ds}(NC)$
$C_{IN} = C_s + 2C_{in}(N) + C_{out} = C_s + 2C_{gs}(N) + 2C_{gb}(N) + C_{gb}(NC) + C_{bd}(NC) + C_{gd}(PC)$
$C_\beta = C_{out} + C_{in}(N) = C_{gs}(N) + C_{gb}(N) + C_{gb}(NC) + C_{bd}(NC) + C_{gd}(PC)$

Table 4. Impedance parameters for the HS integrator.

By analysing in detail the small-signal equivalent model for the complete HS integrator, the value of total integration capacitance C_I can be calculated by Eq.(14). This definition will lead to a simplification of the parasitic pole expressions.

$$C_I = C_{IN} + C_\beta = C_s + 3C_{in}(N) + 2C_{out} \tag{14}$$

Firstly, the differential and common-mode gain can be expressed as follows:

$$A_{DC} \approx \left[(A_N - A_P) + \frac{2g_{ds}(NC)g_{ds}(N)}{g_m(N)(g_M(NC)+g_{ds}(NC))} \right]^{-1} \tag{15}$$

$$A_{CM} \approx -\left[(A_N + A_P) - \frac{2g_{ds}(NC)g_{ds}(N)}{g_m(N)(g_M(NC)+g_{ds}(NC))} \right]^{-1} \approx \frac{-1}{A_N + A_P} \tag{16}$$

In these expressions, the differential negative resistance obtained by the partial positive feedback compensation is shown by means of the difference $\delta g_m=(A_N-A_P)g_m(N)$. The existence of this negative resistance allows the differential dc-gain to be enhanced. Parasitic poles δ_1 and δ_2 can be calculated by means of ratios between α, β and γ. Final expressions are summarized in table 6. The origin of second order effects can be better understood by focusing on their dependence.

$$\alpha \approx ((A_N - A_P)g_m(N))(g_M(NC)+g_{ds}(NC)) + 2g_{ds}(NC)g_{ds}(N) \approx 2g_{ds}(NC)g_{ds}(N) \tag{17}$$

$$\beta \approx G(C_s + 3C_{in}(N) + 2C_{out}) = G\,C_I \tag{18}$$

where considering the dominant term in the transconductance G, the expression can be written as:

$$\beta \approx \left(g_{ds}(N) + g_M(NC) + g_{ds}(NC)\right)C_I \approx g_M(NC)\,C_I \tag{19}$$

Therefore, the parasitic pole δ_1 can be expressed as:

$$\delta_1 = \frac{\alpha}{\beta} \approx \frac{2g_{ds}(NC)g_{ds}(N)}{g_M(NC)\,C_I} \tag{20}$$

Following the same process for the other pole δ_2, we obtain:

$$\gamma \approx C_I\left(C_{gd}(N) + C_{ds}(NC) + C(x)\right) \approx C_I\,C(x) \approx C_I\,C_{gs}(NC) \tag{21}$$

$$\delta_2 = \frac{\beta}{\gamma} \approx \frac{G}{C(X)} = \frac{g_{ds}(N) + g_M(NC) + g_{ds}(NC)}{C_{ds}(N) + C_{bd}(N) + C_{gs}(NC) + C_{bs}(NC)} \approx \frac{g_M(NC)}{C_{gs}(NC)} \tag{22}$$

Similar results can be obtained for the common-mode frequency response:

$$\alpha_{CM} \approx \left((A_N + A_P)g_m(N)\right)\left(g_M(NC) + g_{ds}(NC)\right) \approx (A_N + A_P)g_m(N)g_M(NC) \tag{23}$$

$$\beta_{CM} \approx \beta \approx g_M(NC)\,C_I; \quad \gamma_{CM} \approx \gamma \approx C_I\,C_{gs}(NC) \tag{24}$$

$$\xi_1 = \frac{\alpha_{CM}}{\beta_{CM}} \approx \frac{(A_N + A_P)g_m(N)g_M(NC)}{G\,C_I} \approx \frac{(A_N + A_P)g_m(N)}{C_I}; \quad \xi_2 = \frac{\beta_{CM}}{\gamma_{CM}} \approx \delta_2 \approx \frac{G}{C(X)} \approx \frac{g_M(NC)}{C_{gs}(NC)} \tag{25}$$

4.5 FC transconductance cell frequency response

Considering the previous study of the FC unit cell, a detailed analysis of the frequency response of the complete FC integrator is carried out. The dominant terms of these expressions are obtained in this section. In order to simplify the notation, two small-signal parameters are redefined in table 5.

$G_I = G_s + 2g_{out}$
$C_I = C_s + 3C_{in}(N) + 2C_{out} = C_s + 3C_{gs}(N) + 3C_{gb}(N) + 2C_{gd}(PF)$ $+ 2C_{bd}(PF) + 2C_{gd}(NS) + 2C_{ds}(NS) + 2C_{bd}(NS)$

Table 5. Impedance parameters for the FC integrator.

In accordance with the analysis of the small-signal equivalent model for the complete FC integrator, parameter C_I, defined in table 5, directly represents the total integration capacitance, the expression of which is the same as in the HS integrator. Therefore, the total integrator capacitance of both integrator implementations can be calculated by Eq.(26).

$$C_I = C_s + 3C_{in}(N) + 2C_{out} \tag{26}$$

For the FC integrator, the differential and common-mode gain can be expressed as follows:

$$A_{DC} \approx \left[(A_N - A_P) + \frac{2g_{ds}(NS)(g_{MP}(PF) + 2g_{ds}(P))}{g_m(N)(g_{MP}(PF) + g_{ds}(PF))} \right]^{-1} \approx \left[(A_N - A_P) + \frac{2g_{ds}(NS)}{g_m(N)} \right]^{-1} \tag{27}$$

$$A_{CM} \approx -\left[(A_N + A_P) + \frac{2g_{ds}(NS)}{g_m(N)} \right]^{-1} \approx \frac{-1}{A_N + A_P} \tag{28}$$

In these expressions, the negative resistance obtained by the partial positive feedback compensation is shown again by means of the difference $\delta g_m=(A_N-A_P)g_m(N)$. Parasitic poles, δ_1 and δ_2, are obtained by means of ratios among α, β and γ, as in the HS implementation. The final expressions are summarized in table 6. Consequently, parasitic poles δ_1 and δ_2 can be expressed as:

$$\begin{aligned} \alpha &\approx ((A_N - A_P)g_m(N))(g_{MP}(PF) + g_{ds}(PF)) + 2g_{ds}(NS)(g_{MP}(PF) + 2g_{ds}(P)) \approx \\ &\approx 2g_{ds}(NS)(g_{MP}(PF) + 2g_{ds}(P)) \end{aligned} \tag{29}$$

$$\beta \approx (g + g_{ds}(PF) + g_{MP}(PF))C_I \approx (g_{MP}(PF) + 2g_{ds}(P))C_I \tag{30}$$

$$\delta_1 = \frac{\alpha}{\beta} \approx \frac{2g_{ds}(NS)}{C_I} \tag{31}$$

$$\begin{aligned} \gamma &\approx C_I\left(C_{ds}(PF) + C_{gd}(N) + C\right) \approx C_I C \approx \\ &\approx C_I\left(2C_{gd}(P) + 2C_{ds}(P) + C_{gs}(PF) + C_{bs}(PF)\right) \approx 2\,C_I\,C_{gd}(P) \end{aligned} \tag{32}$$

$$\delta_2 = \frac{\beta}{\gamma} \approx \frac{g_{MP}(PF) + 2g_{ds}(P)}{2C_{gd}(P) + 2C_{ds}(P) + C_{gs}(PF) + C_{bs}(PF)} \approx \frac{g_{MP}(PF) + 2g_{ds}(P)}{2C_{gd}(P)} \tag{33}$$

Similar results can be obtained for the common-mode frequency response:

$$\begin{aligned} \alpha_{CM} &\approx ((A_N + A_P)g_m(N))(g_{MP}(PF) + g_{ds}(PF)) + 2g_{ds}(NS)(g_{MP}(PF) + 2g_{ds}(P)) \approx \\ &\approx (A_N + A_P)g_m(N)(g_{MP}(PF) + g_{ds}(PF)) \end{aligned} \tag{34}$$

$$\beta_{CM} \approx \beta \approx (g_{MP}(PF) + 2g_{ds}(P))\,C_I; \quad \gamma_{CM} \approx \gamma \approx 2\,C_I\,C_{gd}(P) \tag{35}$$

$$\xi_1 = \frac{\alpha_{CM}}{\beta_{CM}} \approx \frac{(A_N + A_P)g_m(N)}{C_I}; \quad \xi_2 = \frac{\beta_{CM}}{\gamma_{CM}} \approx \delta_2 \approx \frac{g_{MP}(PF) + 2g_{ds}(P)}{2C_{gd}(P)} \tag{36}$$

4.6 Comments and discussion

A full analysis has been developed in previous sections in order to draw some design strategies and implement a competitive and robust transconductor cell. Great similarity was found between the two topologies, as reflected in table 6.

The first conclusion regards total integration capacitance, C_I, which has the same definition in both implementations: $C_I=C_s+3C_{in}(N)+2C_{out}$ (Eq. 26), where C_s represents the equivalent

	S_0	S_1	δ_1	δ_2	ξ_1	ξ_2
HS	$\frac{g_m(N)}{C_{gd}(N)}$	$\frac{g_M(NC)+g_{ds}(NC)}{C_{ds}(NC)}$	$\frac{2g_{ds}(NC)g_{ds}(N)}{g_M(NC)C_I}$	$\frac{g_M(NC)}{C_{gs}(NC)}$	$\frac{(A_N+A_P)g_m(}{C_I}$	$\frac{g_M(NC)}{C_{gs}(NC)}$
FC	$\frac{g_m(N)}{C_{gd}(N)}$	$\frac{g_{MP}(PF)+g_{ds}(PF)}{C_{ds}(PF)}$	$\frac{2g_{ds}(NS)}{C_I}$	$\frac{g_{MP}(PF)+2g_{ds}(P)}{2C_{gd}(P)}$	$\frac{(A_N+A_P)g_m(N)}{C_I}$	$\frac{g_{MP}(PF)+2g_{ds}(P)}{2C_{gd}(P)}$

Table 6. Parasitic poles and zeros for the integrator topology.

parallel capacitance between the external capacitance and the Norton capacitance of the input current source. This definition can easily be obtained by taking two different ideas into account. Considering the proposed integrator topology (Fig. 3), the expected integration capacitance will be the external capacitance in addition to the contribution of parasitic capacitors to the input node. For example, looking at the positive branch (the upper one) in this figure, there are three transconductor cell inputs ($g_m(1+)$, $g_m(2+)$ and $g_m(-)$) and two outputs ($g_m(1+)$, $g_m(2-)$) connected to the integrator input and contributing to C_I. Each unit cell forming the complete integrator has been studied in detail and their corresponding models obtained (tables 1 and 2). From this analysis, $C_{in}(N)$ and C_{out} are the equivalent input and output capacitance respectively for each unit cell; which, combined with the previous consideration, leads directly to the definition obtained for C_I.

The low-impedance internal node strategy combined with the use of a negative resistance to enhance the differential dc-gain of the system, while increasing the differential input resistance and reducing the phase error, suggested the use of cascode topologies to us, also taking advantage of their inherent transmission zero reduction. The differential dc-gain obtained in both structures, shown in Eq.(15) and (27) respectively, reflects the existence of this positive feedback compensation, making the ideal infinite dc-gain of the integrator possible. Regarding the common-mode behaviour of the circuit, both implementations are described with the same expression complying with the stability requirement $|A_{CM}|<1$: $A_{CM} \sim -1/(A_N+A_P) \sim -1/2$.

Variations on the negative resistance value are equivalent to considering a mismatching between $M_N(1)$ and $M_N(2)$, which generates a difference between the drain currents of these transistors. Furthermore, the same effect can also be achieved by modifying bias currents I_{BIAS} in these two branches of the circuit. In consequence, the tuning of the value of this negative resistance allows for correction in process deviations and mismatching between the transistors M_N that process the signal.

Finally, cascode stages introduce two dominant parasitic poles and a dominant zero as shown in Eq.(8), which stem from parasitic capacitances mainly associated with the source nodes of cascode transistors. In consequence, an excess phase shift $\Delta\varphi(\omega_t)$ takes place at unity-gain frequency (Eq. 37). Theoretically, by using the minimum length for cascode transistors and minimizing source and drain diffusion areas, the parasitic poles and zero can be located further away from the unity-gain frequency. This consideration will also minimize distributed-channel effects, which cannot be ignored at high frequencies (Tsividis, 1999). However, the minimum channel length is conditioned by the required transistor matching (Pelgrom et al., 1989).

$$\Delta\varphi(\omega_t) \approx \frac{\pi}{2} - \tan^{-1}\left(\frac{\omega_t}{\delta_1}\right) - \tan^{-1}\left(\frac{\omega_t}{\delta_2}\right) - \tan^{-1}\left(\frac{\omega_t}{s_0}\right) \quad (37)$$

These general considerations reflect the benefits of the topology and its frequency limits. Nevertheless, both integrators have been implemented, and the real values drawn from these designs allow a better study of the frequencies related to parasitic effects to be carried out and a comparison to be made between both topologies.

Second-order frequency dependence. The previous study for the frequency behaviour of the integrator was summarized in Eq.(8), and, taking into account the definition for the integrator quality factor (Eq. 4), the general relation in both approximations can be reported as:

$$\frac{1}{Q_{\text{int}}}=\frac{\delta g_m+g_o'}{g_m(N)}-\frac{C_{gd}(N)}{C_I}-\frac{C_p(C)}{C_I}\frac{g_m(N)}{g_m(C)} \tag{38}$$

where g_o' is the sum of output conductances at the input node, C_I is the total integration capacitance, $g_m(C)$ is the transconductance of the cascode element ($g_m(NC)$ and $g_m(PF)$ for the HS and FC approach respectively), and $C_p(C)$ the parasitic capacitance at the low-impedance node associated to the cascode structures (node X in Figs. 8 and 9).

The second-order frequency dependence can be analysed when considering the M_N transistor mismatch A_N-A_P=$\delta g_m/g_m$ and including the frequency dependence for their transconductance $g_m(s)$ (Eq. 39) in the transfer function (Smith et al., 1996; Zele et al., 1996). In this way, the non-quasistatic model for the MOS transistor, required for operation at very high frequencies, is included.

$$g_m(s)=\frac{g_m(N)}{1+s\tau} \quad ; \quad \tau=\frac{2}{5}\frac{C_{gs}(N)}{g_m(N)} \tag{39}$$

By simplifying the result obtained to draw some conclusions, the frequency behaviour of the integrator is described (Eq. 40) reflecting the effect of these additional second-order dependencies.

$$\frac{1}{Q_{\text{int}}}=\left(\frac{g_o'}{g_m(N)}\right)^2+\frac{\delta g_m}{g_m(N)}-\frac{3C_{gs}(N)}{7C_I}-\frac{C_{gd}(N)}{C_I}-\frac{C_p(C)}{C_I}\frac{g_m(N)}{g_m(C)} \tag{40}$$

From this study, a set of parasitic poles and zeros appear at frequencies higher than 15 GHz in the proposed designs, the effect of which is considered negligible in a first-order approximation.

5. Noise

In general, the range of signals that can be accurately driven by electronic devices is limited. For low-signals, electrical noise restricts the minimum amplitude that can be processed. Noise is considered as all the unwanted electrical signals generated within the device or externally and coupled to the output of the system. These signals appear in the system whether input signals are applied or not. Noise signals interfere with the incoming signal and make it impossible to detect with sufficient quality the signals presenting an amplitude comparable to the noise level. Moreover, signals below this level are almost impossible to detect. So, noise in the system represents the lowest level for the incoming signal (Silva-

Martínez et al., 2003). The origins of noise can be classified as intrinsic and extrinsic transistor noise sources, following a similar notation to that used for the parasitic elements of the MOS transistor. There are two dominant intrinsic noise sources in CMOS transistors: channel thermal noise and 1/f or flicker noise. Extrinsic noise is mainly due to the signals produced by the surrounding circuitry and coupled to the device or to the system. The only noise considered in this work is noise generated by the transistor.
The noise of a G_m-C filter is caused by the noise output currents of the transconductors and the mean square noise currents are generally proportional to the corresponding transconductance. Considering the model of a noisy transconductor, the output mean square noise current over a certain frequency interval df is given by:

$$\overline{i_n^2} = 4\,k\,T\,NF\,g_m\,df \tag{41}$$

where k is the Boltzmann constant, T the absolute temperature and NF a noise factor determined by the electronic design of the transconductor. NF=1 corresponds to a transconductor output noise current equal to the thermal noise current of a resistor of value $R=1/g_m$.

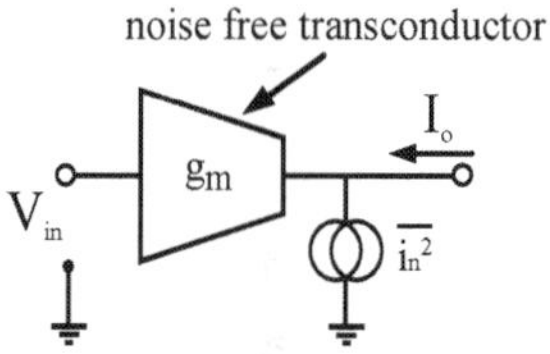

Fig. 10. Transconductor noise model.

To investigate the latter effect in the frequency design of VHF filters, calculating the noise factor and its frequency dependence becomes necessary in both HS and FC transconductor implementations, because filter noise behaviour not only depend on the filter parameters but also on the transconductor noise behaviour. Therefore, the NF-factor appears as a useful quantity to translate the transconductor noise properties to the filter noise properties and, likewise, the properties of each unit cell to the complete transconductor implementation. In consequence, transconductor output noise is determined by the noise properties of the unit cell and by the properties of the transconductor topology.
The transconductor topology analysed is the balanced structure shown in Fig. 3. First, the noise behaviour of each single unit cell is calculated ($H_{HS}(s)$ and $H_{FC}(s)$) and then the noise factor of the complete transconductor in both implementations, $F_{HS}(s)$ and $F_{FC}(s)$.

5.1 Noise model for the unit cell

For high and very-high frequencies, the 1/f or flicker noise can be neglected. Assuming only thermal noise, the drain-current noise of a single transistor can be written as:

$$\overline{i_n^2} = 4\,k\,T\,c\,g_m\,df \tag{42}$$

where k is the Boltzmann constant, T the absolute temperature, g_m the small-signal transconductance of the transistor, and df the frequency interval over which the noise is measured. The constant c value is 2.5 for a short-channel transistor working in the saturation region[2] (Tsividis, 1999). The noise output current of the HS unit cell (Fig. 6(a)) is:

$$\overline{i_n^2}(HS) = 4\,k\,T\,H_{HS}\,df \tag{43}$$

where:

$$H_{HS} = \left(1-\sqrt{\alpha}\right)^2 c(PC)\,g_m(PC) + \alpha\,c(P)\,g_m(P) + \left(1-\sqrt{\beta}\right)^2 c(NC)\,g_m(NC) + \beta\,c(N)\,g_m(N) \tag{44}$$

$$\sqrt{\alpha} = \frac{g_m(PC)+g_{ds}(PC)}{g_m(PC)+g_{ds}(PC)+g_{ds}(P)};\quad \sqrt{\beta} = \frac{g_m(NC)+g_{ds}(NC)}{g_m(NC)+g_{ds}(NC)+g_{ds}(N)} \tag{45}$$

For the FC approach, the noise output current of the unit cell (Fig. 6(b)) can be written as:

$$\overline{i_n^2}(FC) = 4\,k\,T\,H_{FC}\,df \tag{46}$$

where:

$$H_{FC} = \left(1-\sqrt{\alpha}\right)^2 c(PF)\,g_m(PF) + 2\alpha\,c(P)\,g_m(P) + c(NS)\,g_m(NS) + \alpha\,c(N)\,g_m(N) \tag{47}$$

$$\sqrt{\alpha} = \frac{g_m(PF)+g_{ds}(PF)}{g_m(PF)+g_{ds}(PF)+2g_{ds}(P)+g_{ds}(N)} \tag{48}$$

As a result, certain conclusions can be drawn from this study. The main noise sources in both topologies are transistors M_N processing the signal and bias current sources, to which the main contribution applies. The FC unit cell presents a noise factor 2.5 times higher than the HS topology, which is due to the additional current sources required in this topology.

5.2 Noise model for the complete transconductor cell

The noise factor of the complete transconductor can be calculated using the model obtained for the unit cells (Nauta, 1993), or by calculating the complete noise model of the proposed transconductance cell from the start. We use this second option to obtain a more accurate result, modelling the output noise currents of the transistors with noise current sources in parallel with the outputs as in the previous section. The two noise output currents of the differential transconductor are $\overline{i^2_{n,plus}}$ and $\overline{i^2_{n,\text{minus}}}$. Following a similar process as in the noise study on the unit cells, the noise output current of the complete HS transconductor can be written as:

$$\overline{i^2_{n,plus}(HS)} = \overline{i^2_{n,\text{minus}}(HS)} = 4\,k\,T\,F_{HS}\,df \tag{49}$$

[2] The constant c, which models transistor noise behaviour, depends on both its operation region and its dimensions. A theoretical derivation for the transistor working in the linear region leads to 1<c<2; however, c=2/3 for a long-channel device and c=2.5 for a short-channel device are obtained in the saturation region (Tsividis, 1999).

where:

$$F_{HS} = c(PC)\,F_{HS}(PC) + c(P)\,F_{HS}(P) + c(NC)\,F_{HS}(NC) + c(N)\,F_{HS}(N) \quad (50)$$

$$\begin{aligned} F_{HS}(PC) = x_1^{\,2}\,[\,(A_N + A_P - 2)\,g_m(N)\cdot(g_m(NC) + g_{ds}(NC))\cdot x_2 + \\ + 2g_{ds}(P)g_{ds}(PC)x_3 + \; 2g_{ds}(N)g_{ds}(NC)x_2\,]^2 \;\; g_m(PC) \end{aligned} \quad (51)$$

$$F_{HS}(P) = \frac{F_{HS}(PC)}{g_m(PC)}\left(\frac{g_m(PC) + g_{ds}(PC)}{g_{ds}(P)}\right)^2 g_m(P) \quad (52)$$

$$F_{HS}(NC) = \left(g_{ds}(N) + \frac{x_2\,g_m(N)(g_m(NC) + g_{ds}(NC))(g_m(NC) + g_{ds}(NC) - g_{ds}(N))}{\alpha}\right)^2 x_2^{\,2}\,g_m(NC) \quad (53)$$

$$F_{HS}(N) = \left(1 + \frac{x_2^{\,2}(g_m(NC) + g_{ds}(NC))^2 (A_N + A_P)\,g_m^{\,2}(N)}{\alpha^2}\right) x_2^{\,2}(g_m(NC) + g_{ds}(NC))^2\, g_m(N) \quad (54)$$

$$x_1 = \frac{g_{ds}(P)\,x_3}{\alpha}; \quad x_2 = \frac{1}{g_{ds}(N) + g_m(NC) + g_{ds}(NC)}; \quad x_3 = \frac{1}{g_{ds}(P) + g_m(PC) + g_{ds}(PC)} \quad (55)$$

$$\alpha = x_2(g_m(NC) + g_{ds}(NC))((A_N + A_P)g_m(N) - 2g_{ds}(NC)) + 2g_{ds}(NC) + 2g_{ds}(P)g_{ds}(PC)x_3 \quad (56)$$

The noise output current of the complete transconductor implemented with FC unit cells can be written as:

$$\overline{i^2_{n,plus}(FC)} = \overline{i^2_{n,minus}(FC)} = 4\,k\,T\,F_{FC}\,df \quad (57)$$

where:

$$F_{FC} = c(PF)\,F_{FC}(PF) + c(P)\,F_{FC}(P) + c(NS)\,F_{FC}(NS) + c(N)\,F_{FC}(N) \quad (58)$$

$$F_{FC}(PF) = \left(1 - \alpha - \frac{2\alpha(1-\alpha)}{\chi}\right)^2 g_m(PF); \quad F_{FC}(P) = 2\alpha^2\left(1 - \frac{2\alpha}{\chi}\right)^2 g_m(P) \quad (59)$$

$$F_{FC}(NS) = \left(1 - \frac{2\alpha}{\chi}\right)^2 g_m(NS); \quad F_{FC}(N) = \left(1 + \frac{\alpha^2}{\chi^2}(A_N + A_P)\right)\alpha^2\, g_m(N) \quad (60)$$

$$\alpha = \frac{g_m(PF) + g_{ds}(PF)}{g_m(PF) + g_{ds}(PF) + 2g_{ds}(P) + g_{ds}(N)}; \quad \chi = 2(1-\alpha)\frac{g_{ds}(PF)}{g_m(N)} + 2\frac{g_{ds}(NS)}{g_m(N)} + \alpha(A_N + A_P) \quad (61)$$

From this study, the major noise contributions are those due to NMOS transistors in the signal path, M_N, representing almost the total contribution for the FC implementation. The contribution of the bias current sources, which was relevant in the noise analysis of the unit cells, becomes negligible when implementing the complete fully-balanced current-mode

topology. A significant benefit of this structure is that the previous noise-related difference existing between the two unit cells disappears in the complete transconductor, with the same value being obtained mostly for F_{HS} and F_{FC}. Another important conclusion is that the output noise current is independent of the value of the negative resistance δg_m, inherent characteristic of the topology.
Hence, the noise is dominated by the M_N transistors involved in the V to I conversion, with similar noise contributions being achieved for both transconductor implementations. The post-layout simulation results corroborate this idea, with a total input-referred noise of 11 nA_{rms} being obtained for the HS topology and 8 nA_{rms} for the FC approach. This is a very important design conclusion and requires be further considered in the design of high-performance filters.

6. Distortion

The ability to handle large signals with minimum distortion is an important consideration in most linear applications. In this section, the large signal characteristics of the current mode integrator for differential inputs are analysed using first-order square-law MOSFET models, which can provide a sufficiently accurate description of the large signal behaviour for design purposes. The proposed transconductor topology (Fig. 3) has been implemented by using two different cascode architectures: the HS and the FC approach. Transconductor V-I conversion is obtained in the output stage g_m, and a negative resistance, shown in grey in the same figure, is connected at the input in order to improve the topology characteristics.
A theoretical distortion analysis including all the non-linearities would be too complex and provides no practical benefit. Consequently, a separate study of each factor affecting linearity has been developed to obtain simple and efficient design strategies. There are two main contributions to transconductor distortion, the first one will be non-linearities in V to I conversion, and the second non-linearities in the negative resistance. Each effect is analysed separately. The effects of capacitor non-linearities are not taken into account here.

6.1 Non-linearities in V-I conversion

For this study, only the output stage will be considered, i.e., the transconductor cell without the negative resistance. This differential structure will present good linearity in V-I conversion if both paths are perfectly matched. Using a simple model of the MOS transistor, the drain currents of the N- and P-channel MOS transistor in strong inversion can be written as:

$$I_{dn} = \beta_n \left(V_{gsn} - V_{tn}\right)^2; \quad I_{dp} = \beta_p \left(V_{gsp} - V_{tp}\right)^2 \quad with \quad \beta_n = \frac{\mu_n C'_{ox} W_n}{2L_n}; \quad \beta_p = \frac{\mu_p C'_{ox} W_p}{2L_p} \tag{62}$$

Considering $V_{IN}^+=V_C+v_{in}/2$, $V_{IN}^-=V_C-v_{in}/2$ and v_{in} the differential input voltage of the transconductor cell, the differential output current can be written as:

$$I_{out} = i_{out}^+ - i_{out}^- = 2\sqrt{I_{BIAS}\beta_n(N)}\ v_{in} \tag{63}$$

where I_{BIAS} is the bias current per branch and $\beta_n(N)$ the factor associated to the M_N transistor in both implementations. The same result has been obtained with the two possible transconductor structures, the HS and the FC topologies, leading to the same distortion analysis in both situations. This equation is valid as long as the transistors operate in strong inversion and saturation, and show that the differential voltage to current conversion of the transconductor is perfectly linear, making use of the square-law and matching properties of the MOS transistors. However, mobility reduction will cause deviations from the transistor square-law behaviour, generating distortion. The mobility of the NMOS and PMOS transistors can be expressed in the first-order approximation as:

$$\mu_n = \frac{\mu_{on}}{1+\theta_n(V_{gsn}-V_{tn})}; \quad \mu_p = \frac{\mu_{op}}{1-\theta_p(V_{gsp}-V_{tp})} \tag{64}$$

Considering again the output stage of the complete cell (Fig. 3), driven balanced around the common-mode voltage level V_C as previously explained, the differential output current can be calculated as:

$$I_{out} = i_{out}^{+} - i_{out}^{-} \tag{65}$$

Using the previous expressions for mobility reduction and assuming equal transistors operating in saturation, the differential output current can be expanded into Taylor series:

$$I_{out} = \frac{k_{on}V_{on}\left(1+\frac{1}{2}\theta_n V_{on}\right)}{(1+\theta_n V_{on})^2}v_{in} - \frac{\theta_n k_{on}}{(1+\theta_n V_{on})^4}\frac{1}{8}v_{in}^3 - \frac{\theta_n^3 k_{on}}{(1+\theta_n V_{on})^6}\frac{1}{32}v_{in}^5 + \cdots; \tag{66}$$

where: $V_{on} = V_C - V_{tn} = \sqrt{\frac{I_{BIAS}}{\beta}} > 0$ and $k_{on} = \frac{\mu_{on}C'_{ox}W_n}{L_n}$.

The fifth and higher order distortion terms can be neglected, giving the following as a result for distortion calculation:

$$I_{out} = C_1 v_{in} + C_3 v_{in}^3; \quad with \ C_1 = \frac{k_{on}V_{on}\left(1+\frac{1}{2}\theta_n V_{on}\right)}{(1+\theta_n V_{on})^2}; \ C_3 = \frac{-\theta_n k_{on}}{(1+\theta_n V_{on})^4}\frac{1}{8} \tag{67}$$

$$HD_3 = \frac{C_3}{C_1} = \frac{-\theta_n}{8V_{on}(1+\theta_n V_{on})^2\left(1+\frac{1}{2}\theta_n V_{on}\right)} \approx \frac{-\theta_n}{8V_{on}} \tag{68}$$

The expression for HD_3 can be simplified as $\theta_n V_{on} \ll 1$. This expression is the same for both implementations of the transconductor cell. The main conclusion from Eq.(68) is that increasing V_{on}, i.e., the common-mode voltage V_C or the bias current I_{BIAS}, lowers the distortion in the V-I transfer.

6.2 Non-linearities in the negative resistance

Non-linearities in V-I conversion and non-linearities in the negative resistance will introduce non-linear effects in the input resistance of the differential transconductor (Nauta, 1993). The effect of these non-linearities is a signal dependent integrator quality factor, i.e., a phase error; and can cause distortion in filters, especially high-Q filters. As long as the value of the input impedance is high enough and an enhancement of the dc-gain is obtained, the effect of non-linearities will be slight. Considering the equations for a single transistor reported in Eq.(62) and analysing the structure shown in Fig. 3, the differential input voltage can be written as:

$$v_{in} = \sqrt{\frac{I_{BIAS}}{\beta_n}}\left(\sqrt{1+\frac{i}{I_{BIAS}}}-\sqrt{1-\frac{i}{I_{BIAS}}}\right) \tag{69}$$

where (2i) is the differential input current, i.e., $I_i^+ = i$ and $I_i^- = -i$, considering the description shown in Fig. 3. The same relation has been obtained in both implementations, leading again to the same distortion analysis. This expression can be expanded into Taylor series, where $A=i/I_{BIAS}$, $\alpha_1=1$ and $\alpha_3=1/8$.

$$v_{in} = \alpha_1 ACos(wt) + \alpha_2 A^2 Cos^2(wt) + \alpha_3 A^3 Cos^3(wt);\ HD_2 = \frac{\alpha_2 A}{2\alpha_1};\ HD_3 = \frac{\alpha_3 A^2}{4\alpha_1} \Rightarrow HD_3 = \frac{1}{32}\left(\frac{i}{I_{BIAS}}\right)^2 \tag{70}$$

The large signal output characteristics of the current mode integrator present a similar form to that of the standard MOS differential pair. On the other hand, the distortion generated in positive feedback compensation through negative resistance is only dependent on the ratio between the input current and the bias current I_{BIAS}. This analysis is the same for both implementations of the transconductor topology.

6.3 Matching and harmonic-distortion

The process that causes time-independent random variations in physical parameters of identically designed devices is called mismatch, and is a limiting factor in general-purpose analog signal processing. The impact of MOS transistors mismatching becomes very important because the dimensions of the devices are reduced and the available signal swings decrease. To obtain a better circuit design, the physical origin of this effect has been discussed in several studies (Pelgrom et al., 1989), not only for its random but also for its systematic contribution (Gregor, 1992), and also possible measurements for its characterization (Felt et al., 1994).

Thanks to the use of a fully-balanced pseudo-differential topology, with this inherent positive feedback compensation providing the system with an enhancement of the differential dc-gain, distortion resulting from mismatch is small. As mentioned above, the existing negative resistance enables small variations in dimensions of M_N transistors and bias current sources I_{BIAS} to be controlled. In order to obtain a good matching, the minimum channel length related to the considered CMOS technology must be avoided. However, high frequency operation requires short channels, relying on the negative resistance to obtain the adequate matching between transistors in the signal path. In addition, channel length modulation is not considered, as it only has a substantial effect on integrator response linearity at low frequencies, where distortion is suppressed by feedback (Smith et al., 1996; Zele et al., 1996). In consequence, transistors are assumed to be well matched, which will be achieved by means of a lower mismatching sensitive design while obtaining the final layout.

Mismatching between transistors also degrades the outstanding benefits provided by balanced structures, since they depend strongly on the symmetry of the circuit. Therefore, mismatching can also be reflected in unbalanced signal paths. Post-layout simulations have been accomplished regarding sensitivity to this unbalance for both HS and FC transconductor implementations. The starting point was an input signal of 10 MHz and 45 dB of THD. Subsequently, variations of 10% and 20% in magnitude and phase between the input currents I_i^+ and I_i^- were considered while analysing the effect on the THD. Regarding the variations in magnitude: with a shift of 10% between the magnitude of I_i^+ and I_i^-, the THD changes by 1 dB (HS) and 0.5 dB (FC), and for a shift of 20%, the THD changes by 2 dB and 1.5 dB respectively. For deviations between the phase of I_i^+ and the phase of I_i^-, with a shift of 10%, the THD changes by 1 dB for both topologies; and for a shift of 20%, the THD changes by 8 dB (HS) and 6 dB (FC).

As a result, we can conclude that the proposed topology is quite insensitive to transistor mismatching. In addition to this, the effect of common-mode signal mismatching is alleviated by means of feedback compensation, as previously explained, thus supporting the proposed design strategy.

7. Digital programmability

Apart from the usual requirements associated with high frequency CMOS filter design, the issue of programmability brings to the forefront the considerable problem of maintaining performances such as frequency response accuracy, noise and dynamic range across the entire tuning range. Requirements of robust and precise implementation of filtering systems in the VHF range point to programmable G_m-C continuous-time filters as the best option for obtaining a wide programming range (usually 1:5). Due to process and temperature variations, G_m/C time-constants are liable to vary by as much as ±30%. The fact of considering both effects at the same time means that the unity-gain frequency ω_t of each integrator in the filter should be electronically variable over a wide range.

Lower supply voltages required by current digital CMOS technologies make the use of conventional continuous tuning techniques over a wide frequency range very difficult due to their effect on dynamic range and non-linear distortion. These techniques are based on the variation of the transistors biasing points, limiting their application to compensate the inherent changes due to temperature and the technological process. Therefore, discrete tuning is the best option to preserve the dynamic range (DR).

There are three different ways of achieving this wide range of variability: the capacitor, the transconductor or both can be made programmable. At high frequencies, the integrating capacitances are relatively small. If they are replaced by capacitor arrays to obtain C-programmability, the net parasitic capacitances at the terminals of the array can be quite large when the array is implementing the lowest effective capacitance, which is a very difficult problem to solve. In addition to this, switchable array of capacitors provides high precision on filters though the existence of switches in the signal path. So, the constant-C scaling technique is the option considered, leading to the desired programmability by varying G_m discretely while maintaining the noise specifications over the entire frequency range. Furthermore, lower power consumption is achieved at low frequency values of the

programming range. This is the best option for maintaining a trade-off between noise specifications, power consumption and programming range (Pavan et al., 2000).
Two different strategies can be used to extend the tuning range and preserve DR: switchable array of degenerating MOS resistors and parallel connection of identical transconductors switched by a digital word. The first one uses the same transconductor and capacitor throughout the whole frequency range whilst the degeneration resistor R is formed by a parallel connection of MOS triode transistors (Bollati et al., 2001). This technique involves variations over the entire frequency range of the noise factor, which is proportional to the degenerated resistor, generating dynamic range variations. Phase errors will also appear, achieving the worst situation for both undesirable parameters in the opposite ends of the frequency range: minimum DR (maximum R) for the lowest frequency f_{min}, and maximum phase error for the highest frequency f_{max}. However, this strategy leads to the simplest structure, with a small active area and lower power consumption. The second strategy consists of a parallel connection of identical transconductors obtaining a programmable array where the desired time-constant can be digitally tuned (Pavan et al., 2000). This solution is the best option for VHF applications. However, its main drawbacks are power consumption and area, proportional to the number of connected active cells.
Considering all these ideas, the latter strategy is the technique selected for achieving the desired programmability for the proposed topology. Programmability using a parallel connection of conventional differential pairs has been published previously (Pavan et al., 2000); however, these structures are not directly suitable for low-voltage supply. It is worth noting that obtaining a programming range for a transconductor also includes an additional gap of ±30% for the extreme transconductance values G_{min} and G_{max}, in order to compensate the deviation due to temperature and technological process variations. Therefore, the total tunable range will be greater than the nominal one.

7.1 Principle of programmability

Our proposal is to achieve a digitally programmable transconductor, specifically designed for a wide programmability range comprised of parallel connection of unit cells. Fig. 11 shows the conceptual scheme of a 3-bit programmable cell. This topology presents two main drawbacks; the need for additional transistors in the signal path and the variation of parasitic capacitances C_{pin} and C_{pout} depending on the digital word. However, it is necessary to keep the dynamic range constant for each g_m value and the total node parasitic capacitances over the entire programming range.
This solution is adopted for the proposed transconductor topology, giving a HS implementation with a programmable range from 1:7 (Fig. 12(a)) and a FC implementation with a varying range from 1:5 (Fig. 12(b)). Each cascode unit cell (Fig. 6), i.e., cascode amplifier and biasing current source, consists of a parallel connection of equal cells switched by a digital word. The connecting lines of the substrate terminals are not shown on these schematics as the explanation has already been given in previous sections. By driving the gates of the cascode transistors (M_{NC} and M_{PC} in the HS approach, and M_{PF} and M_{NS} in the FC) with modulated digital voltages we can obtain the desired transconductance with no switches in the signal path and power consumption proportional to total transconductance.
The other disadvantage inherent to this topology can be also alleviated with an additional design strategy. When a change in the digital word occurs, some transistors change from saturation to cut-off region and *vice versa*, and different contributions to total input node

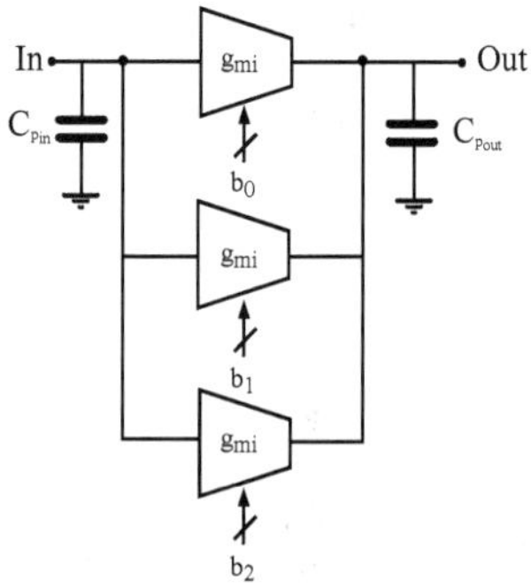

Fig. 11. Topology of the 3-bit programmable transconductor.

parasitic capacitance C_{pin} are obtained. This change can generate a shift in the desired frequency and Q-factor variations, limiting the integrator and filter performance. In consequence, the implementation of each unit cell has been modified by using dummy elements connected at the input, which allow us to make the input capacitance independent of the digital word, maintaining the same parasitic capacitances on the signal processing nodes (Pavan et al., 2000). Note that the total output parasitic capacitance -junction extrinsic capacitance- is also constant because it has almost the same value for cut-off or saturation transistors (Tsividis, 1996; Tsividis, 1999).

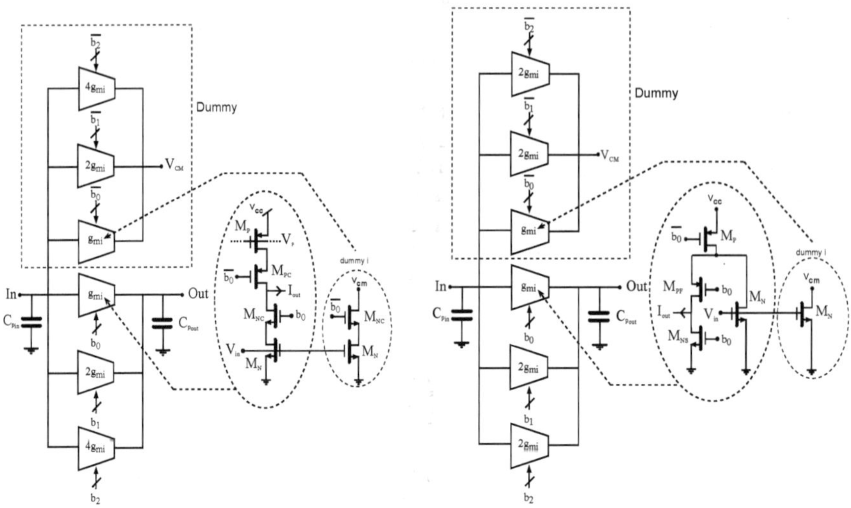

(a) (b)

Fig. 12. Implementation of 3-bit programmable topology: (a) HS and (b) FC transconductor.

As the output conductance is proportional to the transconductance, the differential dc-gain is maintained irrespective of the digital word. Consequently, the relative shape of the frequency response, output noise power and dynamic range are independent of the digital word. Therefore, we obtain the desired programmable transconductor with no switches in the signal path by driving the gates of bias and cascode transistors with a digital word modulated with the adequate analog value. The power consumption is proportional to the necessary transconductance in each frequency range. The digital control word for programming the transconductor is $b_2b_1b_0$, controlling transconductance value from $001 \equiv g_m$ to $111 \equiv 7g_m$ in the HS topology and from $001 \equiv g_m$ to 111≡5gm in the FC topology.

7.2 Implementation results

Two different integrators have been implemented. The first one is based on the HS topology, considering the total input node parasitic capacitance -basically gate-source capacitances- as the total integration capacitance with no need for any external capacitors, in order to reach the maximum operation frequency with moderate power consumption. Therefore, for this situation, maintaining the integration capacitance constant becomes essential and its value can be controlled by means of the dummy-based system.

Fig. 13(a) plots the post-layout simulation results for HS implementation and shows the variation of unity-gain frequency versus digital word value. The expected linear dependence of the transconductance and the constant integration capacitance are observed, and a programming range from 28 to 185 MHz is obtained by varying the digital word. However, a marked phase lag due to parasitic effects (parasitic zero s_0) at high frequency was detected, as expected. A possible compensation scheme, is based on two capacitors, C_C, implemented with dummy MOS transistors and connected in a cross-coupled manner as shown in Fig. 7(a). Consequently, by using this compensating scheme, the phase shift error is effectively reduced and a very efficient scheme endowed with a phase error of less than 3° over the entire frequency range is obtained.

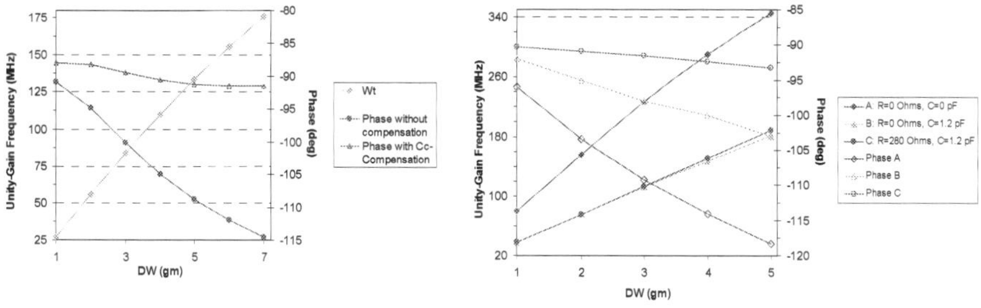

(a) (b)

Fig. 13. Unity-gain frequency and phase vs. digital word value for the: (a) HS integrator with and without compensation scheme; (b) FC integrator with various compensation schemes.

The second implementation is based on the FC topology, considering integration capacitance in this case as the total input node parasitic capacitance together with an additional one (C_{ext}=1.2 pF). Total integration capacitance is once more maintained constant by means of the dummy-based system and Fig. 13(b) plots the post-layout simulation results. The first curve plots the variation of the transconductance as a function of the digital word when no external capacitance is connected at the input; a non-linear response is obtained, due to the expected parasitic poles and zeros. When connecting the external capacitance, the expected linear dependence is obtained, providing the system with coarse

tuning. Nevertheless, a phase shift is obtained even when the compensation scheme based on the two cross-coupled capacitors C_C is used.
Next step involves a resistance R being connected at the input in series with the external capacitance, as shown in Fig. 7(b). This resistor is implemented with a transistor working in the linear region and is the best option to compensate this phase error. Therefore, by varying the digital word, the unity-gain frequency is controlled and the phase error is effectively reduced over the entire programming range. Then, to control the operation frequency and to reduce the phase error, a shunt connection is made at the input between a resistance and the integration capacitance C_I. We obtain a compensation scheme for the FC transconductor based on an RC circuit at the input, leading to a programming range from 40 to 200 MHz by varying the digital word with a phase error of less than 3° over the entire frequency range.
We can define the transconductor input voltage variations around the bias point (V_C) as shown in Fig. 3. The linear input range is constant for digital scaling of the transconductance as shown in Fig. 14. The variation of the g_m as a function of the digital word is presented, providing the system with coarse tuning. In consequence; for the HS topology, ω_t is controlled from 28 to 185 MHz by varying the digital word from 1 to 7; and for the FC topology, ω_t varies from 40 to 200 MHz by varying the DW from 1 to 5. Therefore, by means of a parallel connection of equal transconductors switched by a digital word we guarantee that the DR for each g_m value and the total external node capacitances will be kept constant.

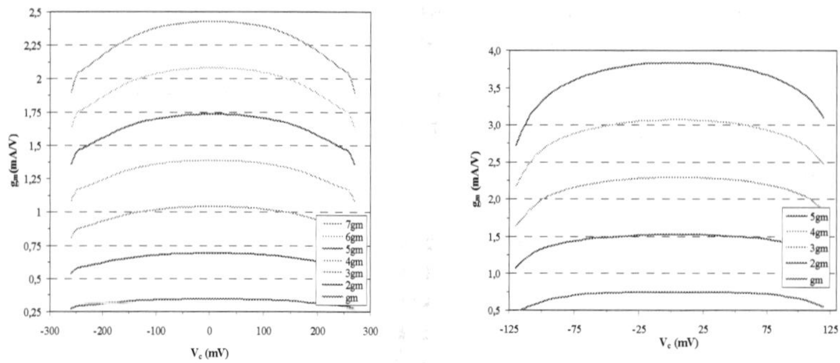

(a) (b)

Fig. 14. Transconductance as a function of the digital word (coarse tuning) for the: (a) HS implementation; (b) FC implementation.

On the other hand, fine tuning can be achieved if necessary, as the transconductance value can be controlled by varying the bias current source for a fixed digital word. Hence, discrete steps are swept by varying the bias current while maintaining the same dynamic range. At the same time, an additional control over the dc-gain can be achieved by modifying the ratio between the bias currents of the negative resistance: M_1/M_2 and M_4/M_5 in both topologies, solving problems associated to mismatching between transistors. Therefore, a complete control of the frequency response can be obtained. The trade-off between transconductance and linear input range is shown in Fig. 15 for both topologies. These figures can also be seen as the fine tuning for the proposed structure since the transconducance value is controlled by varying the bias current source for a fixed digital word: I_{bias} changes from 45 to 180 μA control the HS transconductance from 270 to 452 μA/V, and changes from 40 to 100 μA in the FC topology control the transconductance from 550 to 800 μA/V.

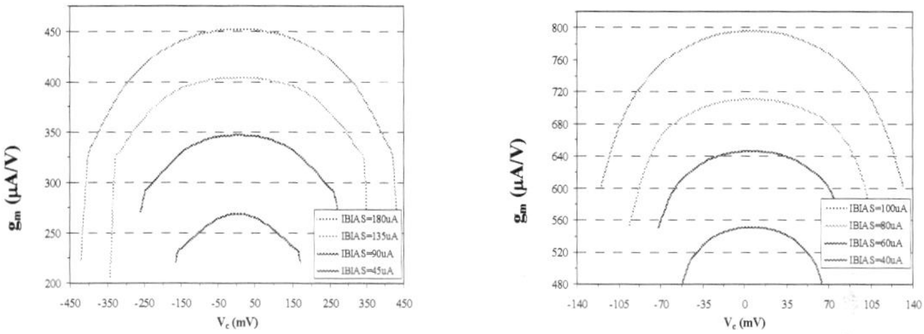

(a) (b)

Fig. 15. Transconductance versus biasisng currents (fine tuning) for the: (a) HS implementation; (b) FC implementation.

To conclude, the proposed structure is a balanced topology aimed at improving immunity to digital noise and linearity. A digitally programmable transconductor has been designed, maintaining the same dynamic range over the entire frequency range. Therefore, it can be used in the design of programmable filters, as the expected characteristics of a programmable cell will be obtained: to maintain Q-factor, noise power and maximum signal swing constant over the entire programming range, leading to a DR independent on the operation frequency. The expected linear dependence of the unity-gain frequency is obtained and the phase error is effectively reduced over the entire programming range in both implementations, with a compensation scheme based on two cross-coupled capacitors for the HS topology and the classical RC circuit connected at the input for the FC approach.

8. Results and discussion

To demonstrate the theoretical advantages of this approach for a programmable transconductor suitable for VHF, two 3-bit programmable integrators have been designed. The HS transconductor has been implemented by using the design kit of an AMI Semiconductor (AMIS) 0.35 μm CMOS technology (P-substrate, N-well, 5-metal, 2-poly) with a 3 V power supply and a nominal bias current of 90 μA per branch; whereas the FC transconductor has been implemented by using the design kit of an AMS (C35B4C3) 0.35 μm CMOS technology (P-substrate, N-well, 4-metal, 2-poly) with a 2 V power supply and a nominal bias current of 100 μA per branch.

The dimensions of the transistors were chosen in order to cover all the design requirements obtained in this chapter, leading to a complete sweep of the discrete step by varying the bias current. In this way, for the HS implementation, the operation point is located at 90 μA and the bias current adjustment is possible from 45-180 μA. However, for the FC implementation, the operating point is located at 100 μA, covering the digital step by varying the bias current from 20-110 μA. In this way, the discrete tunability requirement is obtained but the FC transconductance value at the operation point is maximised.

8.1 Layout strategy

A careful layout has been drawn out to obtain all the characteristics associated with the proposed design accurately and demonstrate the feasibility of the intended approach. As

stated below, we have taken special care to get rid of the unwanted effects related to parasitic elements and mismatching (Baker et al., 1998; Hastings, 2001). All the designs have been carried out taking into account the specific design rules for high frequency operation, which are highly appropriate for obtaining good matching between components. Interdigitized and common-centroid layout techniques have been considered to reduce the variations of threshold voltage, which are associated with gradients in gate-oxide thickness. Guard rings have been included in the design with the aim of reducing substrate noise. Bond-pads have also been carefully laid out and, in this way, input and output pins have been placed as far as possible between them. Balanced structures provide outstanding benefits, but they are strongly dependent on the symmetry of the circuit. Consequently, special care has been taken to outline the paths of the balanced signals, in an attempt to ensure the best matching between them. MOS devices have fragile gates seeing that electrostatic discharges may cause destruction of the device if the oxide breakdown voltage is exceeded. Considering this point, we concluded that it would be advisable to provide the transistors that control the quality factor of the circuit with a path protection system. The scheme chosen to achieve this goal was the anti-parallel diodes configuration. This circuit is very straightforward and simple but is sufficient for the purposes of this work.
Fig. 16(a) shows the drawn layout of the HS test chip with an active area of 0.10 mm². Fig. 16(b) shows the microphotograph of the programmable FC transconductor, with an active area of 0.04 mm² including the compensation RC circuit, where the integration capacitance has been implemented with a double-poly capacitor. The area of the FC active element is 0.03 mm² and a regular and compact arrangement of transistors can be observed.

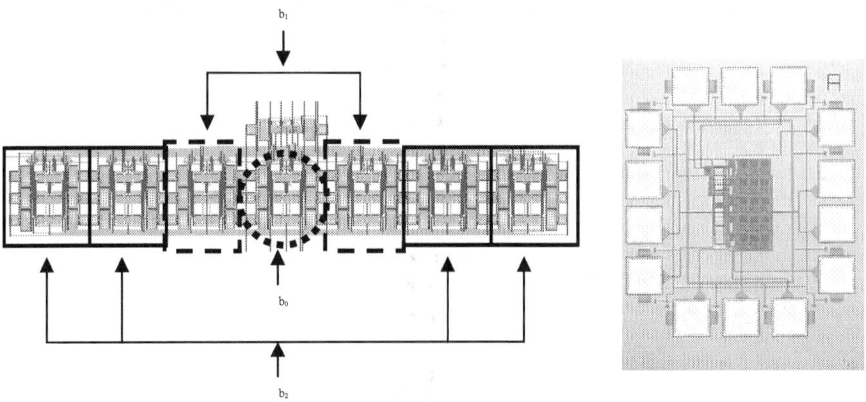

(a) (b)

Fig. 16. (a) Layout of the fully-balanced 3-bit programmable HS integrator.
(b) Microphotograph of the FC integrator, by using double-poly capacitors.

8.2 Experimental results

For the HS approach, a unity-gain frequency of 28 MHz was achieved with a power dissipation of 1.62 mW using a 3 V supply. By varying the digital word from 1 to 7, we expected to control the unity-gain frequency from 28 to 185 MHz and the experimental results lead to a variation between 25 and 185 MHz, as shown in Fig. 17(a). Focusing on the same figure, by varying the bias current source from 45 to 180 μA for a fixed digital word, the transconductance value is modified, providing complementary fine tuning of the frequency. All discrete steps are covered and, in consequence, a frequency span of 25-185 MHz can be provided. The maximum frequency error is obtained at the maximum digital word where a deviation of 6 % is obtained from the 7:1 ratio.

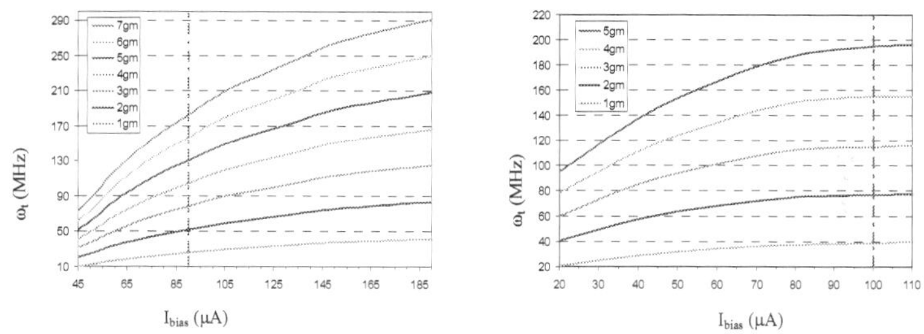

(a) (b)

Fig. 17. Experimental results for coarse and fine tuning of the (a) HS and (b) FC topology. Variation of the unity-gain frequency versus bias currents for all the digital words.

For the FC approach, a unity-gain frequency of 40 MHz is achieved with a power dissipation of 2.4 mW using a 2 V supply, as expected from the post-layout simulation results. By varying the digital word from 1 to 5, the unity-gain frequency is controlled from 40 to 190 MHz, as shown in Fig. 17(b). All discrete steps are swept by varying the bias current from 20 to 110 μA. The maximum frequency error is obtained at the maximum digital word where a deviation of 5 % is obtained from the 5:1 ratio.

The next step is to demonstrate constant linearity by means of a constant THD over the entire programming range. Figs. 18 and 19 show the THD variation as a function of the differential output current for all the digital words. THD was measured for a sine input current of 10 MHz (a) and for the unity-gain frequency (b) in both topologies. These figures show the expected THD dependence, studied above in section §6: lower bias currents or higher input signal amplitudes lead to higher THD values. A corner parameter analysis was carried out following the guidelines provided by the design kit manufacturer of the *'AMI Semiconductor C035M Design-Kit'* and the worst-case analysis for the HS integrator was obtained. This distortion study gave 1 % of THD for a differential input signal of 56 μA and 10 MHz. Experimental results for the design, shown in Fig. 18, lead to a differential input current of 50 μA in the same situation. For the FC approach, the expected value for 1 % of THD was a differential input signal amplitude of 37 μA and 10 MHz; and the experimental results (Fig. 19), give an amplitude of 35 μA.

The post-layout simulated result for the input-referred noise integrated from 0 to 30 MHz in the HS topology was 11.2 nA_{rms}. Hence, the dynamic range, defined as the input signal amplitude at 1 % THD divided by the total noise level integrated over 30 MHz, is 70 dB. In the FC structure, the input-referred noise integrated from 0 to 40 MHz was 8 nA_{rms}. Hence, the dynamic range, defined as the input signal amplitude at 1 % THD divided by the total noise level integrated over 40 MHz, is also 70 dB.

In summary, frequency is adjusted in a coarse discrete way by connecting identical transconductors in parallel and with fine continuous tuning by varying the biasing current.

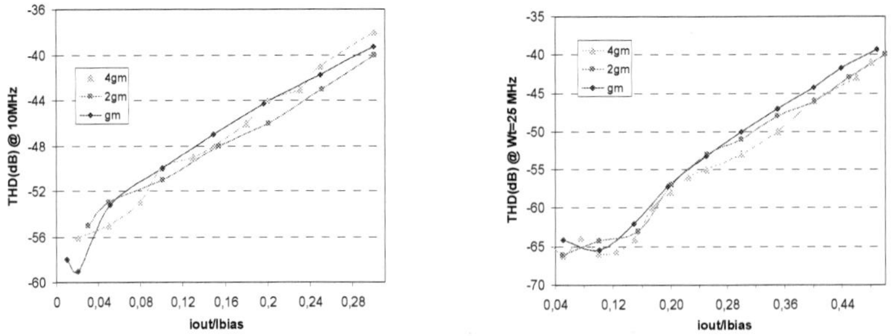

(a) (b)

Fig. 18. THD versus differential output current in the HS integrator for three different digital words: (a) ω(input)=10 MHz, (b) ω(input)= ω_t (25 MHz for $1g_m$).

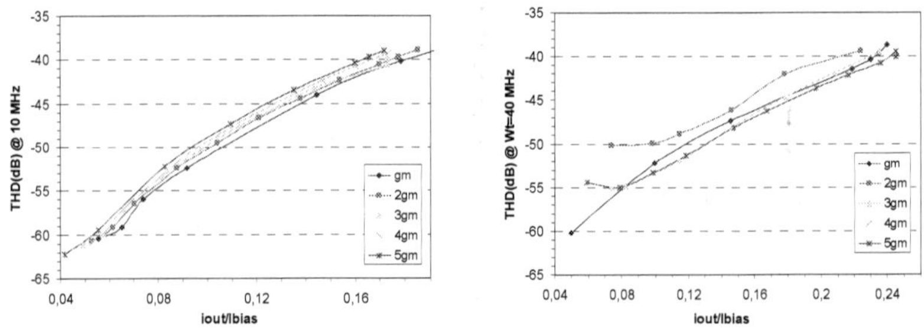

(a) (b)

Fig. 19. THD versus differential output current in the FC integrator for all the digital words: (a) ω(input)=10 MHz, (b) ω(input)= ω_t (40 MHz for $1g_m$).

The feasibility of the programmable array of transconductors has been proven in a 3-bit programmable integrator obtaining frequency scaling as expected. All the specifications in both transconductor implementations are summarized in table 7. The main advantage of the topology proposed was the inherent enhancement of the dc-gain, provided through the existing positive feedback compensation (negative resistance).

The HS design condition was very difficult to achieve because technological process and temperature variations are expected to be greater than the small changes required in this topology. As expected, by varying the external control for this negative resistance, no change was obtained for the dc-gain. The post-layout simulated dc-gain was a variation of 15 dB between the minimum (40 dB) and the maximum (55 dB), with a maximum CMRR of 60 dB. The experimental results lead to a differential dc-gain of 30 dB with no change with the value of the negative resistance and a CMRR greater than 35 dB over the entire frequency range. Therefore, in this case, there is no control on the dc-gain of the system.

The design condition for the FC topology is less restrictive and two different implementations have been fabricated. The post-layout simulation results in both cases showed a dc-gain control of 15 dB from 30 to 45 dB and a maximum CMRR of 50 dB. The first implementation has been designed with the same dimensions for the M_N transistors involved in the negative resistance, and similar results are obtained as in the HS topology. There is no external dc-gain control and an experimental value of 26 dB and CMRR of 33 dB are obtained. In the second one, where a pre-designed mismatching is included between M_N transistors involved in the negative resistance, a variation of 12 dB (from 26 to 38 dB) for the dc-gain is obtained by modifying the value of the negative resistance (Fig. 20). The CMRR is greater than 46 dB over the entire frequency range.

	HS topology	**FC topology**
Power supply voltage	3 V	2 V
Unity-gain frequency	25 MHz	40 MHz
Power dissipation	1.62 mW	2.4 mW
CMRR over the entire pass-band	>35 dB	>46 dB
Active area	0.10 mm^2	0.04 mm^2
Total rms input-referred noise (sim.)	11.2 nA$_{rms}$	8 nA$_{rms}$
Maximum differential input signal current at 1 % THD @ 10 MHz	50 μA (peak)	35 μA (peak)
Dynamic range	70 dB	70 dB

Table 7. Summary of the experimental results for the integrator (1 LSB).

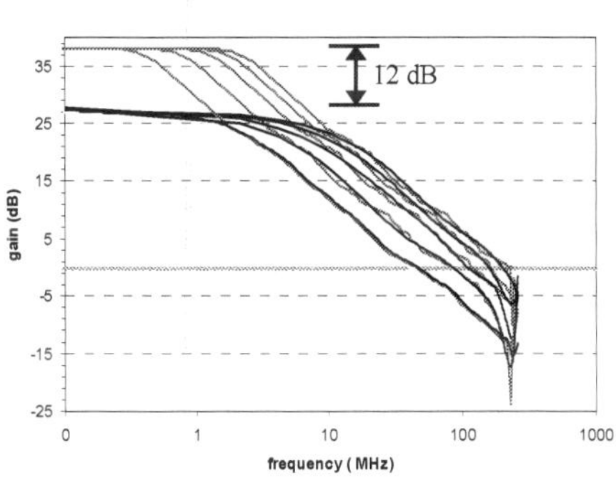

Fig. 20. Experimental dc-gain control for the FC transconductor with a pre-designed mismatching between M_N transistors involved in the negative resistance.

9. Conclusion

This work describes a new approach for implementing digitally programmable and continuously tunable VHF/UHF transconductors compatible with pure digital CMOS technologies and suitable for HDD read channel applications. The cell is suitable for low-voltage operation over an extended frequency range. The programmability exhibited by the transconductor is due to the use of a generic programmable structure that gives a G_m digital control as a parallel connection of unit cells, and the total parasitic capacitances are maintained constant thanks to the specific design of the unit cell: a cascode stage with

dummy elements. This transconductor could be used in any kind of G_m-C filter, thus providing a very wide range of programmable CT filters. The fully-balanced current-mode G_m-C integrator based on this topology exhibits a unity-gain frequency programmability from 25-185 MHz in the HS implementation and 40-200 MHz in the FC approach; with a phase error of less than 4° in both topologies throughout the entire operating frequency range. Total harmonic distortion (THD) of less than 1 % (-40 dB) for a differential input signal of 50 and 35 μA in the HS and FC topology respectively is obtained. The integrator operates over the programming range with 70 dB of dynamic range for 1 % of THD. The cell has been fabricated in a 0.35 μm CMOS process.
The experimental results confirm this approach as an excellent choice to achieve filters exhibiting a good trade-off between tuning capability and dynamic range working in the very high frequency range. The proposed technique can be easily adapted to lower power supply voltages by using folded cascode structures and, in addition, better frequency ranges of operation can be achieved considering current CMOS digital technologies.

10. References

Abidi A. (1988). On the Operation of Cascode Gain Stages. *IEEE Journal of Solid-State Circuits*, Vol. 23, No. 6, 1988, 1434-1437, ISSN: 0018-9200.

Ahn H.T. & Allstot D. J. (2002). A 0.5-8.5 GHz Fully-Differential CMOS Distributed Amplifier. *IEEE Journal of Solid-State Circuits*, Vol. 37, No. 8, August 2002, 985-988, ISSN: 0018-9200.

Baker R.J.; Li H.W. & Boyce D.E. (1998). *CMOS Circuit Design, Layout and Simulation*. IEEE Press Series on Microelectronic Systems, 1998.

Baschirotto A.; Rezzi F. & Castello R. (1994). Low-Voltage Balanced Transconductor with High Input Common-Mode Rejection. *Electronics Letters*, Vol. 30, No. 20, September 1994, 1669-1671, ISSN: 0013-5194.

Bollati G.; Marchese S.; Demicheli M. & Castello R. (2001). An Eight-Order CMOS Low-Pass Filter with 30-120 MHz Tuning Range and Programmable Boost. *IEEE Journal of Solid-State Circuits*, Vol. 36, No. 7, July 2001, 1056-1066, ISSN: 0018-9200.

Croon J.A.; Rosmeulen M.; Decoutere S.; Sansen W. & Maes H.E. (2002). An Easy-to-Use Mismatch Model for the MOS Transistor. *IEEE Journal of Solid-State Circuits*, Vol. 37, No. 8, August 2002, 1056-1064, ISSN: 0018-9200.

Felt E.; Narayan A. & Sangiovanni-Vincentelli A. (1994). Measurement and Modelling of MOS Transistors Current Mismatch in Analog ICs, *Proceedings of the IEEE/ACM International Conference on Computer Aided Design*, pp. 272-277, ISBN: 0-8186-6417-7, San Jose, California, November 1994, Broadway, New York.

Gray P.R. & Meyer R.G. (2001). *Analysis and Design of Analog Integrated Circuits*, 4th Edition, John Wiley & Sons, Inc., 2001.

Gregor R.W. (1992). On the Relationship Between Topography and Transistor Matching in an Analog CMOS Technology. *IEEE Transactions on Electron Devices*, Vol. 39, No. 2, 1992, 275-282, ISSN: 0018-9383.

Hastings A. (2001). *The Art of Analog Layout*, Prentice Hall, Inc., 2001.

Mohan S.S.; Hershenson M.; Boyd S.P. & Lee T.H. (2000). Bandwidth Extension in CMOS with Optimized On-Chip Inductors. *IEEE Journal of Solid-State Circuits*, Vol. 53, No. 3, March 2000, 346-355, ISSN: 0018-9200.

Nauta B. (1993). *Analog CMOS Filters for Very High Frequencies*, Kluwer Academic Publishers, 1993.

Otín A.; Celma S. & Aldea C. (2004). Digitally Programmable CMOS Transconductor for Very High Frequency. *Microelectronics Reliability Journal*, Vol. 44, No. 5, 2004, pp. 869-875, ISSN: 0026-2714.

Otín A.; Celma S. & Aldea C. (2005). A 0.18 μm CMOS 3rd-order Digitally Programmable G_m-C Filter for VHF Applications. *IEICE Transactions on Information and Systems*, Vol. E88-D, No. 7, July 2005, 1509-1510, ISSN: 0916-8532.

Pavan S. & Tsividis Y.P. (2000). *High Frequency Continuous Time Filters in Digital CMOS Processes*, Kluwer Academic Publishers, London, 2000.

Pavan S. & Tsividis Y.P. (2000). Widely Programmable High-Frequency Continuous-Time Filters in Digital CMOS Technology. *IEEE Journal of Solid-State Circuits*, Vol. 35, No. 4, 2000, 503-511, ISSN: 0018-9200.

Pelgrom M.J.M.; Duinmaijer A.C.J. & Welbers A.P.G. (1989). Matching Properties of MOS Transistors. *IEEE Journal of Solid-State Circuits*, Vol. 24, No. 5, October 1989, 1433-1440, ISSN: 0018-9200.

Säckinger E. & Fischer W.C. (2000). A 3 GHz 32 dB CMOS Limiting Amplifier for SONET OC-48 Receivers. *Proceedings of the International Solid-State Circuits Conference, Digest of Technical Papers*, pp. 158-159, ISBN: 0-7803-5855-4, San Francisco CA, February 2000, IEEE Service Center, P.O. Box 1331, Piscataway.

Sansen W.; Huijsing J. & De Plassche R. (1999). *Analog Circuit Design*, Kluwer Academic Publishers, 1999.

Sedra A.S. & Smith K.C. (2004). *Microelectronic Circuits*, Fifth-Edition, Oxford University Press, Inc., New York, 2004.

Silva-Martínez J.; Steyaert M. & Sansen W. (2003). *High-Performance CMOS Continuous-Time Filters*, Kluwer Academic Publishers, 2003.

Smith S.L. & Sánchez-Sinencio E. (1996). Low Voltage Integrators for High-Frequency CMOS Filters using Current Mode Techniques. *IEEE Trans. on Circuits and Systems II: Analog and Digital Signal Processing*, Vol. 43, No.1, 1996, 39-48, ISSN: 1057-7130.

Tsividis Y.P. (1996). *Mixed Analog Digital VLSI Devices and Technology*, McGraw-Hill, New York, 1996.

Tsividis Y.P. (1999). *Operation and Modeling of the MOS Transistor*, 2nd Edition, McGraw-Hill, New York, 1999.

Vadipour M. (1993). A New Compensation Technique for Resistive Level Shifters. *IEEE Journal of Solid-State Circuits*, Vol. 28, No. 1, January 1993, 93-95, ISSN: 0018-9200.

Wakimoto T. & Akazawa Y. (1990). A Low-Power Wide-Band Amplifier Using a New Parasitic Capacitance Compensation Scheme. *IEEE Journal of Solid-State Circuits*, Vol. 25, No. 1, February 1990, 200-206, ISSN: 0018-9200.

Wyszynski A. & Schaumann R. (1994). Avoiding Common-Mode Feedback in Continuous-Time G_m-C Filters by the Use of Lossy-Integrators. *Proceedings of the IEEE*

International Symposium on Circuits and Systems, Vol. 5, pp. 281, Vancouver (Canada), May 1994.

Zele R.H. & Allstot D. (1996). Low-Power CMOS Continuous-Time Filters. *IEEE Journal of Solid-State Circuits*, Vol. 31, No. 2, 1996, 157-168, ISSN: 0018-9200.

9

Impact of Technology Scaling on Phase-Change Memory Performance

Stefania Braga, Alessandro Cabrini and Guido Torelli
Department of Electronics, University of Pavia
Italy

1. Introduction

Nowadays, non-volatile storage technologies play a fundamental role in the semiconductor memory market due to the widespread use of portable devices such as digital cameras, MP3 players, smartphones, and personal computers, which require ever increasing memory capacity to improve their performance. Although, at present, Flash memory is by far the dominant semiconductor non-volatile storage technology, the aggressive scaling aiming at reducing the cost per bit has recently brought the floating-gate storage concept to its technological limit. In fact, data retention and reliability of floating-gate based memories are related to the thickness of the gate oxide, which becomes thinner and thinner with increasing downscaling. The above limit has pushed the semiconductor industry to invest on alternatives to Flash memory technology, such as magnetic memories, ferroelectric memories, and phase change memories (PCMs) (Geppert, 2003). The last technology is one of the most interesting candidates due to high read/write speed, bit-level alterability, high data retention, high endurance, good compatibility with CMOS fabrication process, and potential of better scalability. However, it still requires strong efforts to be optimized in order to compete with Flash technology from the cost and the performance points of view.

In PCMs, information is stored by exploiting two different solid-state phases (namely, the amorphous and the crystalline phase) of a chalcogenide alloy, which have different electrical resistivity (more specifically, the resistivity is higher for the amorphous, or RESET, phase and lower for the crystalline, or SET, phase). Phase transition is a reversible phenomenon, which is achieved by stimulating the cell by means of adequate thermal pulses induced by applying electrical pulses. Reading the resistance of any programmed cell is achieved by sensing the current flowing through the chalcogenide alloy under predetermined bias voltage conditions. The read window, that is, the range from the minimum (RESET) to the maximum (SET) read current, is considerably wide, which allows safe storage of an information bit in the cell and also opens the way to the multi-level approach to achieve low-cost high-density storage. ML storage consists in programming the memory cell to one in a plurality of intermediate resistance (i.e., of read current) levels inside the available window, which allows storing more than one bit per cell (the number of bits that can be stored in a single cell is $n = log_2 m$, where m is the number of programmable levels). The programming power and the read window depend on the electrical properties of the cell materials as well as on the architecture and the size of the memory cell. As the fabrication technology scales down the cell dimensions, new challenges arise to accurately program the cell to intermediate states and discriminate adjacent resistance levels.

In this work, we investigate the impact of technology scaling down on both the program and the read operation by means of a simple analytical model which takes the electro-thermal behavior of the PCM cell and the phase change phenomena inside the chalcogenide alloy into account.

2. Working principle of the PCM cell

The working principle of a PCM cell relies on the physical properties of chalcogenide materials, typically $Ge_2Sb_2Te_5$ (GST), that can switch from the amorphous to the crystalline phase and vice versa when stimulated by suitable electrical pulses. Basically, a PCM cell is composed of a thin GST film, a resistive element named heater (*TiN*), and two metal electrodes, i.e., the top electrode contact (TEC) and the bottom electrode contact (BEC). Only a portion of the GST layer, which is located close to the GST-heater interface and is referred to as active GST, undergoes phase transition when the PCM cell is thermally stimulated. In particular, in this work we focus our attention on the Lance heater geometry (Pellizzer et al., 2006), which is essentially composed of a thin layer of GST alloy and a pillar-shaped heater, as shown in Fig. 1. In the reference Lance heater cell implemented in the 90 *nm* technology node, the GST thickness t is 70 *nm*, the GST-heater contact area A is 3000 nm^2, and the heater height h is 180 *nm*.

The typical V-I characteristic of the PCM cell in the amorphous (RESET) and the crystalline (SET) state is shown in Fig. 2. Consider the case of a cell in its full-SET state: the differential resistance of the cell decreases as the applied voltage increases. This effect is due to the contribution of the crystalline GST to the cell resistance. In fact, the crystalline GST resistivity decreases with increasing electrical field inside the material.

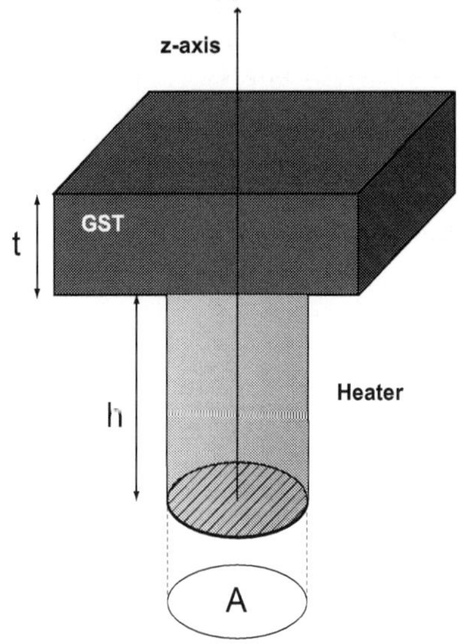

Fig. 1. Conceptual scheme of a PCM Lance heater cell.

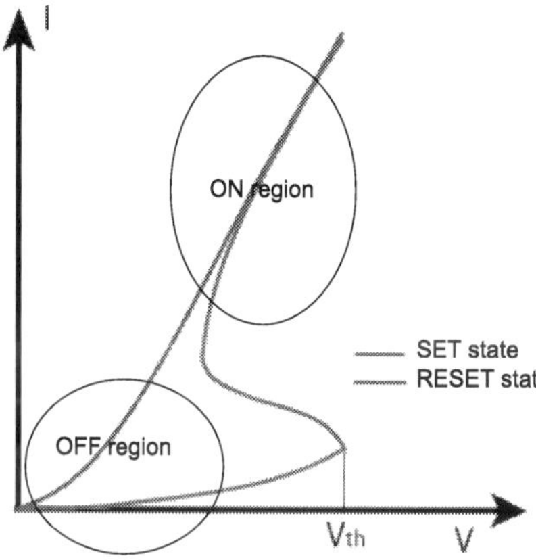

Fig. 2. V-I curve of a PCM device in the SET and the RESET state.

The V-I curve of the cell in its RESET state shows an S-shaped behavior. This effect is due to the threshold switching phenomenon (Adler et al., 1980; Ovshinsky, 1968; Pirovano et al., 2004; Thomas et al., 1976) which consists in a sudden drop of the amorphous GST resistivity as the voltage across the PCM cell exceeds a critical value, typically referred to as threshold voltage, V_{th}. Thus, when low-amplitude voltage pulses are applied to the cell, a low current flows through the device, which is in its high-resistance state (OFF region in Fig. 2). On the other hand, when a high-amplitude voltage pulse is applied to the cell, threshold switching takes place and the device shows a much lower resistance (ON region in Fig. 2). It can be noted that the V-I curves of the cell in the two states (SET and RESET) are almost superimposed in the ON region, while they are substantially different in the OFF region. Thus, readout must be carried out by operating the cell in the OFF region. Typically, a predetermined read voltage is applied to the cell and the current flowing through the device, referred to as read current, is sensed (current sensing approach). The read voltage must be low enough to avoid unintentional modification of the cell contents due to the read pulse. On the other hand, writing is carried out by operating the cell in the ON region, in order to provide the device with enough energy to induce phase change. Since phase transitions are thermally assisted, in PCM devices Joule heating is exploited to raise the temperature inside the chalcogenide material to the required value. The crystalline-to-amorphous phase transition is obtained by applying a high-amplitude electrical pulse to the cell so as to bring the temperature of the active GST material above the melting point T_m (about 600 °C) (Peng et al., 1997), and then quickly cooling the memory cell, in order to freeze the GST material into a disordered (i.e., amorphous) structure. A pulse duration on the order of few tenths of ns is sufficient (Weidenhof et al., 2000). The amorphous-to-crystalline phase transition is obtained by applying an electrical pulse with a lower amplitude and a longer time duration. In this case, the amorphous material is heated to a temperature below the melting point but above the crystallization temperature, that is the temperature necessary to activate the crystallization process in the required time scale (typically an the order of 100 ns). This way, the thermal energy is able to restore the crystalline lattice, which is a minimum-energy configuration. Typical electrical pulses for SET and RESET operations are shown in Fig. 3.

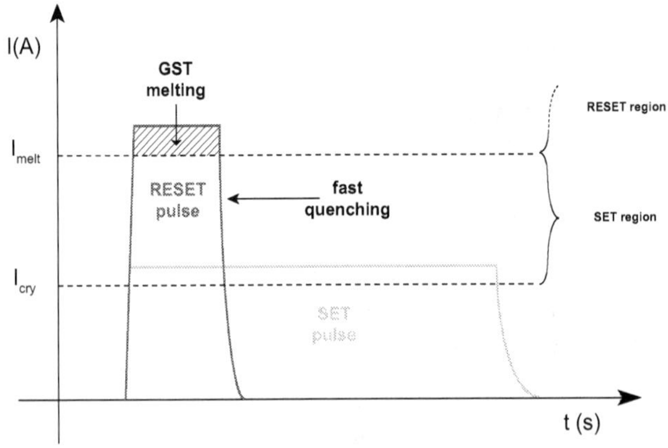

Fig. 3. Standard pulses for bi-level PCM programming.

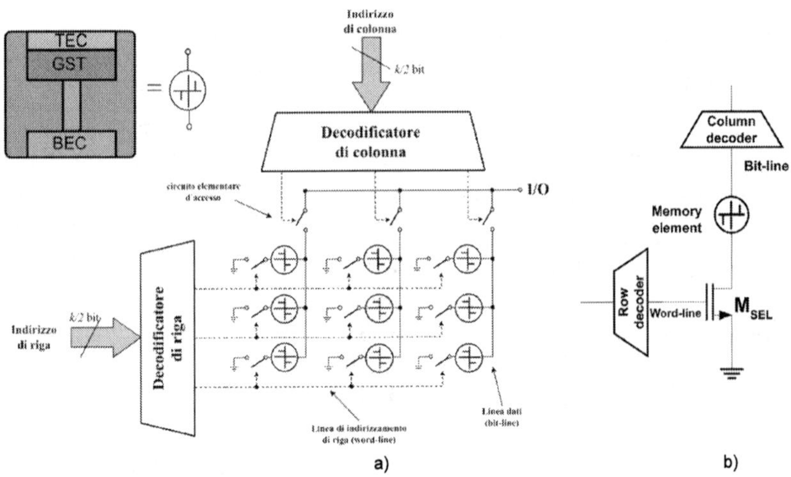

Fig. 4. Architecture of a PCM matrix (a) and schematic of the circuit used to program and read the memory cell (b). Transistors M_{SEL} is the row select transistor.

A PCM memory chip is made of a large number of PCM cells organized in a bi-dimensional array. As opposed to the case of Flash memories, in which the elementary storage consists of a floating-gate transistor, the PCM memory cell is a programmable resistor and, hence, is a two-terminal device. For this reason, a NOR type architecture is adopted (Fig. 4a). As shown in Fig. 4b, each memory cell consists of a PCM storage element connected to a selection transistor M_{SEL} which can be either an MOS or a bipolar device. The gate or the base of all select transistors of the same row are connected to the same word-line, while the TECs of the PCM cells belonging to the same column are connected to the same bit-line. The memory cell is selected by means of row and column decoders that generate the electrical control signals required for read and write operations.

3. Programming operation

We analyzed first the impact of technology scaling on the programming operation, focusing our attention on the electical power (hereinafter referred to as programming power). The maximum programming power is obviously required by the RESET operation, where the highest temperatures are needed to melt the active GST volume. The RESET pulse duration must be higher than the minimum required time for melting \cite{Weidenhof00}, while the cooling time must be short enough to prevent the crystallization process from taking place. The minimum current required to melt a portion of the active GST layer is referred to as melting current, I_m. When the current flowing through the memory cell during a write operation is higher than I_m, the obtained RESET resistance increases with the amplitude of the current pulse. In fact, the maximum temperature inside the cell increases with the pulse amplitude, thus leading to the amorphization of a larger GST volume.

The maximum temperature reached inside a Lance heater cell of given sizes can be estimated by means of an approximated electro-thermal model. In general, the temperature increase in the active GST volume is due to the current flow both through the heater (heater heating) and through the GST layer itself (GST self-heating). Nevertheless, GST self-heating can be neglected when considering high-amplitude RESET pulses. In fact, the resistance of the GST layer (both in the crystalline and in the amorphous state) is negligible with respect to the heater resistance due to high-field effects (the PCM cell is operated in the ON region). Thus, in this case we can estimate the temperature profile inside the PCM cell by considering only the Joule power generated inside the heater when a current I flows through the cell. We assume, for simplicity, a cylindrical geometry of the heater and calculate the temperature along the cell axis. The power generated in a volume $A\delta\tilde{z}$ located at a distance $\tilde{z}$ from the heater-BEC contact is equal to $\delta Q = \frac{I^2\rho_h}{A}\delta\tilde{z}$, ρ_h being the heater electrical resistivity, and contributes to the temperature increase ΔT at the heater-GST interface with a term δT given by

$$\delta T = \left[(R_{th,GST} + R_u(\tilde{z})) \,||\, R_d(\tilde{z})\right]\frac{R_{th,GST}}{R_{th,GST} + R_u(\tilde{z})}\delta Q, \tag{1}$$

where $R_u(\tilde{z}) = \frac{h-\tilde{z}}{\kappa_h A}$ and $R_d(\tilde{z}) = \frac{\tilde{z}}{\kappa_h A}$ (κ_h being the thermal conductivity of the heater material) are the heater thermal resistance from the coordinate $\tilde{z}$ to the heater-GST contact and to the heater-BEC contact, respectively, and $R_{th,GST}$ is the equivalent thermal resistance of the GST layer.

By integrating Eq. (1) along the cell axis from the BEC-heater contact ($\tilde{z}$ = 0) to the heater-GST contact ($\tilde{z}$ = h), we obtain the temperature T at the interface:

$$T = \frac{I^2 \rho_h h}{A} \cdot \frac{R_{th,GST} R_{th,h}}{2(R_{th,GST} + R_{th,h})} + T_0 \tag{2}$$

$$= Q_J \frac{1}{2}(R_{th,GST} \parallel R_{th,h}) + T_0 \tag{3}$$

In the above equations, T_0 is room temperature, $Q_J = \frac{I^2 \rho_h h}{A}$ is the Joule power delivered to the cell during the RESET pulse, and $R_{th,h}$ the thermal resistance of the heater, which can be expressed as $\frac{h}{\kappa_h A}$.

From Eq. (2), taking the expression of $R_{th,h}$ into account, I_m is given by

$$I_m = \sqrt{2 \cdot \frac{(T_m - T_0)}{\rho_h k_h} \frac{(R_{th,GST} + R_{th,h})}{R_{th,GST} R_{th,h}^2}}. \tag{4}$$

In order to estimate the dependence of $R_{th,GST}$ on the geometrical features of the memory cell, we simulated the temperature profile along the cell axis inside the GST layer (Fig. 5a). Fig. 5b shows the simulation results for different values of the GST layer thickness obtained with our previously proposed 3D model (Braga et al., 2008). It can be noticed that the temperature decreases almost linearly inside the GST layer with increasing distance from the GST-heater contact. Moreover, the accuracy of the linear approximation increases as the ratio between the GST layer thickness and the heater radius decreases. Since this behavior suggests that heat flow inside the GST is substantially directed along the cell axis, from the heater-GST interface along the cell axis, a reasonable approximation for the thermal resistance of the GST layer is $R_{th,GST} = \frac{t}{\kappa_{GST} A}$, where κ_{GST} is the thermal conductivity of the GST. Thus, we can rewrite Eq. (4) as

$$I_m = \frac{A}{h} \sqrt{2 \cdot \frac{(T_m - T_0)}{\rho_h} \left(\kappa_h + \kappa_{GST} \frac{h}{t} \right)}. \tag{5}$$

As highlighted by Eq. (5), the melting current depends on the ratios $\frac{A}{h}$ and $\frac{h}{t}$.

Due to fabrication process constraints, heater geometries with a high aspect ratio (i.e., geometries having a high ratio between the GST-heater contact diameter and the heater height), may not be easily manufacturable. Several fabrication solutions have been proposed to overcome lithographic limits and, thus, realize heater structures with minimized contact area (Lam, 2006; Pirovano et al., 2008). In the following, we will consider heater geometries with a high aspect ratio with the purpose of investigating the scaling perspective, even if they may require advanced fabrication techniques. Given a scaling factor $\epsilon < 1$, I_m turns out to be proportional to ϵ in the case of isotropic scaling, where all the linear dimensions are scaled by the same amount, while $I_m \propto \epsilon^2$ in the case of shrinking, where only planar dimensions are scaled. The comparison of melting current reduction in the cases of isotropic scaling and shrinking is shown in Fig. 6.

In order to compare PCM cells having different dimensions, we chose to consider the full-RESET state to be achieved when the maximum temperature inside the PCM cell reaches a

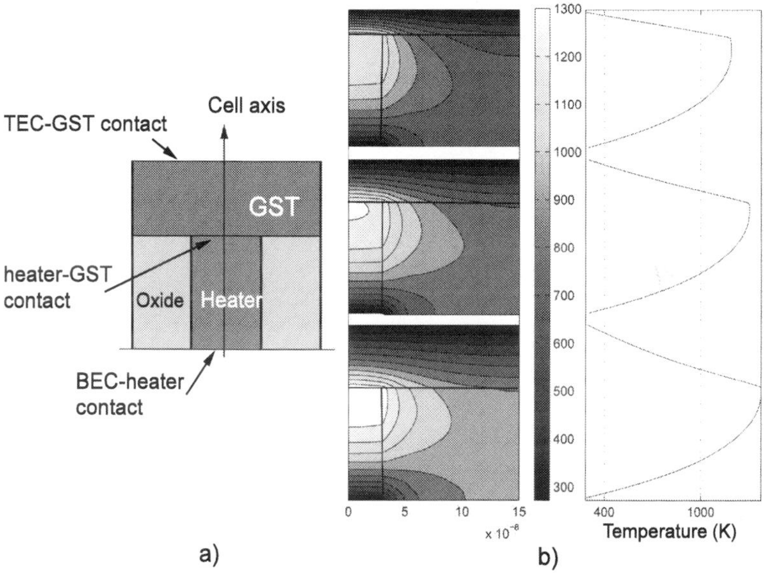

Fig. 5. Cell structure (a) and simulated temperature Maps inside a Lance heater PCM cell with different values of GST layer thickness: 40 nm, 70 nm, and 100 nm (b). Notice that the temperature profile is almost linear inside the GST layer. The maps were obtained by means of our 3D electro-thermal model (Braga et al., 2008).

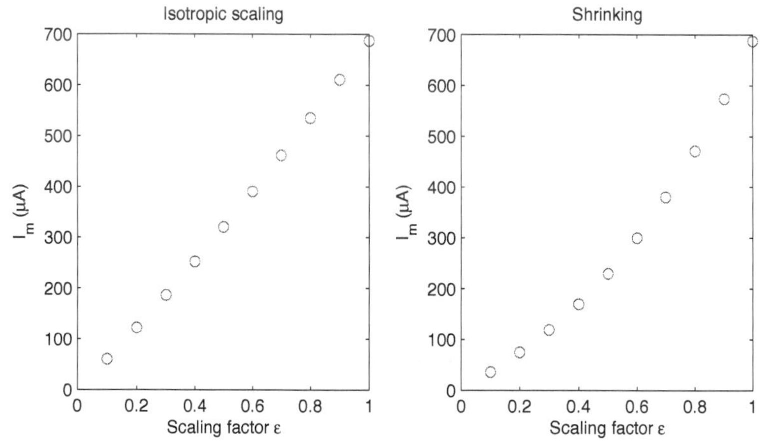

Fig. 6. Melting current reduction in the case of isotropic scaling (left) and shrinking (right). The dimensions are scaled with respect to a reference lance heater cell realized in 90 nm technology

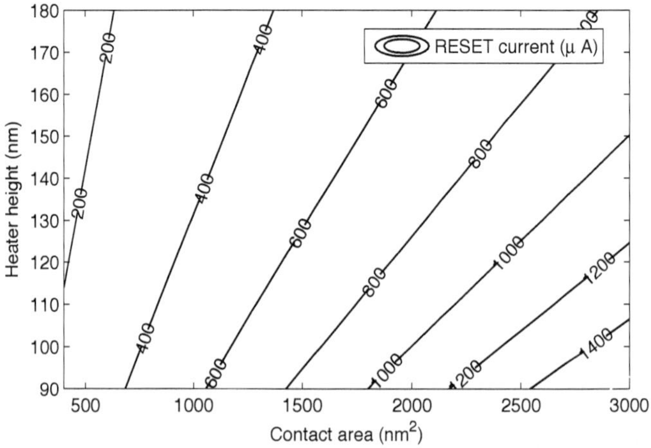

Fig. 7. Map of the RESET current as a function of the GST-heater contact area and the heater height (the GST layer thickness was set to 70 nm).

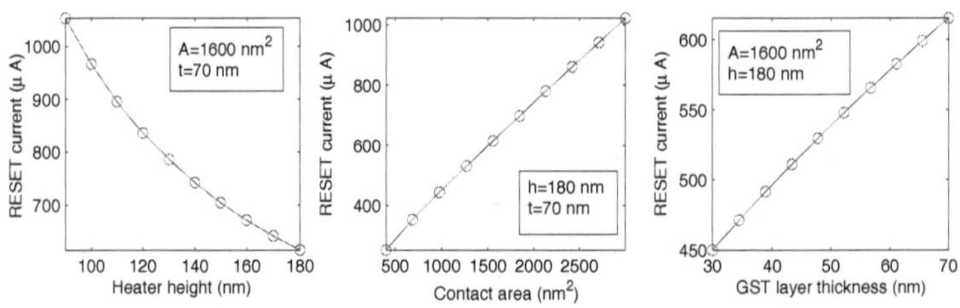

Fig. 8. RESET current dependence on the geometrical parameters of the memory cell.

predetermined value, T_{RST}, which is obtained with a current pulse of amplitude I_{RST}. Typically, I_{RST} is 50% higher than I_m. Different cells require different pulse amplitudes (I_{RST}) to reach T_{RST}, due to the different values of the electrical and the thermal resistance of the device. The dependence of the RESET current on cell sizes obtained by means of Eq. (4) is sketched in Fig. 7 and Fig. 8. The reduction of the heater height leads to a significant increase of I_{RST} due to the decrease of the Joule power and heater thermal resistance. On the contrary, the reduction of the contact area only, that is the shrinking approach, leads to a linear decrease of the RESET current, due to the increase of the Joule power and the thermal resistance of the cell. The same behavior is obtained when considering the scaling of the GST layer thickness.

The values of the electrical and thermal properties used in the above simulations are summarized in Tab. 1. For simplicity, the field dependence of the crystalline GST resistivity was neglected. In order to validate the described analytical compact model, we compared the temperature profiles along the cell axis obtained with this model and our 3D finite-element model (Fig. 9).

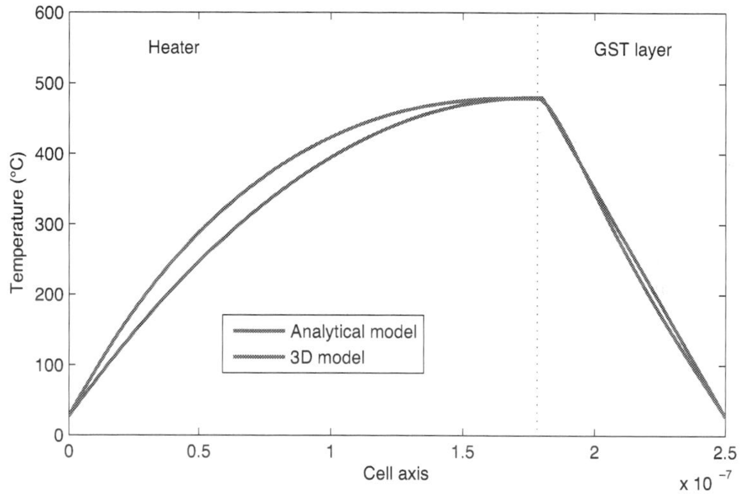

Fig. 9. Comparison of the thermal profile along the cell axis obtained by means of the analytical model and the 3D finite-element model.

Heater thermal conductivity	κ_h	$36\ \frac{W}{m^\circ C}$
GST layer thermal conductivity	κ_{GST}	$0.5\ \frac{W}{m^\circ C}$
Heater electrical resistivity	ρ_h	$30\ \mu\Omega m$
Cryst. GST electrical resist.	ρ_C	$0.1m\ \Omega m$
Amorph. GST electrical resist.	ρ_A	$10m\ \Omega m$

Table 1. Electrical and Thermal Properties of Cell Materials-

A good agreement is observed especially inside the GST layer. The slight temperature disagreement inside the heater is ascribed to the inhomogeneous heat flow in the material that surrounds the heater. To take this thermal evacuation contribution into account, the value of κ_h used in the compact model was set higher than the actual physical value.

4. Read operation

The GST layer undergoes crystalline to amorphous phase transition in the region where the temperature exceeds the melting point. As pointed out above, the temperature profile along the cell axis inside the GST decreases almost linearly with the distance from the GST-heater interface. By approximating the thermal profile inside the GST along the cell axis with a straight line, we derived the analytical expression for the thickness of the amorphous cap x_a obtained when a full-RESET pulse is applied to the cell:

$$x_a = t\frac{(T_{RST} - T_m)}{T_{RST} - T_0}. \tag{6}$$

Thus, the thickness of the amorphous cap obtained by means of the RESET operation is a fraction $f = \frac{(T_{RST} - T_m)}{T_{RST} - T_0}$ of the GST layer thickness (Braga et al., 2009). The volume of amorphous GST determines the value of the GST resistance in the RESET state and, thus, the lower edge of the read window. Since the temperature gradient is much higher along the cell axis than along the other two axis, the ratio between the thickness and the width of the amorphous cap is quite high, thus allowing us to estimate the amorphous GST resistance in the full-RESET state as

$$R_{RST} = \rho_A \frac{ft}{A} + R_h \approx \rho_A \frac{ft}{A}, \tag{7}$$

where ρ_A is the amorphous GST resistivity and R_h has been neglected since it is much lower than the resistance of the GST layer after the full-RESET pulse.
In order to estimate the cell resistance in the full-SET state, by neglecting the current spread inside the crystalline GST, we can write:

$$R_{SET} = \frac{\rho_C t}{A} + R_h, \tag{8}$$

where ρ_C is the resistivity of crystalline GST.
When considering the current sensing approach, we can calculate the minimum and the maximum read current:

$$I_{rd,min} = \frac{V_{read}}{R_{RST}}, \tag{9}$$

$$I_{rd,max} = \frac{V_{read}}{R_{SET}}, \tag{10}$$

where V_{read} is the amplitude of the read voltage. V_{read} must be lower enough to avoid unintended programming during readout. The read current window is affected by both the scaling of V_{read} and the geometrical scaling strategy. It must be pointed out that when V_{read} is kept constant (this approach will be referred to as constant voltage approach), the electrical field E_{read} during readout inside the amorphous GST increases as the size of amorphous cap scales ($E_{read} \approx \frac{V_{read}}{ft}$), thus impacting on the electrical resistivity of the amorphous GST. In this case, in order to calculate the read current, the exponential dependence of the amorphous GST resistance on the electrical field must be taken into account (Ielmini & Zhang, 2007; Kim et al., 2007). For a given PCM cell in the RESET state, neglecting the heater resistance, we have

$$R_{RST} \propto e^{-\frac{E_{read}}{E_{ref}}}, \tag{11}$$

where E_{ref} is the electrical field which activates the electrical resistivity inside the amorphous GST. The value of V_{read} must be chosen so as to ensure that the PCM device is operated in the read region (OFF zone) and the electrical field during readout is below the critical switching field for every considered cell size. In this respect, we chose $V_{read} = 0.3\ V$ and calculated the cell resistance and the read current for both the SET and the RESET state. E_{ref} was set to $30M\ V/m$ (Buckley & Holmberg, 1974).

Several studies (Adler et al., 1980; Buckley & Holmberg, 1974) have shown that V_{th} decreases linearly with the amorphous GST thickness which, in our case, is a fraction of the GST layer thickness. Then, we can scale V_{read} and t consistently, so as to keep the electrical field during readout inside the amorphous GST roughly constant and below the critical value for threshold switching (Buckley & Holmberg, 1974). This scaling approach will be referred to as constant field scaling.

It can be noticed from the simulation results in Fig. 10, that constant voltage approach leads to an increase of the SET read current as the thickness of the GST layer decreases, due to the reduction of the SET resistance. Moreover, a significant increase of the minimum current (RESET state), mainly due to the dependence of amorphous GST resistivity on the electrical field, is apparent. The increase of the RESET read current depends on E_{ref} and is affected by the value of V_{read}. Rather different results are obtained when considering constant

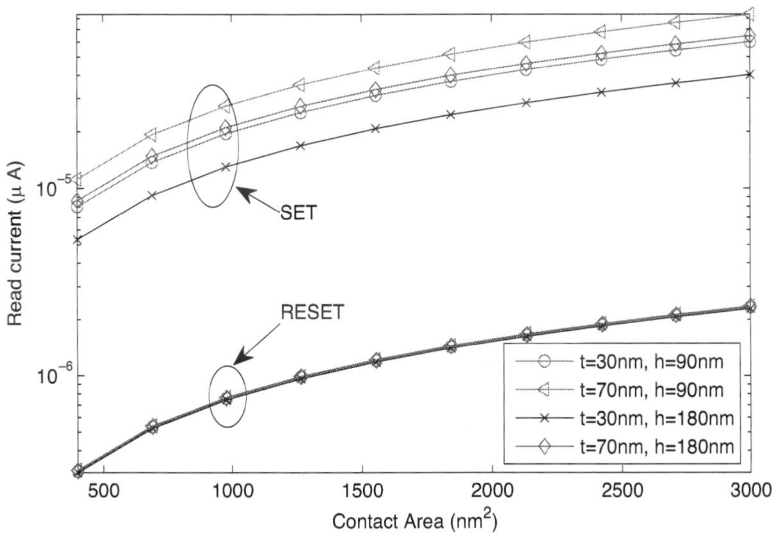

Fig. 10. Constant voltage approach: read current as a function of the contact area A for different values of GST layer thickness t and heater height h. The read voltage is assumed to be 0.3 V.

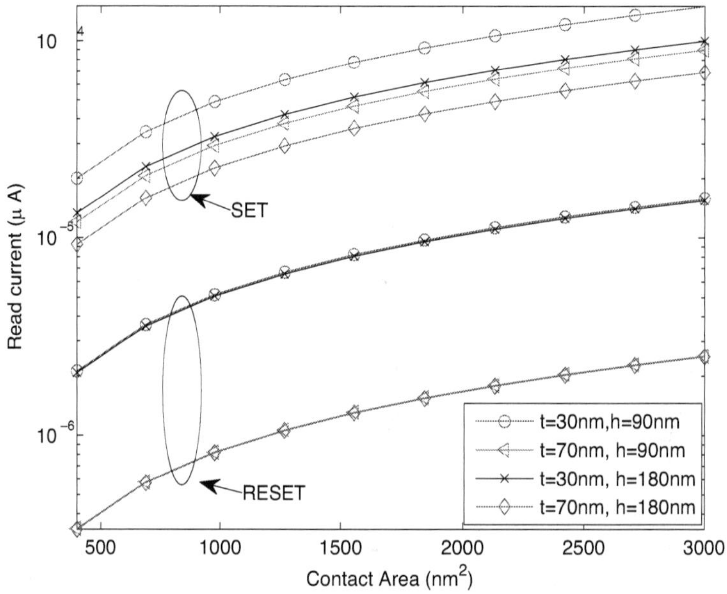

Fig. 11. Constant field approach: read current as a function of the contact area A for different values of GST layer thickness t and heater height h. The read voltage is assumed to be proportional to the thickness of the GST layer (V_{read}= 0.3 V @ t=70 nm).

field scaling. In this case, the current read window scales as shown in Fig. 11. The RESET current is almost independent on t and h, since the read voltage and the cell resistance roughly scale by the same factor. As opposite to the previous approach, in constant field scaling the SET read current decreases with decreasing t due to the fact that R_{SET} is less affected than V_{read} by the reduction of t. The dependence of I_{read} on the contact area is qualitatively similar to the constant voltage case. In both approaches, I_{read} progressively decreases with decreasing A.

5. Conclusions

In this work, we addressed the impact of technology scaling on the performance of phase change memory cells by investigating its effects on both the programming current and the width of the read window. To this end we derived a simplified analytical model of the PCM cell electro-thermal behavior and validate it by means of a 3D finite-elements model of the PCM cell. We considered both constant field and constant voltage scaling approaches. Our study highlights the program-read tradeoffs challenges which aggressive scaling arises and provides analytical insight in the scaling mechanisms.

6. Acknowledgements

This work has been supported by Italian MIUR in the frame of its National FIRB Project RBAP06L4S5.

7. References

Adler, D., Shur, M. S., Silver, M. & Ovshinsky, S. R. (1980). Threshold switching in chalcogenide-glass thin films, *Journal of Applied Physics* 51(6): 3289-3309.

Braga, S., Cabrini, A. & Torelli, G. (2008). An integrated multi-physics approach to the modeling of a phase-change memory device, *Proc. of Solid-State Device Research Conference*, pp. 154–157.

Braga, S., Cabrini, A. & Torelli, G. (2009). Theoretical analysis of the RESET operation in phase-change memories, Semiconductor Science and Technology, 24 (11) 115008 (6pp).

Buckley, W. D. & Holmberg, S. H. (1974). Evidence for critical-field switching in amorphous semiconductor materials, *Phys. Rev. Lett.* 32(25): 1429–1432.

Geppert, L. (2003). The new indelible memories, *IEEE Spectrum* 40(3): 48–54.

Ielmini, D. & Zhang, Y. (2007). Evidence for trap-limited transport in the subthreshold conduction regime of chalcogenide glasses, *Applied Physics Letters* 90(19): 192102.

Kim, D.-H., Merget, F., Först, M. & Kurz, H. (2007). Three-dimensional simulation model of switching dynamics in phase change random access memory cells, *Journal of Applied Physics* 101(6): 064512.

Happ, T.D., Breitwisch, M., Schrott, A., Philipp, J.B., Lee, M.H., Cheek, R., Nirschl, T., Lamorey, M., Ho, C.H., Chen, S.H., Chen, C.F., Joseph, E., Zaidi, S., Burr, G.W., Yee, B., Chen, Y. C., Raoux, S., Lung, H.L., Bergmann, R., Lam, C. (2006). Novel One-Mask Self-Heating Pillar Phase Change Memory, Symposium on VLSI Technology pp. 120–121.

Ovshinsky, S. (1968). Reversible electrical switching phenomena in disordered structures, *Physical Review Letters* 21(20): 1450–1453.

Pellizzer, F., Benvenuti, A., Gleixner, B., Kim, Y., Johnson, B., Magistretti, M., Marangon, T., Pirovano, A., Bez, R. & Atwood, G. (2006). A 90 nm phase change memory technology for stand-alone non-volatile memory applications, *IEEE Symposium on VLSI Technology* pp. 122–123.

Peng, C., Cheng, L. & Mansuripur, M. (1997). Experimental and theoretical investigations of laser-induced crystallization and amorphization in phase-change optical recording media, *Journal of Applied Physics* 82(9): 4183–4191.

Pirovano, A., Lacaita, A. L., Benvenuti, A., Pellizzer, F. & Bez, R. (2004). Electronic switching in phase-change memories, *IEEE Transaction on Electron Devices* 51(3): 452–459.

Pirovano, A., Pellizzer, F., Tortorelli, I., Riganó, A., Harrigan, R., Magistretti, M., Petruzza, P., Varesi, E., Redaelli, A., Erbetta, D., Marangon, T., Bedeschi, F., Fackenthal, R., Atwood, G. & Bez, R. (2008). Phase-change memory technology with selfaligned μtrench cell architecture for 90ănm node and beyond, *Solid-State Electronics* 52(9): 1467 - 1472.

Thomas, C. B., Rogers, B. D. & Lettington, A. H. (1976). Monostable switching in amorphous chalcogenide semiconductors, *Journal of Physics D: Applied Physics* 9(18): 2571–2586.

Weidenhof, V., Pirch, N., Friedrich, I., Ziegler, S. &Wuttig, M. (2000). Minimum time for laser induced amorphization of $Ge_2Sb_2Te_5$ films, *Journal of Applied Physics* 88(2): 657–664.

10

Advanced Simulation for ESD Protection Elements

Yan Han and Koubao Ding
ZJU-UCF Joint ESD Lab, Zhejiang University, Hangzhou 310027, P.R.China

1. Introduction

Electrostatic discharge (ESD) failure is one of the most important causes of reliability problems, therefore the design and optimization of ESD devices have to be done. To achieve very short time to market and reduce the development effort, one tries to make use of the benefit of simulation tools. However, due to the complex physical mechanism of ESD events and the hard mathematic calculation in the snapback region, simulation of the I-V characteristic of ESD protection devices has been proved to be difficult.

This chapter aims at providing a systematic way to ESD simulation, including the process simulation, device simulation and circuit level simulation. Process/device simulation offers an effective way to evaluate the performance of ESD protection structures. However, to prevent the injury of ESD, protection circuits are used sometimes. Therefore circuit level simulation is needed.

There are several process/device simulation tools in the world, the most widely used of which include Tsuprem4/Medici, Athena/Atlas and Dios/Mdraw/Dessis. Tsuprem4, Athena and Dios are process simulators, while Medici, Atlas and Dessis are device simulators. Mdraw is an independent mesh optimization tool, and the similar functions are integrated in device simulation tools, such as Medici and Atlas. The process and device simulation methods introduced in the following will be based on Dios/Mdraw/Dessis, except for the mixed-mode simulation, which is based on Tsuprem4/Medici. And the circuit level simulation will be carried out on the Candence platform.

2. Process simulation

The starting point of ESD simulation is to construct an electronic pattern of the device which can be generated by manual device set-up or process simulation. And obviously, process simulation provides more realistic description of the device. The principle of process simulation is to minimize the errors that might be brought into the following device simulation. Therefore, the physical models used should be carefully chosen. The most important process steps are implantation and diffusion which will be discussed in the following.

Taking Dios for example, this section will introduce physical models used for implantation and diffusion. The implantation models used in Dios consists of analytic implantation models and Monte Carlo implantation model. Monte Carlo implantation model simulates at the atomic level, and it consumes too much time, therefore, in most cases, it is not suitable for ESD simulation. Analytic implantation models are analyzed by series of distribution functions, including Gauss distribution function, Pearson distribution function, Pearson-IV distribution function (P4), Pearson- IV distribution with linear exponential tail function (P4S), Pearson- IV distribution with general exponential tail function (P4K), Gauss distribution with general exponential tail function (GK), Jointed half-Gauss distribution function (JHG), Jointed half-Gauss distribution with general exponential tail function (JHGK). The eight distribution functions are called single primary distribution functions. The complicated expressions of the functions will not be discussed here, and all of them can be found in the DIOS USER'S MANUAL.

The single primary distribution functions describe the relationship between impurity distribution and seven key parameters, which are determined by implantation process step. The seven key parameters are RP (Rp), STDV (σp), STDVSec (σp2), GAMma (γ), BETA (β), LEXP (lexp), LEXPOW (α). The range of parameters that must be specified for each of the single primary distribution functions are shown in Table1. In Table1, x means the parameter must be a real number, x0 means the parameter must be nonnegative, > 0 means the parameter must be positive, and ∅ means the parameter is not allowed for the particular function. Once the implanted element, energy, dose, tilt and rotation of an implantation process step are defined by users, the relevant parameter set will be looked up in implant tables. With proper parameter set, the impurity distribution will be calculated subsequently. If users have data fitted to experiments, the parameter set can be defined in implantation command.

Symbol:	R_p	σ_p	σ_{p2}	γ	β	l_{exp}	α
Keyword:	RP	STDV	STDVSec	GAMma	BETA	LEXP	LEXPOW
Gauss	x	> 0	∅	∅	∅	∅	∅
Pearson	x	> 0	∅ T	x	x	∅	∅
P4	x	> 0	∅	x	x	∅	∅
P4S	x	> 0	∅	x	x	x0	∅
JHG	x	> 0	> 0	∅	∅	∅	∅
GK	x	> 0	∅	∅	∅	> 0	> 0
P4K	x	> 0	∅	x	x	> 0	> 0
JHGK	x	> 0	> 0	∅	∅	> 0	> 0

Table 1. Range of parameter specification for the distribution functions

According to the simulation results, the single primary distribution functions can be divided into 3 groups. Group1 contains Pearson distribution function; group2 contains P4, P4S, P4K distribution functions; group3 contains Gauss, GK, JHG, JHGK distribution functions. Fig.1 (a) shows the 2D impurity distribution with different implantation models; Fig.1 (b) shows the impurity distribution along Y direction. From Fig.1 (a) and Fig.1 (b), we can see that functions in the same group have similar simulation results. Actually, the distribution functions in group3 are usually used in deep implantations, such as WELL implantation in CMOS process; and the distribution functions in group1 and group2 are usually used in shallow implantations, such as drain/source implantation in CMOS process.

In order to obtain more accurate simulation result, we should take ion channeling into consideration. Then the dual primary distribution functions should be used. That is, the profile is divided into two components, the first components representing the profile of ions, which don't channel, and the second one representing the channel ions. A dual primary distribution function is obtained by specifying two single primary functions for the two components mentioned above. It can be defined in the implantation command following the format:

Implantation (..., Function=(function1,function2))

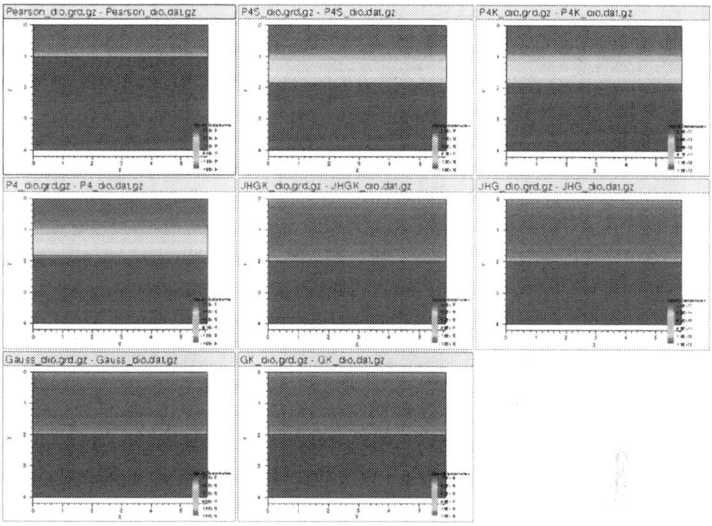

Fig. 1. (a) 2D impurity distribution

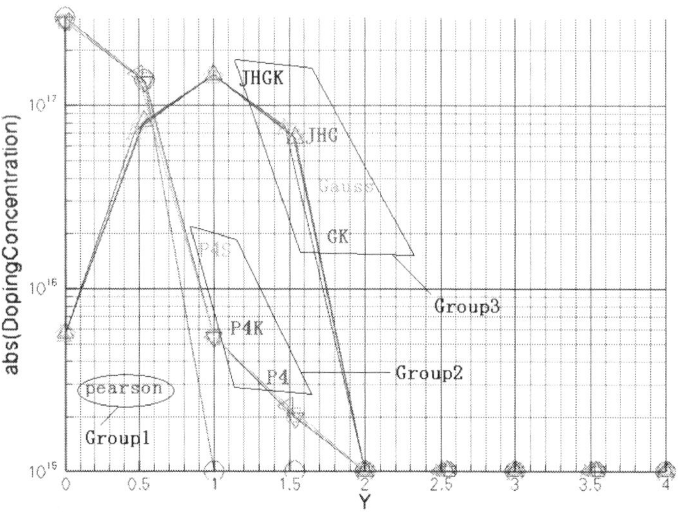

Fig. 1. (b) impurity distribution along Y direction

DIOS provides 5 models for the diffusion process step: Conventional, Equilibrium, Loosely coupled, Semicoupled, and Pairdiffusion. Conventional model is the simplest model but consumes the least time, while Pairdiffusion model is the most accurate model but consumes the most time. In ESD simulation, we'd better select Pairdiffusion model, because it always provides the best boundary shape, which will benefit in convergence problems in the following device simulation.

After selecting proper physical model, the process simulation can be carried out, and the produced electronic pattern of device is then imported into the mesh optimization tool-Mdraw. After the mesh optimization, device simulation is ready.

3. Device simulation

Device simulation is based on solving a set of mathematic and physical equations. And the physical parameters used in these equations are described by different physical models, parts of which are from papers and others are fitted by software engineers. The parameter sets of the physical models are based on the data from several process technologies, and can not cover every process technology. Therefore, to a detailed process technology, some parameters of physical models should be modified. To simulate an ESD event correctly, accurate physical models and proper parameter sets are the most important, no matter which simulation method is chosen.

To account for high electrical field and high temperature effects during an ESD event, the physical models below in ISE TCAD must be included: 1)Fermi-Dirac statistics. When the carrier density exceed 1×10^{19} cm-3, the default Boltzmann statistics becomes not suitable for simulation. 2) Accurate effective intrinsic carrier density model with band gap narrowing and Fermi correction included. 3) A comprehensive mobility model with doping dependence, carrier-carrier scattering, and high field saturation taken into consideration (In MOS devices, surface mobility degradation due to acoustic surface phonons and surface roughness should be also taken into consideration). 4) Recombination model should contain both Shockley-Read-Hall (SRH) model and Auger model, and SRH model should take doping dependence, temperature dependence and field-enhanced recombination into consideration. 5) Avalanche generation. 6) Thermodynamic model considering the self-heating effect. 7) Thermoelectric power model.

Simulating ESD events, three physical parameters are the most important: mobility of carriers (μ), lifetime of free-carrier (τ), and the generation rate (G) dominated by ionization impact.

Mobility is described in ISE TCAD with several degradation models, just as illustrated above. Taking all of these issues into consideration, the mobility is finally formulated as:

$$\mu = \mathrm{f}(\mu_{low}, F) \tag{1}$$

The function is determined by which model is chosen for high field saturation. And μlow in Eq.(1) is formulated as:

$$\mu_{low}^{-1} = \mu_{dop}^{-1} + \mu_{eh}^{-1} + D\mu_{ac}^{-1} + D\mu_{sr}^{-1} \tag{2}$$

In Eq.(2), μ_{dop} represent the doping-dependent mobility degradation mechanism, μ_{eh} is the mobility due to carrier-carrier scattering, μ_{ac} illustrates the surface contribution due to acoustic surface phonons, μ_{sr} is the surface contribution attributed to surface roughness, and D is given by:

$$D = \exp(-x / l_{crit}) \tag{3}$$

where x is the distance from the interface and l_{crit} is a fit parameter. μ_{ac} and μ_{sr} can be ignored in non-surface devices.

We have run simulations using different models, and it is found that Masetti model for doping dependence mobility degradation, Conwell-Weisskopf model for carrier-carrier scattering, and Canali model for high field saturation provide the best result. In Masetti model, μ_{dop} is expressed as:

$$\mu_{dop} = \mu_{\min 1} \exp\left(-\frac{P_c}{N_i}\right) + \frac{\mu_{const} - \mu_{\min 2}}{1+\left(\frac{N_i}{C_r}\right)^{\alpha}} - \frac{\mu_1}{1+\left(\frac{C_s}{N_i}\right)^{\beta}} \tag{4}$$

In Eq.(4), Ni is the total doping concentration, μ_{const} is the mobility in low doping level condition, and other parameters are fit parameters. In Conwell-Weisskopf model, μ_{eh} is expressed as:

$$\mu_{eh} = \frac{D\left(\frac{T}{T_0}\right)^{3/2}}{\sqrt{np}}\left[\ln\left(1+F\left(\frac{T}{T_0}\right)^2 (pn)^{-1/3}\right)\right]^{-1} \tag{5}$$

In Eq.(5), n, p are the electron and hole densities, T_0=300 K, and T denotes the lattice temperature. In Canali model, high field mobility degradation is expressed as:

$$\mu(F) = \frac{\mu_{low}}{\left(1+\left(\frac{\mu_{low}F}{v_{sat}}\right)^{\beta}\right)^{1/\beta}} \tag{6}$$

In Eq.(6), μ_{low} is the low field mobility, v_{sat} and β are temperature dependent parameters, and are expressed as:

$$v_{sat} = v_{sat,0}\left(\frac{T_0}{T}\right)^{v_{sat,\exp}}, \quad \beta=\beta_0\left(\frac{T}{T_0}\right)^{\beta_{\exp}} \tag{7}$$

In Eq.(7), except of T_0 and T, all of the parameters are fit parameters.

Lifetimes of free-carriers are governed by recombination models. SRH recombination rate and Auger recombination rate are given in Eq.(8) and Eq.(9) separately.

$$R_{net}^{SRH} = \frac{np - \gamma_n \gamma_p n_{i,eff}^2}{\tau_p(n+\gamma_n n_1)+\tau_n(p+\gamma_p p_1)} \tag{8}$$

$$R^A = \left(C_n n + C_p p\right)\left(np - n_{i,eff}^2\right) \tag{9}$$

In Eq.(8), $n_{i,eff}$ is the effective intrinsic carrier density, γ_n and γ_p are correction parameters for Fermi statistics, n_1 and p_1 are expressed as:

$$n_1 = n_{i,eff} e^{\frac{E_{trap}}{kT}}, \; p_1 = n_{i,eff} e^{\frac{-E_{trap}}{kT}} \tag{10}$$

where E_{trap} is the difference between defect level and intrinsic level. The silicon default value is E_{trap} =0. In Eq.(8), τ_n and τ_p are temperature and field dependent parameters, expressed as:

$$\tau_c = \tau_{dop} \frac{f(T)}{1 + g_c(F)}, \quad c = n,p \tag{11}$$

The component $[1+g_c(F)]^{-1}$ in Eq.(11) is a field enhancement factor. τ_{dop} and f(T) are expressed as:

$$\tau_{dop}(N_i) = \tau_{min} + \frac{\tau_{max} - \tau_{min}}{1 + \left(\frac{N_i}{N_{ref}} \right)^{\gamma}}, \; f(T) = \left(\frac{T}{300} \right)^{T_\alpha} \tag{12}$$

Except for N_i and T, other parameters in Eq.(12) are all fit parameters.
Auger recombination rate is formulated in Eq.(9), in which the temperature-dependent coefficients C_n and C_p are expressed as:

$$C_i(T) = \left(A_{A,i} + B_{A,i} \left(\frac{T}{T_0} \right) + C_{A,i} \left(\frac{T}{T_0} \right)^2 \right) \left(1 + H_i e^{-\frac{i}{N_{0,i}}} \right), i = n,p \tag{13}$$

Except for T, all other parameters in Eq.(13) are fit parameters.
Another important physical parameter is the ionization impact generation rate G, and it is formulated as $G=a_n n v_n + a_p p v_p$, where $v_{n,p}$ denotes the drift velocity. And $a_{n,p}$ is described by many models, in which vanOverstraeten-deMan model is proved to be the best. In this model, $a_{n,p}$ is formulated as:

$$\alpha(F) = \gamma a e^{-\frac{\gamma b}{F}}, \text{ with } \gamma = \frac{\tanh\left(\frac{h\omega_{op}}{2kT_0} \right)}{\tanh\left(\frac{h\omega_{op}}{2kT} \right)} \tag{14}$$

Two coefficients a and b are used for high and low ranges of electric field. And low electric field and high electric field are distinguished by a parameter E_0 whose default value is 4×10^5 V/cm. In low range of electric field below E_0, the values a(low) and b(low) are applied, while in high range of electric field above E_0, the values of a(high) and b(high) are used. The parameter $h\omega_{op}$ represents the optical phonon energy.
As the physical model has been chosen, the fit parameters mentioned above should be modified. And then the simulation can be carried out. In the simulation, the most difficult problem we may face is the convergence problem. Next, convergence problems and solutions will be proposed.
In our simulation practice, it is found out that convergence problems are mostly caused by five factors: 1) Not enough iteration times. 2) Bad initial guess. 3) Bad mathematic calculation method. 4) Coarse mesh or bad boundary shape. 5) Bad parameter set of physical models.

Fig.2 shows the simulation flow of the device simulator. The parameters, "Notdamped" and "Iterations", dominate when the simulation will be terminated. Therefore, too small values for these two parameters will induce abnormal termination. However, this case rarely happens because the default values for these two parameters are big enough in most times.

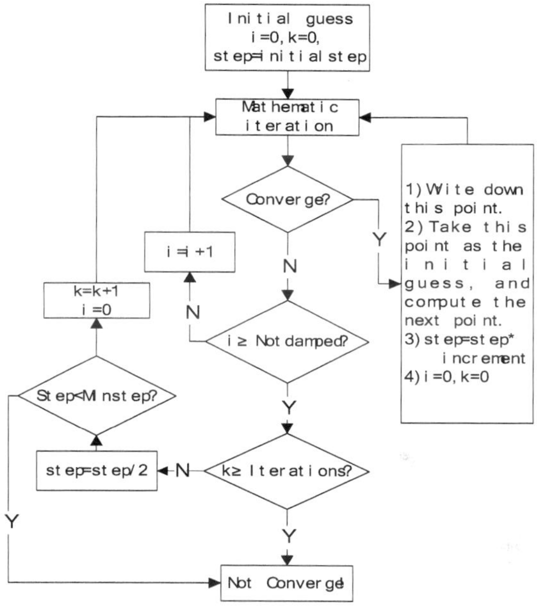

Fig. 2. Device simulation flow

From Fig.2, it is easy to find that all calculations are based on an initial guess. And a bad initial guess will surely induce convergence problem. This case often happens on two occasions. Sometimes, the simulation should be divided into subsections, and in some regions small value for "initialstep" should be used to obtain a good initial guess while in other regions large value for "initialstep" should be used to save time. And a mistaken use of large value for "initialstep" may induce the first point failing to converge. To prevent this convergence problem, the simulation should be divided into subsections in a reasonable way. Meanwhile, large initial voltage imposed on electrodes will also bring on convergence problems. Therefore, another simulation method is necessary. We can set the initial voltage at the electrode to 0 V, and then ramp the voltage to the value we need. In this way, a good convergence will meet. The commands in Fig.3a will cause convergence problems in a great probability while commands in Fig.3b always provide good convergence.

In the snapback region of ESD protection structure, the current increase rapidly. Thus, in the simulation, a small ΔV will induce a large ΔI which induces the simulation failing to converge. Aiming at soling this problem, a particular simulation method is provided in the simulator as shown in Fig.4. A series resistor is put together with the ESD protection structure. Therefore, the current can be written as: $I = (V_{out} - V_{internal})/R$, and in this way, a small ΔI can be gained, which will improve the convergence. In the simulation of ESD events, this method must be included, and generally the value for R is set to be larger than $1\times10^{7}\ \Omega$.

```
Electrode{
        { Name="drain" Voltage=0.0}
        { Name="source" Voltage=0.0 }
       { Name="gate" Voltage=5.0 }
        { Name="sub" Voltage=0.0 }
}
```

(a)

```
Electrode{
        { Name="drain" Voltage=0.0}
        { Name="source" Voltage=0.0 }
     { Name="gate" Voltage=0.0 }
        { Name="sub" Voltage=0.0 }
}
..................
Solve{......}
Goal {name="gate" voltage=5.0V}
```

(b)

Fig. 3. (a) Commands hard to converge, (b) Commands with good convergence

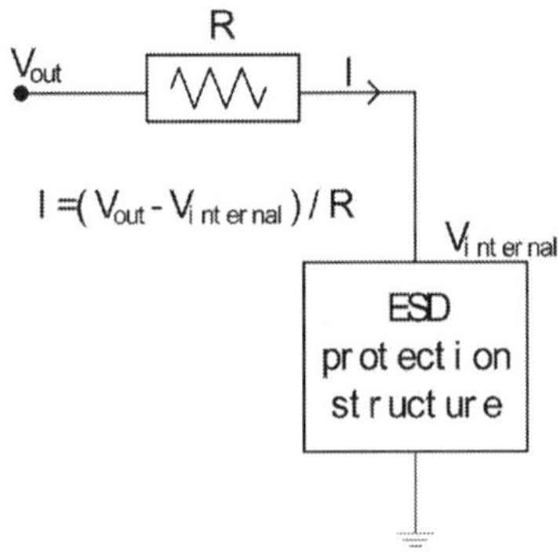

Fig. 4. ESD simulation method

Coarse mesh or bad boundary shape will also cause converge problems. Fig.5 shows the comparison of a bad boundary shape and a good boundary shape. A sharp-angled region can be found in Fig.5a which will cause convergence problem in the later device simulation. It is mainly caused by bad diffusion model and implantation model used in process simulation. It is found that pairdiffusion model used for diffusion and implantation tables based on Crystal-TRIM used for implantation always provide good boundary shape.

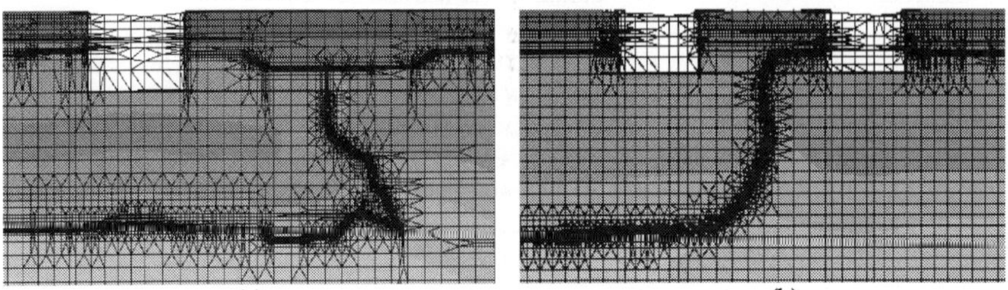

(a) (b)

Fig. 5. (a) Bad boundary shape, (b) Good boundary shape

Another reason for convergence problems is the bad parameter set for device simulation. A small value for the parameter "α" in Eq.(14) and a large value for the parameter "τ_{max}" in Eq.(12) may result in convergence problem, the current failed to increase near the breakdown region. In addition, a great difference between the values of "α" in low field region and high field region may result the simulation failed to converge after it snapbacks, just as shown in Fig.6. When the curve snapbacks, the simulation will change from the high field condition to low field condition, and the sudden change of the value for "α" finally result in the convergence problem. Therefore, when modifying the parameters, great difference between *a*(low) and *a*(high), *b*(low) and *b*(high) is forbidden.

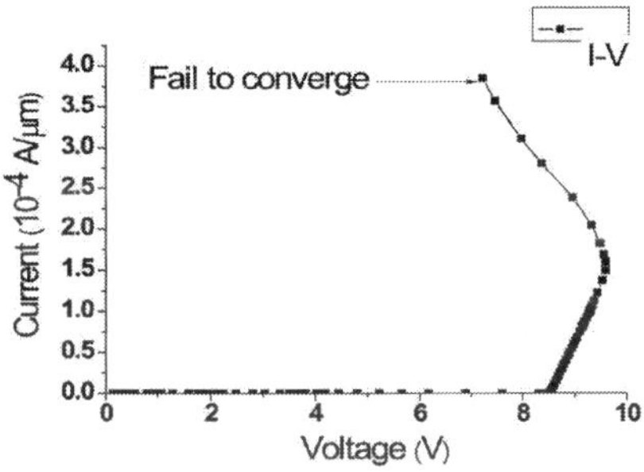

Fig. 6. Simulation fails to converge after the snapback happens

4. ESD simulation methods

There are three main methods to simulate the I-V characteristic of the ESD protection device: DC simulation, TLP simulation and mixed mode simulation. DC simulation provides the fastest simulation speed while it is confronted with the most serious convergence problem. TLP simulation method and mixed mode simulation method can both reflect transient characteristic of devices. In this section, DC simulation and traditional TLP simulation and their limitations will be illustrated. Then a new simulation method based on the traditional TLP simulation method is proposed, which can predict key parameters of ESD protection devices precisely. Mixed mode simulation will be illustrated separately, which is carried out in TSUPREM4/MEDICI environment, and the method to evaluate the effectiveness, the robustness, the speed, the transparency of ESD protection devices is proposed.

To illustrate DC simulation and TLP simulation method, a traditional LSCR (Lateral Silicon-controlled rectifier) shown in Fig.7 is considered, in which D1 is 1.5 μm, D2 is 0.5 μm, D3 is 0.6 μm, and D4 is 1 μm. Fig.8 is the doping profile which is simulated by DIOS, and the total concentration of different layers is shown in Table 2.

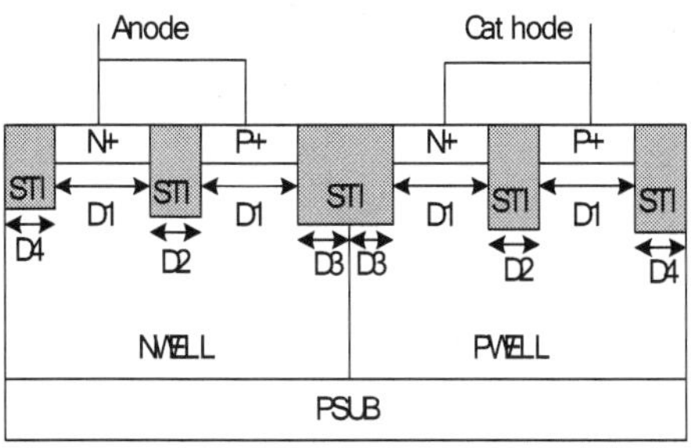

Fig. 7. A cross section of LSCR

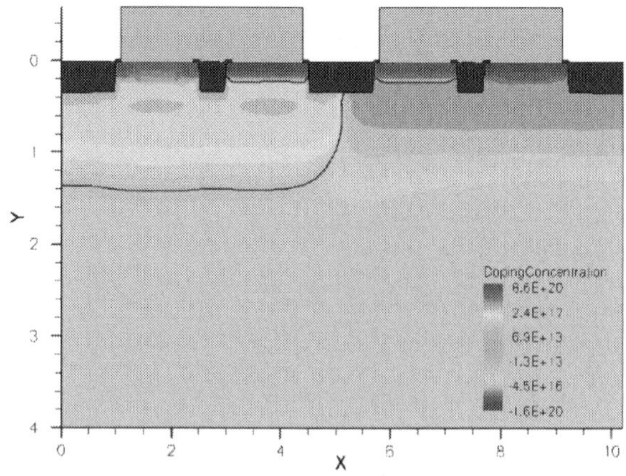

Fig. 8. Doping profile of LSCR

	PSUB	NWELL	PWELL	N+	P+
Total Concentration	1×10^{15}	3.7×10^{17}	2.6×10^{17}	5.1×10^{20}	2.4×10^{20}

Table 2. Total concentration of varies layers

Then, the structure obtained from the process simulation is imported into the device simulator. And the device simulation can be carried out in two ways. To evaluate the trigger voltage (V_{t1}), the holding voltage (V_h), and the second breakdown current (I_{t2}) precisely, selecting proper physical models and parameters is the key point. Table 3 lists the parameters modified in the simulation, and the parameters not mentioned in the table remain default. The value for parameter α mentioned in Eq.(14) determines V_{t1}, while the values for μ mentioned in Eq.(1) and τ mentioned in Eq.(11) are crucial for V_h.

Parameter	Value	Value for electron	Value for hole	Mentioned in Eq.
b(low)	-	9.85×10^{5}	$1.629\text{x}10^{6}$	Eq.(13)
b(high)	-	9.85×10^{5}	$1.354\text{x}10^{6}$	Eq.(13)
F	1×10^{13}	-	-	Eq.(5)
Cr	-	9×10^{16}	$1.5\text{x}10^{17}$	Eq.(4)

Table 3. Parameter set in the simulation

Actually, traditional TLP simulation can not evaluate DC characteristic of ESD protection devices, due to the voltage overshoot. Fig.9 (a) shows the current pulse imposed on the devices simulated, and Fig.9 (b) shows the corresponding I-V curve, comparing with the TLP test result. From Fig.9 (b), we can see that the simulation result deviates from the test result a lot.

DC simulation can evaluate V_{t1} and V_h, but it can not evaluate I_{t2} precisely. DC simulation is based on the solving of thermal equilibrium equations, but in fact, there is no thermal equilibrium established in the structure when the ESD event happens. Therefore, DC simulation can no longer evaluate the characteristic of ESD events when the temperature becomes much more than 300K. The non-equilibrium can only be described by a transient simulation. Fig.10 shows the result of DC simulation, together with the TLP test result.

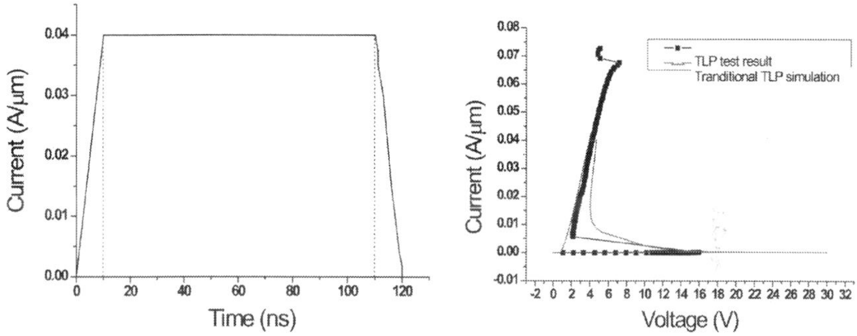

(a) (b)

Fig. 9. (a) Current pulse imposed on the simulated structure (b) I-V characteristic obtained from TLP test and traditional TLP simulation method

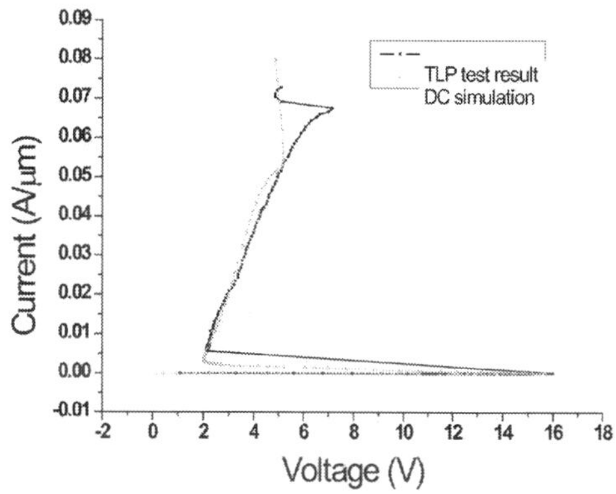

Fig. 10. Comparison of DC simulation and TLP test result

To evaluate the performance of ESD protection devices, Vt1, Vh, and It2 are all indispensable. Based on traditional TLP simulation, we propose a novel TLP simulation method, which can simulate all of the three parameters precisely. Firstly, we should make sure that this method can evaluate V_{t1} and V_h. As the novel TLP simulation begins, series of current pulses are imposed on the structure as shown in Fig.11 (a). The obtained voltage vs. time curves are shown in Fig.11 (b). Then average current value in the range of 70%~90% time for each I-t curve is calculated, and so is the average voltage value, the same as the TLP measurement works. Then each pair of voltage and current is plotted as a point in Fig.12. After connect these points together, comparing it with the tested results, it is found that they meet very well.

Table 4 lists the TLP test results and simulation results with DC simulation method and the novel TLP simulation method. We can see that DC simulation method and the novel simulation method provide almost the same result in terms of evaluating V_{t1} and V_h.

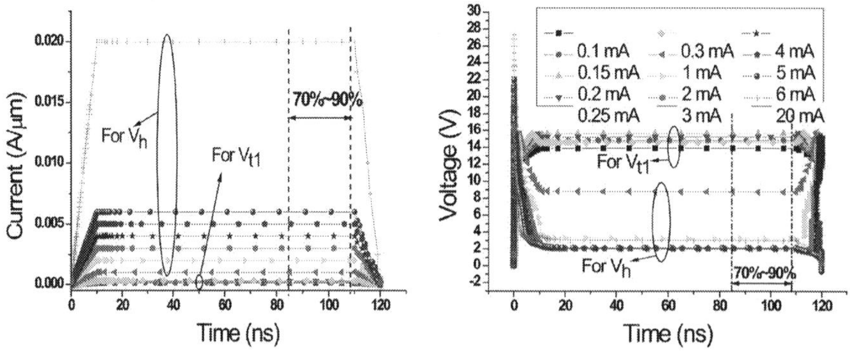

(a) (b)

Fig. 11. (a) Series of current pulses are imposed on the structure simulated, and average currents of the 70%~90% section of each curve are calculated, (b) Voltage vs. time curves are obtained from the simulation. And the average voltage of the 70%~90% section of each curve is calculated.

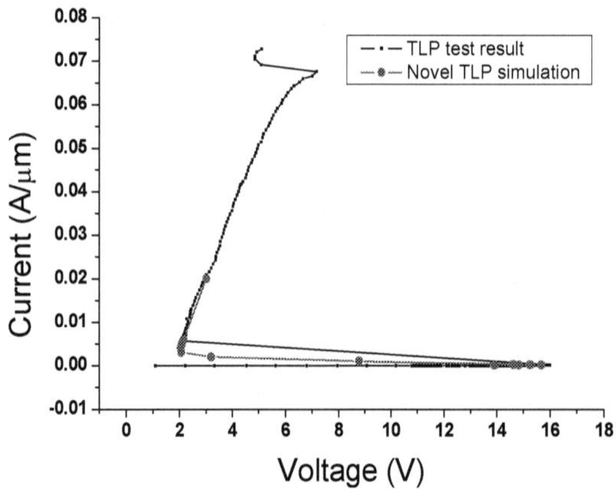

Fig. 12. Comparison of TLP test result and the novel TLP simulation result

	V_{t1}(V)	Absolute crror (V)	Relative crror	V_h(V)	Absolute crror (V)	Relative crror
TLP test	16	-	-	2.16	-	-
Novel TLP simulation	15.69	0.31	1.94%	2.03	0.13	6.02%
DC simulation	15.69	0.31	1.94%	2.02	0.14	6.48%

Table 4. Test result and simulation results

To evaluate I_{t2}, current pulses whose peak values are 0.04A, 0.05A, 0.06A, 0.066A, 0.068A, 0.07A, 0.08A, 0.09A are imposed on the structure, and several points obtained from simulation, together with the points obtained before, the whole curve is shown in Fig.13, from which we can see that that as the current arrive 0.066A, the voltage comes back. And this current is treated as I_{t2}.

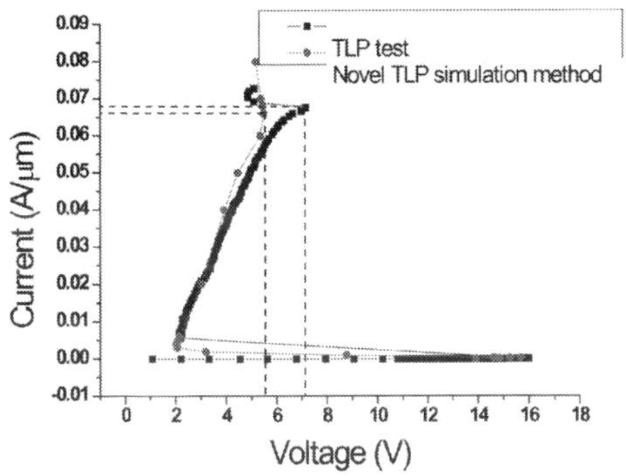

Fig. 13. I_{t2} obtained from novel TLP simulation and that from TLP test

We can also evaluate I_{t2} by the maximum temperature in the structure, as thermal breakdown is caused by high temperature ultimately. After the simulation, we can obtain T_{max} vs. time curves, as shown in Fig.14. When the maximum value of T_{max} exceeds the melting point of Si (1687 K), it can be judged that thermal breakdown happens. From Fig.14, we can see that I_{t2} is about 0.064 A.

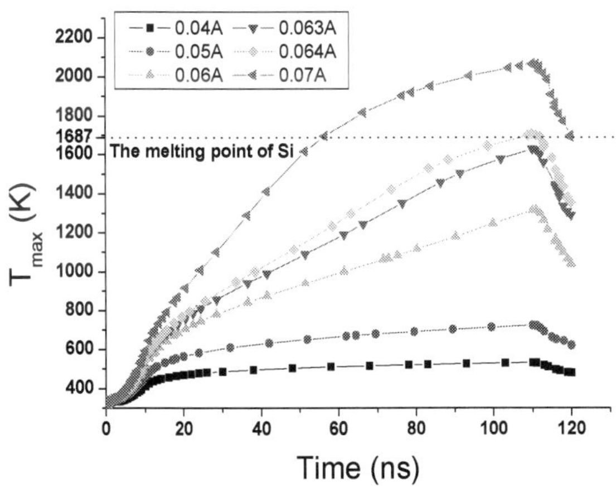

Fig. 14. Maximum temperature in the structure vs. time curves when series of current pulses are imposed on the structure.

Table 5 lists the test result, the result simulated with the novel TLP simulation method and judged by the voltage's snapback, and the result simulated with the novel TLP simulation method and judged by the maximum temperature in the structure.

	I_{t2}(A/μm)	Absolute error(A/μm)	Relative error
TLP test	0.068	-	-
Judged by voltage's snapback	0.066	0.002	2.94%
Judged by maximum temperature	0.064	0.004	5.88%

Table 5. Test and simulation results

From the discussion above, we can conclude that the most effective and fastest way to evaluate the performance of ESD protection devices is to evaluate V_{t1} and V_h with DC simulation method, and evaluate I_{t2} with the novel TLP simulation method introduced above.

Next, the mixed mode simulation method is introduced, taking the CDM model for example. The equivalent circuit of CDM model is shown in Fig.15. The device to be evaluated is a MLSCR, as shown in Fig.16, and the doping profile gained by simulation with TSUPREM4 is shown in Fig.17.

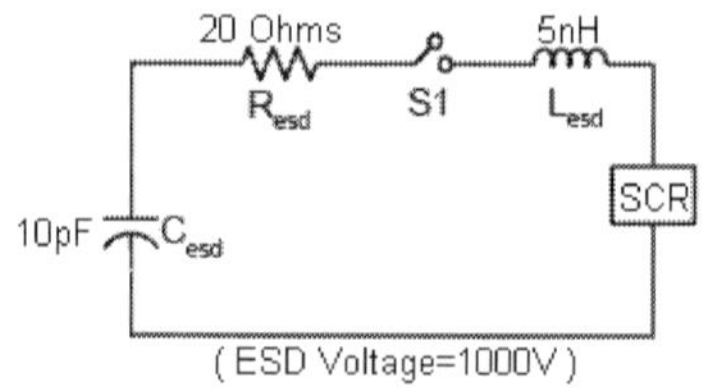

Fig. 15. Equivalent circuit of CDM Model

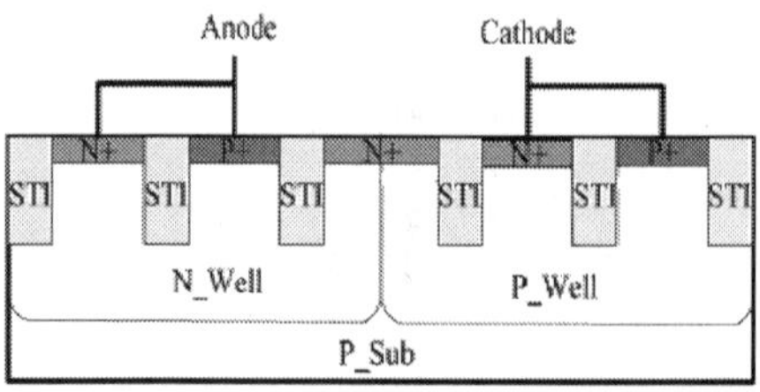

Fig. 16. A cross section of MLSCR

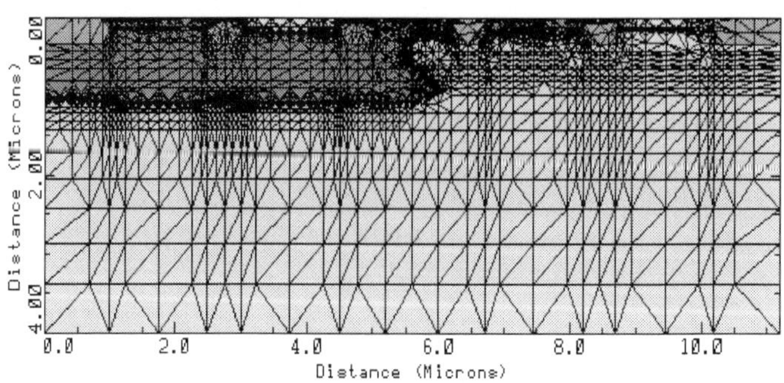

Fig. 17. Doping profile of MLSCR

4.1 Effectiveness evaluation

From the current vs. time curve gained from the mixed mode simulation, as shown in Fig.18, we can see that the ESD current is completely released through the device in 2.5 ns. This time and the peak current at the T_{imax} point reflect the effectiveness of the device. Smaller value of the time and larger peak current mean that the device can release larger current in smaller time, in other words, the device is more effective.

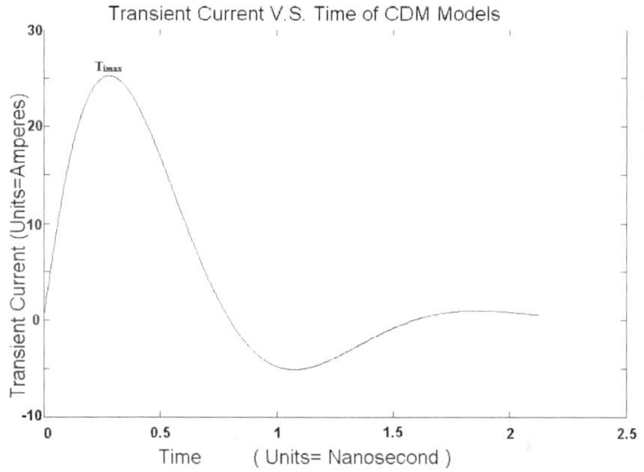

Fig. 18. Current vs. time curve

4.2 Speed evaluation

From the voltage vs. time curve shown in Fig.19, we evaluate the speed using the recover time. The recover time is defined as the time that the device voltage quickly rises and then returns to the normal working voltage, which is described as the $T_{recover}$ in Fig.19. The smaller value of $T_{recover}$ shows that the ESD protection device can make faster reaction to the electrostatic signal.

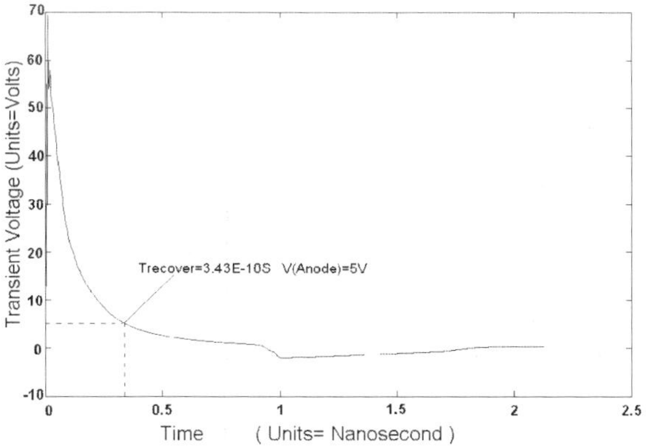

Fig. 19. Voltage vs. time curve

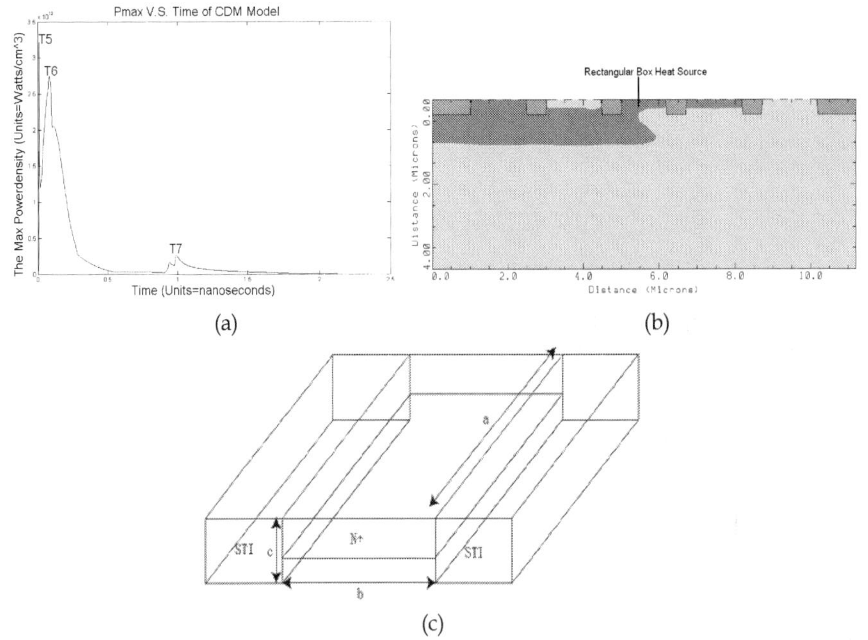

Fig. 20. (a) Pmax-t, (b) Rectangular box heat source model (Zoom out), (c) Rectangular box heat source model (Zoom in)

4.3 Robustness evaluation

There are mainly two aspects should be considered when evaluating the robustness: the first one is to inspect whether the electro thermal characteristics become uncontrollable, when the instantaneous power of ESD comes to the maximum (P_{max}); the second one is to inspect the power distribution in the ESD protection device when the ESD event happens. Taking advantage of the P_{max}-t curve in Fig.20 (a) and the rectangular box heat source model of Ajith Amerasekera, a modified rectangular box heat source model is proposed to evaluate the robustness of the SCR protection device. In the modified model, the power is supposed to be concentrated in a cuboid whose three side lengths are a, b and c respectively, as shown in Fig.20 (b) and Fig.20 (c). Define $P_{normalized}(t)$ as $(\int_{t=0}^{T=t} P_{max}(t)\partial t)/t$, the power instilled into the SCR device is $P(t)=abcR(t)P_{normalized}(t)$, where R(t) is a fitting parameter (0<R(t)<1), and $R(t)P_{normalized}(t)$ is the average power density of the rectangular source heat source. The relationship between the temperature difference ΔT(t) (at this time, the highest temperature T_{max}=T0+ΔT, T0 is the initial temperature, T_{max} is the highest temperature) and P(t) is a subsection function depicted in equations (15) to (18):

$$P = \frac{\rho abcC_p\Delta T}{t} \quad (0 \le t < t_c) \tag{15}$$

$$P = \frac{ab\sqrt{\pi K \rho C_p}\Delta T}{\sqrt{t} - \sqrt{t_c}/2} \quad (t_c \le t < t_b) \tag{16}$$

$$P = \frac{4\pi K a \Delta T}{\log_e(t/t_b) + 2 - c/b} \quad (t_b \le t < t_a) \tag{17}$$

$$P = \frac{2\pi K a \Delta T}{\log_e(a/b) + 2 - c/2b - \sqrt{t_a/t}} \quad (t \ge t_a) \tag{18}$$

In these equations, K is the thermal conductivity, C_p is the specific heat capacity, D= $K/\rho C_p$, ρ is the density of silicon, $t_c=c^2/4\pi D$, $t_b=b^2/4\pi D$, $t_a=a^2/4\pi D$, and K, C_p, and $\rho$is dependent on the process. Therefore we can calculate the highest temperature at every time point, and then calculate the heat produced carriers n_d caused by highest temperature. If n_d extends the background impurity concentration, the robustness of this device cannot meet the need. The transform equation is depicted in Eq.(19):

$$n_d = 1.69\times 10^{19} \exp\left(\frac{-6.377\times 10^3}{T\max}\right)\cdot \left(\frac{T\max}{300}\right)^{3/2} \tag{19}$$

The method to estimate whether the device enters electro thermal uncontrollable condition through the curve of P_{max}-t, as mentioned above can also be quickly implemented by mathematic project software such as Matlab.

The inside power distribution profiles of the ESD protection device when ESD event happens can reflect the robustness of the device. An ESD protection device with strong robustness should spread the inner power as dispersive as possible, especially when the power extremum is very large. Fig.21 shows the power distribution when the power comes to its peak.

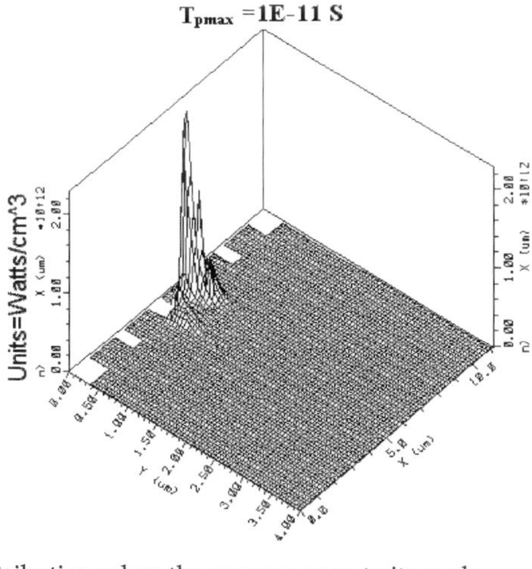

Fig. 21. The power distribution when the power comes to its peak

4.4 Transparency evaluation

We can inspect the leak currents on 0 to 1.2 VDD bias voltages when evaluating DC transparency (depicted in Fig.22 (a)). We need to inspect the leak current under I/O signal frequency when evaluating the transparence of AC signal. (Take 100K rectangular wave as example, see Fig.22 (b)). The leak current under frequency signal is larger than that under DC voltage, which is mainly caused by high frequency couple effect.

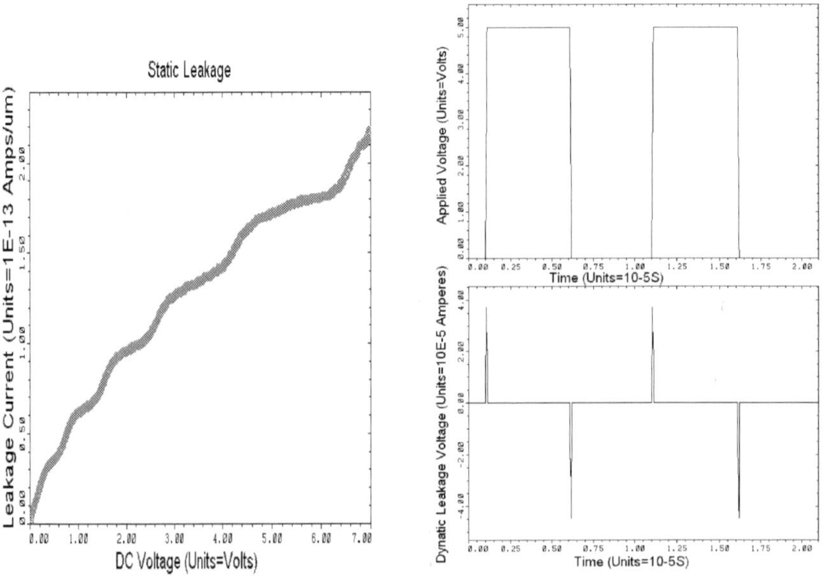

Fig. 22. (a) DC leakage current of the SCR-based ESD protection device, (b) Leakage current of the SCR-based ESD protection device under 100K frequency signal

4.5 Overall evaluation

At the last, we can obtain the transient curve [I(t),V(t)] which describes the entire ESD event as shown in Fig.23, from which we can make a comprehensive evaluation on the effectiveness, speed, robustness and transparency of the ESD protection device. $T0 < T3 = T5 < T6 < T7 < T1 < T_{recover} < T4 < T2$. The current value at T1 reflects the effectiveness of the ESD protection device. $T_{recover}$ reflects the trigger speed of the ESD protection device. The hyperbola family in this figure represents the power of the ESD protection device, and the distance from the hyperbola family to the origin reflects the robustness of the ESD protection device. Besides, the power density extremum also reflects the robustness of the ESD protection device. When time is 1E-11 S, the max power density of the device comes to the peak. The current when the device first comes to 5V in an ESD event reflects the transparency of the ESD protection device. An ideal transient curve of an ESD protection device should be close to the vertical axis with most of the points staying on the left of the line V=VDD.

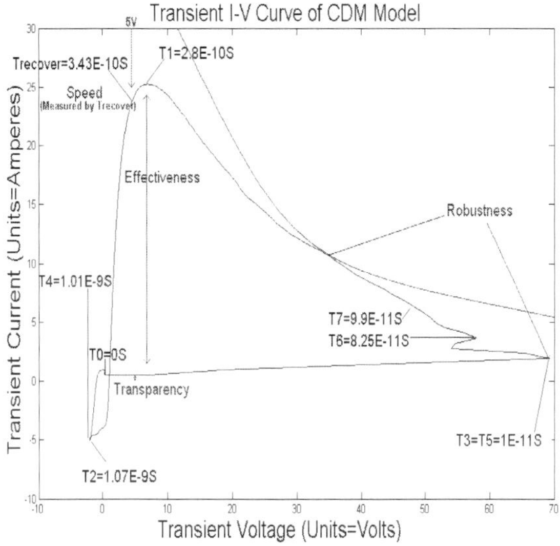

Fig. 23. Ransient I(t) versus transient V(t) of SCR-based ESD protection device

5. ESD protection element characteristic evaluation based on SPICE simulation

5.1 SPICE Simulation based design-transient power clamp

As technology is scaling down, the gate oxide is shrinking and becoming more vulnerable to ESD. The resistance of the routing rail metal increases apparently with the technology advances. Traditional rail-based static ESD power clamp protection (Fig.24) is more challenge. Transient power clamp, which consists of a RC network based detection circuit and the main ESD device NMOS (Fig.25), is becoming more and more attracting for their fast turn-on speed and low turn-on voltage. The key advantage of the transient power clamp is the capability with the SPICE simulation, which enables the optimization in the pre-silicon phase. A major drawback of the transient power clamp is the large RC network, needed to trigger the main protection device, will response any fast event on the power rails.

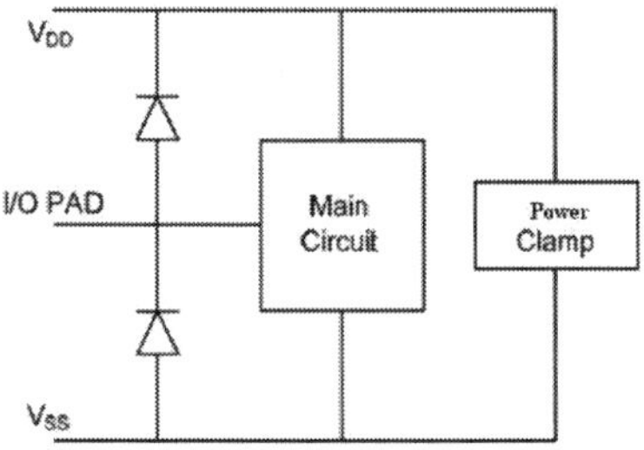

Fig. 24. Rail-based ESD protection scheme with power clamp

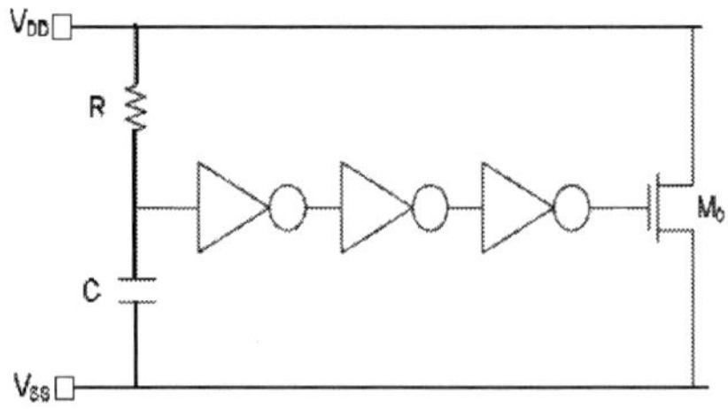

Fig. 25. Three-stage inverter based transient ESD power clamp

The transient power clamp uses the RC network to detect the ESD event and turns on the main ESD protection device NMOS (Fig.25), to shunt the ESD event on the supply pin. The main NMOS conducts the ESD current through the channel and this can be simulated in the SPICE. As the peak current of the HBM is around the orders of amperes, the main NMOS needs to be large enough to shunt the ESD current safely. It is always about millimeter. In normal condition, the gate of the NMOS is low and the main protection device is off. The rise time of ESD event is between 100ps and 60ns.However, the rise time of power up is about millisecond range. In order to keep the main protection device on, the RC constant is set to larger than the duration of the ESD event, which is about 1μs for HBM ESD stress, and shorter than the rise time of power on. The typical value of RC time constant is 1μs. The large RC time constant not only consumes large silicon area but also leads susceptibility to the power bus noise.

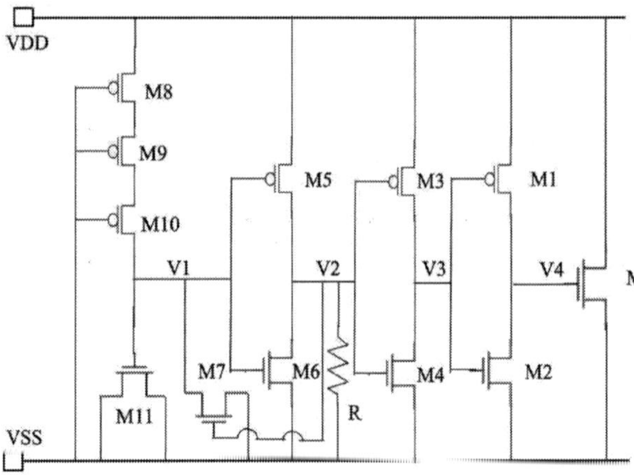

Fig. 26. Proposed three-stage inverter based ESD power clamp with feedback

The M0 is the main protection NMOS to shunt the ESD current.M1~M6 consist of the three stage inverter. The signal at the node V1 transfers through the three stage inverter to control the gate of main device M0. M8~M10 consist of the resistor M11 is the NMOS capacitor. M7 is the feedback NMOS and R is the pull-down resistor. In normal conditions, the node V1 charge up to VDD and V2 is low. The pull-down resistor R confirms the node to couple to VSS. This ensures the feedback NMOS is in its off state. And the voltage at node V2 transfers through two stage inverter to ensure the node V4 is Low. And the M0 is in off. The low voltage at the node V4 enables the reduction in the leakage of M0.In ESD conditions, because of the RC delay, the voltage at the node V1 is low. The M5 is on and the node V2 is charge to VDD. The high voltage in V2 enables the feedback NMOS M7.The M7 pulls the node of V1 to VSS. And the low voltage at the node V1 enhances the pull-up of the POMS M5.The high voltage at node V2 transfers through two stage inverter and enables the M0.The main protection device M0 shunts the ESD current. The feedback significantly increases the time to keep V4 in high voltage. So the RC time constant can be reduced significantly which translates into reduction in the silicon area. The most advantage is the smaller RC time constant reduces the susceptive to the fast transient event on the power lines. In the design, the specific dimension of the RC network is list in Table 6.

Device	Dimension
M8	W/L=7.12um/0.4um
M9	W/L=7.12um/0.4um
M10	W/L=7.12um/0.4um
M11	W/L=1.4um/3.5um

Table 6. RC network device dimension

The power clamp is simulated in the Cadence Specture environment. A simplified RC network (Fig.27) is to simulated the HBM ESD event. The switch SW1 and SW2 are voltage controlled switch. When SW2 is on and SW1 is off, the C1 is charge through the voltage source V2 before 1ns.After 1ns, the switch SW1 is on and SW2 is off, the capacitor discharge through the 1.5k resistor R2 to the power clamp.

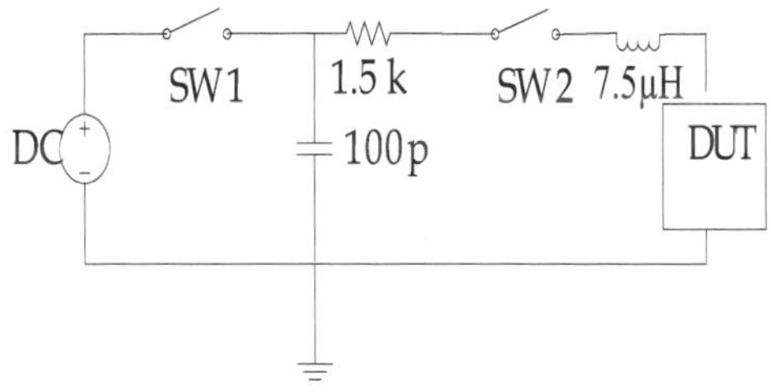

Fig. 27. The simplified RC network to simulated HBM ESD event.

The simulated result of the transient power clamp under a 5kV HBM ESD event in 90nm process is shown in Fig.28.The width of the main protection device M0 is 3000μm. The breakdown voltage of gate oxide for 1.0V core device is about 5V in DC condition. The transistor in the power clamp is 1.8V devices to reduce the leakage. The breakdown voltage of gate oxide for 1.8V device is about 9.5V in DC condition. From the simulated results, the voltage at the gate of the M0 is smaller than the breakdown voltage 9.5V. And the NMOS keeps on state at almost 1μs.The voltage at the VDD rail is also smaller than 9.5V.The NMOS can safely shunt the 5KV HBM ESD current.

To evaluate the immunity to the fast transient, a fast power on 100μs pulse with a rise time of 10μs and a fall time of 10μs is applied at the power clamp. The pulse voltage is 1.8V. The voltage response is shown is Fig.29.The peak voltage at node 4 is 0.05V and it keep almost

0V at most time. So the main NMOS in is off state. And the power clamp is immunity to the fast transient power on.

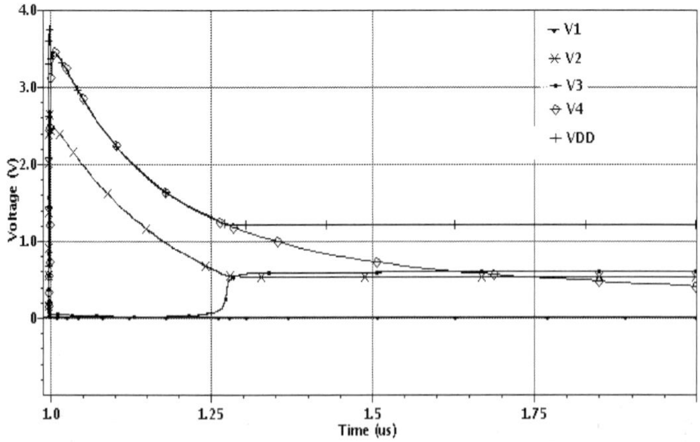

Fig. 28. Simulated voltage at the different node under 5KV HBM ESD event

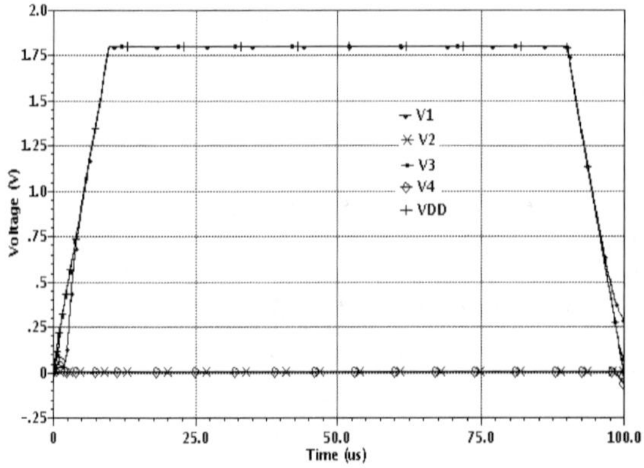

Fig. 29. Simulated voltage at the different node at fast power on state

TLP like pulse with rise time of 10ns and fall time of 10ns and pulse with 100ns is stressed at the power clamp. The pulse voltage is 1.8V. The results are shown in Fig.30. The voltage at node V4, which transfers after three-stage inverter, is a square like pulse. This ensures the main NMOS is on in the pulse width and can shunt the ESD current safely.

The SPICE simulation based transient power clamp is compatibility with the normal SPICE simulation. This enables an early optimization phase in a pre-silicon state. The transient power clamp responds to any fast transient event. An example of the transient power clamp is introduced in the 90nm CMOS process to show the design flow. The susceptibility to fast power on issue is addressed in the example. From the simulation result, the power clamp can achieve a level of 5KV HBM ESD without suffering mistriggering from fast power on.

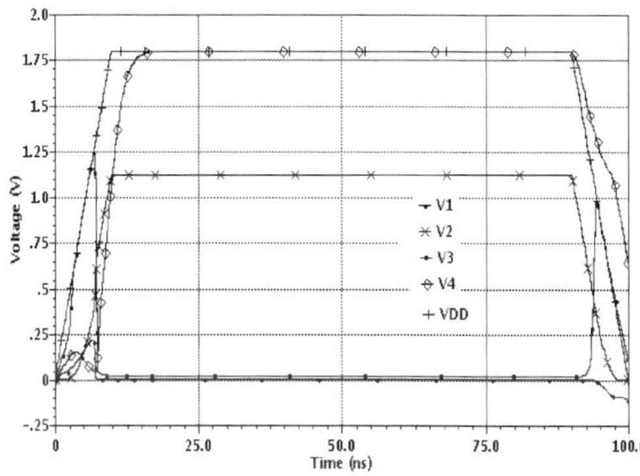

Fig. 30. Simulated voltage at the different node at TLP like pulse

5.2 Triggering characteristic evaluation

SCR is an efficient ESD protection device in integrated circuit area. In order to estimate the ESD device performance, including trigger voltage (Vt1), holding voltage (Vh), failure current (It2), a lot of research are spent base in TCAD simulation. However, a precise evaluation method does not exist as the high ESD current model is not support in spice model. Therefore, a desirable technique is in need to evaluating the ESD device performance in ESD protection device design process. In this section, a new technique is proposed to evaluate the trigger voltage of SCR base in spice simulation.

5.2.1 SCR triggering characteristic evaluation

The equivalent schematic of SCR is showed in Fig.31, which consists of Bipolar junction transistor PNP and NPN. The left part of Fig.31 is an ESD voltage pulse generation circuit. There are different ways to trigger a SCR, including voltage-triggering by slowly stepping up Vac(voltage of anode to cathode) or using a dV/dt transient, and current-triggering by injecting seeding currents from the base of PNP or NPN. A current source is employed to regard as the base current of NPN when the SCR occurring avalanche breakdown. The SCR will turn to latch up state once the base current reaches a value which induces the inside feed back of SCR occurring. The simulation results are showed in Fig.32. As Fig.2 shows, the SCR reaches latch up state when the base current of NPN is 1.3mA.

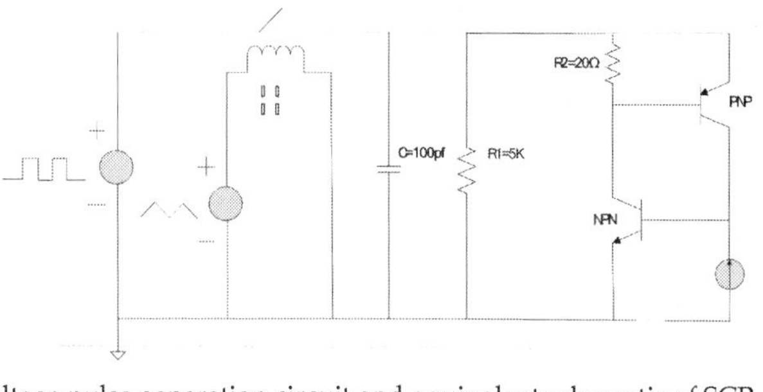

Fig. 31. ESD voltage pulse generation circuit and equivalent schematic of SCR

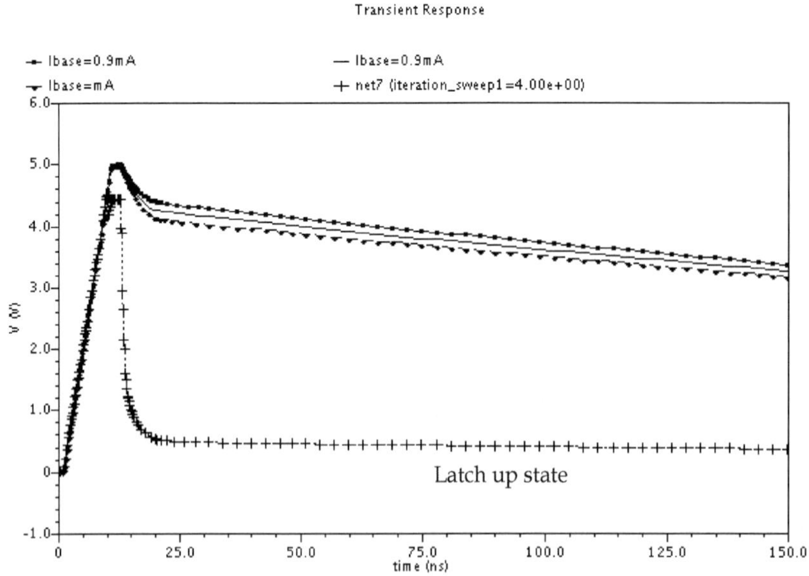

Fig. 32. Simulation results of normal SCR triggering characteristic

5.2.2 Darlington SCR triggering characteristic evaluation

Increasinig the common-base current gains $\beta$of PNP and NPN can make for reducing the trigger voltage of SCR. A Darlington SCR configure is showed in Fig.33. The Q2 and Q3 form to a Darlington transistor, which equates to a NPN transistor here. A current source is also employed to emulate base current as above SCR simulation. The simulation results are showed in Fig.34. SCR turns to latch up state when the base current achieves 0.37mA which is almost one third of normal SCR. In other words, the Darlington configured SCR needs less base current to trigger the SCR into latch up and, therefore, low breakdown voltage to keep the NPN operation. The triggering characteristics of normal SCR and Darlington SCR are showed in Fig.35 when the base current of NPN is 0.37mA.

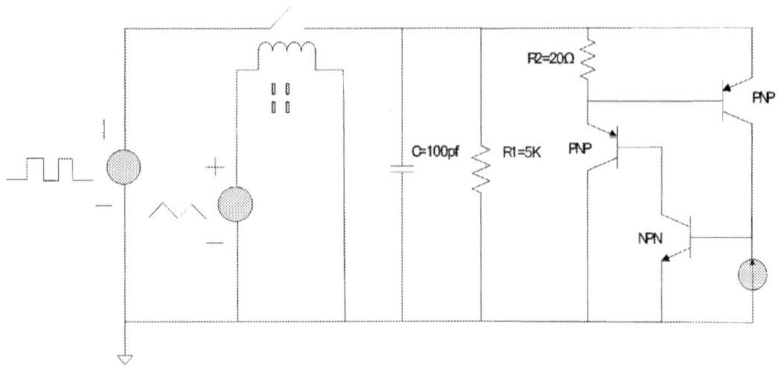

Fig. 33. ESD voltage pulse generation circuit and equivalent schematic of Darlington SCR

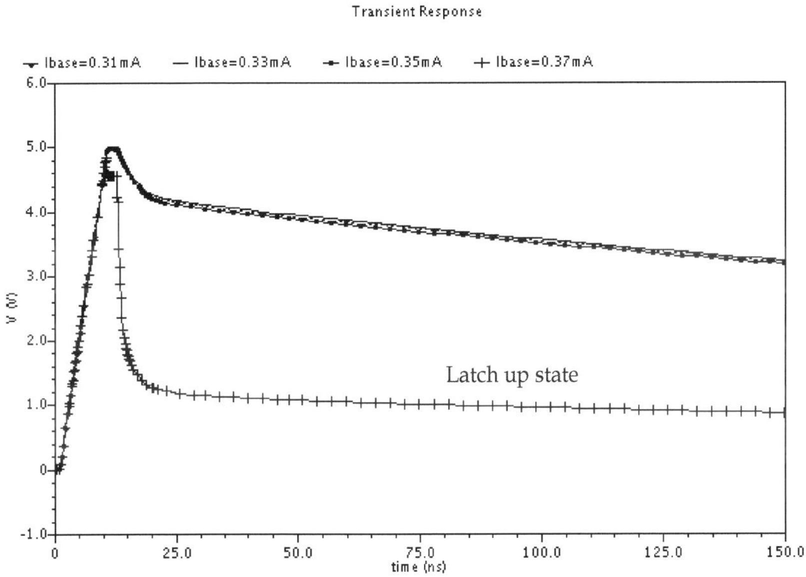

Fig. 34. Simulation results of Darlington SCR triggering characteristic

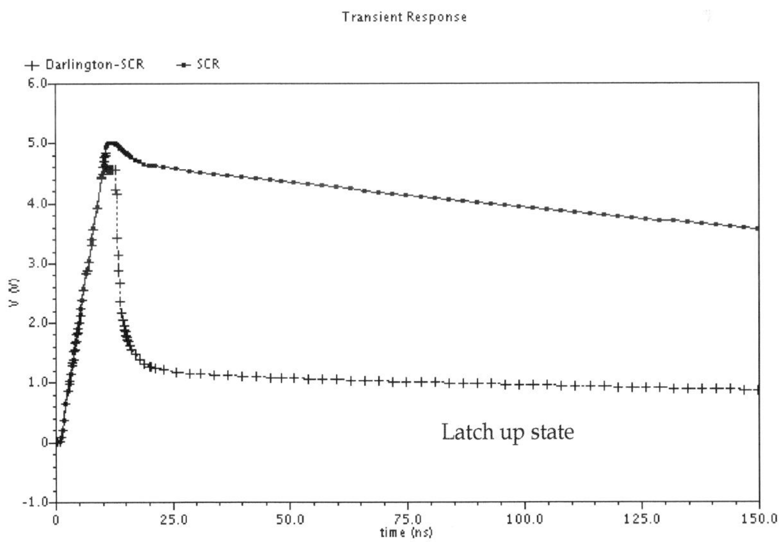

Fig. 35. Trigger characteristic comparison of normal SCR and Darlington SCR when the base current is 0.37mA

6. Acknowledgements

The authors would like express thanks to our students: M.S.E candidates Dahai Huang, Mingliang Li and D.E. candidate Bo Song, Qiang Cui for their help in writing the first draft.

7. References

[1] Ajith Amerasekera (2002). *ESD in Silicon Integrated Circuits.* JOHN WILEY & SONS

[2] Bin Li (2006). An ESD Protection Device Simulation Design Methodology based on MEDICI. *ICSICT 2006*, 2:1358-1360.

[3] Amerasekera A. (1991) .Characterization and modeling of second breakdown in NMOST's for the extraction of ESD-related process and design parameters. *ElectronDevices, IEEE Transactions*,38: 2161-2168

[4] Javier A. Salcedo (2007). TCAD Methodology for Design of SCR Devices for Electrostatic Discharge (ESD) Applications. *Electron Devices, IEEE Transactions* .54:822-832

[5] Salamero C (2006) . TCAD Methodology for ESD Robustness Prediction of Smart Power ESD Devices. *Device and Materials Reliability, IEEE Transactions*. 6:399-407

[6] Salamero, C (2005). Efficient TCAD methodology for ESD failure current prediction of smart power ESD protection. *Power Semiconductor Devices and ICs, 2005.* 115-118

[7] Amerasekera. A (1993). Prediction of ESD robustness in a process using 2D device simulations. *Reliability Physics Symposium*, 161-167

[8] Hillkirk (2003). Investigation of thermal breakdown mechanism in 0. 13um technology ggNMOS under ESD conditions. *Simulation of Semiconductor Processes and Devices*,125-128

[9] Tremouilles D. (2002). TCAD and SPICE modeling help solve ESD protection issues in analog CMOS technology, *Microelectronics, 2002. MIEL 2002. 23rd International Conference* 2: 749 - 752

[10] Amerasekera A (1994). Electrothermal behavior of deep submicron nMOS transistors under high current snapback (ESD/EOS) conditions. *Electron Devices Meeting, 1994. Technical Digest.*, 455 - 458

[11] Cui, Q., Dong, S.R., Liou, J.J., Han, Y., 2008. Robustness Evaluation of ESD Protection Devices in NEMS Using a Novel TCAD Methodology. Proceedings of the 3rd IEEE Int. Conf. on Nano/Micro Engineered and Molecular Systems., Sanya, China.

[12] DESSIS User's Manual, Integrated Systems Engineering AG, Switzerland, 2004.

[13] DIOS User's Manual, Integrated Systems Engineering AG, Switzerland, 2004.

[14] Esmark, K., 2001. Device Simulation of ESD Protection Elements. PhD Thesis, University of Hamburg, Germany.

[15] Fankhauser, B. and B. Deutschmann (2004). Using device simulations to optimize ESD protection circuits. Electromagnetic Compatibility, 2004. EMC 2004. 2004 International Symposium on.

[16] Feng, H.G., Chen, G., Zhang, R.Y., Wu, Q., Guan, X.K., Xie, H.L., Wang, A.Z., Gafiteanu, R., 2003. "A mixed-mode ESD protection circuit simulation-design methodology." Solid-State Circuits, IEEE Journal of 38(6): 995-1006. [doi: 10.1109/JSSC.2003.811978]

[17] Fichtner, W., Esmark, K., Stadler, W., 2001. TCAD Software for ESD On-Chip Protection Design. Electron Devices Meeting, 2001. IEDM Technical Digest. International. Washington, DC, USA. [doi: 10.1109/IEDM.2001.979493]

[18] Huang, C. Y., 2005. "Simulation prediction of electrothermal behaviors of ESD N/PMOS devices." Solid State Electronics 49(12):1925-1932.[doi:10.1016/j.sse.2005.08.016]

[19] Jiang, L.,Yang, X., Wang, J.J., 2001. Simulation models of ESD event in ICs. Solid-State and Integrated-Circuit Technology, 2001. Proceedings. 6th International Conference on.

[20] Li, B., Li, R.Y., Luo, H.W., 2006. An ESD Protection Device Simulation-Design Methodology based on MEDICI. Solid-State and Integrated Circuit Technology, 2006. ICSICT '06. 8th International Conference on.

[21] Salcedo, J.A., Liou, J.J., Liu, Z.W., Vinson, J.E., 2007. TCAD Methodology for Design of SCR Devices for Electrostatic Discharge (ESD) Applications. IEEE Transactions on Electron Devices. [doi: 10.1109/TED.2007.891251]

[22] Salaméro, C., Nolhier, N., Gendron, A., Bafleur, M., Besse, P., Zécri, M., 2006. TCAD Methodology for ESD Robustness Prediction of Smart Power ESD Devices. IEEE TRANSACTIONS ON DEVICE AND MATERIALS RELIABILITY. [doi: 10.1109/TDMR.2006.881483]

[23] Stricker, A.D., Mettler, S., Wolf, H., Mergens, M., Wilkening, W., Gieser, H., Fichtner, W., 1999. Characterization and optimization of a bipolar ESD-device by measurements and simulations. Microelectronics Reliability. [doi: 10.1109/EOSESD.1998.737049]

[24] Voldman, S., W. Anderson, et al., 2001. "A strategy for characterization and evaluation of ESD robustness of CMOS semiconductor technologies." Microelectronics Reliability 41(3): 335-348. [doi:10.1109/EOSESD.1999.819064]

[25] Wang, A. Z., Tsay, C., Lele, A., Deane, P., 1998. "A study of NMOS behavior under ESD stress: simulation and characterization." Microelectronics Reliability 38(6-8): 1183-1186. [doi: 10.1016/S0026-2714(98)00117-6]

[26] K. Esmark, H. Gossner, and W. Stadler, Advanced simulation methods for ESD protection development: Elsevier Science, 2003.

[27] R. Merrill and E. Issaq, "ESD design methodology," in *Proc EOS/ESD Symp*, 1993, pp. 233-237.

[28] S. Dabral, R. Aslett, and T. Maloney, "Core clamps for low voltage technologies," in *Proceedings of the Electrical Overstress/Electrostatic Discharge (EOS/ESD) Symposium*, 1994, pp. 141-149.

[29] C. A. Torres, J. W. Miller, M. Stockinger, M. D. Akers, M. G. Khazhinsky, and J. C. Weldon, "Modular, portable, and easily simulated ESD protection networks for advanced CMOS technologies," *Microelectronics Reliability*, vol. 42, pp. 873-885, 2002.

[30] J. C. Smith and G. Boselli, "A MOSFET power supply clamp with feedback enhanced triggering for ESD protection in advanced CMOS technologies," *Microelectronics and Reliability*, vol. 45, pp. 201-210, 2005.

[31] M. D. Ker and C. C. Yen, "Investigation and Design of On-Chip Power-Rail ESD Clamp Circuits Without Suffering Latchup-Like Failure During System-Level ESD Test," *IEEE Journal of Solid-State Circuits*, vol. 43, p. 2533, 2008.

[32] H. Sarbishaei, O. Semenov, and M. Sachdev, "A transient power supply ESD clamp with CMOS thyristor delay element," in *29th Electrical Overstress/Electrostatic Discharge Symposium (EOS/ESD), 2007. EOS/ESD*, 2007, pp. 7A.3-1-7A.3-8.
[33] Ajith Amerasekera, ESD in silicon integrated circuits, John wiley & Sons, LTD, 1992.
[34] Albert Z.H.Wang, On-chip ESD protection for integrated circuits, Springer, 2002.

11

Directional Tuning Control of Wireless/Contactless Power Pickup for Inductive Power Transfer (IPT) System

Jr-Uei William Hsu, Aiguo Patrick Hu and Akshya Swain
The University of Auckland
New Zealand

1. Introduction

Inductive Power Transfer (IPT) systems have successfully been developed and used to replace traditional conductive power transfer systems where physical connection is either inconvenient or impossible, such as biomedical implants, undersea vehicles, and contactless battery chargers of robots, for providing power to movable or detachable loads (Kim et al., 2001; Feezor et al., 2001; Harrison, 2007). As IPT systems extend to more fields, better control methods are required to cope with various operating environments to satisfy users' needs. Difficulties in controlling the power flow in a wireless/contactless power pickup using IPT technologies can arise from several factors, which include but not limited to load and circuit parameter variations, magnetic field coupling variations between the primary and secondary coils, the operating frequency drift of the primary power supply, etc (Jackson et al., 2000; Chao et al., 2007). These factors can cause the output voltage of the secondary power pickup to deviate significantly from the original designed value, resulting in an undesirable characteristic for applications where a stable output voltage is required. Hence, there is a need to develop controllers under various operating conditions.

Practical power flow control of an IPT sytem can generally be categorized into three different types: namely, primary power supply control, secondary power pick-up control, and coordinated control of both primary and secondary circuits. Among these three, direct power flow control at secondary power pickups is most commonly used to stabilize the output voltage, paricularly for multiple power pickup applications (Hu et al., 2007; Wang et al., 2006; Gao, 2005). This chapter presents the basic theory and control algotithm of an improved directional tuning control method for power flow control of secondary contactless/wireless power pickup circuits.

2. Background of Inductive Power Transfer (IPT) system

The basic structure of an IPT system is shown in Fig. 1 (Wang et al., 2000; Wang et al., 2005; Bieler et al., 2002). The system comprises two electrically isolated parts: the primay power supply and the secondary power pickup. The primary power supply is normally stationay and consists of a resonant power supply and an elongated conductive path for producing a constant AC track current. The secondary movable part, also called the power pickup, is mutually coupled with the primary track and moves with respect to the track loop as the operation requires. Since the system is often loosely coupled between its primary and secondary side, the induced voltage source is usually unsuitable for direct use in applications. As a result, proper tuning and control are essential in the system design for providing a constant DC voltage to the load.

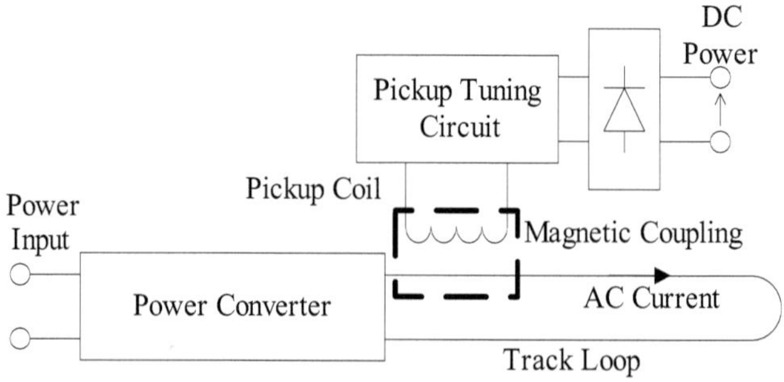

Fig. 1. Basic structure of an IPT system with uncontrolled power pickup.

Figure 2. shows the structure of a typical IPT power pickup. L_S and C_S represent the secondary pickup coil inductance and tuning capacitance respectively, a parallel tuning configuration is adopted here for boosting the induced open circuit voltage.

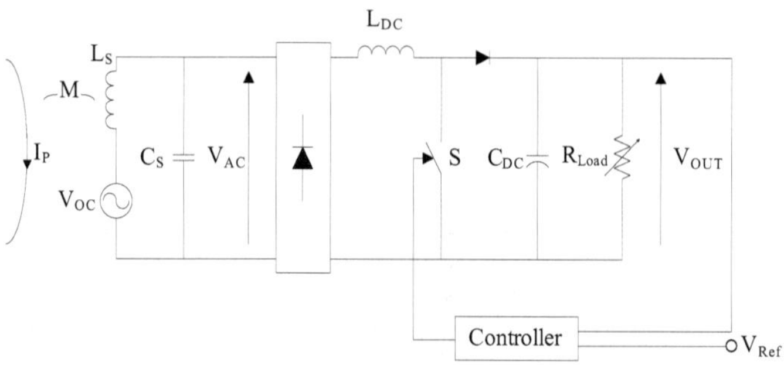

Fig. 2. Basic structure of an IPT power pickup with shorting-control.

The open circuit voltage V_{OC} and short circuit current (I_{SC}) of the pickup coil are governed by the following equations:

$$V_{OC} = j\omega M I_P \tag{1}$$

$$I_{SC} = V_{OC} / j\omega L_S \tag{2}$$

In Figure 2 the voltage V_{AC} after tuning is converted from AC to DC through rectifiers to provide a DC output voltage V_{OUT}. To simply the analysis, the rectifier and load can be represented with the equivalent AC resistor R_{AC}. The transfer function of the system given in (3) can be derived from the simplified second order system shown in Fig. 3, and it also can be seen that at steady state the pickup provides a current source to the load when it is fully-tuned.

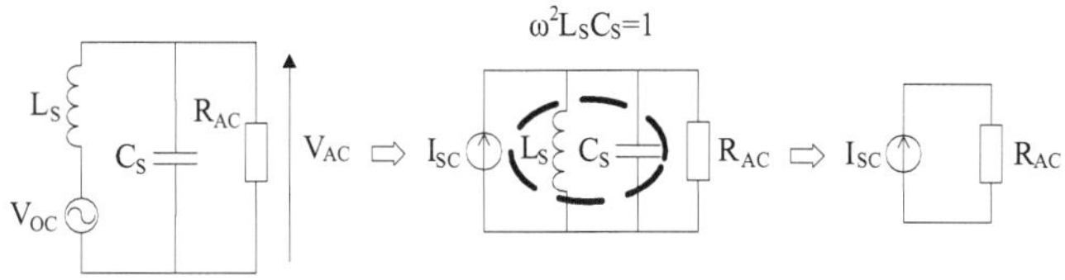

Fig. 3. Simplified second order tuning circuit of power pickup.

$$\frac{V_{AC}}{V_{OC}} = \frac{1/L_S C_S}{s^2 + 1/R_{AC} C_S \, s + 1/L_S C_S} \tag{3}$$

where ω is the system operating frequency. The maximum voltage boost-up factor of power pickup is gorverned by Q factor of the tuning circuit, and under fully-tuned condition it can be expressed as:

$$k_v = \left|\frac{V_{AC}}{V_{OC}}\right| = \frac{R_{AC}}{\sqrt{\left[R_{AC}\left(1-\omega^2 L_S C_S\right)\right]^2 + (\omega L_S)^2}} \tag{4}$$

$$Q = k_{v\max} = \frac{R_{AC}}{\omega L_S}, \quad if \quad \omega^2 L_S C_S = 1 \tag{5}$$

The AC equivalent load resistance R_{AC} is given by:

$$R_{AC} = \frac{\pi^2}{8} R_{Load} \tag{6}$$

where R_{Load} is the DC load resistance. A DC inductor L_{DC} is normally added after the rectifier to maintain a continuous current flow, so that the available power of the secondary pickup can be fully delivered to the load. The output voltage regulation is normally achieved by using a well-known control technique called "Shorting-Control" (Boys et al., 2000; Elliot et al., 1995; Raabe et al., 2007). Its working principle is similar to a boost converter. The constant output voltage is maintained by controlling the average current flowing through the load by switching a semiconductor device (S, shown in Fig. 2) on and off using either hysteresis or PWM control. However, this controller cannot maintain the full-tuning condition of the secondary power pickup circuit. Therefore, the maximum power which can be transferred may be significantly reduced if the circuit parameters vary. And due to the fact that the short circuit current of the pickup coil has to flow through the switch during shorting period, which causes high power losses particularly under light loading conditions, this shortcoming also decreases the potential capability of the primary power supply to operate with more pickups due to unnecessary power loss and possible circuit mistuning.

An alternative method that has been investigated to further improve the power flow control is the dynamic tuning/detuning technique (Hu et al., 2004; Si et al., 2006). Figure 4 shows the general structure of dynamic tuning/detuning control scheme. The fundamental concept of this control method is to dynamically change the tuning condition of the power pickup according to the actual load demands. This helps to maintain maximum power transfer

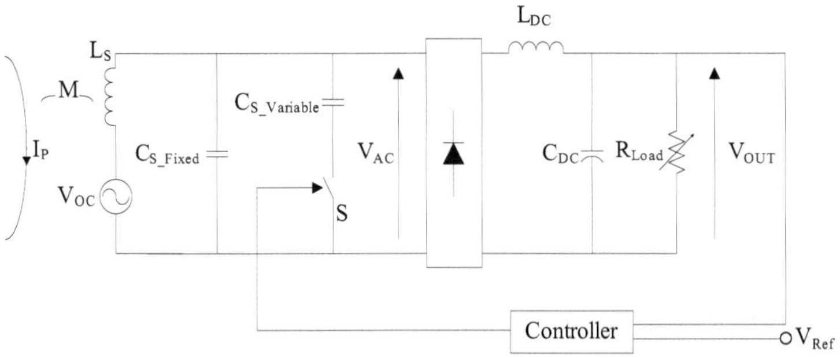

Fig. 4. Basic structure of an IPT power pickup with dynamic tuning/detuning control.

capacity, improve the overall efficiency of the system under light loading condition while keeping the output voltage to be constant. The control strategy is achieved by using a PI controller to control the on/off time of a soft-switched tuning inductor/capacitor to obtain the desired values of equivalent inductance/capacitance in the resonant tank. However, because the relationship between the tuning components and the output voltage is bell-shaped (shown in Fig. 5), there are two possible operating points with one in the over-tuned region and the other in under-tuned region. If the operating point has been accidentally shifted to the other region due to variations of circuit parameters, the desired equivalent values may be tracked in the wrong direction and consequently fail to control the output voltage.

To overcome the problems associated with existing control methods of power pickups such as shorting control, dynamic tuning/detuning control, etc., an LCL (Inductor-Capacitor-Inductor) based power pickup with directional tuning control (DTC) algorithm is proposed and has been discussed in detail in this chapter. Its working principle is similar to the dynamic tuning/detuning control technique. However, instead of using the traditional PI controller to perform the tracking process, it uses the present and previous control results to determine the correct tracking direction in the next step, and retune the circuit to deliver the required power (Hsu et al., 2006). Such an approach covers the full-tuning curve, so dual-side (full-range) control can be achieved. The proposed controller can provide reliable

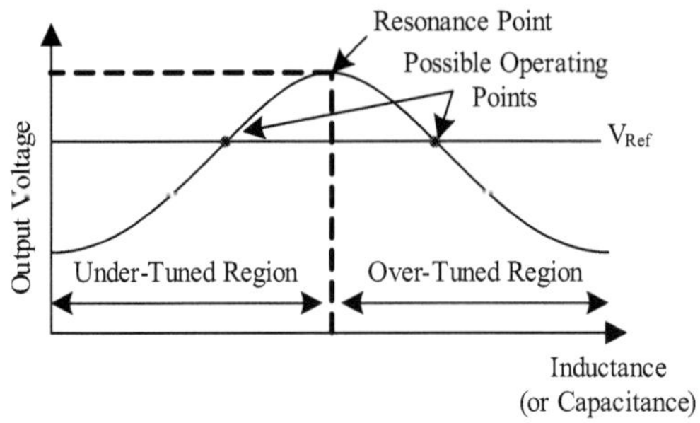

Fig. 5. Relationship between tuning inductance/capacitance and output voltage of IPT power pickup.

constant output voltage under various circuit parameter variations, thus eliminating the need for tedious fine-tuning process required by traditional IPT pickups. As a result, it is more cost-effective for mass production with reduced tuning and component tolerance requirements.

3. Effects of power pickup parameter variations on output voltage

In practical operations, the pickups are often deviated from its designated operating point due to the variation of circuit parameters. Since the deviation of output voltage may not be regulated by the general controller, especially under full-tuning range, the effect of each parameter variation on the output voltage is therefore need to be individually examined so the control range based on the given maximum tolerance to pickup parameters can be better understood (Hsu et al., 2007). The considered circuit parameters include: system operating frequency, magnetic coupling between the primary and secondary side, load resistance and tuning capacitance. Figure 6 shows the structure of the proposed secondary power pickup. An LCL tuning configuration is being used here to provide a constant output voltage to the load under resonant conditions, and a magnetic amplifier in the tuning circuit serves as a variable inductor for changing the tuning condition of the power pickup. The DC current (I_{MA}) which controls the magnetic amplifier is varied through a transistor operating in linear mode which essentially functions as a variable resistor. The equivalent inductance of L_{S2} is adjusted through changing the output signal V_{ctrl} from the DTC algorithm, which allows the power pickup to deliver the right amount of power required by the load (Hsu et al., 2009).

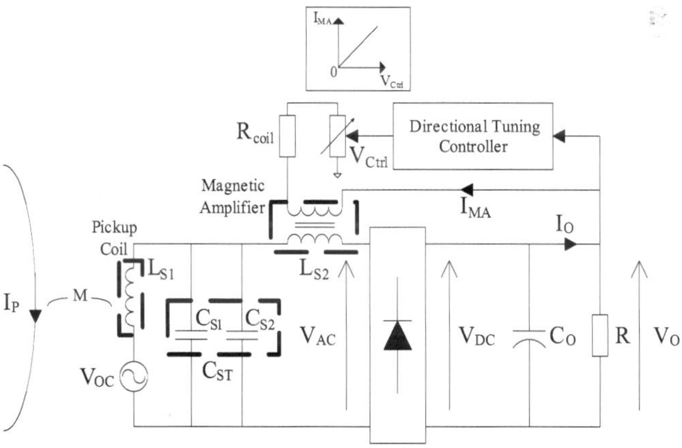

Fig. 6. The proposed LCL power pickup with directional tuning control.

The boost-up factors for ac voltage (V_{AC}) and current (I_{AC}) of the LCL tuning circuit can be determined from the following two transfer functions.

$$k_v = \frac{V_{AC}}{V_{OC}} = \frac{R/L_{S1}L_{S2}C_{ST}}{s^3 + \frac{R}{L_{S2}}s^2 + \frac{L_{S1}+L_{S2}}{L_{S1}L_{S2}C_{ST}}s + \frac{R}{L_{S1}L_{S2}C_{ST}}} \quad (7)$$

$$k_i = \frac{I_{AC}}{I_{SC}} = \frac{sL_{S1}}{R} \cdot \frac{V_{AC}}{V_{OC}} \tag{8}$$

As shown in Fig. 6, the value of C_{ST} can be separated into C_{S1} and C_{S2} which resonate respectively with L_{S1} and L_{S2} *i.e.* $j\omega L_{S1}C_{S1}= j\omega L_{S2}C_{S2}=1$. The ac voltage boost-up factor k_r under full resonant condition can be expressed as:

$$k_r = \frac{V_{AC}}{V_{OC}} = \frac{L_{S2}}{L_{S1}} = \frac{C_{S1}}{C_{S2}} \tag{9}$$

With the considered circuit parameters, the magnitude of AC boost-up factor k_v in (7) can be further expressed as:

$$k_{vm} = \frac{\alpha_v \alpha_f \alpha_r k_r R}{\sqrt{\left\{\alpha_r R\left[\alpha_f{}^2\alpha_c(k_r+1)-k_r\right]\right\}^2 + \left\{\omega\left[k_r L_{S1} - \left[\alpha_f{}^2\alpha_c(k_r+1)-k_r\right]L_{S2}\right]\right\}^2}} \tag{10}$$

where α_v, α_r, α_f, and α_c is the per unit variation of open circuit voltage, load resistance, primary operating frequency, and tuning capacitance, respectively and these are equal to unity when they are at their nominal values. For example if the open circuit voltage increases or decreases by 10%, the value of α_v is set to 1.1 or 0.9 respectively. By rearranging (10) into a quadratic equation of L_{S2}, the solution can be obtained as:

$$L_{S2} = \frac{k_r L_{S1}}{\alpha_f{}^2\alpha_c(k_r+1)-k_r} \pm \frac{\alpha_r R\sqrt{(\alpha_v\alpha_f k_r)^2 - \left\{k_{\min}\left[\alpha_f{}^2\alpha_c(k_r+1)-k_r\right]\right\}^2}}{\omega\, k_{\min}\left(\alpha_f{}^2\alpha_c(k_r+1)-k_r\right)} \tag{8}$$

where k_{min} is defined as the required minimum ratio between V_{AC} and V_{OC}, reflecting the required AC voltage boost-up capability under all possible variations in α_v, α_r, α_f, and α_c.

3.1 System operating frequency variation

Depending on the design of primary power supplies, the operating frequency may drift which often causes significant power loss due to the mismatch in the resonant frequency between the primary and secondary sides. This is particularly a major concern in wireless power transfer systems using resonant variable frequency converters.

Figure 7 shows the effects of system operating frequency variation on AC voltage of the power pickup. It can be seen from the graph that the operating frequency is drifted with the variation so the tuned-point (T-P) is shifted accordingly. As for the magnitude of V_{AC}, it is also changed due to the tuning circuit requires different value of L_{S2} to achieve resonant condition and therefore resulted in various k_r. Note that there are two possible operating points for L_{S2} to compensate for the variations, and both of them are able to keep V_{AC} constant. However, depending on the design specifications, designer can choose to either work with the lower or higher inductance point.

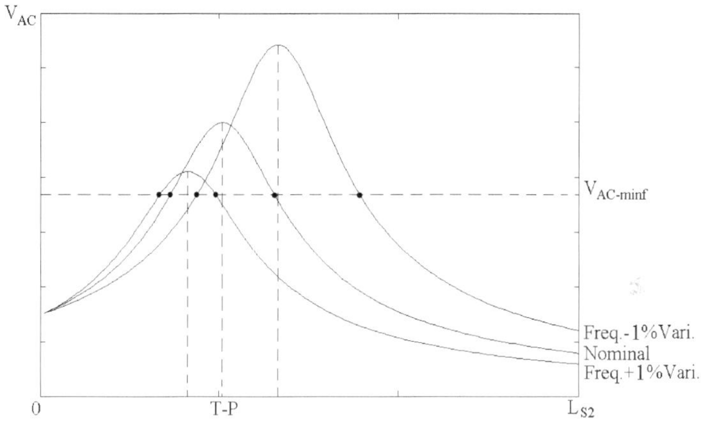

Fig. 7. The effect of system operating frequency variation on AC voltage of LCL power pickup.

3.2 Magnetic field coupling variation

The IPT system is normally involved in loosely coupled applications which allow free movements between the primary and secondary sides. In such applications, fluctuating open circuit voltage of the pickup coil is usually caused by coupling variations due to the free movements, and hence it needs to be compensated for keeping the output voltage constant.

Effect of the magnetic field coupling variation on AC voltage of the power pickup is shown in Fig. 8. It can be seen that the tuned-point and shape of the tuning circuit have both remained the same. Only the magnitude of open circuit voltage of the pickup coil has been changed and therefore resulted in different peak value of V_{AC}.

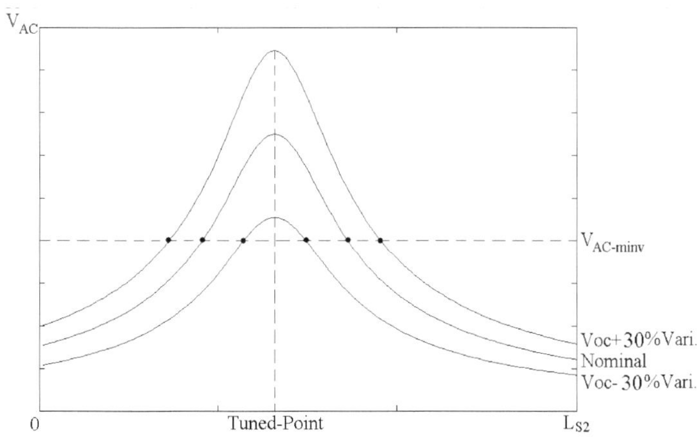

Fig. 8. The effect of magnetic coupling variation on AC voltage of LCL power pickup.

3.3 Load resistance variation

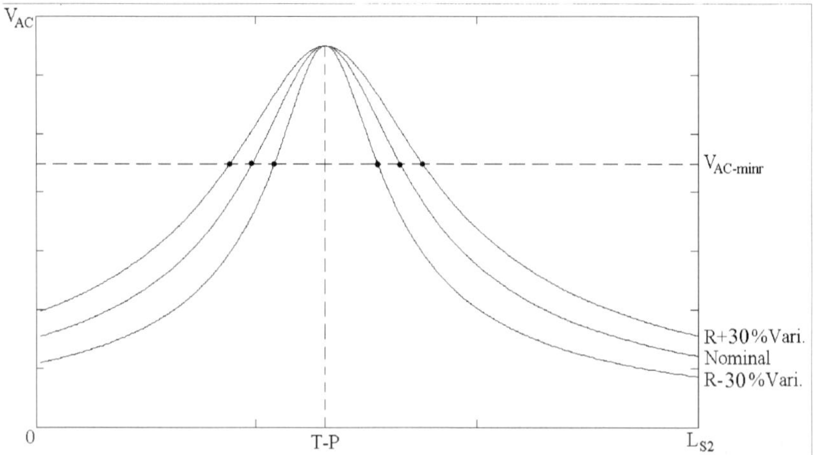

Fig. 9. The effect of load resistance variation on AC voltage of LCL power pickup.

Another variable whose effects need to be studied is the load resistance which varies as the loading condition changes. Fig. 9 shows the effect of load variation on V_{AC}. It can be seen from Fig. 9 that when the load increases, the sensitivity of V_{AC} with respect to L_{S2} decreases. On the contrary, when the load decreases, V_{AC} becomes very sensitive to the change of L_{S2}. These two results have indicated that when the power pickup is operating at extreme loading conditions, either L_{S2} will not be able to compensate for the variation, or the tuning circuit will be too sensitive with respect to L_{S2}.

3.4 Tuning capacitance variation

Unwanted variations of the tuning capacitor such as the variation caused by temperature change may result in undesired tuning condition change and affect the output voltage. This is particularly severe when the seondary system is working with high Q factor since the circuit becomes extremely sensitive to parameter variations.

Similar to the operating frequency variation, both the magnitude of peak V_{AC} and the T-P have been changed and shifted to different places after the variation as can be seen from Fig. 10. Note that as the tuning capacitance decreases/increases, the corresponding L_{S2} also needs to be increased/decreased to keep the circuit tuned, and this consequently causes the pickup to have different peak V_{AC} (or k_r).

3.5 Determination of range of the tuning inductance

In practical operations, the system operating frequency, magnetic coupling, and load resistance as well as other parameters may vary simultaneously. To design the variable capacitor and its controller properly, the worst-case maximum and minimum values of L_{S2} should be identified based on the integrated effect of concerned parameter variations to cover the full control range. Given the maximum allowed tolerance for each variation, the desired maximum and minimum inductance can be calculated by using (8), with the following conditions:

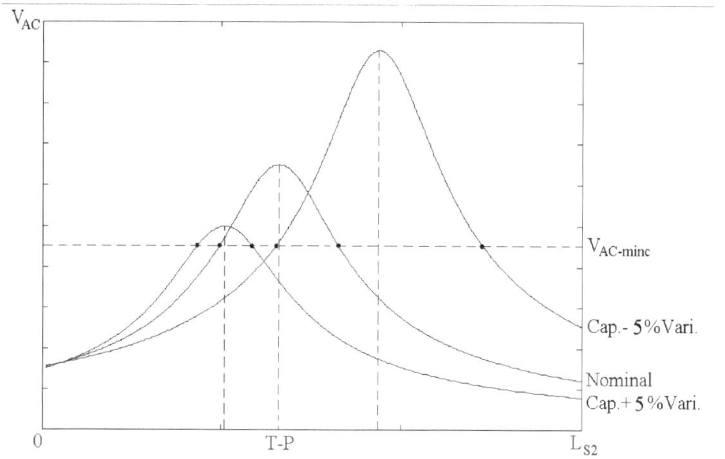

Fig. 10. The effect of tuning capacitor variation on ac voltage of power pick-up.

1. Maximum Inductance
   - Open circuit voltage, operating frequency, tuning capacitor, and load resistance are all at *Nominal value - maximum allowed tolerance.*
2. Minimum Inductance
   - Open circuit voltage, operating frequency, tuning capacitor, and load resistance are all at *Nominal value + maximum allowed tolerance.*

The method presented here can be extended to other possible parameter variations in the system for calculating the range of L_{S2} in worst-case scenario.

4. Design of Directional Tuning Control (DTC) algorithm

In both the shorting-control and dynamic tuning/detuning control method, traditional PI controller has been employed for their output voltage regulation and proven to be effective when the power pickup operates under single-side tuning condition. Nevertheless, it is practically difficult to maintain single-side operation, particularly for high Q systems. The system parameter variations may force the pickup to traverse from one operating region to the other region of the tuning curve and fail to control the output voltage. Directional Tuning Control (DTC) algorithm has been proposed to overcome the problems associated with full-range tuning of the power pickup. The fundamental concept of DTC is based on comparing the present value of control input with its immediate past value, and then use this result to determine the next control action. Instead of depending only on the output error detection as the traditional controllers do, the proposed controller generates the control signal based on the memory of previous control action following the procedure outlined in the flow chart of Fig. 11.

4.1 Standard procedure of DTC algorithm

The flow chart of DTC algorithm is shown in Fig. 11. Standard procedures of the DTC algorithm start with initializations. In this process, the controller initializes the settings according to the user specifications, which include sampling time of the controller and initial state of each processing block. Since the algorithm is designed for controlling the power pick-up to focus on the steady state control, variation of the circuit time constant caused by other system parameter variations must be specified in the initial time delay of the program to avoid inaccurate sampling. After the initialisations, the output voltage at present-state V_{OUT}^k will be sampled, stored, and used to compare with a voltage reference V_{ref} and its previous stored value V_{OUT}^{k-1} for generating logic signals $S_1(k)$ and $S_2(k)$, respectively. These control signals are then collected by the next processing block to check with a predetermined truth table (Table 1) for determining the next-state control signal $S_4(k)$. Note that the memory block after the decision block stores the present control signal as $S_3(k)$, so it can later be used in the next execution for validity checking of the present control action.

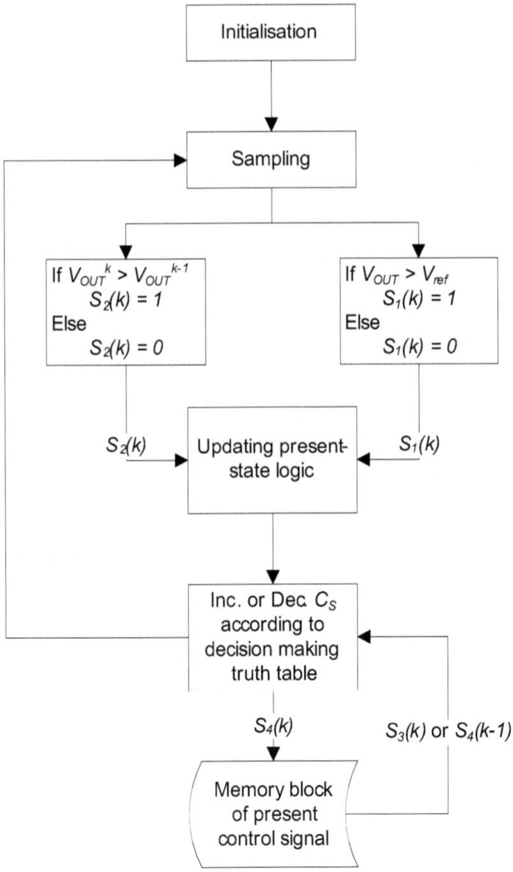

Fig. 11. Flow chart of the directional tuning control algorithm.

S_1	S_2	Previous -State (S_3)	Next-State (S_4)
0	0	0	1
0	0	1	0
0	1	0	0
0	1	1	1
1	0	0	0
1	0	1	1
1	1	0	1
1	1	1	0

Table 1. Truth table for L_{S2} increasing direction determination.

The simplified Boolean expression corresponding to Table I and the actual output signal of the controller can be expressed by:

$$S_4 = S_3(S_1 \oplus S_2) + \overline{S_3}(S_1 \equiv S_2) \tag{9}$$

$$U(k) = U(k-1) + (-1)^{S_4+1} \cdot \Delta h \tag{10}$$

where $U(k)$ is the present-state control signal, $U(k-1)$ is the previous-state control signal, and Δh is the step-size of the adjustment.

4.2 Fuzzy logic control for automatic selection of tuning step-size *Δh*

Despite the fact that the DTC algorithm can effectively control the output voltage of the pickup, the control quality is still restrained by the predefined tuning step-size. A larger step change in the inductance often causes chattering of the output voltage. Although the chattering effect can be reduced by using smaller step change in the inductance, it causes the overall response to be sluggish. To overcome the difficulties associated with the chattering problems and to make the overall response fast, a fuzzy logic controller is integrated with the classical DTC algorithm to further improve the performance of the controller (Hsu et al., 2008). The objective of the fuzzy logic controller is to dynamically determine the step change Δh of the tuning inductance in (10).

4.2.1 Fuzzification

Design of the fuzzy controller consists of fuzzification, formulation of control rule base, and defuzzification. In the process of fuzzification, operating region of the controller is designed to allow error and rate of error to lie inside a predetermined interval (-*L*, *L*). The inputs to the fuzzy PI controller are given as:

$$GE \cdot e(n) = GE \cdot [y_r(n) - y(n)] \tag{11}$$

$$GR \cdot r_1(n) = GR \cdot [e(n) - e(n-1)] \tag{12}$$

$$GR \cdot r_2(n) = GR \cdot [|e(n)| - |e(n-1)|] \tag{13}$$

where $y(n)$ is the output voltage, $y_r(n)$is the reference signal, $e(n)$ is the error signal, *GE* and *GR* are scaling factors for the error and the rate of error respectively. Since the rate of error is calculated from values of output voltage at two consecutive sampling instances i.e. n and n-1, the rate of error $r(n)$ has been further separated into two different variables $r_1(n)$ and $r_2(n)$, where $r_1(n)$ represents the rate of error when the output voltage at both these sampling instances i.e. y(n) and y(n-1) lie either above or below the reference value and $r_2(n)$ represents the rate of error when the output voltage at these two instances lie in different regions with respect to the reference value. The membership functions for error positive (e_p), error negative (e_n), rate positive (r_p), and rate negative (r_n) can be claculated from the following expressions:

$$\mu_{ep} = \frac{L + GE \cdot e(n)}{2L}, \quad \mu_{en} = \frac{L - GE \cdot e(n)}{2L} \tag{14}$$

$$\mu_{rp} = \frac{L + GR \cdot r(n)}{2L}, \quad \mu_{rn} = \frac{L - GR \cdot r(n)}{2L} \tag{15}$$

However, a simple fuzzy PI controller will fail to eliminate the chattering effect at the output voltage since the positive and negative errors calculated using (14) could be the same and cancel out with each other. Therefore a D controller is introduced here with a new set of inputs given by:

$$GD \cdot y_d(n) = GD \cdot |y_r(n) - y(n)| = GD \cdot |e(n)| \tag{16}$$

$$GM \cdot \Delta y(n) = GM \cdot |y(n) - y(n-1)| \tag{17}$$

where $y_d(n)$ is the absolute value of the error, $\Delta y(n)$ is the absolute value of the rate of output, GD and GM are scaling factors for the absolute error and the absolute rate of output respectively. The membership functions for absolute error large (y_{dl}), absolute error zero (y_{dz}), absolute rate of output large (Δy_l), and absolute rate of output zero (Δy_z) are given as:

$$\mu_{ydl} = \frac{GD \cdot y_d(n)}{L}, \quad \mu_{ydz} = 1 - \frac{GD \cdot y_d(n)}{L} \tag{18}$$

$$\mu_{\Delta yl} = \frac{GM \cdot \Delta y(n)}{L}, \quad \mu_{\Delta yz} = 1 - \frac{GM \cdot \Delta y(n)}{L} \tag{19}$$

4.2.2 Control rule base

The control rules for the normal tuning operation are as follows:

R_1: If $GE\ e(n)$ is e_p and $GR\ r(n)$ is r_{1p} Then $\Delta u_{PI}(n)$ is o_l.

R_2: If $GE\ e(n)$ is e_p and $GR\ r(n)$ is r_{1n} Then $\Delta u_{PI}(n)$ is o_z.

R_3: If $GE\ e(n)$ is e_n and $GR\ r(n)$ is r_{1p} Then $\Delta u_{PI}(n)$ is o_z.

R_4: If $GE\ e(n)$ is e_n and $GR\ r(n)$ is r_{1n} Then $\Delta u_{PI}(n)$ is o_l.

An extra set of four control rules for reducing the output chattering are:

R_5: If *GE* $e(n)$ is e_p and *GR* $r(n)$ is r_{2p} Then $\Delta u_{PI}(n)$ is o_l.

R_6: If *GE* $e(n)$ is e_p and *GR* $r(n)$ is r_{2n} Then $\Delta u_{PI}(n)$ is o_z.

R_7: If *GE* $e(n)$ is e_n and *GR* $r(n)$ is r_{2p} Then $\Delta u_{PI}(n)$ is o_l.

R_8: If *GE* $e(n)$ is e_n and *GR* $r(n)$ is r_{2n} Then $\Delta u_{PI}(n)$ is o_z.

The D controller considered here has only four control rules since it only takes the absolute value of the error and the rate of output as its inputs.

R_9: If *GD* $y_d(n)$ is y_{dl} and *GM* $\Delta y(n)$ is Δy_l Then $\Delta u_D(n)$ is o_z.

R_{10}: If *GD* $y_d(n)$ is y_{dl} and *GM* $\Delta y(n)$ is Δy_z Then $\Delta u_D(n)$ is o_z.

R_{11}: If *GD* $y_d(n)$ is y_{dz} and *GM* $\Delta y(n)$ is Δy_l Then $\Delta u_D(n)$ is o_l.

R_{12}: If *GD* $y_d(n)$ is y_{dz} and *GM* $\Delta y(n)$ is Δy_z Then $\Delta u_D(n)$ is o_l.

In the above rules, $\Delta u_{PI}(n)$ and $\Delta u_D(n)$ stands for crisp incremental output of the fuzzy PI controller and the fuzzy D controller respectively.

4.2.3 Defuzzificaction

Defuzzification of the output for fuzzy PI and fuzzy D controller is carried out by using center of gravity algorithm and are expressed as:

$$\Delta\mu_{1PI} = \frac{H\cdot S(\mu_{R_1R_4}) + 0\cdot S(\mu_{R_2R_3})}{S(\mu_{R_1R_4}) + S(\mu_{R_2R_3})} \tag{20}$$

$$\Delta\mu_{2PI} = \frac{H\cdot S(\mu_{R_5R_7}) + 0\cdot S(\mu_{R_6R_8})}{S(\mu_{R_5R_7}) + S(\mu_{R_6R_8})} \tag{21}$$

$$\Delta\mu_{D} = \frac{H\cdot S(\mu_{R_{11}R_{12}}) + 0\cdot S(\mu_{R_9R_{10}})}{S(\mu_{R_{11}R_{12}}) + S(\mu_{R_9R_{10}})} \tag{22}$$

where the membership of output fuzzy sets for control rules R_1R_4, R_2R_3, R_5R_7, R_6R_8, R_9R_{10}, and $R_{11}R_{12}$ are obtained from Lukasewicz fuzzy logic, or, i.e. $\mu_{R_1R_4} = \min(\mu_{R_1} + \mu_{R_4}, 1)$. The function $S(\mu)$ is computed using Mamdani reference.

$$S(\mu) = \mu(2-\mu)H \tag{23}$$

The actual output of the controller which determines the tuning step-size for the variable capacitor is given by:

$$GU\cdot\Delta\mu = GU\cdot(\Delta\mu_{PI} - \Delta\mu_D) \tag{24}$$

where *GU* is a scaling factor for the crisp incremental output of the fuzzy PID controller.

5. Simulation results

To illustrate the effectiveness of the proposed fuzzy based DTC algorithm, a power pickup model has been created in MATLAB Simulink and PLECS.

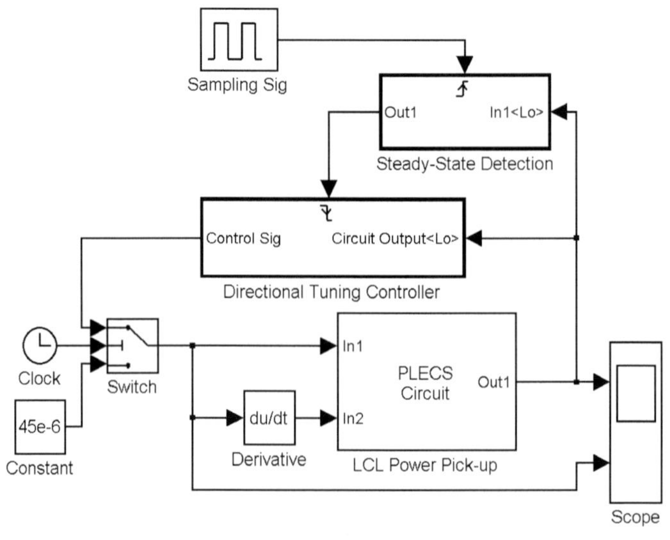

Fig. 12. Simulink model of LCL based power pickup with DTC.

The secondary power pickup model with DTC is shown in Fig. 12. Operating conditions of the power pickup can generally be categorized into four different cases such as: Under-Tuned with Low Start-up Voltage (UT-LSV), Under-Tuned with High Start-up Voltage (UT-HSV), Over-Tuned with Low Start-up Voltage (OT-LSV), and Over-Tuned with High Start-up Voltage (OT-HSV). However, their results are similar to each other during the control process and therefore only two of them are presented here.

The simulation result of V_{OUT}, and L_{S2}, are shown in Figure 13(a) and (b) respectively when the power pickup is operating under UT-LSV. The simulation was started from the circuit start-up with a predetermined delay of 0.05s (for separating the initialization and the actual control process, easing the observation) until it reaches the desired output voltage level (5V). As the error gets reduced, the step change in the tuning inductance also decreases to remove the output chattering effect.

Figure 14 shows the simulation results of the controlled power pickup operating under OT-HSV. As can be seen from the results, both UT-LSV and OT-HSV give similar outcome for providing a constant voltage at the output.

From the results of simulation studies of the controlled power pickup under different operating conditions, it was observed that the proposed controller is capable of controlling the output voltage to the desired value with a response time of 0.1~0.25s. However, the sampling frequency of the controller has to be selected carefully to achieve a more efficient output voltage regulation.

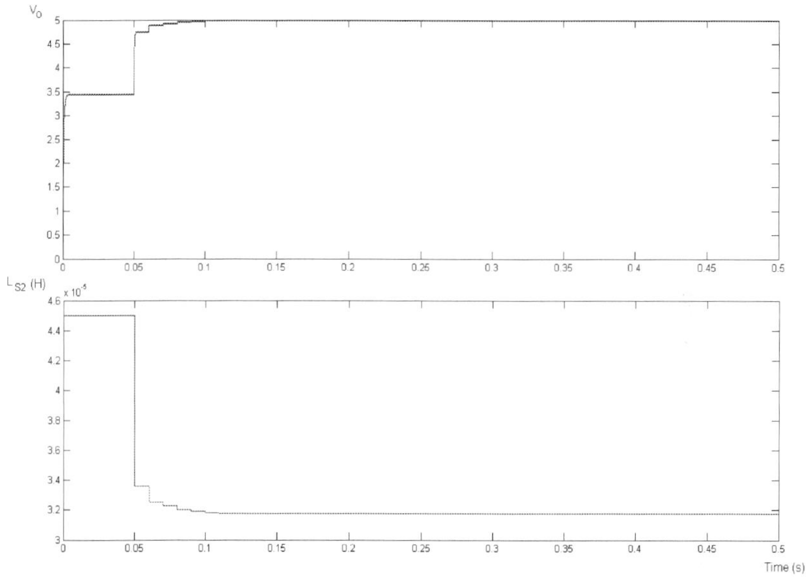

Fig. 13. Waveform of: a) output voltage of power pickup and b) tuning inductance, with Fuzzy based DTC algorithm controlled power pickup operating under UT-LSV.

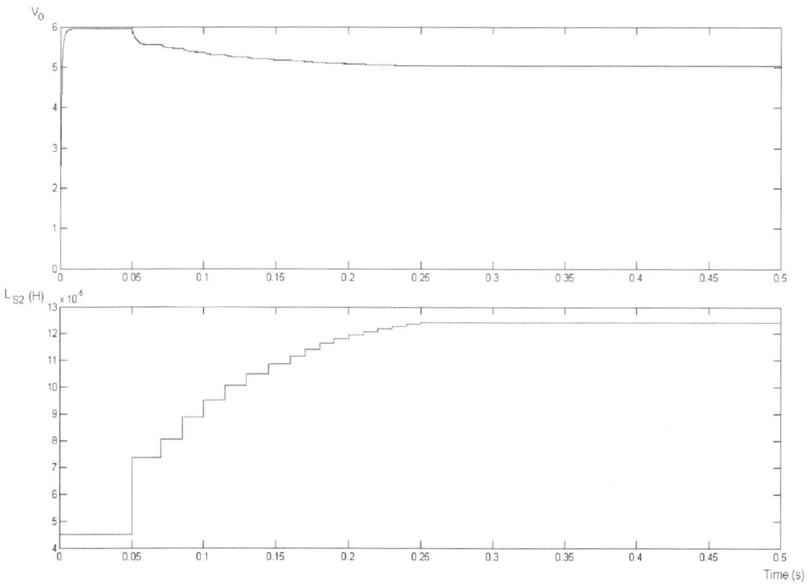

Fig. 14. Waveform of: a) output voltage of power pickup and b) tuning inductance, with Fuzzy based DTC algorithm controlled power pickup operating under OT-HSV.

6. Conclusions

A fuzzy based controller tuning step-size adjuster has been integrated with directional tuning controller to automatically determine the tuning step-size and to effectively regulate the output voltage of the power pickup for inductive power transfer system. The integrated controller has solved the directional tracking problem of the traditional PI dynamic tuning/detuning controller and hence achieved full-range power flow control of the secondary power pickup. The simulation performed by MATLAB Simulink and PLECS have demonstrated the effectiveness of the controller under different testing conditions and it has been shown that a desired constant output voltage can be maintained using the proposed controller without chattering effect. Within certain allowable tolerance of the pickup circuit parameters, the controller can automatically find the correct tuning directions. This helps to ease the circuit component selection in design and eliminates the tedious fine-tuning process in practical implementation.

7. Future research

As the fuzzy based directional tuning control algorithm is developed in discrete-time domain, sampling frequency becomes a very important factor which often affects the performance of the controller. Although the power pickup system will never go unstable since the output voltage is confined by the tuned-point, the true control result of each control action and the response time of the controller are still significantly affected by the sampling frequency. Two different aspects e.g. the magnitude of voltage variation after each control action and the time constant of the DC filter of the power pickup have been preliminarily investigated. However, a clear relationship between these two variables has not yet been found and therefore needs to be further explored.

8. References

T. Bieler, M. Perrottet, V. Nguyen, and Y. Perriard, "Contactless power and information transmission," *IEEE Transactions on Industry Applications*, vol. 26, no. 5, pp. 1266-1272, 2002.

J.T. Boys, G.A. Covic, and A.W. Green, "Stability and control of inductively coupled power transfer systems," *IEE Proceedings of Electric Power Applications*, vol. 147, no. 1, pp. 37-43, 2000.

Y.-H. Chao, J.-J. Shieh, C.-T. Pan, W.-C. Shen, and M.-P. Chen, "A primary-side control strategy for series-parallel loosely coupled inductive power transfer systems," in *ICIEA 2007 2nd IEEE Conference on Industrial Electronics and Applications*. May 23-25 2007.

G.A.J. Elliott, J.T. Boys, and A.W. Green. "Magnetically coupled systems for power transfer to electric vehicles," in *International Conference on Power Electronics and Drive Systems*, Feb 21-24 1995.

M.D. Feezor, F.Y. Sorrell, and P.R. Blankinship, "An interface system for autonomous undersea vehicles," *IEEE Journal of Oceanic Engineering*, vol. 26, no. 4, pp. 522-525, 2001.

J. Gao, "Inductive power transmission for untethered micro-robots," in *IECON 2005 32nd IEEE Annual Conference of Industrial Electronics Society*. Nov 6-10 2005.

R.R. Harrison, "Designing efficient inductive power links for implantable devices," in *ISCAS 2007 IEEE International Symposium on Circuits and Systems*. May 27-30 2007.

J.-U.W. Hsu, A.P. Hu, A. Swain, X. Dai, and Y. Sun, "A new contactless power pick-up with continuous variable inductor control using magnetic amplifier," in *PowerCon 2006 International Conference on Power System Technology*. Oct 2006.

J.-U.W. Hsu, A.P. Hu, and A. Swain, "Fuzzy based directional tuning controller for a wireless power pick-up," in *TENCON 2008 IEEE Region 10 Conference*, Nov 19-21 2008.

J.-U. Hsu, A.P. Hu, "Determining the variable inductance range for an LCL wireless power pick-up," in *EDSSC 2007 IEEE Conference on Electron Devices and Solid-State Circuits*, Dec 20-22 2007.

J.-U.W. Hsu, A.P. Hu, and A. Swain, "A wireless power pick-up based on directional tuning control of magnetic amplifier," in *IEEE Transactions on Industrial Electronics*, vol. 56, no. 7, pp. 2771-2781, 2009.

A.P. Hu, I.L.W. Kwan, C. Tan, and Y. Li, "A wireless battery-less computer mouse with super capacitor energy buffer," in *ICIEA 2007 2nd IEEE Conference on Industrial Electronics and Applications*. May 23-25 2007.

A.P. Hu and S. Hussmann, "Improved power flow control for contactless moving sensor applications," *IEEE Power Electronics Letters*, vol. 2, no. 4, pp. 135-138, 2004.

D.K. Jackson, S.B. Leeb, and S.R. Shaw, "Adaptive control of power electronic drives for servomechanical systems," *IEEE Transactions on Power Electronics*, vol. 15, no. 6, pp. 1045-1055, 2000.

C.-G. Kim, D.-H. Seo, J.-S. You, J.-H. Park, and B.H. Cho, "Design of a contactless battery charger for cellular phone," *IEEE Transactions on Industrial Electronics*, vol. 48, no. 6, pp. 1238-1247, 2001.

S. Raabe and G.A.J. Elliott, "A quadrature pickup for inductive power transfer systems," in *ICIEA 2007 2nd IEEE Conference on Industrial Electronics and Applications*. May 23-25 2007.

P. Si, A.P. Hu, S. Malpas, and D. Budgett, "Switching frequency analysis of dynamically detuned ICPT power pick-ups," in *PowerCon 2006 International Conference on Power System Technology*, Oct 2006.

C.-S. Wang, O.H. Stielau, and G.A. Covic, "Load models and their application in the design of loosely coupled inductive power transfer systems," in *PowerCon 2000 International Conference on Power System Technology*. Dec 4-7 2000.

C.-S. Wang, O.H. Stielau, and G.A. Covic, "Design considerations for a contactless electric vehicle battery charger," *IEEE Transactions on Industrial Electronics*, vol. 52, no. 5, pp. 1308-1314, 2005.

L. Wang, M. Chen, and D. Xu, "Increasing inductive power transferring efficiency for maglev emergency power supply," in *PESC 2006 37th IEEE Power Electronics Specialists Conference*. Jun 18-22 2006.

12

A 7V-to-30V-Supply 190A/μs Regulated Gate Driver in a 5V CMOS-Compatible Process

David C. W. Ng[1], Victor So[1], H. K. Kwan[1], David Kwong[1] and N. Wong[2]
[1]*The Hong Kong Applied Science and Technology Research Institute (ASTRI)*
[2]*The University of Hong Kong*
Hong Kong, China

1. Introduction

The growing markets of electronic components in automotive electronics, LCD/LED drivers and TV sets lead to an extensive demand of high-voltage integrated circuits (HVICs), which are normally built by HV-MOSFETs. These HV-MOSFET devices generally occupy large die areas and operate at low speed due to large parasitic capacitance and small trans-conductance (g_m). There are two types of HV-MOSFET devices, namely, thick-gate and thin-gate oxide devices. Thick-gate oxide devices can sustain a high gate-to-source voltage, V_{GS}, but suffer from a reduced g_m, poor threshold voltage V_T control in production and higher cost due to the need of extra processing steps. Thin-gate devices have a larger g_m, smaller parasitic capacitance, less processing steps and a lower cost. These properties make the thin-gate HV-MOSFETs attractive, though they face severe limitation on V_{GS} swing. There are two main concerns when thin-gate HV-MOSFETs are used. The first is how to achieve high current driving capability to drive capacitive loads in high-voltage (HV) application, whereas the second is how to protect the thin-gate oxide from HV stress breakdown. For current-driving capability, Bales (Bales, 1997) proposed a class-AB amplifier using bipolar technology which consumes a high quiescent current and is expensive due to a large die area and complicated masking. Lu & Lee (Lu & Lee, 2002) proposed a CMOS class-AB amplifier which can only drive around 6mA and does not meet the driver requirements of large and fast current responses (Hu & Jovanovic, 2008). Mentze et al. (Mentze et al., 2006) proposed a HV driver using pure low-voltage (LV) devices but this architecture requires an expensive silicon-on-insulator (SOI) process to sustain substrate breakdown in HV application. Tzeng & Chen (Tzeng & Chen, 2009) proposed a driver that consumes a large die area with all transistors inside the circuit being HV transistors. On the other hand, transistor reliability becomes a serious issue in HV thin-gate oxide transistor circuits. Chebli et al. (Chebli et al., 2007) proposed the floating gate protection technique. The voltage range under protection will change according to the ratio of capacitors and the HV supply, V_{DDH}. This technique, however, cannot limit the voltage across the nodes of gate and source well when the variation of the supply voltage is large. Riccardo et al. (Riccardo et al., 2001) proposed a method which requires an extra Zener diode to protect the thin-gate oxide transistors, so a special process and higher cost are incurred. Declercq et al. (Declercq et al., 1993) suggested a HV-MOSFET op-amp driver with a clamping circuit to protect the thin-gate oxide, but it consumes a significant amount of die area as all devices are HV-MOSFETs. To overcome these drawbacks, the main aims of the proposed driver architecture are:

a. to minimize the number of HV devices so as to save die area in HV application.
b. to develop a HV driver with fast transient responses.
c. to develop reliable thin-gate protection circuitry in HV application, so as to enjoy cost saving from reduced processing steps and take advantages of better V_T process control and high current gain g_m comparing to the thick-gate HV-MOSFET counterparts.

As a result, a HV high-speed regulated driver is developed using mostly LV-MOSFETs with the minimum number of thin-gate HV-MOSFETs.

In this chapter, we present a high-speed CMOS driver that operates with a HV 7V-to-30V supply delivering an output drive up to 190A/µs at a regulated 4.8V output voltage. It is particularly suitable for HV applications such as LCD/LED/AC-DC drivers loaded with power (MOS)FETs. The circuit consists of only 5V LV devices and two thin-gate HV asymmetrical MOS transistors (HV-MOSFETs) fully compatible with standard CMOS technology. The design features a small-area cost-effective solution, measuring only 650µm×200µm in a 0.5µm standard 5V/40V (V_{GS}/V_{DS}) CMOS process. The approach of the regulated output driver can adjust itself to the desired V_{GS}, helping to fully utilize the effect of V_{GS} on minimizing the on-resistance, $R_{DS\text{-}ON}$, of the power FET. Novel thin-gate protection circuits, based on source-follower (SF) configurations, have been deployed to limit the V_{GS} swing to within 5V for the HV-MOSFETs. A dual-loop architecture provides an extremely fast slew rate and transient response under a low quiescent current of 90µA in its static state and 860µA during switching. A dead-time circuit is included to eliminate the power loss incurred by shoot-through current, saving 75mW under a 30V HV supply. Moreover, stability analysis and compensation techniques are described in details to ensure stable operation of the driver in both loaded and un-loaded conditions. Lab measurements are in good agreement with simulations. A comparison with existing works then demonstrates the efficacy and superiority of the proposed design.

In this chapter, Section 2 introduces the use of LV devices to build HV high-speed regulated driver, together with stability analyses for both cases when the power FET load is ON or OFF. Section 2 also discusses the power saving techniques in driving HV-MOSFETs. In Section 3, simulation and lab measurement results are shown which confirm the merits of the proposed design. Finally, the conclusion is drawn in Section 4.

2. Principles of operation

2.1 Circuit structure and basic operation

Fig. 1(a) shows the high-level block diagram of the proposed driver. It consists of a LV error amplifier, a HV thin-gate protection circuit, a feedback resistor network with pole-zero cancellation and a fast transient regulated driver with dead-time control. A HV nMOS, hvn01, is connected to the node V_{reg} in a SF configuration. The driver requires a LV supply, V_{DDL}, as well as a HV supply, V_{DDH}. We first develop an internal regulator, which gives a 4.8V DC voltage, V_{reg}, through hvn01. The drain of hvn01 is connected to V_{DDH}, which is 30V in our design. The V_{reg} acts as a supply voltage to a chain of inverter buffers, which in turn drive the output load at the node V_{out}. The switching activities are started from V_{in} all the way to V_{out}. The output load here is the gate of a 1A on-chip thin-gate power (MOS)FET. The equivalent gate capacitance is around 270pF. The driver provides a 4.8V output and therefore protects the thin gate of the loading power FET by limiting its V_{GS} right below 5V.

The node V_{out} can also be connected externally to drive external power FETs. The approach of the regulated output driver can always adjust itself to the desired V_{GS}, helping to fully utilize the effect of V_{GS} on the on-resistance, R_{DS-ON}, of the power FET. In this connection, and with reference to (1) and (2) (Gray et al., 1990), the on-resistance and die area of the on-chip power FET can be minimized:

$$I_{DS} \approx \frac{\mu_n C_{ox}}{2}\left(\frac{W}{L}\right)(V_{GS} - V_T)^2 \;, \tag{1}$$

$$R_{DS\text{-}ON} \approx \frac{1}{\mu_n C_{ox}\left(\frac{W}{L}\right)(V_{GS} - V_T)} . \tag{2}$$

Equation (1) describes the behavior of a (HV) nMOS in saturation region, while (2) approximates the turn-on resistance of a (HV) nMOS in the linear region. I_{DS} is the current flowing from the drain to source of a MOSFET. V_{GS} is the gate-to-source voltage. V_T is the threshold voltage to turn on the MOSFET. Also, μ_n is the mobility of electrons and C_{ox} is the gate-oxide capacitance per unit area, whereas W and L are the width and length of the transistor, respectively.

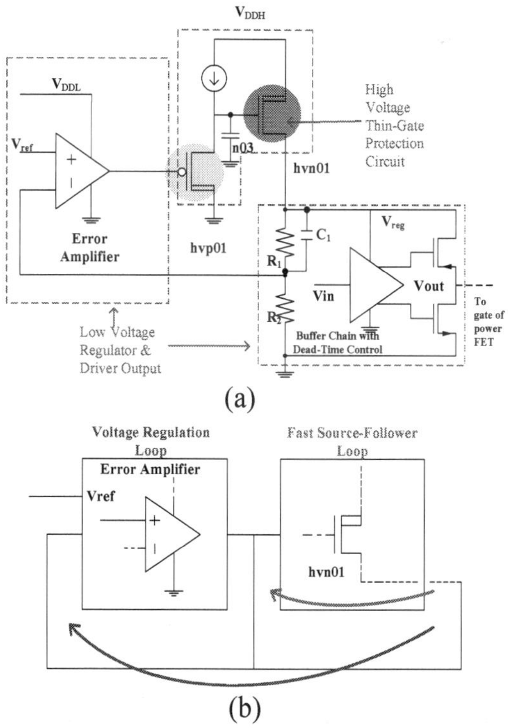

Fig. 1. (a) Architecture of the proposed driver; (b) Dual-loop structure in the driver

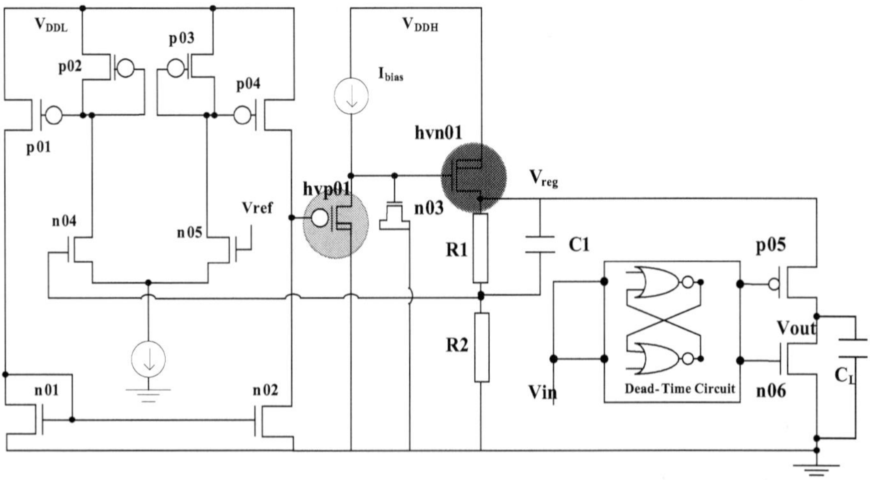

Fig. 2. Detail schematic of the proposed driver

	$C_L \approx$ 0pF (power FET OFF)	$C_L \approx$ 0pF (power FET OFF with Zero)	$C_L = C_{powerFET} \approx$ 270pF (power FET OFF)	$C_L = C_{powerFET} \approx$ 270pF (power FET ON with Zero)
Loop gain(dB)	45	45	45	45
Phase Margin (deg)	83.7	109	57.5	77.1
Unity-Gain Frequency (UGF) (kHz)	27.1	33.8	24.5	28.1
Gain Margin (dB)	35.7	31	18.2	21.9
Gain Margin Frequency (kHz)	623.6	918.3	94.6	184.6
Source Follower Unity-Gain Frequency (MHz)	>100 (Gray et al., 1990)	>100	>100	>100

Table I. Summary of frequency responses of the driver with output = high and output = low

2.2 Regulated driver with fast transient response

2.2.1 Fast dual-loop operation

As shown in Figs. 1 & 2, there are two loops in the driver, namely, the voltage-regulation (VR) loop and the source-follower (SF) loop to achieve fast transient responses. Firstly, for the VR loop, the error amplifier senses the V_{reg} through the resistor network and amplifies the error signal between the scaled V_{reg} and the reference voltage V_{ref}. The error signal is then shifted up to a higher voltage through the thin-gate protection circuit and regulates hvn01 to correct the error, thereby generating a steady and accurate V_{reg}. Secondly, for the SF loop, the SF configuration of hvn01 itself is a fast feedback loop. Referring to (1) and Fig. 1, the feedback mechanism is obvious: When the node V_{reg} goes down due to load current change, the gate-to-source voltage of hvn01, $V_{GS\text{-}hvn01}$, increases and sources a larger output current to charge up the node V_{reg} again. The main function of the VR loop is to provide a regulated voltage of around 4.8V in the steady state, while the fast SF loop provides an immediate response when there is a sudden load change.

2.2.2 Loop gain analysis with the power FET being ON/OFF

We first analyze the SF loop and later the VR loop. For the SF loop, it is well known for its fast response with its unity-gain frequency (UGF) in the 100MHz to 1GHz range (Gray et al., 1990). Its pole effect is generally beyond the UGF of the VR loop and therefore negligible. For the VR loop, there are two scenarios in the stability analysis: the power FET ON and the power FET OFF. When it is ON, $C_L = C_{L\text{-}ON} = C_{powerFET} \approx 270\text{pF}$, and when it is OFF, $C_L = C_{L\text{-}OFF} \approx 0\text{pF}$. Here $C_{powerFET}$ is the equivalent gate capacitance of the power FET. The AC simulation with and without the power FET is shown in Table I and Fig. 3. The phase margin of the VR loop is larger when the power FET is OFF. This can be explained by the following loop gain analysis:

$$T(s) = A(s)\left(\frac{R_2}{R'}\right)\left(\frac{1+\frac{s}{z_{hvp01}}}{1+\frac{s}{p_{hvp01}}}\right)\left(\frac{1+\frac{s}{z_{hvn01}}}{1+\frac{s}{p_{hvn01}}}\right)\left(\frac{1+\frac{s}{z_f}}{1+\frac{s}{p_f}}\right), \quad (3)$$

where z_{hvp01}, z_{hvn01}, p_{hvp01} and p_{hvn01} are the zeros and poles from hvp01 and hvn01, respectively. A(s) is the transfer function of the error amplifier. $R' = R_1 + R_2$. The zeros and poles are defined as

$$z_{hvp01} = \frac{g_{m,hvp01}}{C_{gs,hvp01}},\ z_f = \left(\frac{1}{R_1C_1}\right),\ p_{hvp01} = \frac{1+g_{m,hvp01}r_{o,Ibias}}{C_{gs,hvp01}r_{o,Ibias}},\ z_{hvn01} = \frac{g_{m,hvn01}}{C_{gs,hvn01}},$$
$$p_{hvn01} = \frac{1+g_{m,hvn01}R'}{(C_{gs,hvn01}+C_L)R'},\ p_f = \frac{R'}{C_1R_1R_2}, \quad (4)$$

where $g_{m,hvp01}$, $g_{m,hvn01}$, $C_{gs,hvp01}$, $C_{gs,hvn01}$ are the trans-conductances and gate capacitances of hvp01and hvn01, respectively, whereas $r_{o,Ibias}$ is the output impedance from the current source I_{bias}. We assume the gains of the SF configurations formed by hvp01 and hvn01 are unity. Also, p_{hvn01} is the pole contributed by hvn01 where $p_{hvn01} = p_{hvn01\text{-}ON}$ and $p_{hvn01} = p_{hvn01\text{-}OFF}$ when the power FET is ON and OFF, respectively. Typically, the zeros are located at higher frequencies than poles in the SF configuration except for p_{hvn01}. As $C_{L\text{-}ON} \approx 270\text{pF} >> C_{L\text{-}OFF} \approx 0$, the pole $p_{hvn01\text{-}ON} << p_{hvn01\text{-}OFF}$. A double-pole effect before the UGF happens and may lead to instability when $C_L = C_{L\text{-}ON} = C_{powerFET}$ when the power FET is ON.

To avoid instability, we designed a feedback-resistive network which creates a medium frequency zero for warranting the stability. Referring to R_1, R_2 and C_1 in Fig. 2,

$$\frac{v_{g,n04}}{v_{reg}} = \left(\frac{R_2}{R'}\right)\left(\frac{1+\dfrac{s}{z_f}}{1+\dfrac{s}{p_f}}\right), \tag{5}$$

where V_{reg} and $V_{g,n04}$ are the voltages at the nodes at V_{reg} and gate of n04, respectively. The frequency of the zero, z_f, is lower than the pole frequency, p_f, and this zero can be used to cancel the pole effect of $p_{hvn01\text{-}ON}$.

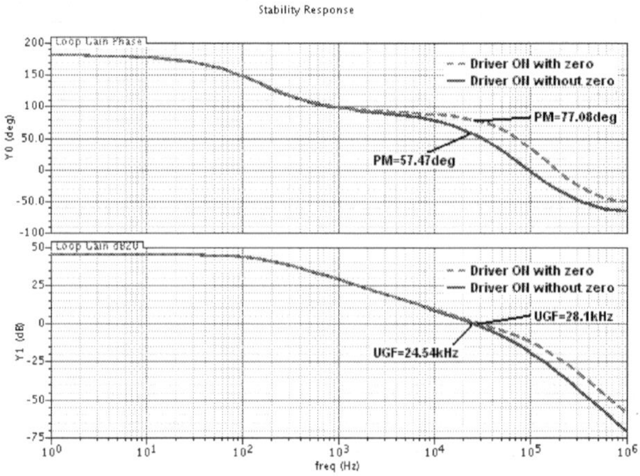

(a)

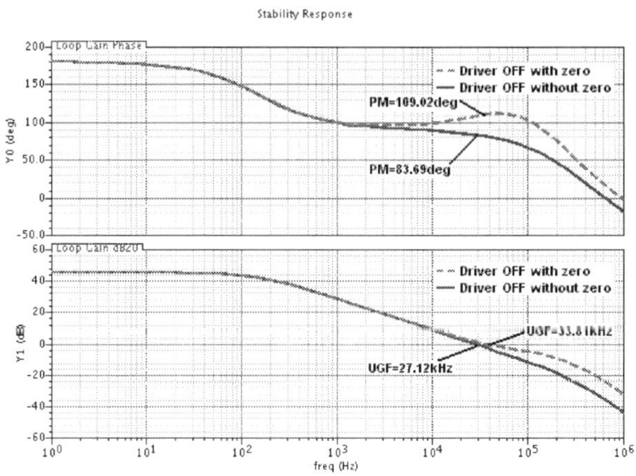

(b)

Fig. 3. (a) Simulated loop gain of the proposed driver with power FET ON; (b) Simulated loop gain of the proposed driver with power FET OFF

In order to have $z_f << p_f$, R_2 should be much smaller than R_1. From Fig. 3, the phase margin is very good even when the power FET is ON. However, if C_1 is not inserted, the double-pole effect will be significant. Results in Table I clearly show the pole-zero cancellation. When the power FET is OFF, the phase margins are around 109° and 83° with and without the zero z_f, respectively. When the power FET is ON, the phase margins are around 77° and 57° with and without the zero z_f, respectively. The differences in phase margin, with and without the zero z_f, are around 20° to 25° in both cases. With the pole-zero cancellation technique, the unity gain frequencies are also larger in both the power FET ON/OFF cases. These differences are significant in stability and transient analyses. The larger the phase margin, the less the ringing is. As the phase margin is larger, the settling time is faster also (Gray et al., 1990). The lab measurement results in Section 3 will demonstrate the steady and fast transient responses of the driver, thereby verifying the usefulness of the pole-zero cancellation technique in this type of regulated gate driver.

2.3 Power-saving: LV devices in HV application

HV devices differ from the normal LV ones in several ways. The size of a HV transistor is much larger than that of a LV transistor (Murari et al., 1995). There are several problems in using HV devices as inverter chains to drive power FETs, namely,

a. Large parasitic capacitance: The larger size HV transistors result in larger parasitic capacitance. The dynamic power, which is the product of the capacitance (C) and the square of the voltage (V), CV^2, is directly proportional to the parasitic capacitance. As a result, the total power consumption of a HV inverter is much higher than that of the LV one. The number of stages also trades off with the rise and fall times of the driver output and subsequently the delay of the driver output signal.

b. Severe V_{GS} limitation for thin-gate devices: Though LV devices are preferred, there is a gate-to-source V_{GS} swing limitation when LV devices are used in HV application. If the gate-to-source voltages of the pMOS and nMOS inside the inverters are above 5V, we must use thick-gate devices. The gate capacitance of the thick-gate devices are large and therefore will slow down the rise and fall times and the propagation delay. It also increases the cost as an extra processing step for thick-gate is needed.

c. Significant power loss in shoot-through current: During switching of the inverter chain, there is a shoot-through current flowing from the V_{reg} node to ground. Such dynamic current causes the V_{reg} voltage to drop (Heydari & Pedram, 2003). Since the operating voltage is 30V, the power of the shoot-through current still contributes much to the power loss.

d. Large die area: using HV-MOSFETs will occupy huge die areas and hence increase the wafer cost.

In the following, we propose solutions to solve the above problems by employing LV devices in HV driver application.

2.3.1 Power saving & thin-gate protection in the regulated driver

In the proposed design, we use all LV transistors (5V) in HV (30V) applications except two HV thin-gate transistors. This approach results in low dynamic power consumption and a small die area. We use LV devices to construct the inverter chain. The supply voltage of the inverters is given by the internal regulator at the V_{reg} node which maintains a 4.8V supply. This node is connected to the source of hvn01 whose drain is connected to V_{DDH}. This connection ensures that the internal regulator can give sufficient current to the inverter chain to drive the load. The maximum current is limited by the size of hvn01, or the internal supply voltage V_{reg} will go down if the loading current is too large. This regulated driver approach helps protect the thin-gate oxide of the power FET from damage by HV stresses.

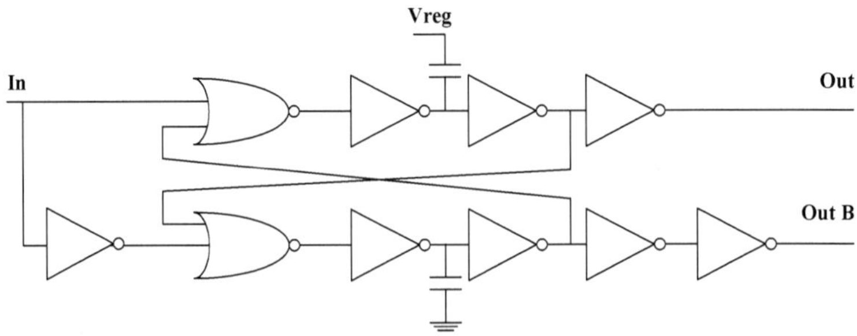

Fig. 4. Dead-time circuit

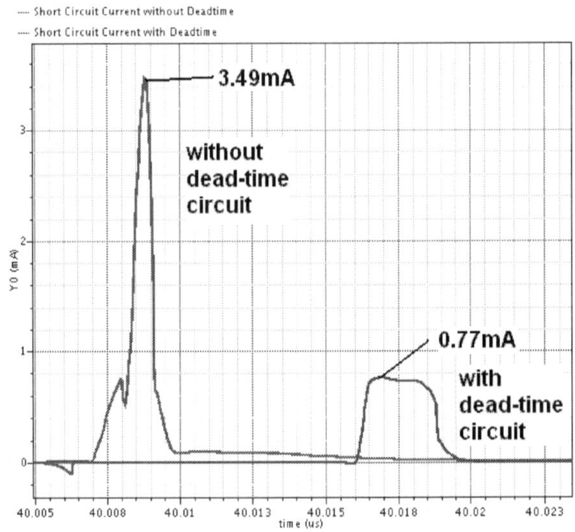

Fig. 5. Shoot-through current

2.3.2 Power saving via dead-time circuit

There are several ways to reduce the shoot-through current. In the proposed circuit, a dead-time control circuit is added for this purpose. This dead-time circuit prevents the flow of shoot-through current by a break-before-make logic. Fig. 4 shows the dead-time circuit and Fig. 5 shows the current going from the V_{reg} node to ground when the driver is charging up the load. Driver with the dead-time circuit only peaks up to 0.77mA, which is one-fifth of the driver without dead-time circuit. The 0.77mA current is mainly due to the switching activity of the dead-time logic. Since the dead-time circuit eliminates the shoot-through current of the final-stage driver, the driver possesses a higher slew rate and higher efficiency to drive the output capacitive load. The original driver has a 97ns rise time and 39.41V/μs slew rate, while the one with the dead-time circuit is 73ns and 52.06V/μs, respectively. The slew rate is improved by 32% owing to a larger portion of current charging up the output load C_L instead of being shunted to ground as shoot-through current.

2.3.3 Thin-gate protection circuit

Using SF configuration as thin-gate protection circuitry for HV-MOSFET is one of the innovations in this design. Referring to Fig. 2, the gate voltage of hvn01, $V_{G\text{-}hvn01}$, is limited by the SF configuration, where the gate voltage of hvn01, $V_{G\text{-}hvn01} \approx (V_{DDL} - V_{DS\text{-}p04} + V_{GS\text{-}hvp01}) \approx (4.5 - 0.2 + 1.2) = 5.5\text{V}$. The gate-to-source voltage of hvn01, $V_{GS\text{-}hvn01} \approx V_{G\text{-}hvn01} - V_{reg} = 5.5 - 4.5 \approx 1\text{V}$. The gate-to-source voltage of hvp01 is limited by $V_{DS\text{-}n02} + V_{t\text{-}hvp} \approx 0.2 + 1.2 \approx 1.4\text{V}$. The gate-to-source voltages of both hvp01 and hvn01 are therefore well limited below 5V. In other words, we utilize the SF characteristic where the source voltage tracks the gate voltage and subsequently protects the thin-gate oxide.

3. Simulations and lab. measurements

3.1 High current drive

Fig. 6 shows V_{DDH} vs V_{reg} with V_{DDL} fixed at 5V. Measurement result shows that V_{reg} becomes regulated when V_{DDH} exceeds 7V. The line regulation of V_{reg} from V_{DDH} at 7V to 30V is 0.113mV/V. Fig. 7 shows the transient simulations of the driver. The corresponding lab measurements are shown in Figs. 8-11, and the die photo is shown in Fig. 12. Obviously, the measurement agrees with the simulation results. When the power FET turns ON, the transient output current rises from 0 to 100mA in 525ps, i.e., about 190A/μs. When the power FET turns OFF, the output sinking current is about 120mA. With the large current driving capability, the output can charge a 270pF load within 100ns. That is, the driver is able to operate up to 10MHz even under heavy loading.

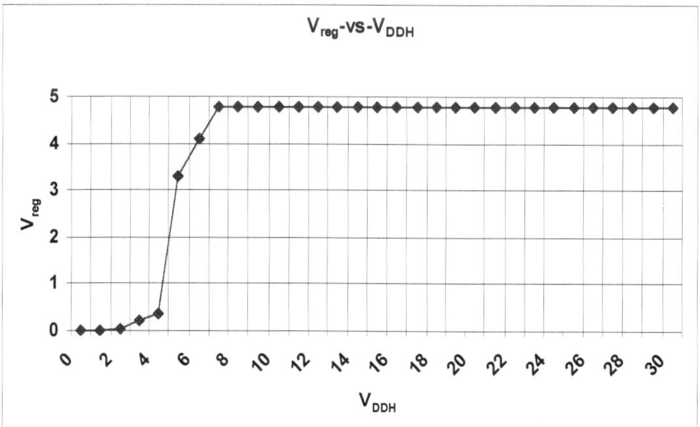

Fig. 6. V_{reg} vs V_{DDH} with V_{DDL}=5V

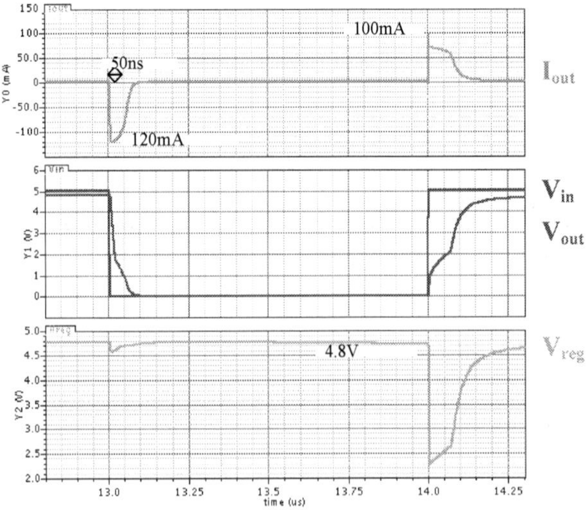

(a)

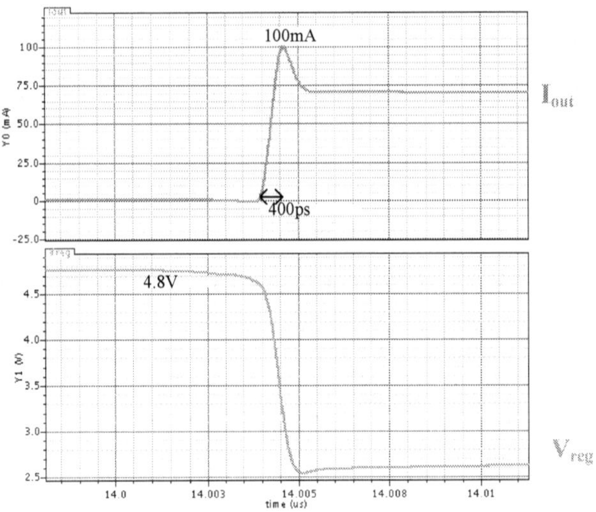

(b)

Fig. 7. Simulated transient responses with V_{DDH} = 30V: (a) overall waveforms (b) transient current I_{out} and V_{reg} when charging up output capacitor (gate capacitor of power FET)

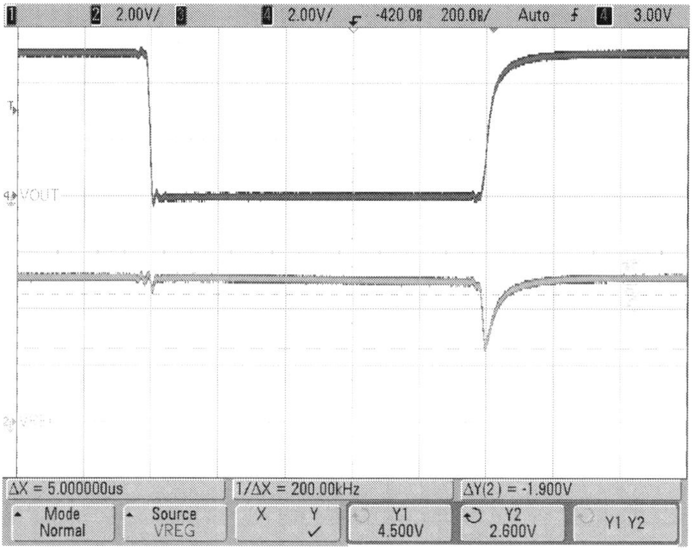

Fig. 8. V_{DDH}=30V; C_L=270pF; Red: V_{out}; Green: V_{reg}. Transient responses of V_{reg} when V_{out} is driving output capacitor (gate capacitor of power FET) (Zoomed in)

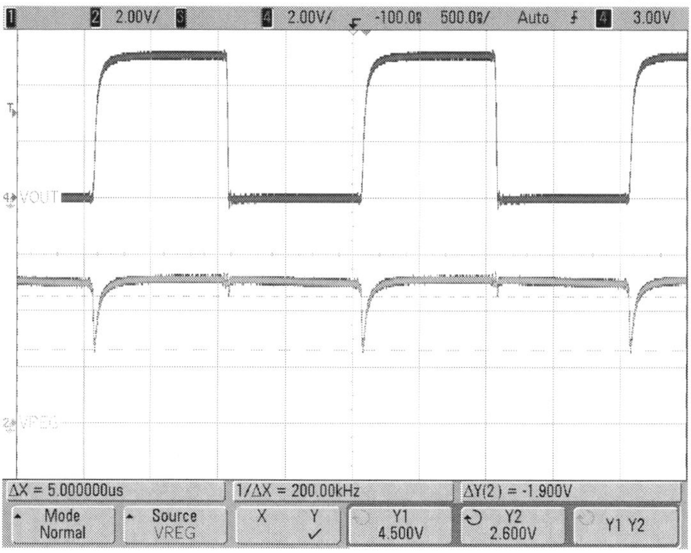

Fig. 9. V_{DDH}=30V; C_L=270pF; Red: V_{out}; Green: V_{reg}. Transient responses of V_{reg} when V_{out} is driving output capacitor (gate capacitor of power FET) (Zoomed out)

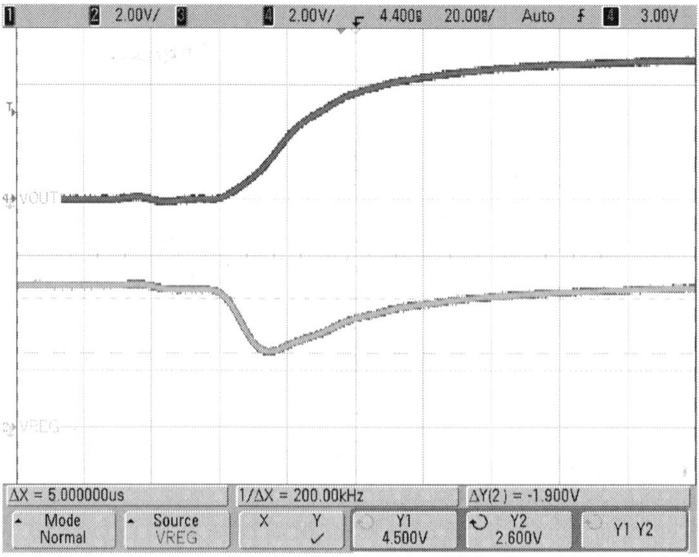

Fig. 10. Rise Time V_{DDH}=30V; C_L=270pF; Red: V_{out}; Green: V_{reg}

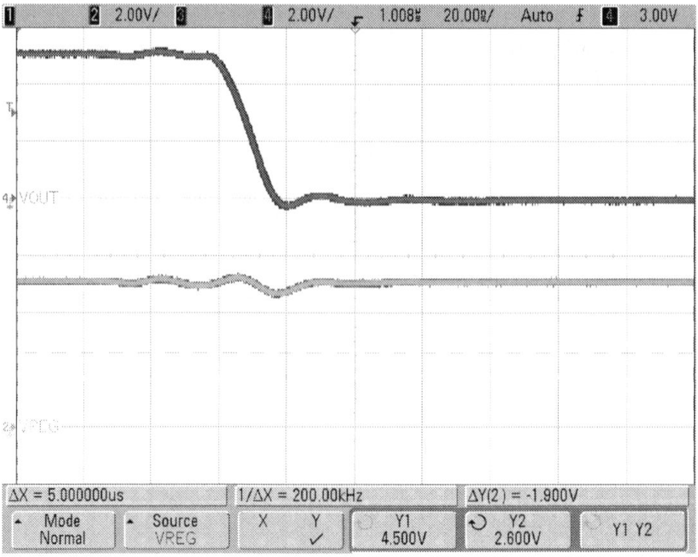

Fig. 11. Fall Time: V_{DDH}=30V; C_L=270pF; Red: V_{out}; Green: V_{reg}

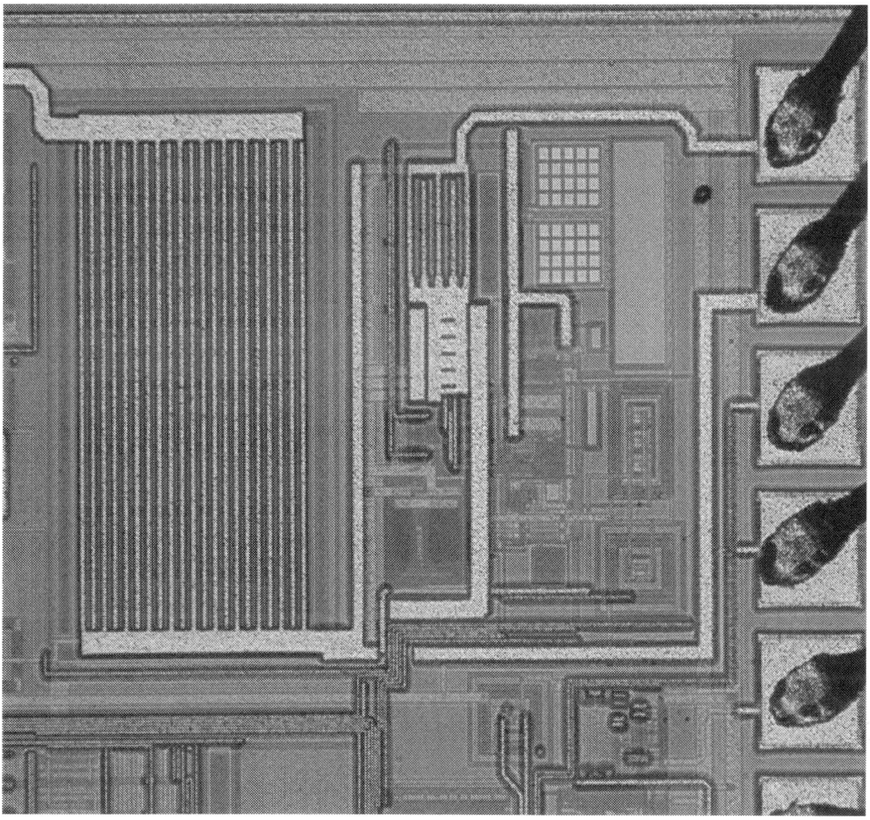

Fig. 12. Die photo of the proposed driver

3.2 Table of comparison

The left part of Table II shows the performance comparison between the proposed driver and other HV circuits with thin-gate protection. Our work features small die area (650μm × 200μm), high slew rate (52V/μs), fast transient current (190A/μs), fast rise (73.8ns) and fall time (17.5ns). The right part of Table II shows the comparison of our work and other drivers, including high-speed LV ones. Our work still features the smallest die area, highest slew rate, and fastest rise and fall times among all CMOS implementations. The bipolar implementation only shows fast rise and fall times under unloaded measurement, and its bipolar nature makes it unattractive for implementation due to high cost.

	This Work	Floating Gate Protection Technique (Chebli et al., 2007)	Low to High Voltage Digital Interface (Declercq et al., 1993)	High-Voltage CMOS OpAmp (Declercq et al., 1993)	Class AB Output Stage Op-Amp (Bales, 1997)	Class AB buffer amp with Slew Rate Enhancement (Lu & Lee, 2002)	Regulated Gate Driver (Tzeng & Chen, 2009)
Process	0.5μm	0.8μm	2.0μm	2.0μm	Bipolar	0.6μm	0.5μm
Die Area	0.13mm^2	0.9mm^2	N/A	N/A	0.8 mm^2	N/A	0.72mm^2
Dead-time circuit	√	×	×	×	×	×	×
Load	270pF	100pF	30pF	1000pF	N/A	680pF	2400pF
Slew Rate	52V/μs	N/A	–	15V/μs	N/A	2.41V/μs	N/A
Rise time	73.8ns	474ns	80ns	–	7ns	1.6μs	≈ 670ns
Fall time	17.5ns	445ns	80ns	–	7ns	1μs	N/A
Maximum output current	100mA @charge 120mA @discharge	N/A	N/A	20mA @charge	100mA @charge	N/A	N/A
HV supply, V_{DDH}	30V	60V	75V	75V	N/A	N/A	30V
LV supply, V_{DDL}	5V	5V	5V	N/A	5V	N/A	N/A
Thin-gate oxide Protected ?	Yes	Yes	Yes	Yes	N/A	N/A	N/A
Power	4mW	0.55mW	N/A	N/A	>7.5mW	1mW	546mW

Table II. Performance comparison between this work and similar works

4. Conclusion

A 7V-to-30V high-speed CMOS regulated driver for on-chip thin-gate power MOSFET has been developed. A small die area is achieved by minimizing the number of HV devices. The

HV devices are all thin-gate type and the corresponding V_{GS} and driver output voltages are constrained below 5V by the SF circuit technique. The driver is capable of delivering a fast transient current response of up to 190A/μs when charging up the output capacitor. The maximum charging and discharging currents are 100mA and 120mA, respectively, while keeping the quiescent current to below 90μA at static state for 7V to 30V application. A dead-time circuit is incorporated to reduce 75mW power loss due to shoot-through current. In short, we have developed a thin-gate-protected fast driver using standard HV CMOS process for HV applications with a small die area. This topology is applicable at 30V supply voltage or higher. The stability analysis for compensating this type of regulated driver is also presented to provide useful insights and guidelines for driver IC design.

5. References

Bales J. (1997). A low-power, high-speed, current-feedback op-amp with a novel class AB high current output stage, *IEEE J. Solid-State Circuits*, vol. 32, no. 9, pp. 1470–1474, Sep. 1997.

Chebli, R. et al., (2007). High-voltage DMOS integrated circuits with floating gate protection technique, in *Proc. of IEEE International Sym. on Circuits and Systems (ISCAS)*, pp. 3343–3346, 2007.

Declercq, M. et al., (1993). 5V-to-75V CMOS output interface circuits, in *Proc. 36th Int. Solid-State Circuits Conf. (ISSCC)*, pp. 162–163, 1993.

Gray, P. et al., (1990). *Analysis and Design of Analog Integrated Circuits*. New York: John Wiley & Sons, Inc., 1990.

Heydari, P. & Pedram, M. (2003). Ground bounce in digital VLSI circuits, *IEEE Trans. VLSI Syst.*, vol. 11, no. 2, pp. 180–193, Apr. 2003.

Hu, Y. Q. & Jovanovic, M. M. (2008). LED driver with self-adaptive drive voltage, *IEEE Trans. Power Electron.*, vol. 23, no. 6, pp. 3116–3125, Nov. 2008.

Lu, C. & Lee, C. (2002) A low-power high-speed class-AB buffer amplifier for flat-panel-display application, *IEEE Trans. VLSI Syst.*, vol. 10, no. 2, pp. 163–168, Apr. 2002.

Ma, D. et al., (2008). Adaptive on-chip power supply with robust one-cycle control technique, *IEEE Trans. VLSI Syst.*, vol. 16, no. 9, pp. 1240–1243, Sep. 2008.

Mentze, E. et al., (2006). A scalable high-voltage output driver for low-voltage CMOS technologies, *IEEE Trans. VLSI Syst.*, vol. 14, no. 2, pp. 1347–1353, Dec. 2006.

Murari, B. et al., (1995). *Smart Power ICs Technologies and Applications*. Springer-Verlag, 1995.

Parpia, A. & Salama C. A. T. (1990). Optimization of RESURF LDMOS transistors: an analytical approach, *IEEE Trans. Electron Devices*, vol. 37, no. 3, pp. 789–796, Mar. 1990.

Parpia, Z. et al., (1987). Modeling and characterization of CMOS-compatible high-voltage device structures, *IEEE Trans. Electron Devices*, vol. 34, no. 11, pp. 2335–2343, Nov. 1987.

Riccardo, D. et al., (2001). High voltage level shifter for driving an output stage, *US patent 6236244*, May 2001.

Tzeng, R. H. & Chen, C. L. (2009). A low-consumption regulated gate driver for power MOSFET, *IEEE Trans. Power Electron.*, vol. 24, no. 2, pp. 532-539, Feb. 2009.

13

Millimeter-Wave CMOS Impulse Radio

Ahmet Oncu[1,2] and Minoru Fujishima[1]
[1]*Hiroshima University*
[2]*The University of Tokyo*
Japan

1. Introduction

Millimeter waves are electromagnetic waves with wavelengths of 1 to 10 mm in vacuum, and they were discovered experimentally in the 19th century (Wiltse, 1984). In 1946, the most unique feature of millimeter waves, oxygen absorption at 60 GHz, was reported, which results in the rapid attenuation of electromagnetic waves in the air (Beringer, 1946). Although the oxygen absorption makes long-distance wireless communication difficult, it enables us to allocate a wide frequency band, which realizes ultra-high-speed communication greater than 1Gbps (gigabits-per-second). Recently, the well-known feature of millimeter-wave communication has attracted attention again because millimeter-wave circuits have been realized with advanced CMOS technologies, and the recent 60GHz band license-free regulations with license-free bandwidths of 9GHz in Europe and 7GHz in Japan, USA, Canada and Korea. In academic conferences and journals, many studies on millimeter-wave CMOS circuits were reported in the past few years, and consumer devices are expected to be available soon.

Here, for realizing the consumer application of millimeter waves, the reduction of power consumption is the most important issue. It is noted that the power-hungry building blocks in a transceiver are the local oscillator (LO) based on the phase-locked loop (PLL), and analog-to-digital and digital-to-analog converters (ADC and DAC) as shown in Fig. 1(a) (Marcu, 2009). If these blocks can be eliminated partially or completely in a transceiver, power consumption will be considerably reduced. From this viewpoint, we have studied millimeter-wave pulse communication for high-performance CMOS wireless transceivers as shown in Fig. 1(b) and Fig. 1(c). In this study, low-power direct pulse generators, high-speed switches and receivers, which are the most important building blocks in millimeter-wave pulse communication, are discussed for high-speed wireless communications using the 60 GHz band. In conclusion, the prospects for millimeter-wave pulse communication will be addressed.

2. 60GHz CMOS pulse transmitter

In this section, three low-power 60GHz CMOS pulse transmitter circuits are presented. The first one is a carrier-less direct pulse generator circuit, (Badalawa, 2007). The second design presents an 8Gbps millimeter-wave CMOS switch used for an Amplitude Shift Keying (ASK) modulator (Oncu, 2008, b) and the last one presents a design of a low-power 10Gbps

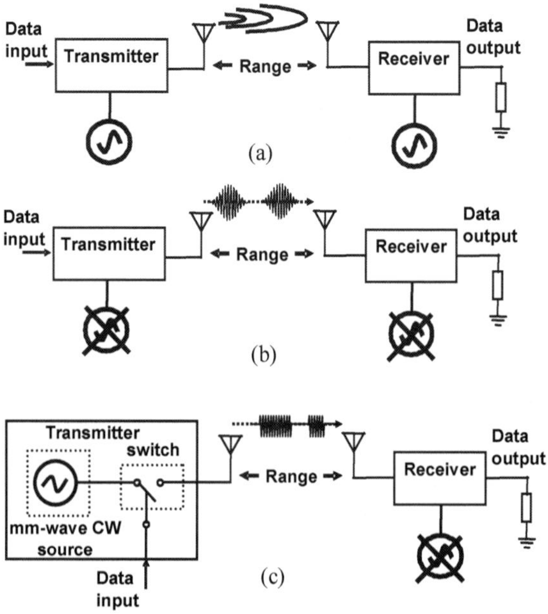

Fig. 1. Block diagram of wireless communication based on (a) carrier modulation, (b) direct pulse generator without oscillator, (c) pulse generator with millimeter-wave oscillator.

CMOS transmitter for a 60GHz millimeter-wave impulse radio, where a 60GHz millimeter-wave continues-wave (CW) source and ASK modulator circuits are embedded on the same silicon substrate.

2.1 60GHz CMOS pulse generator design

The circuit topology of the proposed pulse generator (PG) is shown in Fig. 2. This circuit has a monopulse generator (MPG) cell is composed of two CMOS inverters to contribute the delay and two NMOS transistors to produce the pulses by combing edges as shown in Fig. 3(a). The inverter A is driven by falling edges of baseband data. Just before the falling edge, NMOSFET C is "off" and NMOSFET D is "on". When the signal passes through inverter A, NMOSFET C is turned "on" and the output node is discharged. Next, when the input signal passes through inverter B, NMOSFET C is turned "off" and the output node is charged by a pulling-up inductor. At this moment, one pulse is produced according to the propagation delay of inverter B. The transmitter can be implemented with a low power consumption using this topology, because the circuit is activated only when falling edges of the input signal are fed from the baseband data. Since no power is consumed at other times, consumed power has a linear relationship with the input data rate.

To fit the delay time per inverter to 8ps, which is being equal to half the reciprocal of the carrier frequency, it is essential to reduce the load capacitance of the transistors that are connected to each inverter output node. To obtain a short delay time, the gate widths of NMOS and PMOS transistors in the inverter should be increased to obtain a large drain

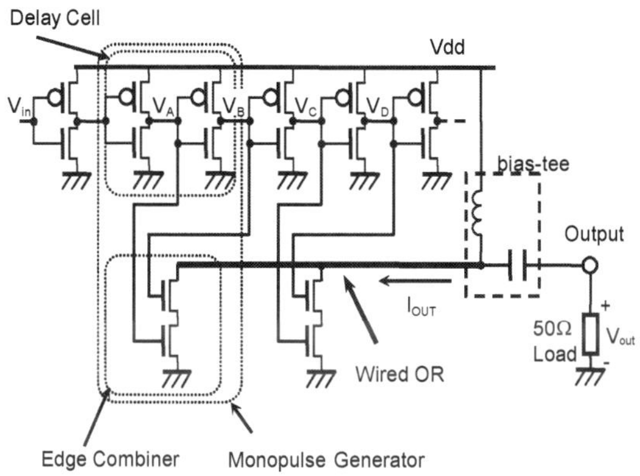

Fig. 2. Circuit topology of a 60GHz CMOS pulse generator.

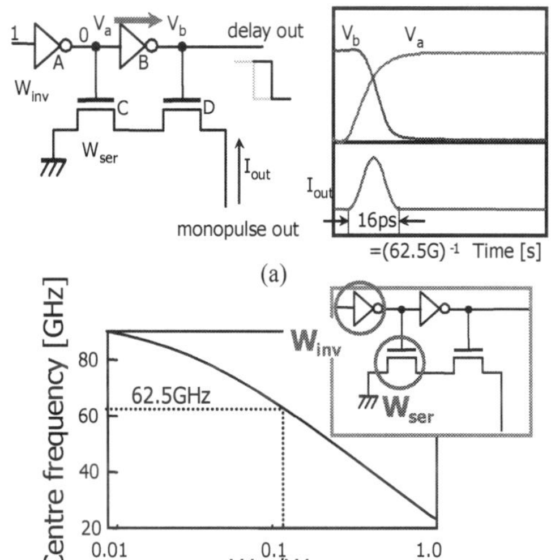

Fig. 3. (a) An edge combiner comprising MOSFETs C and D has to generate a 16ps pulse. (b) Centre frequency as a function of NMOS width Wser over inverter width Winv.

current. Since the load capacitance connected to the inverter output node is varied favorably or unfavorably with the inverter delay, the relationship between the size of the inverter and the edge combiner NMOSFET is essential to obtain a carrier frequency of 62.5GHz. Figure 3(b) shows the relationship between centre frequency when the fan-out is varied from 0.01 to 1. To realize a 60GHz PG using this circuit topology, the fan-out should be set to 0.1. Not only the optimization, but also selecting of CMOS process with small threshold voltage is one of the key points to implement 60GHz pulse generator as mentioned above. Here, we choose the 9metal TSMC CMOS 90nm process, which has the 1/2 times of small threshold voltage of the CMOS process used in 22-29GHz UWB CMOS pulse generator circuit in (Fujishima, 2006).

The power spectrum must fit into a spectrum mask to meet regulations as shown in Fig. 4. Here, filtering is employed (Maruhashi, 2005) to satisfy the regulations while increasing the power consumption. To solve these problems, an all-digital low-power CMOS pulse generator with 14 delay stages, which generates a pulse width of 224ps, is adopted. To satisfy the power spectrum regulations without any filters, monopulse amplitudes within a single pulse are adjusted to four levels to approximate the ideal Gaussian power spectrum by sizing the edge-combiner NMOSFET as shown in Fig. 5.

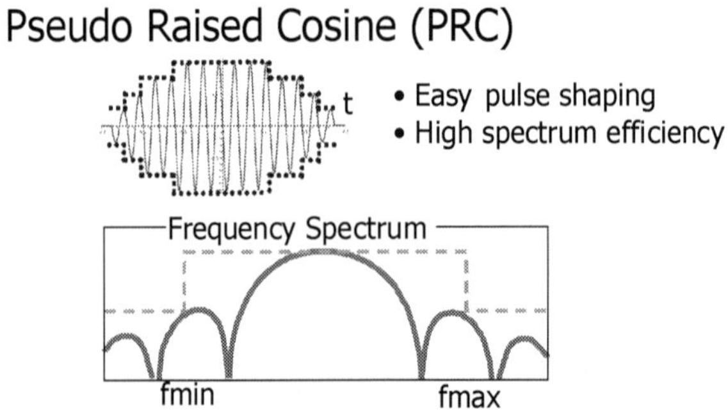

Fig. 4. Pseudo-raised cosine pulse for satisfying specified.

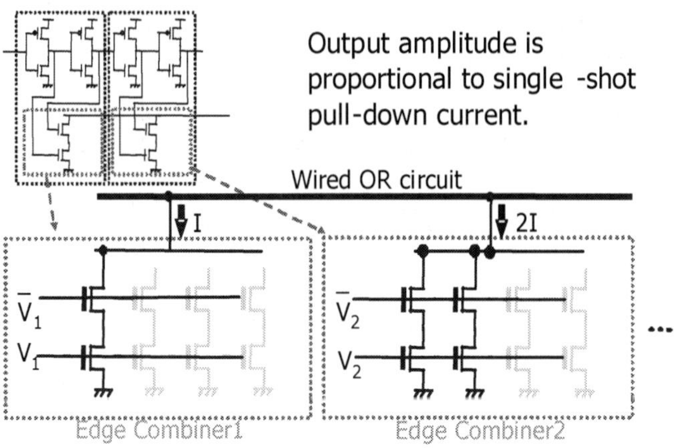

Fig. 5. MOSFET sizing for generating pseudo-raised cosine pulse.

Figure 6 shows a chip micrograph of the CMOS pulse generator with a die area of 590×380μm², where a 90nm CMOS process with nine metal layers was used. The time-domain response of the pulse generator is shown in Fig. 7, where the 62.5GHz operating frequency is observed at a supply voltage of 1.15V, and the four-level approximation is confirmed.

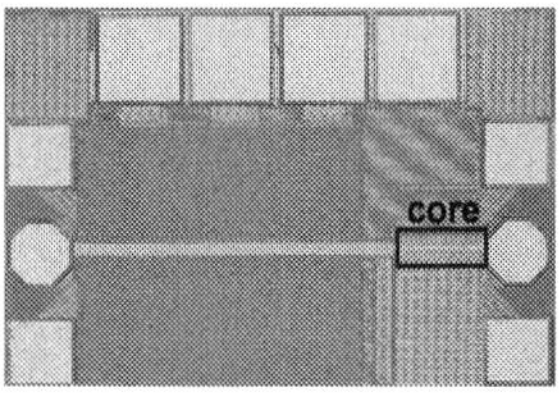

Chip micrograph CMOS 90nm
(nine metals one ploy)

Technology	90nm CMOS
Frequency Band	59-66GHz
Modulation	PPK
Supply Voltage	1.15V
Input Data Rate	1.5Gbps
Power Consumption	11mW@ 1.5Gbps
Core Size	90μm×20μm

Fig. 6. Chip micrograph and specification of 60GHz CMOS pulse generator.

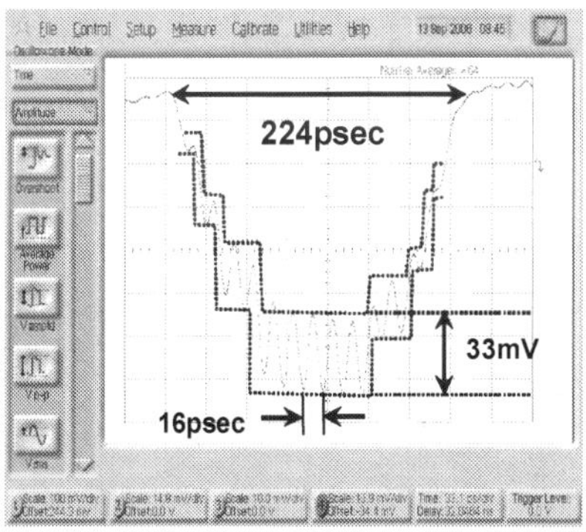

Fig. 7. Measured transient response of 60GHz CMOS pulse generator.

Figure 8 shows carrier frequencies and output powers as function of supply voltage, and also shows power consumptions as function of input data rate. The carrier frequency increased with supply voltage with inverse proportional relationship, while output power is almost unchanged when supply voltage is higher than 0.7V. The linear dependence of power consumption on input data rate is confirmed by the measurement data. Since power is only consumed at rising edges of the input signal, a low average power consumption is observed at 1.5Gbps compared with those in (Maruhashi, 2005; Nakakita, 1997). The power consumption for the proposed pulse generator is 11.5mW at a supplies voltage of 1.15V.

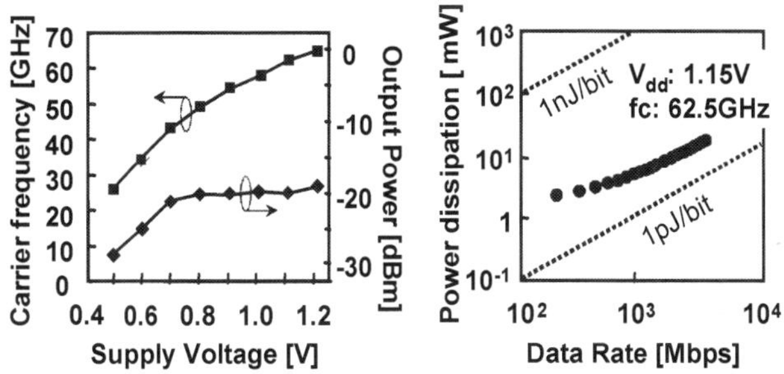

Fig. 8. (a) Carrier frequency and output power as a function of supply voltage and (b) power dissipation as a function of input data rate.

In this section millimeter-wave pulse generator was studied. By designing pulse generators in digital circuits, a 60GHz millimeter-wave pulse can be generated without using a power-hungry LO. As a result, the pulse generator consumes a small amount of power proportional to input data rate. However, this architecture strictly depends on the used technology to achieve higher RF power. We concluded that shorter channel advanced CMOS processes would provide better speed and RF power performance. In the following sections, we study the pulse generator architectures consisting of a low-power millimeter-wave ASK modulator and a 60GHz oscillator in standard CMOS process which is generally used for digital processor design.

2.2 8Gbps 60GHz CMOS ASK modulator

A millimeter-wave CMOS impulse radio with ASK modulator, as shown in Fig. 9, is promising for low-cost and low-power wireless communication, in which a digital switch controls a millimeter-wave CMOS ASK modulator in the transmitter. This architecture will have less sensitivity to the used CMOS technology than that of a direct millimeter-wave pulse generator. The receiver receives 60GHz pulses and converts them to a digital signal (Oncu, 2008, a; Lee, 2009). In this section, we study a design of an 8Gbps CMOS ASK modulator for a 60GHz millimeter-wave impulse radio.

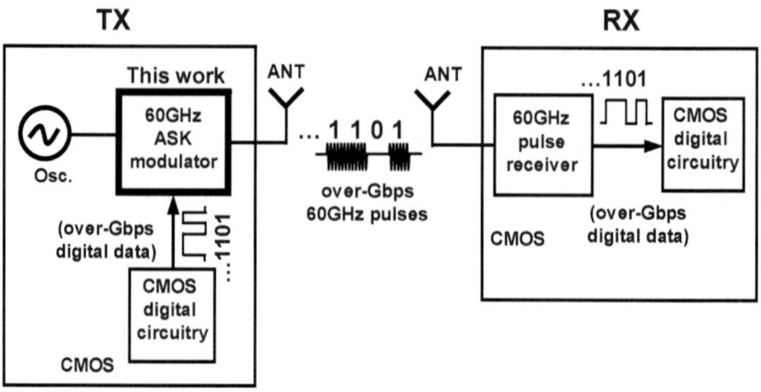

Fig. 9. Block Diagram of millimeter-wave impulse radio with a 60GHz ASK (Amplitude Shift Keying) modulator.

Figure 10(a) shows a conventional millimeter-wave ASK modulator in CMOS (Chang, 2007). It consists of an oscillator and a buffer. Millimeter-wave pulses are obtained by turning the biasing on and off. Although this architecture has high isolation when the biasing is turned off, the switching speed is limited by the stored energy in the oscillator tank. High-speed conventional distributed traveling-wave millimeter-wave ASK modulators in compound semiconductors have been reported (Mizutani, 2000; Ohata, 2000; Ohata, 2005; Kosugi, 2003; Kosugi, 2004). They were realized using distributed shunt switches between the signal and the ground line of a transmission line as shown in Fig. 10(b). In this architecture, when the switches are off the input signal is transferred to the output and the ASK modulator is in the ON state. On the other hand, when the switches are turned on, no input signal is transferred to the output and the ASK modulator is in OFF state. The distributed structure requires a large number of switches since the resistances of the switches in the OFF state should be small to realize a lossy transmission line.

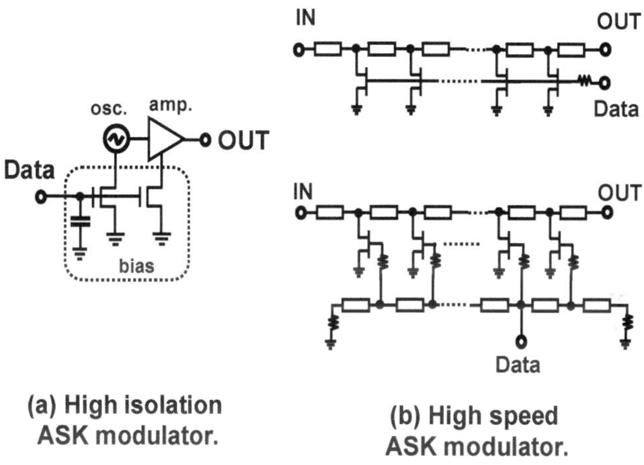

Fig. 10. Architectures of conventional (a) high-isolation and (b) high-speed ASK modulators.

2.2.1 Millimeter-wave CMOS ASK modulator design

A possible distributed CMOS modulator is shown in Fig. 11(a). However, low-quality parasitic capacitances in the switches, which are located on a silicon substrate, are expected to degrade the transmission line characteristics. In this study, a reduced-switch architecture is used for a high-speed millimeter-wave CMOS ASK modulator as shown in Fig. 11(b). Note that the isolation characteristics become degraded upon reducing the number of switches since each switch has a leakage to the output. To achieve high isolation with a reduced number of switches, the transmission line length between switches is adjusted. When the millimeter-wave signal travels from the source to the load, the switches do not only dissipate the incident signal, but they also reflect and leak it as shown in Fig. 12. Note

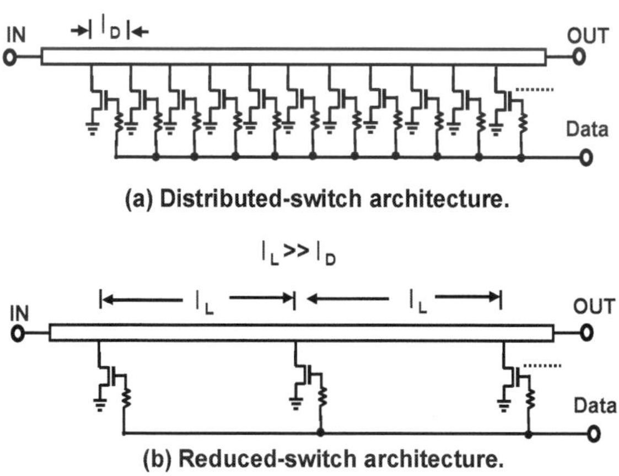

Fig. 11. Architectures of (a) distributive and (b) reduced-switch ASK modulators in CMOS process.

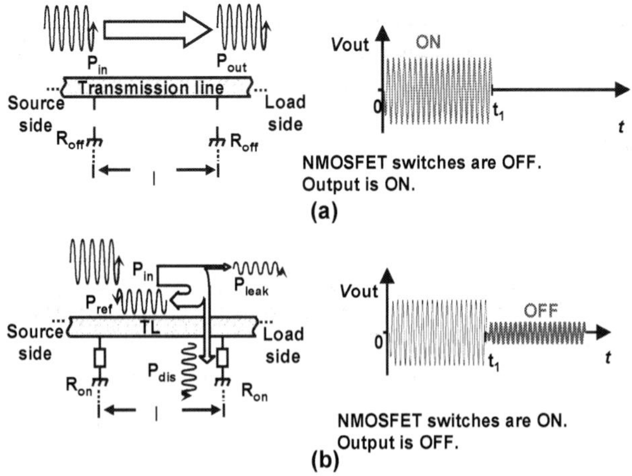

Fig. 12. Illustration of transmitted, reflected, dissipated and leaked signals of a switch in the (a) ON and (b) OFF states of the modulator when the millimeter-wave signal travels from source to the load.

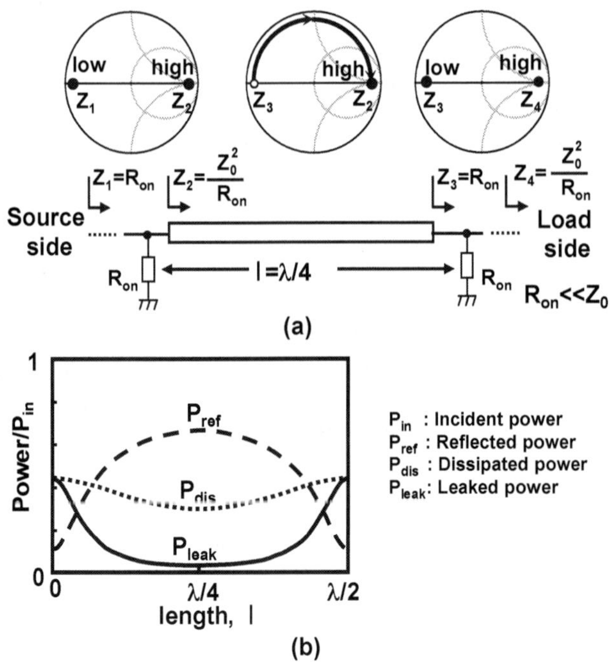

Fig. 13. (a) Impedance transformation along the modulator and (b) calculated reflected, dissipated and leaked powers as a function of the transmission line distance between switches.

that, in a transmission line, impedance transformation between the two terminals occurs as shown in Fig. 13(a). In Fig. 13(b), the calculated leaked, reflected and dissipated powers are shown as a function of the distance between switches. Since the dissipated power in the switches is insensitive to the transmission line length, reflection should be maximized to minimize the leakage. To obtain maximum reflected power and minimum leaked power, the switches are separated by a quarter-wavelength distance. In this case, the isolation is maximized with a lower number of switches.

A 60GHz CMOS ASK modulator is designed with three NMOSFET switches and two quarter-wavelength transmission lines as shown in Fig. 14. When the digital input is 0V, the NMOSFET switches are turned off. Since the parasitic capacitance of each switch in the OFF state is negligible, the input impedance of each transmission line is equal to the load impedance and the input power is transferred to the output. When the digital input is 1V, the switches are turned on. The transmission line with a quarter wavelength transforms the low impedance of the switch to a high impedance and reflection is maximized. In this case, the leaked power to the output is minimized and high isolation is achieved.

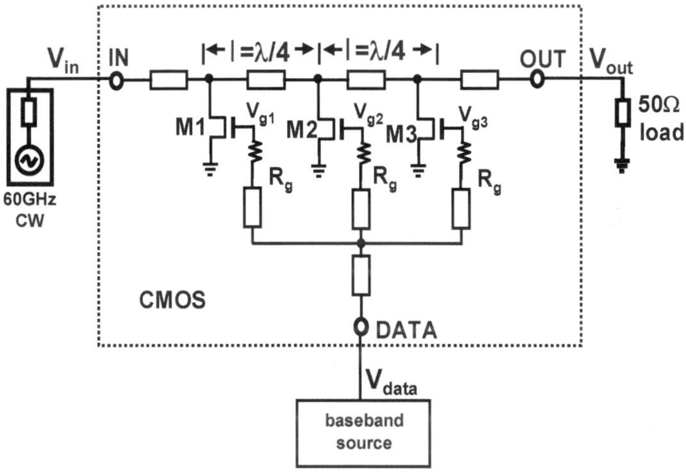

Fig. 14. Circuit schematic of the CMOS ASK modulator for 60GHz wireless communication.

Millimeter-wave NMOSFET models are established by extracting the parasitic components based on on-wafer measurements (Doan, 2005). The slow-wave transmission line (SWTL) (Cheung, 2003) shown in Fig. 15 is used for implementing the quarter-wavelength transmission lines and the networks between the circuit and the pads to reduce the size of the modulator. In the SWTL, a slotted ground shield under the signal line is laid orthogonal to the direction of the signal current flow. This structure results in the propagating waves having lower phase velocity; thus, the corresponding wavelength at a given frequency is reduced. A quarter wavelength is obtained using a 450-μm-long SWTL. Note that the quarter wavelength would be 850μm if a microstrip line (MSL) was used.

200Ω gate resistors are inserted to ensure operation with sufficient high-speed. Transient internal waveforms are simulated as shown in Fig. 16. A 200ps pulse is applied from the data port to analyze the response of the circuit. The total time of the rising and falling gate

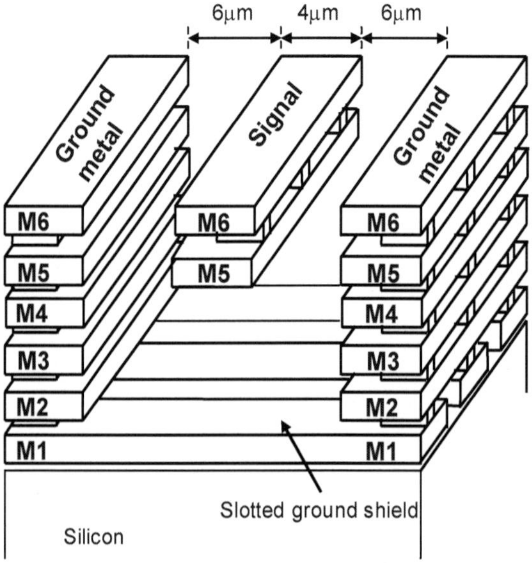

Fig. 15. Structure of the slow-wave transmission line used in the circuit.

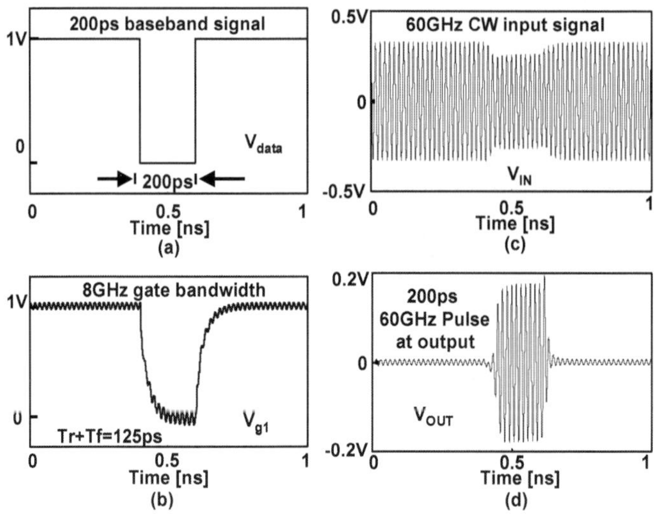

Fig. 16. Transient simulation; (a) 200ps applied data pulse, and responses of (b) the gate voltage of the NMOSFET switch, and (c) input and (d) output signals.

voltages is estimated as 125ps, which corresponds to the maximum data rate of 8Gbps. The 60GHz millimeter-wave ASK modulator is fabricated by a 6-metal 1-poly 90nm CMOS process. The cutoff frequency f_T and the maximum operation frequency of the nMOSFET are 130GHz and 150GHz, respectively. Figure 17 shows a micrograph of the fabricated ASK modulator. The size of the chip is 0.8mm × 0.48mm including the pads. The core size is 0.61mm × 0.3mm.

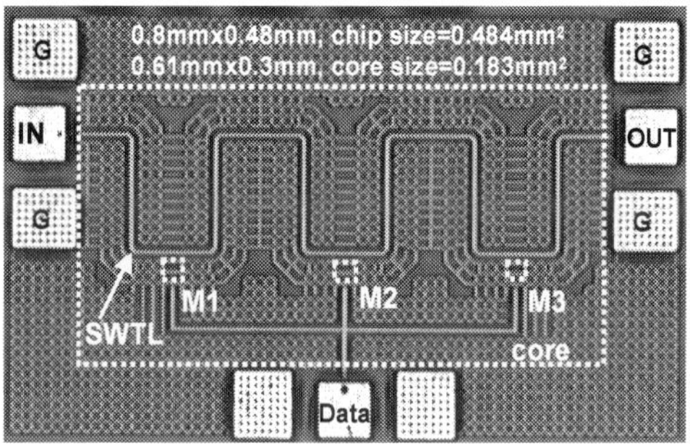

Fig. 17. Micrograph of the fabricated chip.

2.2.2 Experimental result and discussion

On-wafer two-port measurements were performed up to 110-GHz with Anritsu ME7808 network analyzer with transmission reflection modules for the ON and OFF states by applying 0V and 1V DC voltages to the gate terminal, respectively. The measured and simulated insertion losses of the modulator for the two states are shown in Fig. 18(a) for comparison. The insertion losses in the ON and OFF states are 6.6dB and 33.2dB, respectively, at 60GHz. Isolation is defined as the insertion loss difference between the ON and OFF states, which is 26.6dB. The isolation is nearly flat from 20 to 80GHz, although the maximum isolation is measured at 60GHz. As a result, shorter transmission lines may be adopted to reduce the insertion loss caused by the SWTL in the ON state of the modulator. The simulated isolation is shown at frequencies up to 350GHz in Fig. 18(b) to demonstrate

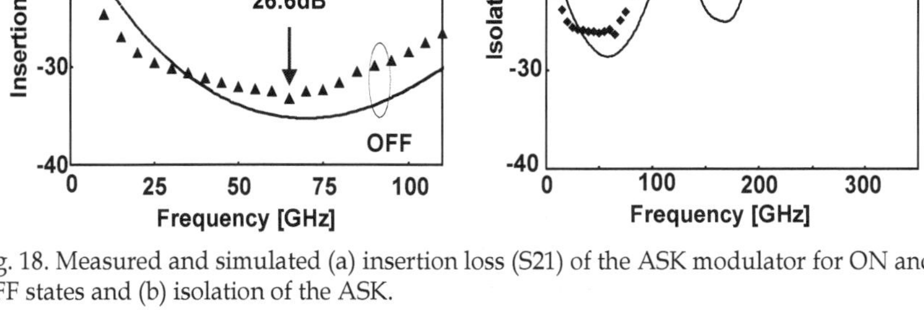

Fig. 18. Measured and simulated (a) insertion loss (S21) of the ASK modulator for ON and OFF states and (b) isolation of the ASK.

the frequency behaviour of the modulator. The minimum isolation appears at 60GHz when the electrical length of the transmission lines is $\lambda/4$, where λ is the wavelength. Local maxima in the OFF-state insertion loss occur at 180GHz and 300GHz, which correspond to $3\lambda/4$ and $5\lambda/4$, respectively.

The time-domain response is measured using a 70GHz sampling oscilloscope, a 60GHz millimeter-wave source module and a pattern generator. No external filters are applied in the measurement. A 60GHz continuous wave is applied to the RF input and the modulator is controlled by the pattern generator. The rising and falling times of the applied baseband signal are 6ps and 8ps, respectively. The output response for the maximum data rate is shown in Fig. 19(a). In Fig. 19(b), the output response is shown for a 125ps single-baseband pulse by reducing the scale to 20ps.

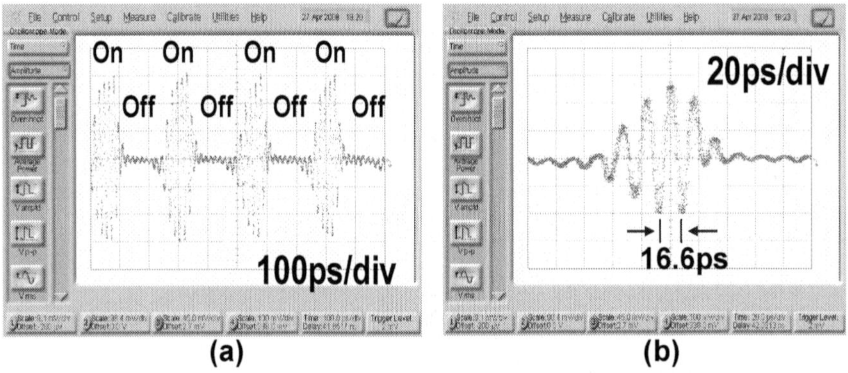

Fig. 19. Measured output response of the modulator for (a) an 8Gbps data train and (b) a single 125ps data pulse.

The maximum data rates as a function of the isolation of the millimeter-wave ASK modulators are shown in Fig. 20. It can be seen that the isolation and the maximum data rate have a tradeoff relationship. The product of the maximum data rate and the isolation of this modulator is 170GHz, which is the highest value among multi-Gbps ASK modulators.

2.3 12.1mW 10Gbps pulse transmitter for 60GHz wireless communication

In this section, we present a design of a low-power 10Gbps CMOS transmitter (TX) for a 60GHz millimeter-wave impulse radio, where a 60GHz millimeter-wave CW source and ASK modulator circuits are embedded on the same silicon substrate as shown in Fig. 21. An 8Gb/s CMOS ASK modulator for 60GHz wireless communication is studied in Section 2.2. This single-pole-single-throw (SPST) reduced NMOSFET switch architecture is capable of high-speed operation without DC power dissipation. Its isolation was maximized by a quarter-wave length transmission line which results in a long transmission lines, therefore the insertion loss becomes high. Figure 22(a) shows TX configuration which consists of an off-chip 60GHz millimeter-wave CW source and an on-chip CMOS modulator. Off-chip millimeter-wave source module will increase the size, the total power consumption and the cost of the TX system. The oscillator should be embedded in the CMOS chip for a practical application. The millimeter-wave CMOS oscillators are commonly designed in differential

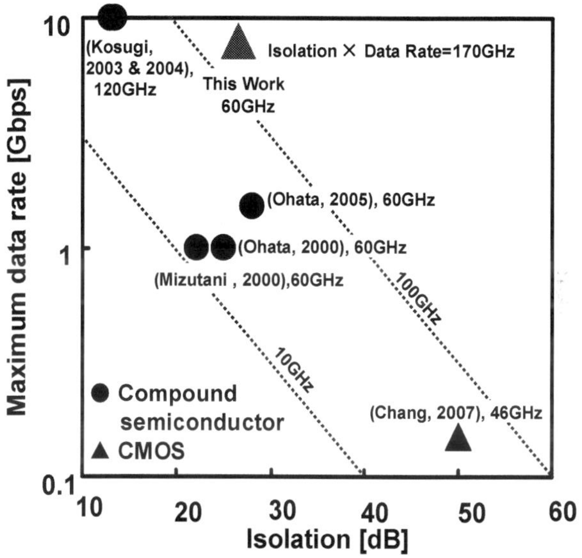

Fig. 20. Maximum data rates as a function of isolation of the ASK modulators.

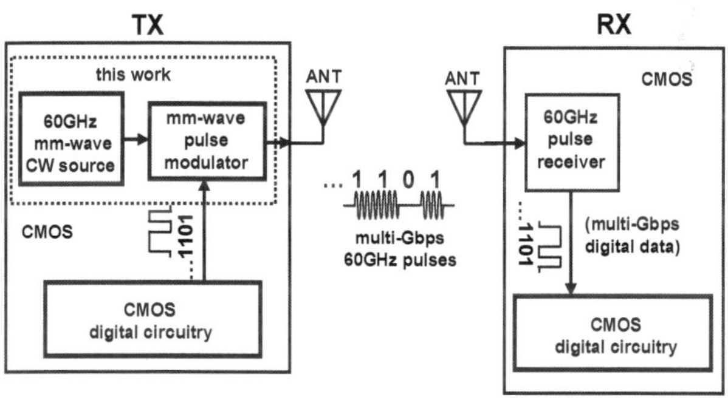

Fig. 21. Block diagram of a Giga-bit millimeter-wave wireless pulse communication in CMOS.

ended (Huang, 2006). In this design a differential ended CMOS oscillator was designed for a 60GHz CW source. To utilize the differential-ended output signal, a double-pole-single-throw (DPST) switch was proposed for modulator as shown in Fig. 22(b).

2.3.1 60GHz pulse transmitter design

2.3.1.1 60GHz CMOS CW Signal Source Design

Figure 23 shows the schematic of the on-chip 60GHz CW source circuit which consist of two sub-blocks, a 60GHz oscillator and a buffer. The oscillator generates a 60GHz CW signal and the buffer drives the ASK modulator. The 60GHz oscillator contains an on-chip transmission

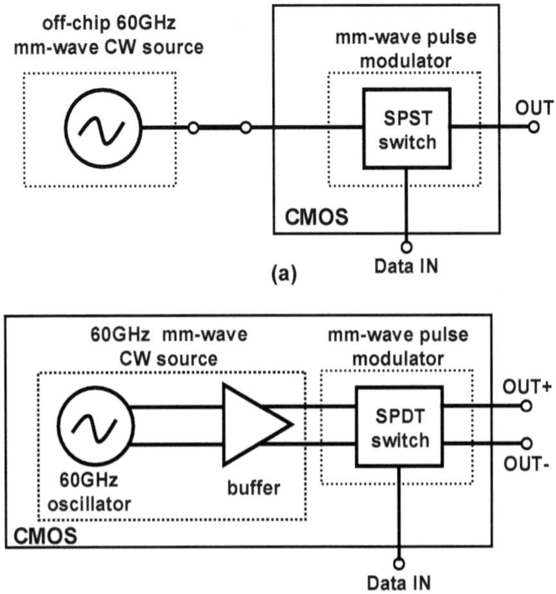

Fig. 22. Architecture of (a) a single-ended millimeter-wave pulse transmitter with off-chip 60GHz CW source and (b) a proposed differential-ended pulse transmitter with on-chip 60GHz CW source.

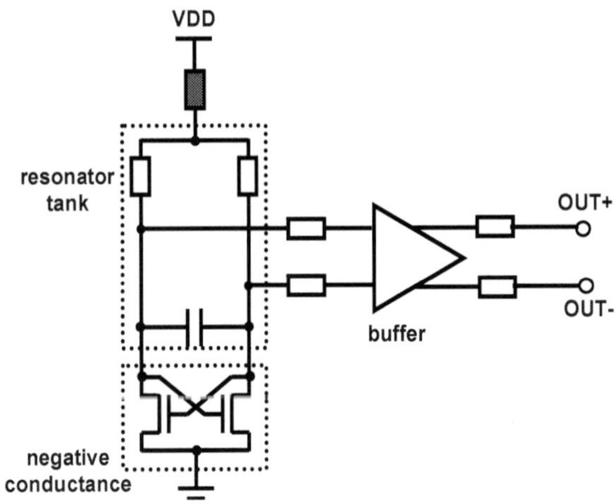

Fig. 23. Circuit schematic of a 60GHz millimeter-wave continues-wave (CW) source.

line resonating tank with a MOS capacitor and two cross-coupled MOSFETs which realize a negative conductance in parallel with the tank. The size of the devices was chosen by considering the parasitic and the process variations to keep the resonation at the 60GHz millimeter-wave band. The active device and the MOS capacitor models were obtained from the foundry. The transmission lines were characterized by a 3D full-wave electromagnetic field simulation using high-frequency structure simulator (HFSS).

The bias voltage does not only affect the negative conductance but also power consumption. High supply voltage results in a high-power dissipation. Even though a maximum 1.2V supply voltage is allowed in this CMOS process, it is simulated in spectre RF that the oscillation starts when the supply voltage is approximately 0.9V. 0.1V was decided as a margin and the supply voltage was set to be 1V for low-power operation.

2.3.1.2 Millimeter-wave Differential Ended CMOS ASK Modulator Design

Figure 24 shows the 60Hz differential ended CMOS ASK modulator. It is designed by a DPST switch consisting of a parallel connected two SPST switches. The inputs are connected to the complementary outputs of the on-chip 60GHz signal source. The gates of the switches are controlled by binary data. Each SPST switch is designed with two NMOSFET switches and a transmission line, TL1 as shown in Fig. 24. When the digital input is 0V, the NMOSFET switches are turned off. Since the parasitic capacitance of each switch in the OFF state is negligible, the input impedance of each transmission line is equal to the load impedance and the input power is transferred to the output as shown in Section 2.2 Fig. 12(a). When the digital input is 1V, the switches are turned on. The transmission line transforms the low impedance of the switch to high impedance and reflection is increased. In this case, the leaked power to the output is reduced and isolation is improved as shown in Section 2.2 Fig. 12(b).

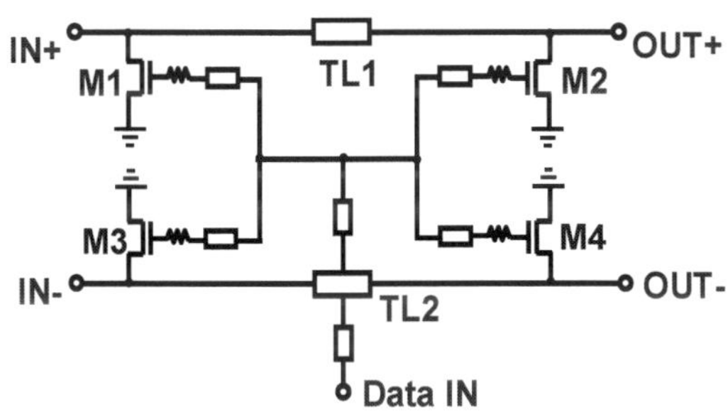

Fig. 24. Circuit schematic of the differential-ended ASK modulator for 60GHz millimeter-wave pulse transmitter.

The isolation is theoretically maximized when the switches are separated by a quarter-wavelength transmission line however long transmission lines result higher insertion loss. The isolation was maximized with two quarter-wavelength transmission lines whose total length is 900μm which results in 6.6dB insertion loss in Section 2.2. The isolation is nearly flat from 20 to 80GHz, although the maximum isolation is measured at 60GHz. As a result, shorter transmission lines may be adopted to reduce the insertion loss caused by the on-chip transmission line in the ON state of the modulator. In this CMOS technology, the length of a quarter-wavelength transmission line is 600μm. We designed the switch with a 300μm long transmission line where the isolation will slightly degrade but the insertion loss will improve.

2.3.2 60GHz pulse transmitter measurement and discussions

The proposed pulse transmitter, a 60GHz millimeter-wave source and an ASK modulator test circuits were fabricated by an 8-metal-1-poly 90nm CMOS process with a rewiring layer fabricated by a wafer-level chip-scale package (W-CSP). Figure 25 shows the micrographs of the pulse transmitter chip. In this design, the pitch of radio frequency and the biasing pads are designed 150μm.

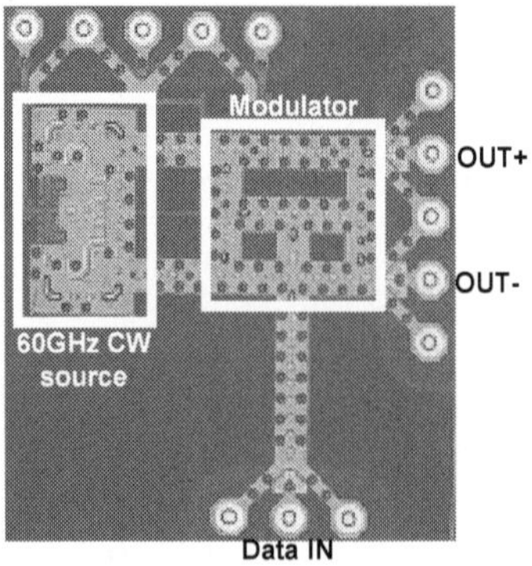

Fig. 25. Micrograph of the fabricated 60GHz pulse transmitter chip.

2.3.2.1 60GHz CW signal source

The spectrum of the 60GHz CW signal source was measured using an Agilent E4407B spectrum analyzer and an Agilent 11970V 50-75GHz harmonic mixer. A 60GHz continues-wave signal was measured at the output of the circuit whose spectrum is shown in Fig. 26. In this measurement setup, the total power loss of the probe, cables, connecters and harmonic mixer is approximately 42dB. It was observed that the fabricated chip starts to oscillate when the bias voltage is larger than 0.7V. The measured operating frequency as a function of supply voltage is plotted in Fig. 27(a). Figure 27(b) shows the power dissipation and millimeter-wave RF power as a function of the supply voltage from 0.7V to 1.4V. As the supply voltage increases, the power dissipation rapidly increases. However, the millimeter-wave output power saturates when the supply voltage reaches near to 1V. The power

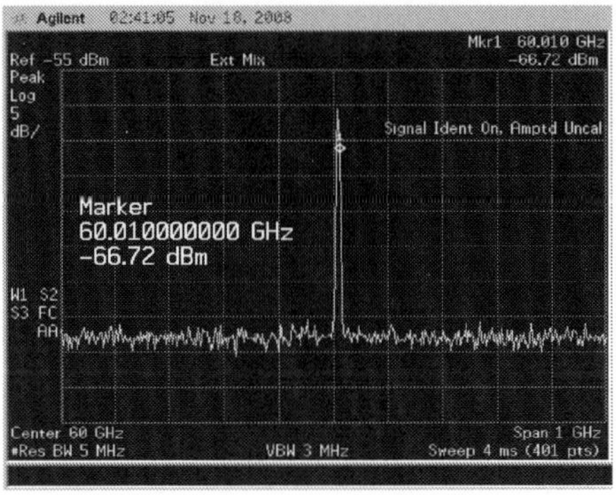

Fig. 26. Measured output spectrum of the 60GHz CW source.

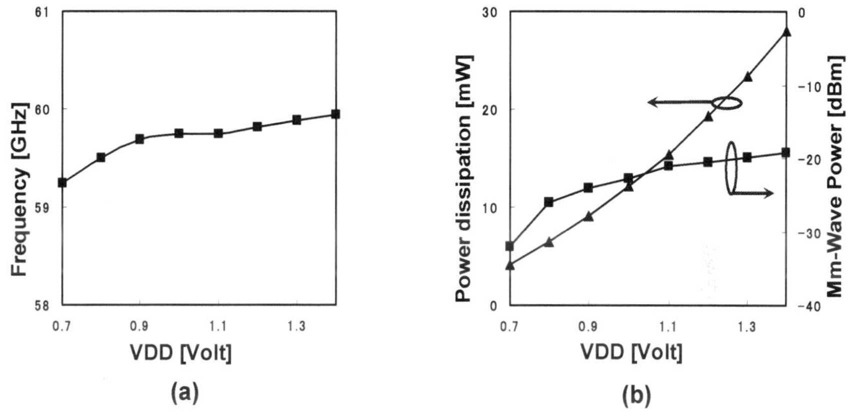

Fig. 27. Measured (a) operating frequency of the oscillator and (b) power dissipation and output millimeter-wave power of the oscillator as a function of supply voltage.

dissipation was measured to be a 19.2mW at a maximum allowed supply voltage of 1.2V. We reduced to the supply voltage to 1V for low-power operation where the millimeter-wave output power was measured to be -20.7dBm and power dissipation of 12.1mW. In this study, we found out that our layout versus schematic verification software had not been functioning properly while we had been designing the circuit using this 90nm CMOS technology first time. The core of the oscillator operates properly; however, because of the verification error in the layout, we noticed that the buffer attenuates the generated millimeter-wave signal by 18dB although it was designed to have 10dB gain.

2.3.2.2 Millimeter-wave CMOS ASK Modulator

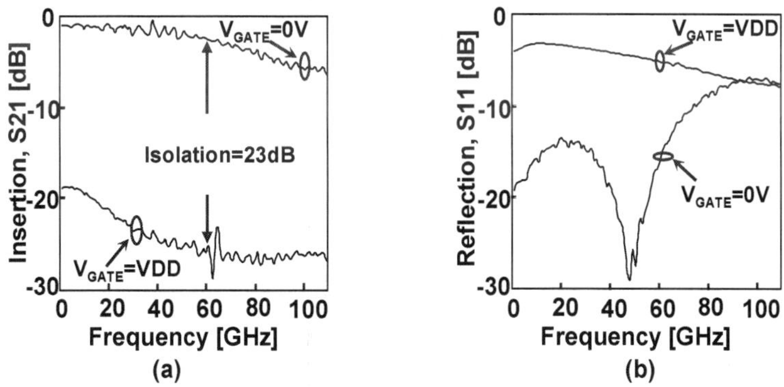

Fig. 28. Measured (a) insertion loss (S21) and (b) reflection loss (S11) of the ASK modulator for ON and OFF states.

The scattering parameters of the ASK modulator test circuit were measured on-wafer up to 110GHz with Anritsu ME7808 network analyzer with transmission reflection modules for the ON and OFF states, respectively. The measured insertion losses of the modulator for the two states are shown in Fig. 28(a). When the gate voltage is 0 volt, the insertion loss was measured to be a 2.3dB at 60GHz. When the gate voltage was increased to VDD, the insertion loss became 25.8dB therefore isolation was calculated to be 23.5dB at 60GHz, which is defined as the insertion loss difference between the ON and OFF states. Figure 28(b) shows the measured reflection of loss of the modulator for the two states. When the modulator is ON, S11 is lower than -10dB up to 75GHz and it was measured to be a -16.2dB at 60GHz where it was matched to 50Ω system. When the modulator was turned on by increasing the gate voltage, the S11 became -5.2dB. The maximum data rates as a function of the isolation of the millimeter-wave ASK modulators are shown in Fig. 29. It can be seen that the isolation and the maximum data rate have a tradeoff relationship. The product of the maximum data-rate and the isolation of this modulator is slightly less than the previous work in Section 2.2 but its maximum data is increased by 2Gbps and the insertion loss is improved by 4.3dB.

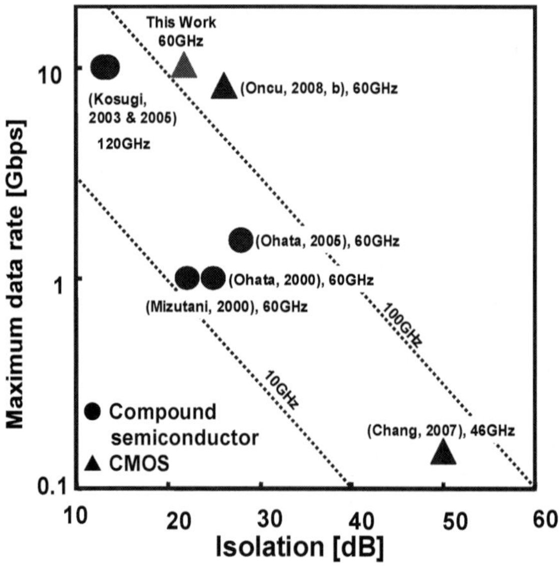

Fig. 29. Maximum data rates as a function of isolation of the ASK modulators.

2.3.2.3 60GHz Pulse Transmitter

The time-domain response of the pulse transmitter was measured using an Agilent Infiniium DCA 86100B wide-bandwidth oscilloscope with an Agilent 86118A 70GHz remote sampling module. The chip was measured by on-waver. The output is connected to the sampling oscilloscope by on-wafer probe and cables. The measurements were performed without any external filters at the output. The internal impedance of the measurement equipment is equal to a 50Ω. Figure 30(a) and Fig. 30(b) show the output response for 1Gbps and 10Gb/s respectively. Due to the high-speed binary base-band signal leakage from the gate, the baseline varied. Especially the leakage became stronger at 10GHz but it will not distort the transmitted millimeter-wave signal since the base-band leakage will be filtered out in the 60GHz band antenna. The RF power can be measured from the time-domain response shown in Fig. 31. The maximum peak-to-peak voltage was measured to be 45mV for a 50Ω load impedance. It corresponds to -23dBm peak power. By using this circuit up 10Gbps short-range wireless or proximity communication can be realized a power dissipation of 12.1mW. Our study showed us that with a proper buffer design and improved layout verifications, the output RF power would be increased up to a few dBm with an additional cost of a few tens of mW power dissipation for longer range applications.

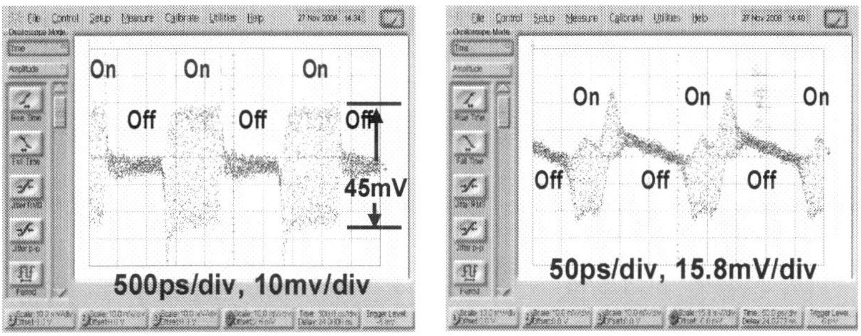

(a) 1Gbps (b) 10Gbps

Fig. 30. Measured output response of the transmitter for (a) a 1Gb/s and (b) a 10Gb/s data trains.

3. 60GHz CMOS pulse receiver

In the past few years, millimeter-wave quadrature amplitude modulator (QAM) receiver circuits in the short-channel standard CMOS process have been reported with a several Gbps data rate and a better energy-per-bit efficiency than WLAN and UWB (Pinel , 2007). Conventional QAM receivers downconvert the received millimeter-wave signal to baseband using one or two voltage-controlled oscillator (VCO) and phase-locked loop (PLL) circuits. However, these building blocks consume several tens of mW. Additionally, total power consumption further increases using an analog-to-digital converter and a high-speed modulator, particularly when the data rate exceeds 1Gbps. By removing these power-hungry building blocks, 2Gbps and 5Gbps millimeter-wave CMOS impulse radio receivers were developed with a better power efficiency. The 2Gbps receiver detects millimeter-wave single-ended pulses using a single-ended CMOS envelope detector, and high-speed data is only processed using a limiting amplifier. The second receiver design contains a differential envelope detector, a voltage control amplifier, a current mode offset canceller and the data is processed using a high-speed comparator with hysteresis. In this section, 2Gbps and 5Gbps millimeter-wave CMOS impulse radio receivers will be studied.

3.1 19.2mW 2Gbps CMOS pulse receiver

The general architecture of conventional millimeter-wave QAM receivers is shown in Fig. 31(a), where the received signal is downconverted using a local oscillator (LO) consuming a power of several tens of mW (Razavi, 2007; Mitomoto, 2007). Also, total power dissipation will even increase using a high-speed analog-to-digital converter (ADC) and a high-speed demodulator (DMOD), particularly for the multi-Gbps data rate. Instead of using an LO, an ADC and a DMOD, a low-power CMOS pulse receiver is proposed in this work for multi-Gbps wireless communication, as shown in Fig. 31(b). The architecture is adopted from that of optical communication receivers due to the similarity between an optical pulse and a millimeter-wave pulse. In the following sections, the pulse receiver design and the measurement results are presented.

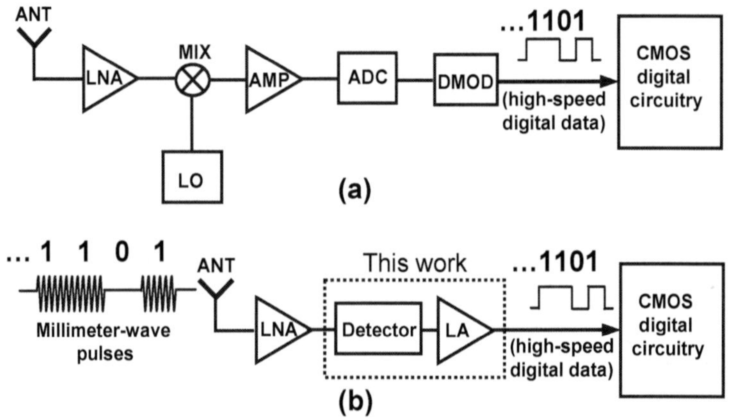

Fig. 31. Architectures of (a) a conventional 60GHz receiver and (b) the proposed 60GHz pulse receiver.

3.1.1 19.2mW 2Gbps CMOS pulse receiver design

Multi-Gbps communication will have low power consumption when a received signal is detected without using a high-frequency LO and high-speed data are processed using only a limiting amplifier (LA), as shown in Fig. 31(b). Figure 32(a) shows the widely used optical receiver architecture (Narasimha, 2007; Le, 2004). By adopting a similar principle, a 60GHz-band CMOS pulse receiver used for investigating the above concept is shown in Fig. 32(b). Here, a low-noise amplifier (LNA) is not implemented in this work to determine the inherent features of the millimeter-wave pulse receiver. As a result, the receiver consists of a nonlinear amplifier (NLA), a five-stage LA, an off-set canceller and an output buffer. To detect the millimeter-wave pulses, a metal-insulator-insulator-metal (MIIM) diode (Rockwell, 2007) or a Schottky diode (Sankaran, 2005) was conventionally used. However, the MIIM diode is used in special CMOS process, thus increasing the cost of the pulse receiver. And a Schottky diode is not always available in general design rules. To overcome this issue, a common-source amplifier, utilizing a square-law relationship between the drain current I_d and the gate voltage V_g of an NMOSFET, is used as a detector. In the NLA, V_g is adjusted to maximize $\partial^2 I_d/\partial V_g^2$ to detect the envelope of the millimeter-wave pulses efficiently. At the output of the NLA, the base-band signal is generated as shown in Fig. 33. The remainder of the circuitry is designed in the same way as for similar types of optical receivers.

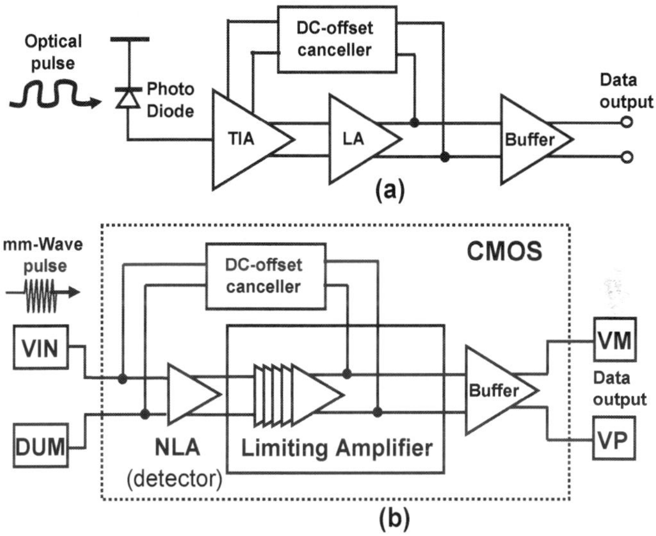

Fig. 32. (a) Typical optical receiver architecture and (b) diagram of receiver block in this work to realize the proposed pulse receiver.

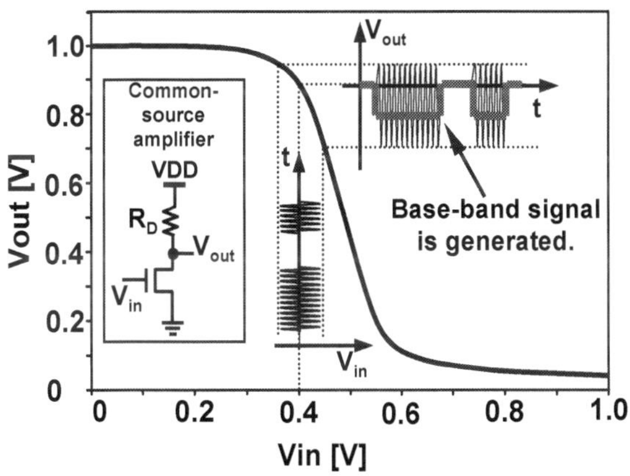

Fig. 33. Nonlinear pulse detection using a common-source amplifier.

3.1.2 Measurement and discussions

The receiver was fabricated by a 90nm CMOS process. A micrograph of the receiver is shown in Fig. 34. The millimeter-wave switch in Section 2.2 was used for measurement. A 60GHz continuous-wave (CW) signal applied to the switch input is modulated using a pattern generator in a bit-error-rate tester (BERT). To filter out base-band fluctuations due to switching, a V-band waveguide is inserted between the transmitter and the receiver. Before applying the pulses to the receiver input, the average pulse power is measured using a millimeter-wave power meter. The 60GHz pulses and the demodulated digital signals transmitted at a data rate of 2Gbps are shown in Fig. 35. The eye diagram and bit-error rate (BER) of the receiver are obtained using 2^{31}-1 bits of pseudo-random data. The eye diagram of the receiver is shown in Fig. 36 for the data rates of 1 and 2Gbps. In both cases, clear eye openings are observed. The output was 313mV peak to peak. The measured BER with respect to the average pulse power is plotted in Fig. 37 for 1 and 2Gbps data rates. The theoretical BER curves for the case of square-law detection are fitted to the measured data, the shapes of which agree with the square-law detection theory. The BER of the pulse receiver decreases more rapidly with increasing input power than that of a linear-detection receiver.

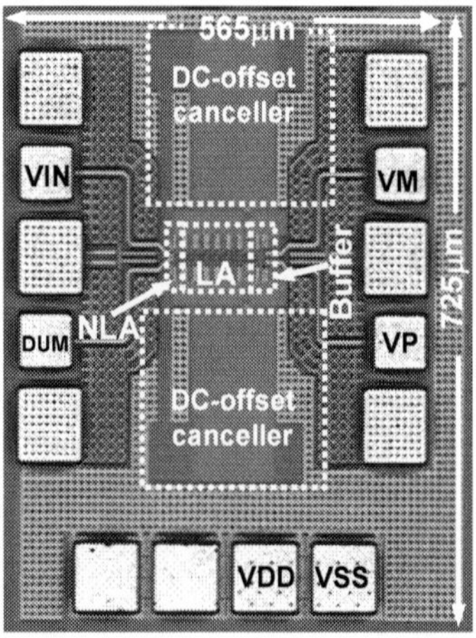

Fig. 34. Micrograph of the pulse receiver.

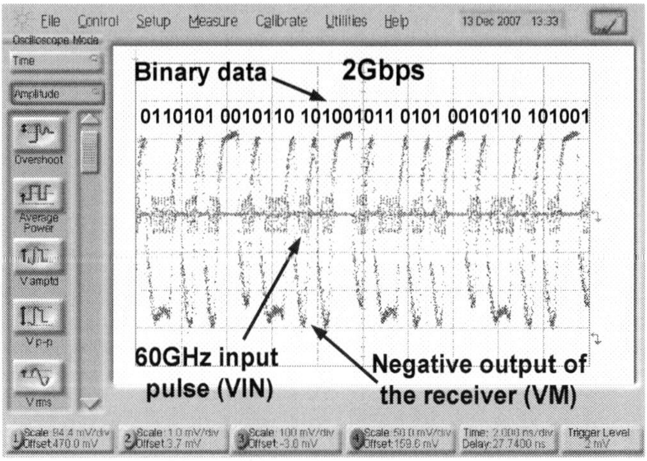

Fig. 35. Receiver input and output waveforms for pseudo-random data.

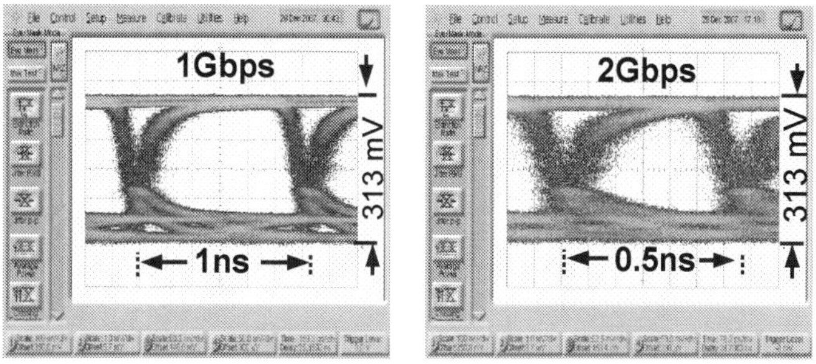

Fig. 36. Eye diagram with 2^{31}-1 random bits of data at 1 and 2Gbps data rates.

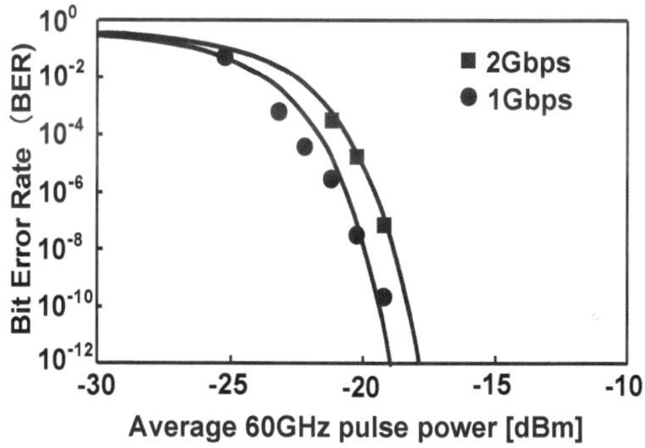

Fig. 37. Bit error rate with 2^{31}-1 random bits of data at 1 and 2Gbps data rates.

The total power consumption of the pulse receiver including the buffer is 19.2mW. To compare between this receiver and optical receivers, a figure of merit FOM is determined as G•DR/P_{DC}, where G is the power gain, DR is the data rate, and P_{DC} is the power consumption. The product of G and DR is plotted as a function of PDC, as shown in Fig. 38, where the FOMs are given by the slope. The FOM of this receiver is a slightly better than

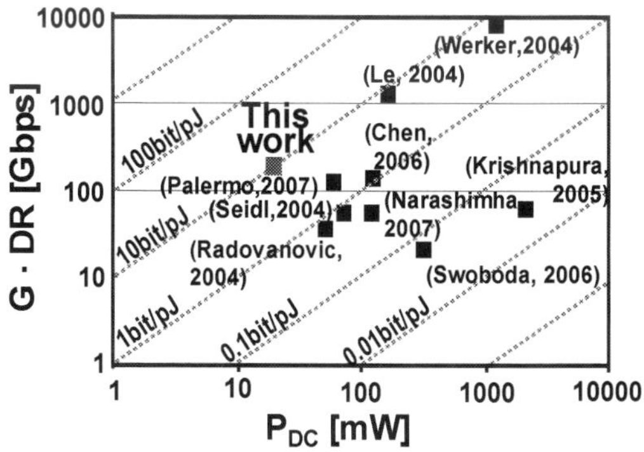

Fig. 38. Product of gain and data rate as a function of power dissipation for the receivers in this work and previously reported optical receivers.

those of other reported optical receivers. It was shown by measuring the scattering parameters that suitable input matching would increase the power gain by 4.9dB. The receiver is also compared with recently reported millimeter-wave receivers in Table 1. Note that digital codes are provided at the output with only 19.2mW of the power consumption using the proposed pulse receiver.

A low-power 60GHz-band CMOS pulse receiver was proposed for multi-Gbps wireless communication. Using a 90nm 1P6M standard CMOS process, the proposed pulse receiver achieved a 2Gbps data rate with a total power dissipation of 19.2mW, which consumes less power than recently reported 60GHz receivers. The performance of this pulse receiver indicates the possibility of new low-power multi-Gbps wireless communication at the 60GHz band.

	DC Power	Missing Building Blocks
This Work	19.2mW	LNA
(Afshar, 2008)	24mW	PLL, DMOD
(Parsa, 2008)	36mW	DMOD
(Scheir, 2008)	65mW	DMOD
(Razavi, 2007)	80mW	DMOD
(Mitomoto, 2007)	144mW	DMOD

Table 1. Comparison of 60GHz receivers.

3.2 49mW 5Gbps CMOS receiver

The receiver circuit in Section 3.1 operates up to a 2Gbps data rate with a total power dissipation of 19.2mW, consuming less power than conventional 60GHz millimeter-wave QAM receivers. However, it suffers from input common-mode noise, sensitivity to supply voltage, and an insufficient data rate for 4.5Gbps wireless high-definition multimedia interface applications. To overcome these issues, a fully differential 5Gbps millimeter-wave CMOS impulse radio receiver in an 8M1P 90nm standard CMOS process was realized. The receiver contains an on-chip matching circuit, a fully differential envelope detector, a voltage-controlled amplifier (VGA), a current-mode offset canceller, a high-speed comparator with hysteresis.

3.2.1 49mW 5Gbps CMOS receiver design

A block diagram of the proposed receiver is shown in Fig. 39. The on-chip matching network is used for 50Ω impedance matching and also helps reject the off-band signals. The envelope detector detects the envelope of the received pulses; the VGA amplifies the received signal to the required level, and then the high-speed comparator processes the signal. The current-mode offset canceller circuit both cancels the offset due to the mismatching of the differential amplifiers through the receiver chain and drives the NMOSFETs of the fully differential envelope detector.

Input signals are first given to the fully differential envelope detector through the input matching circuit. In practical applications an LNA will be included at the input of the receiver. Unlike the single-ended LNA, the differential LNA is superior in terms of common-mode noise rejection (Sun, 2006). The degradation of the common-mode noise will be stronger for an impulse radio receiver since the analog front-end and the logic circuits share the same substrate. To solve this issue, a fully differential CMOS envelope detector is designed. The fully differential envelope detector (FDD) is shown in Fig. 40, along with a conventional single-ended detector (SED) for comparison. The SED used in Section 3.1 only detects single-ended pulses. In the proposed FDD, the differential signals are applied to the gates of two parallel NMOSFETs with the same size. Also, an active balun is used for generating a differential output and common-mode rejection as shown in Fig. 40. The FDD rejects common-mode noise from the substrate and power line.

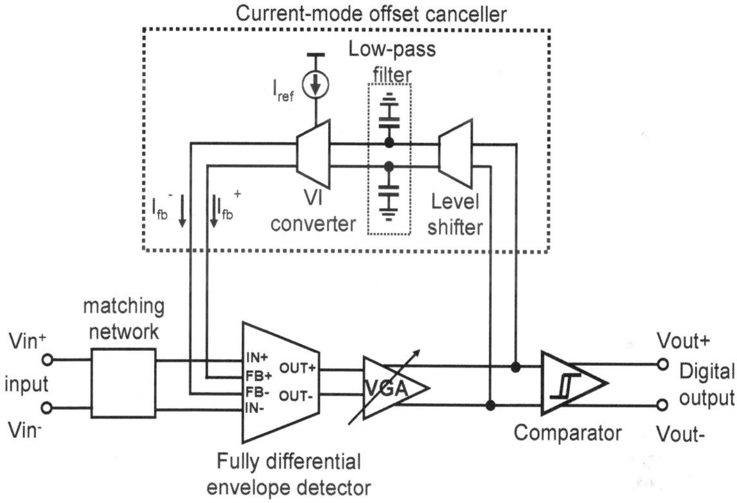

Fig. 39. Block diagram of fully differential 60GHz band millimeter-wave CMOS impulse radio receiver.

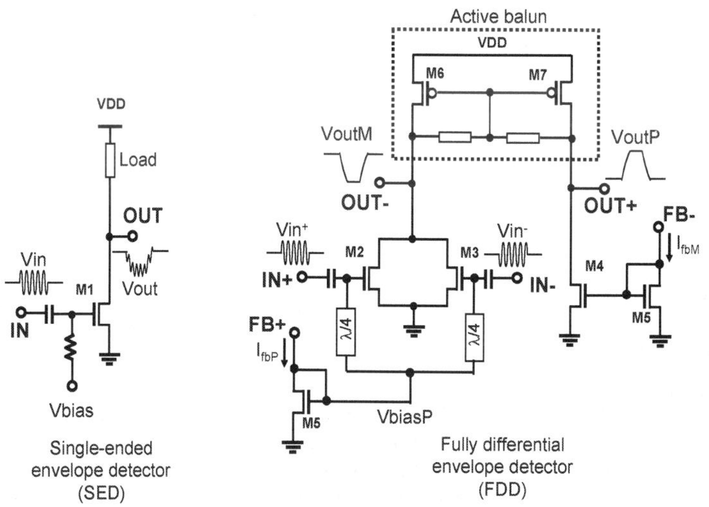

Fig. 40. Millimeter-wave CMOS envelope detector circuits.

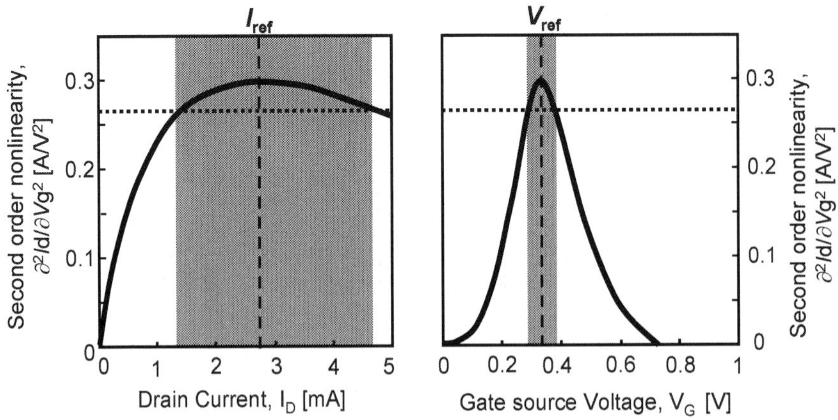

Fig. 41. Second-order nonlinearities with respect to drain current and to gate voltage.

To improve the immunity of PVT variations, current-mode offset canceller is proposed. The envelope detector circuits, driven by the offset canceller as well as 60GHz input pulses, detect the envelope of the pulses using the square-law relationship between the drain current I_d and the gate voltage V_g of the NMOSFETs. In (Oncu, 2008, a), V_g was adjusted to maximize $\partial^2 Id/\partial V_g^2$ to detect the envelope of the millimeter-wave pulses efficiently, where V_g is determined by the output common-mode voltage of the limiting amplifier. Here, the simulated second-order nonlinearity with respect to I_d is shown in Fig. 41, along with that with respect to V_g for comparison. The maximum nonlinearity is obtained when the transistor is biased in the moderate inversion region in both cases. However, since the peak characteristics of the nonlinearity with regard to I_d are flatter than that with regard to V_g, the nonlinearity is insensitive to the deviation from the maximum point due to the PVT variations when the drain current I_d is adjusted with respect to a reference current I_{ref} and the envelope of the millimeter-wave pulses is efficiently detected. To utilize this advantage, the current-mode offset canceller is used, which contains a level shifter, a low-pass filter, a voltage-independent reference current generator, and a VI converter.

A high-speed comparator with hysteresis is used in this design to process the input signal with rejecting a noise. Its circuit schematic is shown in Fig. 42. It has three subcirctuis: a positive-feedback decision circuit, a predriver, and a line driver. In the positive-feedback decision circuit, a differential driver and a positive-feedback load are composed of NMOSFETs to realize high speed with moderate bias current. No stacking transistor is used in the load to maximize an output voltage swing. Two current mirrors by PMOSFETs are used between the driver and the load. Since the operating speed of the PMOSFET current mirrors has to be improved to realize high-speed operation, higher overdrive voltage is applied to the PMOSFETs than to the NMOSFETs. The predriver utilizes a PMOSFET differential pair to obtain sufficient bias voltage since the output common-mode voltage of the positive-feedback decision circuit is reduced. The CMOS line driver is used for the final stage. The comparator test circuit is measured at a data rate up to 6Gb/s with 500mVpp output voltage swing at a supply voltage of 1.2V and a current of 11.9mA, where the power consumption of the line driver is included.

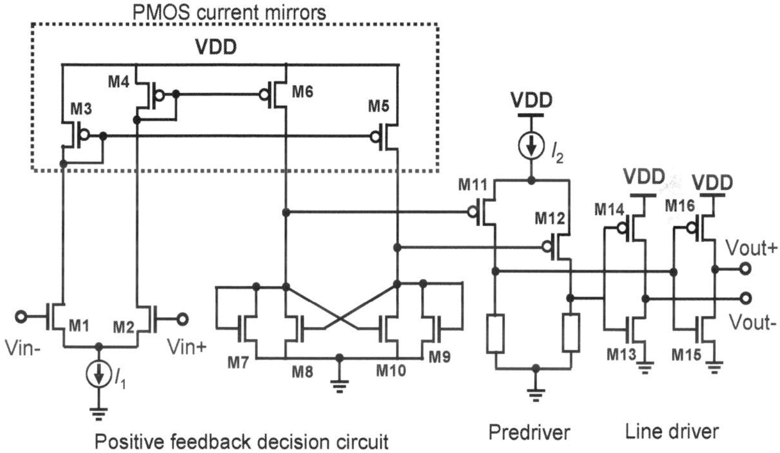

Fig. 42. High-speed CMOS comparator with hysteresis.

3.2.2 Measurement and discussion

The fabricated receiver is measured using an on-wafer probe station. The chip micrograph is shown in Fig. 43, where the chip size is 950μm × 750μm. The input reflection coefficient of the receiver was measured using a 4-port network analyzer. $S11_{dd}$ is less than -10dB at frequencies from 60GHz to 64GHz. Using an 8Gbps ASK CMOS modulator in Section 2.2 millimeter-wave pulses are generated to characterize the dynamic behaviour of the receiver. 62GHz differential ended pulses are applied to the input of the receiver using a magic tee, and the receiver is also tested using single-ended pulses. The receiver can receive 62GHz short-pulses in a time as short as 200ps. The measured receiver sensitivity is approximately -20dBm, which is suitable for high-speed millimeter-wave proximity communication applications. An LNA and a high-gain antenna will improve the sensitivity for long-range applications. An eye diagram of the receiver is obtained using 2^{31}-1 pseudorandom bits of data. The eye diagram obtained at a data rate of 5Gb/s data requires a total power consumption of 49mW. Measured results of the receiver performance are summarized in Fig. 44. The power consumptions of recently reported wireless digital receivers are compared in Fig. 45. The slope shows the figure of merit and the energy per bit. The graph shows that millimeter-wave receivers have better power efficiency than WLAN and UWB (Nathaward, 2008; Zheng, 2008). The millimeter-wave impulse radio receiver consumes the lowest energy per bit. The impulse receiver in Section 3.1 and the present impulse receiver have approximately the same energy-per-bit consumption of 9.8pJ/bit. However, this receiver is 2.5 times faster than that in Section 3.1. It is verified that millimeter-wave pulse receivers require low-power for high-speed communication. The 60GHz millimeter-wave band pulse communication can be used for low-power several Gbps wireless multimedia communication applications using a standard CMOS process.

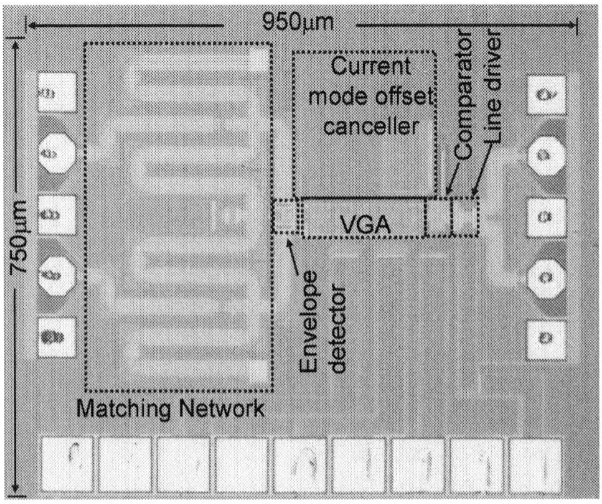

Fig. 43. Chip micrograph.

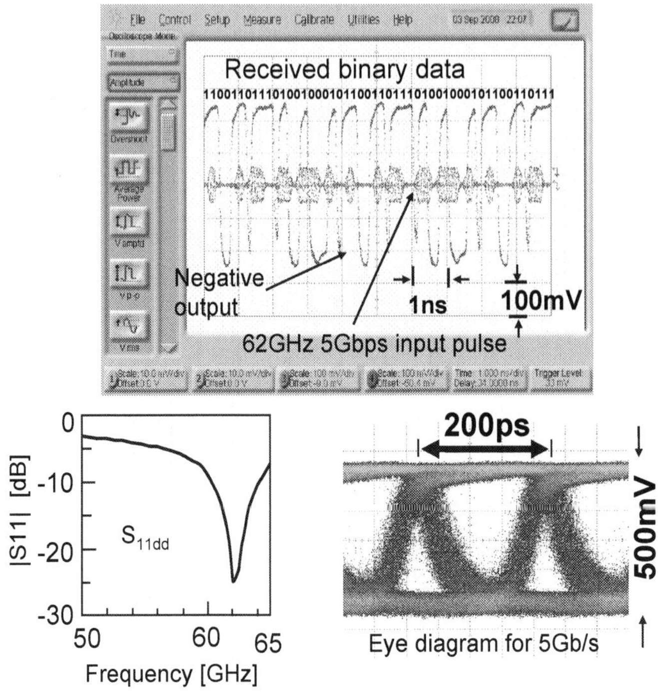

Fig. 44. Summary of measured results of the receiver.

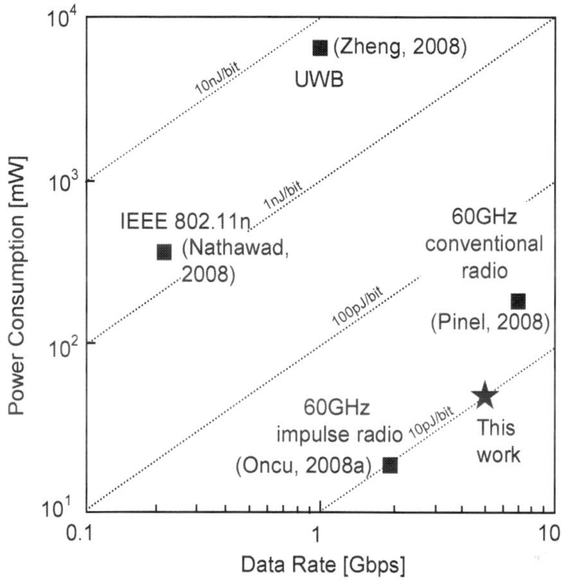

Fig. 45. Comparison of power consumption with respect to data rate of recently reported wireless communication devices.

4. Conclusion

Millimeter-wave impulse radio for low-power high-speed wireless communication was studied. Because of the several GHz license free bandwidth of the 60GHz band, the millimeter-wave impulse radio was optimized to operate at 60GHz band. To study the important building blocks of the millimeter-wave impulse radio, five prototype CMOS circuits, operating at 60GHz band, were successfully realized using 90nm standard CMOS processes from various foundries. A millimeter-wave CMOS pulse generator, a high-speed millimeter-wave ASK modulator, a 60GHz pulse transmitter circuit, 2 and 5Gbps millimeter-wave CMOS pulse receivers are studied for a realizing low-power and high-speed millimeter-wave impulse radio.

A carrier-less 60GHz CMOS pulse generator was fabricated using a 6-metal 1-poly 90nm CMOS process. By designing pulse generators in digital circuits, a millimeter-wave pulse can be generated without using a power-hungry LO. As a result, the pulse generator consumes a small amount of power proportional to input data rate.

After that to provide a better RF performance using available CMOS technologies, pulse transmitter circuits containing a high-speed millimeter-wave ASK modulator and a 60GHz oscillator were studied. A 60GHz millimeter-wave band ASK modulator was successfully fabricated using a 6-metal 1-poly 90nm CMOS process. The maximum isolation at 60GHz was obtained by adjusting the transmission line length. The isolation and maximum data rate of the switch were measured to be 26.6dB and 8Gbps, respectively. The ASK modulator does not consume DC operating power. Results indicate that a very high data-rate can be obtained at a 60GHz millimeter-wave band using a standard CMOS process.

Then, a 60GHz pulse transmitter circuit and to study its building blocks, a 60GHz millimeter-wave CW signal source and a millimeter-wave ASK modulator circuits were successfully fabricated by an 8-metal 1-poly 90nm CMOS process. The RF power of the 60GHz CW signal source circuit was measured to be -20.7dBm. The isolation of the ASK modulator was measured to be 23.5dB at 60GHz. The insertion loss of the modulator is 2.3dB which is 4.3dB better than that of the previous ASK modulator. The data-rate and output peak-to-peak voltage on a 50Ω load of the transmitter was measured up to 10Gb/s and 45mV respectively. The total power dissipation of the transmitter is 12.1mW. The results indicate that a short-range, multi-Gb/s data-rate and low-power 60GHz millimeter-wave band wireless communication can be realized using a sub-100nm CMOS technology.

In this study, a low-power 60GHz-band CMOS pulse receiver was proposed for multi-Gbps wireless communication. To investigate low-power and high speed pulse receivers, at first a prototype of a 60GHz pulse receiver was realized using a 90nm 1poly-6metal standard CMOS process. The proposed pulse receiver achieved a 2Gbps data rate with a total power dissipation of 19.2mW, which consumes less power than recently reported 60GHz receivers. The performance of this pulse receiver indicates the possibility of new low-power over-Gbps wireless communication at the 60GHz band.

Then, to suppress the input common-mode noise, sensitivity to supply voltage, and reach a sufficient data rate for 4.5Gbps wireless high-definition multimedia interface (HDMI) applications, a prototype of a differential ended 5Gbps 60GHz pulse receiver was successfully realized in a 1poly-8metal standard 90nm CMOS process. It receives up to 5Gbps millimeter-wave pulses with a power consumption of 49mW. Both pulse receivers have approximately same energy-per-bit consumption but the second one operates 2.5 times faster than the first one. It is verified that millimeter-wave pulse receivers require low-power for high-speed communication.

Millimeter-wave pulse transmitter and receiver architectures were discussed in this chapter, where pulse signals can be received without using an LO nor an ADC by adopting asynchronous detection, which will lead to the realization of a low-power millimeter-wave wireless transceiver system. The study of CMOS millimeter-wave impulse radio will encourage the widespread adoption of consumer millimeter-wave applications.

5. References

Afshar, B., Wang, Y. & Niknejad, A.M. (2008). A Robust 24mW 60GHz Receiver in 90nm Standard CMOS, *ISSCC Dig. Tech. Papers*, pp. 182-183, ISBN: 978-1-4244-2010-0, San Francisco, Feb. 2008.

Badalawa, B.B.M.W. & Fujishima, M. (2007). 60 GHz CMOS pulse generator, *Electronics Letters*, January 18 2007, Vol. 43, Pages. 100 - 102, ISSN: 0013-5194.

Beringer, R. (1946). The absorption of one-half centimeter electromagnetic waves in oxygen., *Phys. Rev.*, July 1 1946, Vol. 70, No. 1-2, Pages. 53-57.

Chang, H.Y.; Lei, M.F.; Lin C.S.; Cho, Y.H.; Tsai, Z.M.; Wang H. (2007). A 46-GHz Direct Wide Modulation Bandwidth ASK Modulator in 0.13-μm CMOS Technology, *IEEE Microwave and Wireless Components Letters*, Vol. 17, No. 9, Sept. 2007, pp. 691-693, ISSN: 1531-1309.

Chen W.Z. & Gan R.M. (2006). A Single Chip 2.5 Gbps CMOS Burst Mode Optical Receiver, *Symposium on VLSI Circuits Digest of Technical Papers*, pp. 120-121, ISBN: 1-4244-0006-6, Honolulu, June 2006.

Cheung, T.S.D.; Long, J.R.; Vaed, K.; Volant, R.; Chinthakindi, A.; Schnabel, C.M.; Florkey, J.; Stein, K. (2003). On-Chip Interconnect for mm-Wave Applications Using an All-Copper Technology and Wavelength Reduction, *ISSCC Dig. Tech. Papers*, pp. 396-397, ISBN: 0-7803-7707-9, San Francisco, Feb. 2008.

Chowdhury, D.; Reynaert, P. & Niknejad, A.M. (2008). A 60GHz 1V +12.3dBm Transformer-Coupled Wideband PA in 90nm CMOS, *ISSCC Dig. Tech. Papers*, pp. 560-561, ISBN: 978-1-4244-2010-0, San Francisco, Feb. 2008.

Doan, C.H.; Emami, S.; Niknejad, A.M.; Brodersen, R.W. (2005). Millimeter-wave CMOS design, *IEEE Journal of Solid-State Circuits*, Jan. 2005, Vol. 40, No. 1, pp. 144-155, ISSN: 0018-9200.

Fujishima, M.; Badalawa, B.B.M.W.; Oncu, A.; Wang, T. (2006). 22-29GHz CMOS Pulse Generator for Ultra-Wideband Radar Application, Proceedings of the 32nd European Solid-State Circuits Conference, pp. 279-282, ISBN: 1-4244-0303-0, Montreux, Sept. 2006.

Huang, D.; Hant, W.; Wang N.Y.; Ku, T.W.; Gu, Q.; Wong, R.; Chang, M.-C.F. (2006). A 60GHz CMOS VCO Using On-Chip Resonator with Embedded Artificial Dielectric for Size, Loss and Noise Reduction, *ISSCC Dig. Tech. Papers*, pp. 314-315, ISBN: 1-4244-0079-1, San Francisco, Feb. 2006.

Kosugi, T.; Shibata, T.; Enoki, T.; Muraguchi, M.; Hirata, A.; Nagatsuma, T.; Kyuragi, H. (2003). A 120-GHz millimeter wave MMIC chipset for future broadband wireless access applications, *IEEE MTT-S Int. Dig.*, vol.1, pp. 129-132, ISBN: 0-7803-7695-1, Jun 2003.

Kosugi, T.; Tokumitsu, M.; Enoki, T.; Muraguchi, M.; Hirata, A.; Nagatsuma, T. (2004). 120-GHz TX/RX chipset for 10-Gbit/s wireless applications using 0.1-mm-gate InP HEMTs, *Proceedings of Compound Semiconductor Integrated Circuit Symposium*, pp.171-174, ISBN: 0-7803-8616-7, Oct. 2004.

Krishnapura, N.; Barazande-Pour, M.; Chaudhry, Q.; Khoury, J.; Lakshmikumar, K.; Aggarwal, A. (2005). A 5Gb/s NRZ transceiver with adaptive equalization for backplane transmission, *ISSCC Dig. Tech. Papers*, pp.60-61, ISBN: 0-7803-8904-2, San Francisco, Feb. 2005.

Le, Q.; Lee, S.G.; Oh, Y.H.; Kang, H.Y.; Yoo, T.H. (2004). Burst-mode receiver for 1.25Gb/s Ethernet PON with AGC and internally created reset signal, *ISSCC Dig. Tech. Papers*, pp. 474-475, ISBN: 0-7803-8267-6, San Francisco, Feb. 2004.

Lee, J.; Huang, Y.; Chen, Y.; Lu, H.; Chang C. (2009). A Low-Power Fully Integrated 60GHz Transceiver System with OOK Modulation and On-Board Antenna Assembly,

ISSCC Dig. Tech. Papers, pp. 316-317, ISBN: 978-1-4244-3458-9, San Francisco, Feb. 2009.

Marcu, C.; Chowdhury, D.; Thakkar, C.; Ling-Kai K.; Tabesh, M.; Jung-Dong P.; Yanjie W.; Afshar, B.; Gupta, A.; Arbabian, A.; Gambini, S.; Zamani, R.; Niknejad, A.M.; Alon, E. (2009). A 90nm CMOS Low-Power 60GHz Transceiver with Integrated Baseband Circuitry, *ISSCC Dig. Tech. Papers*, pp. 314-315, ISBN: 978-1-4244-3458-9, San Francisco, Feb. 2009.

Maruhashi, K.; Kishimoto, S.; Ito, M.; Ohata, K.; Hamada, Y.; Morimoto, T.; Shimawaki, H. (2005). Wireless uncompressed-HDTV-signal transmission system utilizing compact 60-GHz-band transmitter and receiver, *Proceedings of IEEE MTT-S International Microwave Symposium*, pp.1867-1868, ISBN: 0-7803-8845-3, June 2005.

Mitomoto, T.; Fujimoto, R.; Ono, N.; Tachibana, R.; Hoshino, H.; Yoshihara, Y.; Tsutsumi, Y.; Seto, I. (2007). A 60-GHz CMOS Receiver with Frequency Synthesizer *Symposium on VLSI Circuits Digest of Technical Papers*, pp. 120-121, ISBN: 978-4-900784-05-5, Kyoto, June 2007.

Mizutani, H. & Takayama, Y. (2000). DC-110-GHz MMIC traveling-wave switch, *IEEE Transactions on Microwave Theory and Techniques*, Vol.48, No.5, May 2000, pp. 840-845, ISSN: 0018-9480.

Nakakita, H. ; Suginoshita, F. & Yazawa, N. (1997), 60-GHz-band ultracompact transmitter for HDTV, *Proceedings of IEEE MTT-S International Microwave Symposium*, pp.1143-1146, ISBN: 0-7803-3814-6, Denver, June 2005.

Narasimha, A.; Analui, B.; Yi Liang; Sleboda, T.J.; Gunn, C. (2007). A Fully Integrated 4×10Gb/s DWDM Optoelectronic Transceiver in a standard 0.13µm CMOS SOI, *ISSCC Dig. Tech. Papers*, pp. 42-43, ISBN: 1-4244-0853-9, San Francisco, Feb. 2007.

Nathawad, L.; Zargari, M.; Samavati, H.; Mehta, S.; Kheirkhahi, A.; Chen, P.; Gong, K.; Vakili-Amini, B.; Hwang, J.; Chen, M.; Terrovitis, M.; Kaczynski, B.; Limotyrakis, S.; Mack, M.; Gan, H.; Lee, M.; Abdollahi-Alibeik, S.; Baytekin, B.; Onodera, K.; Mendis, S.; Chang, A.; Jen, S.; Su, D.; Wooley, B. (2008). A Dual-Band CMOS MIMO Radio SoC for IEEE 802.11n Wireless LAN, *ISSCC Dig. Tech. Papers*, pp. 358-359, ISBN: 978-1-4244-2010-0, San Francisco, Feb. 2008.

Oncu A. & Fujishima M. (2008). 19.2mW 2Gbps CMOS pulse receiver for 60GHz band wireless communication, *Symposium on VLSI Circuits Digest of Technical Papers*, pp. 158-159, ISBN: 978-1-4244-1804-6, Honolulu, June 2008.

Oncu, A.; Takano, K. & Fujishima, M. (2008). 8Gbps CMOS ASK Modulator for 60GHz Wireless Communication, *A-SSCC Dig. Tech.*, pp. 125-128, ISBN: 978-1-4244-2604-1, Fukuoka, Nov. 2008.

Ohata, K.; Maruhashi, K.; Matsuda, J.; Ito, M.; Domon, W.; Yamazaki, S. (2000). A 500Mbps 60GHz-Band Transceiver for IEEE 1394 Wireless Home Networks, *European Microwave Conf. Proc.*, pp. 1-4, Oct 2000.

Ohata, K.; Maruhashi, K.; Ito, M.; Nishiumi, T. (2005). Millimeter-wave Broadband Transceivers, *NEC Journal of Advanced Technology*, Vol. 2, No. 3, Summer 2005, pp: 211-216.

Palermo, S.; Emami-Neyestanak, A. & Horowitz, M. (2007). A 90nm CMOS 16Gb/s transceiver for optical interconnects, *ISSCC Dig. Tech. Papers*, pp. 44-45, ISBN: 1-4244-0853-9, San Francisco, Feb. 2007.

Parsa, A. & Razavi, B. (2008), A 60GHz CMOS Receiver Using a 30GHz LO, *ISSCC Dig. Tech. Papers*, pp.190-191, ISBN: 978-1-4244-2010-0, San Francisco, Feb. 2008.

Pinel, S.; Sarkar, S.; Sen, P.; Perumana, B.; Yeh, D.; Dawn, D.; Laskar, J. (2008). A 90nm CMOS 60GHz Radio, *ISSCC Dig. Tech. Papers*, pp. 130-131, ISBN: 978-1-4244-2010-0, San Francisco, Feb. 2008.

Radovanovic, S.; Annema, A.-J. & Nauta, B. (2004), 3Gb/s monolithically integrated photodiode and pre-amplifier in standard 0.18μm CMOS, *ISSCC Dig. Tech. Papers*, pp.472-743, ISBN: 0-7803-8267-6, San Francisco, Feb. 2004.

Razavi, B. (2007). A mm-Wave CMOS Heterodyne Receiver with On-Chip LO and Divider, *ISSCC Dig. Tech. Papers*, pp. 188-189, ISBN: 1-4244-0853-9, San Francisco, Feb. 2007.

Rockwell, S.; Lim, D.; Bosco, B.A.; Baker, J.H.; Eliasson, B.; Forsyth, K.; Cromar, M. (2007). Characterization and Modeling of Metal/Double-Insulator/Metal Diodes for Millimeter Wave Wireless Receiver Applications, *Proceedings of IEEE Radio Frequency Integrated Circuits Symposium*, pp. 171-174, ISBN: 1-4244-0530-0, June 2007.

Sankaran, S. & Kenneth, K. O. (2005). Schottky Barrier Diodes for Millimeter Wave Detection in a Foundry CMOS Process, *IEEE Electron Device Letters*, July 2005, vol. 26, no. 7, pp. 492-494, ISSN 0741-3106.

Scheir, K.; Bronckers, S.; Borremans, J.; Wambacq, P.; Rolain, Y. (2008). A 52GHz Phased-Array Receiver Front-End in 90nm Digital CMOS, *ISSCC Dig. Tech. Papers*, pp.184-184, ISBN: 978-1-4244-2010-0, San Francisco, Feb. 2008.

Seidl, C.; Knorr, J. & Zimmermann, H. (2004). Single-stage 378MHz 178kΩ transimpedance amplifier with capacitive-coupled voltage dividers [optical fiber receiver applications], *ISSCC Dig. Tech. Papers*, pp.470-741, ISBN: 0-7803-8267-6, San Francisco, Feb. 2004.

Sun Y.; Glisic S. & Herzel F. (2006). A Fully Differential 60 GHz Receiver Front-End with Integrated PLL in SiGe, *European Microwave Integrated Circuits Conference Dig. Papers*, pp. 198-201, ISBN: 2-9600551-8-7, Manchester, Sep. 2006.

Swoboda, R. & Zimmermann, H. (2006). 11Gb/s monolithically integrated silicon optical receiver for 850nm wavelength, *ISSCC Dig. Tech. Papers*, pp. 904-911, ISBN: 1-4244-0079-1, San Francisco, Feb. 2006.

Werker, H.; Mechnig, S.; Holuigue, C.; Ebner, C.; Mitteregger, G.; Romani, E.; Roger, F.; Blon, T.; Moyal, M.; Vena, M.; Melodia, A.; Fisher, J.; de Mercey, G.; Geib, H. (2004), A 10 Gb/s SONET-compliant CMOS transceiver with low cross-talk and intrinsic jitter, , *ISSCC Dig. Tech. Papers*, pp.172-173, ISBN: 0-7803-8267-6, San Francisco, Feb. 2004.

Wiltse, J.C. (1984). History of Millimeter and Submillimeter Waves, IEEE Transactions on Microwave Theory and Techniques, Sep. 1984, Vol. 32, No: 9, pp. 1118-1127, ISSN: 0018-9480.

Zheng Y.; Arasu M. A.;Wong K. W.; The Y. J.; Suan A.P.H.; Tran D.D.; Yeoh W.G.; Kwong D.L. (2008). A 0.18µm CMOS 802.15.4a UWB Transceiver for Communication and Localization, *ISSCC Dig. Tech. Papers*, pp. 118-119, ISBN: 978-1-4244-2010-0, San Francisco, Feb. 2008.

14

CMOS Integrated Switched-Mode Transmitters for Wireless Communication

Ellie Cijvat
Lund University
Sweden

1. Introduction

Power amplifiers (PAs) determine much of the efficiency and linearity of transmitters in wireless communication systems, both on the base station side and in the handset device. With the move to third-generation (3G) communication systems as well as other systems such as Ultra-Wideband (UWB), a higher linearity is required due to envelope variations of the radio frequency (RF) signal. The traditional way of guaranteeing sufficient linearity is backing off the PA; however, this results in a significant drop in efficiency, and thus in reduced battery lifetime for the handheld device and increased cooling requirements for the base station. With the current energy costs, and increased density of base stations, this is fast becoming an important issue.

A second issue in current wireless communication systems is the requirement for a certain range of transmitter output power control, e.g. for 3G systems. Depending on the distance to the base station, a difference in handset output power in the range of tens of dB may occur. If the PA efficiency is peaking for maximum output power, and is reduced considerably for lower output power, the average efficiency of the transmitter calculated over its full output power range of operation will be low. Thus, efficiency improvement for lower output power is an important aspect in transmitter design.

Moreover, current wireless communication handsets require a multi-band/multi-standard approach, so that several communication standards are incorporated in one device. Ideally this would all be achieved by one PA, but current standard is that multiple PAs are used for multiple standards, in worst case each with its bulky, costly output filter.

In order to address efficiency and linearity issues, different transmitter architectures have been proposed and implemented throughout the years, such as for instance Envelope Elimination and Restoration (EER) or Envelope Tracking (ET), varieties of polar transmission where the envelope and phase of the signal are processed separately. Also, different PA architectures have been used, such as Doherty and switched mode amplifiers, often complemented with linearity-improving measures such as digital predistortion or feedback.

With the coming of age of handset production, cost effectiveness has driven wireless communication transceiver design to higher levels of integration. As many building blocks as possible are integrated on the same chip, and the use of external bulky filters is avoided if possible. CMOS technology has been the main choice for this development, due to the possible integration of digital, mixed-signal and analog circuits. However, CMOS was not suitable for PA design due to frequency, output power, efficiency and linearity requirements. Thus, the stand-alone PA has long been manufactured in III-V technologies or specialized technologies such as LDMOS.

In recent years however, CMOS technology has evolved for radio frequencies in two ways: (1) Decreasing device dimensions have resulted in higher clocking frequencies, thus e.g. providing the opportunity for clocking speeds of several times the RF frequency; (2) The technology provides special RF properties such as thick top metal, allowing for e.g. integrated inductors or transformers with high quality factor. These two technology trends have enabled a higher level of transmitter integration. In combination with the use of switches, for which CMOS devices are extremely suitable, so-called digitally assisted RF transmitters have been designed, that is, transmitters where building blocks are switched on or off by means of digital control signals, or biasing settings are changed based on digital signals.

Recently transmitter design research has taken the next step: increasingly using digital techniques for the full transmitter. A fully integrated GSM radio has been presented with all-digital phase and amplitude signal paths, including an all-digital phase-locked loop. Other examples are a class-E switched mode PA with pulse-width and pulse-position modulation (PWPM) implemented with all-digital blocks, an array of power mixers, controlled by digital logic, and an array of digitally controlled cascode transconductance stages not unlike current-steering digital-to-analog converters, referred to as digital-to-RF conversion. However, efficiency over a wide power range is still a major concern, as will be shown.

In this chapter an overview of switched-mode power amplifiers will be presented. This will be followed by an overview of transmitter architectures suitable for switched-mode transmitters; direct modulation as well as polar and Cartesian modulation will be described by looking at traditional architectures and recent developments, with focus on switched-mode implementations, resulting in a future outlook for integrated transmitter design for wireless communication.

2. Power amplifier technology issues

Generally a switched-mode (SM) amplifier consists of one or more transistors that are behaving as a switch, that is, having an on- and an off-stage, ideally without on-resistance and near-zero raise- and fall time. These conditions can be approximated by heavily overdriving the transistor input, and by operating the device at significantly lower frequencies than the device's f_t. The SM transistor is thus used differently than normal amplifier transistors, which are generally used as either current, voltage or transconductance amplifying elements.

Overdriving the transistor input, however, has certain consequences: the device will act non-linearly, and small-signal models are not always valid. Moreover, for wireless communication applications the difference between operating frequency and device unity gain frequency f_t is rather small - this in contrast to e.g. audio applications, where switched-mode techniques have been used extensively. In this section we will first discuss power amplifier technology issues, and then address losses in switched-mode power amplifiers.

2.1 PA technology aspects

It is only fairly recent that CMOS technology has come up as an alternative for integrated circuit power amplifier design, as CMOS previously was not suitable for PA design due to frequency, output power, efficiency and linearity requirements. Thus, stand-alone PAs have long been manufactured in III-V technologies such as GaAs or GaN, or specialized technologies such as LDMOS or SiGe bipolar junction transistors.

Largely driven by the drive for integrating more digital functionality on the same chip area, CMOS devices have continued to shrink in device dimensions, basically following Moore's law. Accordingly, transistor f_t and f_{max} are expected to rise to several hundreds of GHz, thus allowing for circuit operation in excess of 100GHz (Niknejad et al., 2007).

However, the trend of shrinking device dimension comes with certain distinct disadvantages for analog circuit design, and more specifically for PA design. Due to shrinking oxide thickness, the breakdown voltage of the devices is reduced, implying that supply voltages must be reduced for safe operation. This has implications for CMOS PAs, as the maximum output power, assuming load-line matching, is then given by

$$P_{out} = V_{DD}^2/2R_l \quad (1)$$

such that in a 50Ω system, and a supply voltage of 1V, the output power is limited to 10mW or 10dBm. Thus, impedance transformation must be used so that the amplifier sees a lower impedance. This is practically limited to 1-5Ω; Having such a low impedance makes the PA efficiency very sensitive to parasitic series resistance in the output network, because of conduction losses: A 0.1mΩ parasitic resistance in series with a load resistance of 1Ω gives a loss of 10%.

Due to these increasing technology limitations, in modern CMOS deep-submicron technologies special transistors are provided having a thicker gate oxide and thus allowing for higher supply voltage.

2.2 Losses in switched-mode amplifiers

Looking at RF power amplifiers, we want to have an output signal at the frequency of interest - usually the fundamental frequency, sometimes a harmonic - but no disturbing output signals at other frequencies. In other words, some filtering must be performed in order to use a switch in a power amplifier.

The ideal waveforms for a switched-mode (SM) transistor in a PA, assuming a broadband load, are shown in Fig. 1. From this figure it can be seen that the voltage and current are ideally never non-zero simultaneously, thus no power is consumed, and ideally a 100% efficiency can be achieved. However, considerable power is generated at harmonic frequencies. Thus the maximum theoretical efficiency for this broadband SM PA is slightly larger than 80%, achieved at a 50% duty cycle.

In order to reduce the power present in harmonic frequencies, a tuned amplifier can be used. This can be implemented in several ways. One way is by introducing harmonic shorts in parallel to R_l in Fig. 1, so that harmonics other than the desired frequency are grounded. The maximum theoretical efficiency now reaches 100%, however, for relatively low duty cycles (and thus very short pulses and low output power) (Cripps, 1999, p. 153).

Another strategy is to have a resonance circuit in series with R_l, to make sure that only the desired frequency signal is passed on. This issue will be explored more in the section on class-F amplifiers.

Device and switching losses

Aside from the harmonic losses discussed in the previous section, some other losses can be identified in a SM amplifier/transistor (El-Hamamsy, 1994). First of all, the transistor will

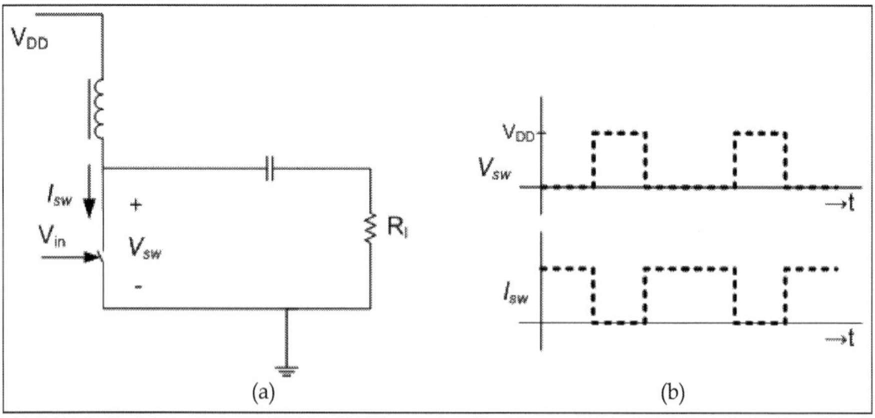

Fig. 1. An ideal switched-mode (SM) power amplifier, (a). Schematic, (b). Voltage and current waveforms.

suffer from non-idealities, of which one is a non-zero on resistance. This will cause a non-zero voltage drop and thus so-called conduction loss, resulting in reduced efficiency.
Secondly, the transistor will have non-zero rise- and fall times, potentially causing the current and voltage to be non-zero simultaneously. Also CMOS subthreshold current will contribute to this.
Thirdly, dynamic losses due to charging and discharging of parasitic capacitors must be taken into account - the switching losses. These are proportional to the switching frequency f, and will likely dominate for RF applications.

Other losses

External elements such as output networks may cause losses as well, for example a tuning or impedance transformation network consisting of on-chip or discrete passive elements. These inductors and capacitors will include parasitics such as capacitances or series resistances. These may cause power dissipation and thus reduce the amplifier efficiency.
A MOSFET is very suitable as a switch, toggling between the *off* mode for low gate-source voltage V_{GS}, and the triode region for high V_{GS}. The *on* resistance of the device is then given by

$$R_{on} = (L/W) \cdot (k'(V_{GS} - V_t - V_{DS}))^{-1} \quad (2)$$

where L is the transistor length, W the transistor width, k' the transistor gain factor, V_t the threshold voltage, and V_{GS} and V_{DS} the gate-to-source and drain-to-source voltage, respectively.
The *on* resistance can thus be minimized by choosing a large ratio W/L. Having a low resistance decreases the conduction losses caused by the switch. Other considerations of interest for PA design are the current density capacity and parasitic capacitances. The former is important if high output power is desired and the supply voltage is low. A larger width increases the current capacity. The parasitic capacitance may, however, cause increased dynamic losses, thus potentially decreasing the efficiency especially at high frequencies.

3. CMOS switched-mode power amplifiers

Now that general technology issues have been discussed, SM amplifiers for radio frequencies will be addressed in this section, and an overview will be given of specific CMOS implementations.

3.1 Switched-mode amplifier classes

In amplifier theory, several different switched-mode types are established: the classes D, E and F (Cripps, 1999; Raab, 2001). They will briefly be addressed below, before looking into CMOS implementations in the next section.

Class-D

Class-D amplifiers use a double-switch structure, with a series resonance circuit (see Fig. 2). The output current is alternatingly supplied by each switch, similar to a push-pull configuration. The simplest implementation for the two switches is an inverter. The maximum theoretical efficiency is 100%, with a square-wave voltage and a half-wave rectified sine wave current in each device. In that case the voltage contains only odd harmonics, and the current even harmonics.

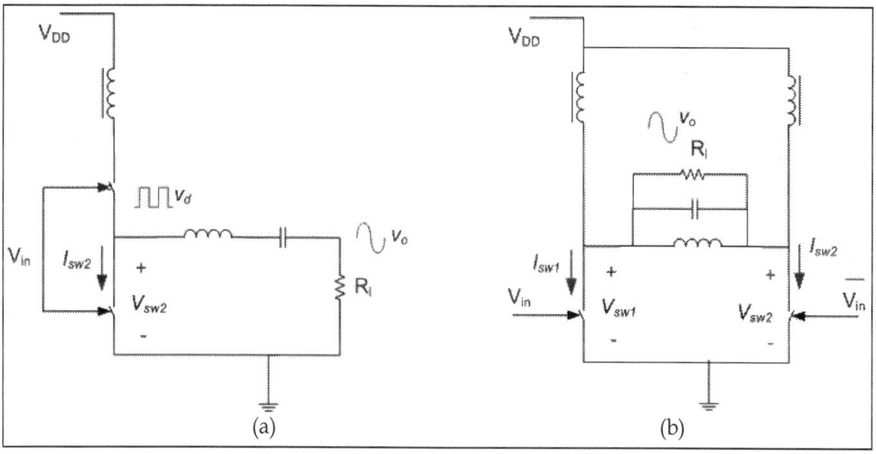

Fig. 2. Simplified schematic of a class-D amplifier, (a). A voltage-mode amplifier, (b). A current-mode amplifier.

This amplifier may also be implemented as current-mode (see Fig. 2b). Instead of having a series resonance circuit in series with the load, a parallel resonance circuit is then used at the output of the amplifier. In that case the current approximates a square-wave, containing odd harmonics, while the drain voltage for each device approximates a half-wave rectified sine wave. It has been shown that a high efficiency can be achieved, assuming the amplifier can be designed for Zero Voltage Switching (Long et al., 2002; Kobayashi et al., 2001).

Class-E

A class-E amplifier consists of a single switching device with a carefully tuned output network. The voltage derivative, close to the timing point when the device is switched off, is designed to be very small (so-called Zero Voltage Switching, ZVS) so that potential static losses are kept very low. Also for this amplifier the theoretical maximum efficiency is 100%. One of the characteristics of class-E is that large voltage peaks occur; thus, care must be taken to avoid high voltages across the CMOS device, as the breakdown voltage of CMOS devices is relatively low.

Class-F

A class-F amplifier is basically an amplifier with a current that approaches a half-wave rectified sine wave, and a voltage that approaches a maximally flat shape. Tuning a limited number of odd-order harmonics of the fundamental signal is used to achieve this. Two different structures are in use for class-F design, depending on which harmonics are seen at the drain: Regular class-F for odd-order harmonics, that is, the voltage is approximately maximally flat, and inverse class-F for even harmonics, i.e. a half-wave rectified sine wave-shaped drain voltage and a maximally flat shaped drain current (Raab, 2001). It must be noted that the inherent pulse shaping makes this amplifier less suitable for e.g. Pulse Width Modulated (PWM) input signals (Sjöland et al., 2009).

All three amplifier classes depend to some extent on a frequency-selective output network. Thus, their operation cannot be considered broadband. Either they can only be used in a narrow, specific frequency range, or each amplifier's behavior may show significant differences depending on the frequency of operation.

Research is progressing into variable output networks, where digital control signals are used to e.g. change the frequency of operation, or reconfigurable PAs, as well as output networks allowing for concurrent multi-band operation (Colantonio et al., 2008). In such digitally assisted systems the use of CMOS technology, also for the PA, may lead to a higher level of integration. This will be addressed more extensively in the section on transmitter architectures.

3.2 CMOS PA implementations

By the mid-1990s, the first publications on integrated CMOS PAs for RF appeared. These works initially focused on more or less linear amplifier structures such as class A, AB, B or C, but research has since then focused more on the switched-mode class-D, E and F, as higher clocking or switching speeds became available with improvements in CMOS technology.

Su and McFarland (1997) presented a 0.8μm CMOS SM amplifier consisting of four stages with the final stage in switched-mode. A Power-Added Efficiency (PAE) of 42% was achieved at 850MHz with a 2.5V supply, and largely off-chip input and output matching networks were used. Yoo and Huang (2001) presented a 0.25μm CMOS class-E PA, using a finite DC feed inductor to reduce the peak voltage over the device, as well as Common Gate (CG) switching instead of the more usual Common Source (CS) structure. These strategies allow for a higher supply voltage to be used, thus reducing the necessity for a low load impedance.

Reynaert and Steyaert (2005) have presented a fully integrated 0.18μm CMOS class-E PA, consisting of three stages and including supply modulation to provide amplitude variation. A PAE of 34% was achieved for an output power of 23.8 dBm, using a supply voltage of 3.3 V and extra thick gate oxide for the final stage.

As limited supply voltage is one of the major challenges in CMOS PA design, other strategies have been used to effectively add the output voltages, such as using a transformer to combine output power (Aoki et al., 2008; Haldi et al., 2008) or stacking devices, making sure that the voltage over each device stays below the maximum (Stauth & Sanders, 2008; Jeong et al., 2006). However, generally this slightly impairs the efficiency, counteracting the intended advantage of a higher supply voltage. Apart from voltage stacking, current combining has been implemented (Kavousian et al., 2008; Kousai & Hajimiri, 2009), as well as the switching in of several parallel stages (Walling et al., 2008). The latter two will be covered more in the section on transmitter architectures.

Reference	Class	Technology	Supply voltage	Output power	Efficiency (PAE)	Frequency
Su et al., 1997	D?	0.8μm CMOS	2.5 V	30 dBm	42%	850 MHz
Tsai et al., 1999	E	0.35μm CMOS	2.0 V	30 dBm	48%	1.9 GHz
Yoo et al., 2000	E	0.25μm CMOS	1.9 V	30 dBm	41 %	900 MHz
Kuo et al., 2001	F	0.2μm CMOS	3.0 V	32 dBm	43 %	900 MHz
Sowlati et al., 2003	?	0.18μm CMOS	2.4 V	24 dBm	42 %	2.4 GHz
Reynaert et al., 2005	E	0.18μm CMOS	3.3 V	24 dBm	34 %	1.75 GHz
Stauth et al., 2008	D	90nm CMOS	2.0 V	20 dBm	38.5%	2.4 GHz

Table 1. An overview of CMOS integrated switched-mode power amplifiers.

4. Transmitter architectures

As we have seen before, one of the basic requirements for power amplifiers in modern wireless communication systems is to accommodate envelope variations and to provide variable output power. Wireless communication standards have moved from constant-envelope, low- channel bandwidth to more complex signal shapes in order to increase data rates in limited bandwidth, resulting in variable envelope RF signals and larger channel bandwidths in the range of tens of MHz.

In SM amplifiers output power variation can be achieved by varying the supply voltage, by varying the duty cycle of the signal, by varying the load, or by a combination of these. In this section some transmitter architectures will be discussed that adopt such strategies; only the strategy of varying the load impedance will not be addressed here.

4.1 Supply variation

On the transmitter architecture level, one of the classical methods of varying the output power is based on polar modulation, where a baseband Cartesian signal $v_{RF}(t)$ is first converted into its polar form, separating envelope (amplitude) and phase information, which are then processed separately and combined before being transferred to the antenna:

$$\begin{aligned} v_{RF}(t) &= I(t)\cos(2\pi f_0 t)\cdot + Q(t)\sin(2\pi f_0 t) \quad \text{(Cartesian)} \\ &= A(t)\cos(2\pi f_0 t + \varphi(t)) \quad \text{(polar)} \end{aligned} \tag{3a}$$

where

$$\begin{aligned} A(t) &= \sqrt{(I(t)^2 + Q(t)^2)} \quad \text{(amplitude)} \\ \varphi(t) &= \tan^{-1}(I(t)\,/\,Q(t)) \quad \text{(phase)} \end{aligned} \tag{3b}$$

Polar modulation is recently gaining more and more interest due to its potential to maintain linearity while having a relatively high efficiency even for lower output power, thus improving the average efficiency over a wide output power range.
One of the most well-known polar schemes is Envelope Elimination and Restoration (EER), brought to attention by Khan (Khan, 1952; Wang et al., 2006; Su & McFarland, 1998). The envelope is used to control the PA supply level, while the phase signal is upconverted to RF and transformed to a constant envelope signal, driving the PA input. Thus, a non-linear PA can be used. Su and McFarland (1998) have demonstrated a CMOS implementation of an EER system, including a delta-modulated supply, a limiter, and envelope detectors, driving a switched-mode PA, resulting in significant linearity and efficiency improvements.

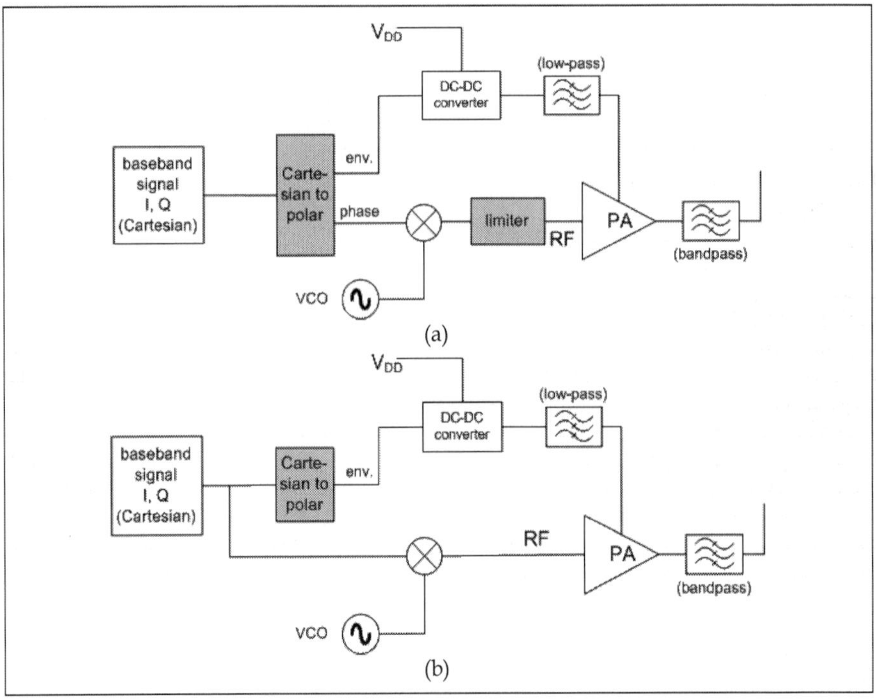

Fig. 3. Simplified representation of the Envelope Elimination and Restoration (EER) and Envelope Tracking (ET) transmitter architectures.

Envelope tracking (ET) describes a transmitter architecture where the Cartesian RF signal is amplified by means of a linear amplifier, with its supply controlled by the envelope of the signal (Hanington et al., 1999; Takahashi et al., 2008). One of the main advantages is that the bandwidth of the PA input signal is not expanded, but a linear amplifier generally has a lower efficiency than a SM amplifier. However, requirements on the envelope signal and timing are less stringent (Wang et al., 2006). So-called hybrid EER architectures have been demonstrated, where the ET linear amplifier is replaced by a SM amplifier, however, still driven by the full Cartesian RF signal (Wang et al., 2006).

Both the EER, ET and hybrid EER depend on utilizing an efficient power supply modulator, that must be able to handle the bandwidth of the envelope signal. For this, a boost dc-dc converter, a Buck dc-dc converter, or a switched-mode low-frequency amplifier can be used, controlled by a Pulse Width Modulator (PWM), a Sigma-Delta modulator (ΣΔM) or a Delta modulator (ΔM) (Kitchen et al., 2007). Generally, independent of supply modulator type, a bulky low-pass filter must be used to filter out undesired signals such as noise or harmonics.

4.2 Changing the duty cycle

If the duty cycle D of a square-wave signal is changed, the output power at the fundamental frequency will be changed according to

$$P_{out}(f_0) = (4V_{DD}^2/\pi^2R_l)\ sin^2(\pi D) \tag{4}$$

assuming ideal frequency selection at the output. This can be used to accommodate the envelope and power variations in a polar transmitter, by changing the amplifier's threshold voltage. Implementations exist with discrete steps as well as continuous change (Yang et al., 1999; Cijvat et al., 2008; Smely et al., 1998). A major advantage of these strategies is that no DC-DC converter is necessary; A disadvantage is that linearity may be worse compared to an amplifier where the supply voltage is changed, possibly resulting in tougher requirements for digital predistortion. Moreover, the efficiency drops rapidly at small duty cycles (Cijvat et al., 2008).

Smely et al. (1998) combined discrete supply voltage steps with changing the drain current of the output stage of a class-F stage by means of varying the GaAs MESFET gate voltage, depending on the amplitude of the input signal. Yang et al. (1999) focused on improving the efficiency of a class-A amplifier, by using variable bias to change the current in the output stage as well as changing the supply voltage.

Variable gate bias was used (Cijvat et al., 2008) for CMOS class-D amplifiers, with the goal of creating a PWM signal at the output of the amplifier. The proposed architecture uses the envelope signal to control the gate bias, and the RF signal is assumed to be sinusoidal, containing only the phase information.

For this amplifier structure, loss mechanisms as discussed in section 2 cause a drop in drain efficiency for lower output powers. It is likely that switching and harmonic losses are significant; the amplifier switches as often as for full output power, thus having roughly the same switching loss, and the harmonic content of a PWM signal increases for duty cycles other than 0.5, thus increasing harmonic losses. As can be seen in Fig. 4.b, the amplifier aimed for higher output power, having larger output devices and thus larger parasitic capacitances, reaches a lower maximum drain efficiency as a result.

As was addressed by Sjöland et al. (2009), one of the challenges of polar modulation is the sharp notch in amplitude variation which causes fast amplitude variations that are difficult to track for a DC-DC converter with limited bandwidth. Thus, a combination of EER and Pulse Width Modulation is proposed. This is applied to the aforementioned 130 nm CMOS class-D inverters, and simulation results are presented in Fig. 5.

It can be seen from this figure that efficiency gains of EER and PWM combined are minimal in this case, compared to EER-only. Moreover, combining the two strategies will lead to greater transmitter complexity; the additional power that is required is not taken into account in the simulations. However, as was mentioned earlier, this solution may address the bandwidth limitations of EER.

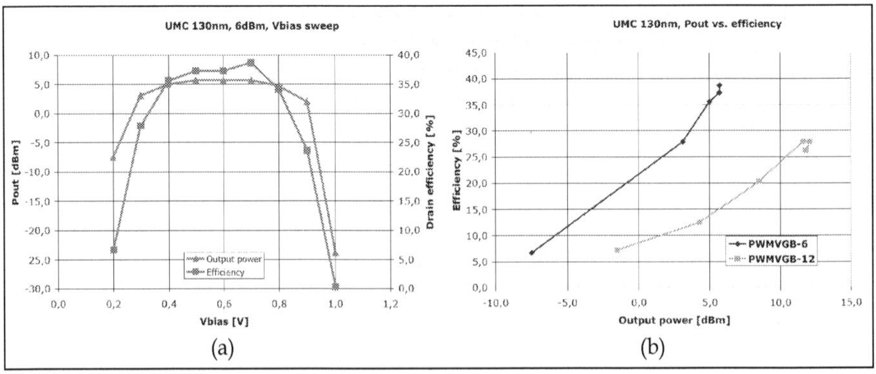

Fig. 4. (a). Measured output power and efficiency of a 6 dBm 130nm CMOS class-D inverter chain, using gate bias variation to create a pulse width modulated inverter output voltage (Cijvat et al., 2008). (b). Efficiency versus output power of two amplifiers, one with 6dBm and one with 12 dBm output power. The supply voltage was 1.2 V. The 6 dBm amplifier operated at 1.5 GHz, the 12 dBm amplifier at 1 GHz.

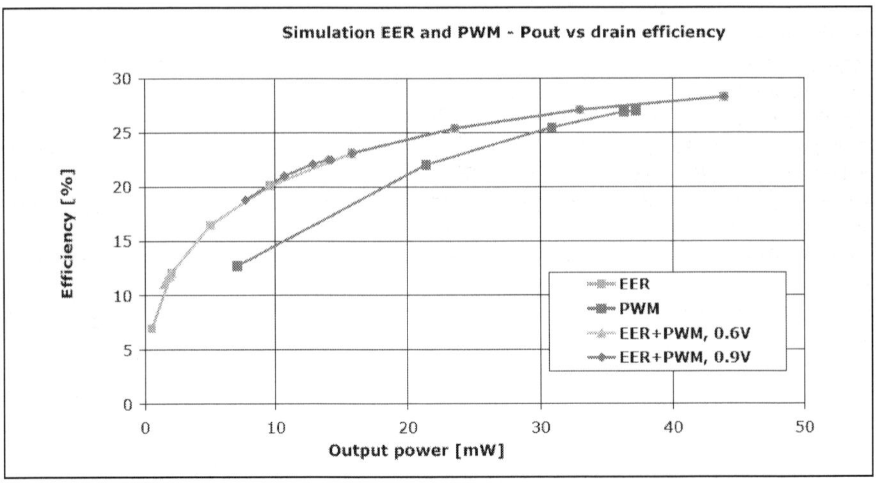

Fig. 5. Simulated PA drain efficiency versus output power, combining EER modulation for high amplitudes and PWM for lower amplitudes. The voltage where EER takes over is varied; one curve shows results for a border value of 0.6V and the second curve for a border value of 0.9V.

4.3 Burst-mode transmitters

A third method for varying the output power is so-called burst mode transmission. Effectively the RF signal is turned on and off by means of a bit stream. The envelope signal may be digitized e.g. by means of a $\Sigma\Delta$ or a Pulse Width Modulator (Jeon et al., 2005; Berland et al., 2006; Stauth & Sanders, 2008).

A burst-mode pulsed power oscillator to be used as a final stage in a transmitter was presented by Jeon et al. (2005). The oscillator is turned on and off by a PWM representation of the low-frequency envelope signal, thus resulting in the high-frequency RF signal multiplied by the PWM signal, appearing as bursts at the oscillator output. An isolator and bandpass filter are used to prevent reflected power to return into the oscillator and filter out undesired frequency components.

Berland et al. (2006) analyzed two varieties of using a one-bit signal to be multiplied with the slightly modified Cartesian signal. The one-bit signal was derived from the envelope signal by utilizing a Pulse Width Modulator and a Sigma-Delta Modulator, respectively. A high operating frequency of several GHz is, however, necessary to reach sufficient performance.

A polar modulator using a baseband ΣΔM and an RF Pulse Density Modulator (PDM) were used to drive a class-D amplifier with a 1-bit signal (Stauth & Sanders, 2008). This solution, basically all-digital, was implemented in 90nm CMOS and the cascade PA operated from a 2V supply. The PA performance can be seen in Table 1. The Bluetooth 2.1+EDR spectral mask was met for an output signal in the range of 10dBm, including a bandpass filter at the output.

4.4 Digitally controlled TX

In analogy to current-steering Digital-to-Analog converters (Zhou & Yuan, 2003), a fourth strategy to control output power has recently gained attention, which is switching in parallel stages. One example is the work by Kavousian et al. (2008), where the low-frequency envelope of the polar signal was transformed into a thermometer code used to switch on and off unit stages, while the constant-envelope RF phase signal drives the input of each stage. The authors refer to this as digital-to-RF conversion.

Shameli et al. (2008) used 6 control bits to both switch in a number of parallel output stages and at the same time change the supply voltage with a ΣΔ modulator. A 62 dB power control range was achieved, as well as a 27.8dBm maximum output power and an average WCDMA efficiency of 26.5%.

Current summing was also used by Kousai and Hajimiri (2009), utilizing 16 parallel power mixers and a transformer at the output. The phase information modulates the high-frequency digital LO signal. Linearization could be chosen to be analog, by sensing and feeding back the signal level for each mixer core, or digital, by using a thermometer code for the envelope signal, switching on and off mixer cores. Both the baseband and the LO signal where controlled digitally with a number of bits. A 16-QAM (Quadrature Amplitude Modulation) signal at 1.8 GHz and a symbol rate of 4 MSym/s was reproduced with an output power of 26 dBm, a PAE of 19% and an EVM (Error Vector Magnitude) of 4.9%.

Presti et al. (2009) used 7-bit thermometer + 3 bit binary weighted current summing combined with analog input power control for low-power levels. Relative broadband operation, 800-2000 MHz, and a 70dB power control range is achieved. With Digital Pre-Distortion (DPD) both WCDMA, EDGE and WiMAX specifications are met.

In these architectures no supply voltage modulator is used. Sufficient resolution to achieve a high linearity or amplitude accuracy is achieved by increasing the number of parallel stages. However, the efficiency of these current-summing amplifiers follows a class-B curve (Presti et al., 2009):

$$\eta \propto \sqrt{P_{out}} \tag{5}$$

Walling et al. (2008) used control bits to generate a suitable Pulse Width/Pulse Position (PWPM) signal, which was then provided to four class-E quasi-differential stages. In a 65nm technology, a maximum output power of 28.6 dBm and PAE of 28.5% is achieved at 2.2 GHz with the output stage using a supply voltage of 2.5 V. For a 192kHz symbol rate, non-constant envelope π/4-DQPSK (Differential Quadrature Phase Shift Keying) modulated signal, an output power of 27 dBm is achieved with an EVM of 4.6%.

4.5 Direct RF modulation

A third strategy to process the signal is to directly modulate the RF signal into the SM amplifier. For instance, a Pulse Width/Pulse Position modulator (PWPM) or a Sigma-Delta (ΣΔ) modulator can be used (Nielsen & Larsen, 2008; Wagh & Midya, 1999). This is depicted in Fig. 6. A major disadvantage however is that generally a high sampling or operating frequency is necessary, typically at least $4f_{RF}$, in order to achieve the desired resolution. This implies a large power consumption in the modulator, as this is directly proportional to the frequency. Moreover, since the PA switches more often, more switching loss will occur, reducing the efficiency.

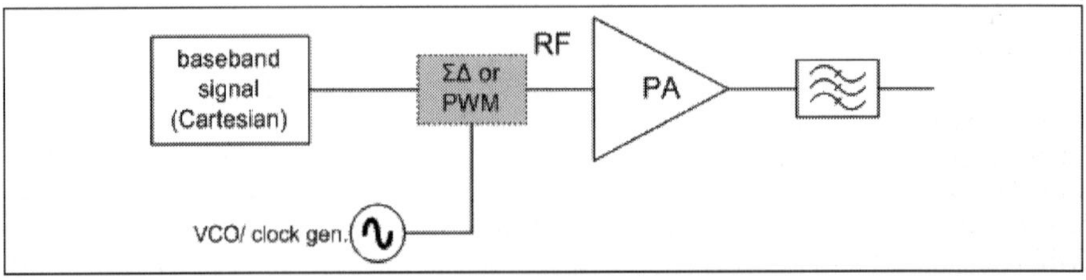

Fig. 6. Direct modulation of the RF signal by means of Sigma-Delta (ΣΔ) or Pulse Width Modulation (PWM).

Wagh and Midya (1999) presented the concept of Pulse Width Modulation for RF. Nielsen and Larsen (2008), utilizing GaAs technology, used a feedback circuit and a comparator to generate an RF PWM signal. The signal's adjacent channel power ratio stayed well below the UMTS spectrum mask, allowing for some non-linearity from a subsequent PA.
Direct modulation was also proposed by Jayaraman et al. (1998), utilizing a bandpass ΣΔ modulator, simulated with GaAs HBT technology. Discussions on efficiency were presented, and it was indicated that the linearity demands were moved from the PA to the ΣΔM.

4.6 Cartesian modulation

Even though polar modulation has some distinct efficiency advantages, as an alternative Cartesian modulation may be used, that is, the I and Q baseband signal that differ 90° in phase are each processed in the transmitter and then summed either directly before the PA, or alternatively, each signal is amplified and the two signals are combined after the amplifiers. An advantage is that the signal is not transformed into its amplitude- and phase component, a non-linear transformation putting tough requirements on the delay and recombination of the two signals.
Bassoo et al. (2009) have proposed a combination of Cartesian and polar modulation, where the SMPA input signal is a SD modulated Cartesian signal divided by the amplitude signal, which may be more or less bandlimited (see Fig. 7). Analysis showed that the envelope

signal can be limited to 75% of the channel bandwidth without impairing the efficiency, still keeping OFDM clipping limited and EVM very low. Thus, a combination of EER and PWPM can be used to have a high efficiency over a wide range of output power while avoiding the bandwidth expansion of polar modulation.

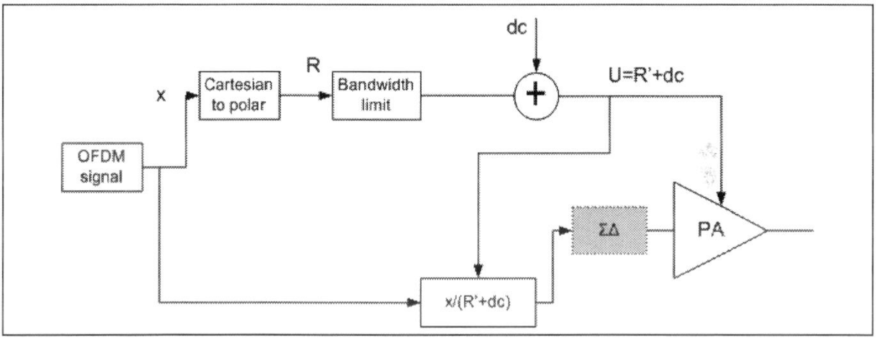

Fig. 7. Simplified architecture presented in Bassoo et al. (2009) for a combined polar and Cartesian modulator.

4.7 Efficiency comparison

Simulations have been performed on a 130nm CMOS class-D switched-mode amplifier, in order to compare the drain efficiency versus output power of the different architectures that have been discussed in the previous sections, such as Envelope Elimination and Restoration (EER), Envelope Tracking (ET), and Pulse Width Modulation by Variable Gate Bias (PWMVGB). Moreover, hypothetical curves for class-A and class-B operation have been drawn (see Fig. 8), with the peak efficiency as starting point. Class-A represents linear amplifier operation while class-B can be said to represent current-summing architectures.
Not unexpectedly the EER and ET architectures perform best, showing the highest efficiency for lower output power ranges. It may thus be concluded that the use of supply modulation is desirable for high average efficiencies. However, it can also be seen that efficiency remains a challenging aspect, especially taking into account numerous other requirements such as linearity, channel bandwidth, multi-mode/multi-standard operation and output power control range.

5. Summary

It is only fairly recent that CMOS technology has become a competitive alternative for integrated circuit power amplifier design for wireless communication handsets, as CMOS previously was not suitable for PA design due to frequency, output power, efficiency and linearity requirements. Thus, stand-alone PAs have long been manufactured in specialized technologies. Nowadays however CMOS has evolved to operating frequencies far into the GHz range, and many of the limitations, such as efficiency when used as linear amplification element, can be compensated by more digital control. Thus, a higher level of integration and more complex transmitter design result. However, the trend in CMOS technology development is to reduce device dimensions and as a consequence breakdown voltage. This complicates CMOS power amplifier design.

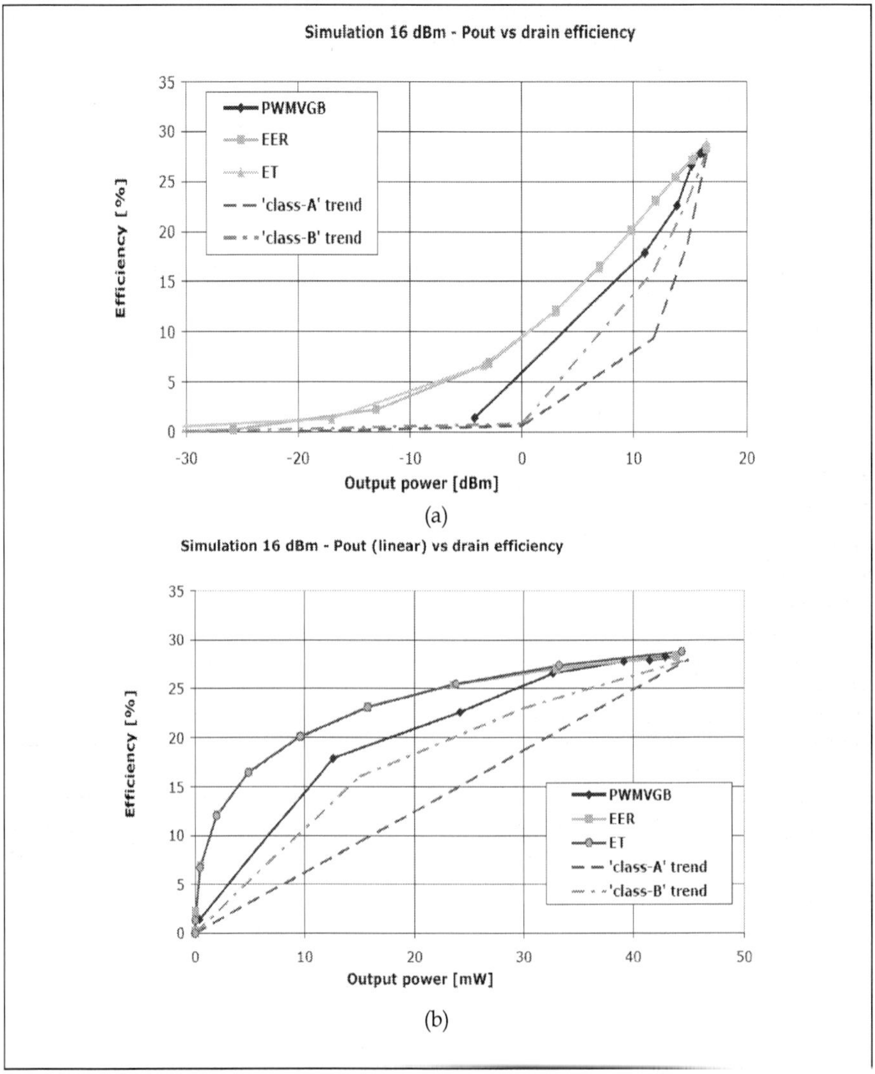

Fig. 8. Simulated drain efficiency for a CMOS class-D amplifier in different architectures, such as Envelope Elimination and Restoration (EER), Envelope Tracking (ET), and Pulse Width Modulation by Variable Gate Bias (PWMVGB). Class-A and class-B curves serve only as an illustration. The amplifier operated on a 1.2V supply and the input signal had a frequency of 2 GHz. (a). The output power (x-axis) represented in dBm, (b). The output power in mW.

Transmitter architectures using polar signals have gained in popularity, as splitting the Cartesian signal into a low-frequency envelope signal and a high-frequency phase signal provides excellent opportunity for efficiency improvements because a non-linear power amplifier can be used. A number of different polar architecture implementations exist, both digital and analog. However, signal bandwidth and supply requirements are challenging aspects of such designs. Other strategies have thus been used to avoid supply voltage modulation, such as switched control of the supply voltage or variable gate bias. Moreover, direct RF modulation can be used, implemented as a sigma-delta or pulse width modulator at high operating frequency. Recently, design strategies such as current steering have gained interest for use in PA and transmitter design. Digital control bits are used to generate a scaled output current, providing a high output power without straining the devices.

However, efficiency over a wide range of output power is still a challenging aspect of transmitter design, especially if other requirements such as linearity, power control, multi-mode/multi-band operation and channel bandwidth must be fulfilled simultaneously.

5.1 Future outlook

As CMOS technologies continue to develop to dimensions well below 65nm, special devices suitable for high supply voltage will likely continue to be provided, for example using high-K metal gate material. Such devices can be used on the same chip as digital circuits with clocking speeds of several GHz. Moreover, other substrate types may be used more extensively, such as Silicon-on-Isolator substrates. As they are less lossy, this may provide efficiency improvements.

On the other hand, performance requirements will continue to rise with the development and maturing of wireless communication systems, especially because of the desire to cover more and more standards in one handset (multi-mode/multi-standard operation). Digital control may be used to accommodate greater flexibility, reconfigurability and on-chip calibration in transmitter design. Moreover, techniques may be used to increase the adaptivity of components such as antennas, duplexers, filters and matching networks.

CMOS will continue to expand into the millimeter-wave range, with operating frequencies beyond 60 GHz. However, other technology developments may play an important role in future integrated circuit design for wireless communication, such as integrated RF MEMS (microelectromechanical systems). Also devices such as carbon nanotubes may be used for wireless applications. But such technologies have some way to go until they reach the level of integration that current CMOS technology has.

6. References

Aoki, I., Kee, S., Magoon, R., Aparicio, R., Bohn, F., Zachan, J., Hatcher, G., McClymont, D., Hajimiri, A. (2008). A Fully-Integrated Quad-Band GSM/GPRS CMOS Power Amplifier. *IEEE Journal of Solid-State Circuits*, Vol. 43, no. 12, December 2008, pp. 2747-2758.

Bassoo, V., Tom, K., Mustafa, A.K., Cijvat, E., Sjöland, H., and Faulkner, M. (2009). A Potential Transmitter Architecture for Future Generation Green Wireless Base Station. *EURASIP Journal on Wireless Communications and Networking*, August 2009.

Berland, C., Hibon, I., Bercher, J.F., Villegas, M., Belot, D., Pache, D. and Le Goascoz, V. (2006). A Transmitter Architecture for Nonconstant Envelope Modulation. *IEEE Transactions on Circuits and Systems-II: Express Briefs*, Vol. 53, no. 1, January 2006, pp. 13-17.

Cijvat, E. and Sjöland, H. (2008). Two 130nm CMOS Class-D RF Power Amplifiers suitable for Polar Transmitter Architectures, in *Proceedings of the 9th International Conference on Solid-State and Integrated-Circuit Technology (ICSICT)*, pp. 1380-1383, Beijing, China, October 2008, IEEE, Piscataway, NJ, USA.

Colantonio, P., Giannini, F., Giofrè, R. and Piazzon, L. (2008). A Design Technique for Concurrent Dual-Band Harmonic Tuned Power Amplifier. *IEEE Transactions on Microwave Theory and Techniques*, Vol. 56, no. 11, November 2008, pp. 2545-2555.

Cripps, S.C. (1999). *RF Power Amplifiers for Wireless Communication*. Artech House, ISBN 0-89006-989-1, Norwood, MA, USA.

El-Hamamsi, S-A. (1994). Design of High-Efficiency RF Class-D power Amplifier. *IEEE Transactions on Power Electronics*, Vol. 9, no. 3, May 1994, pp. 297-308.

Haldi, P., Chowdhury, D., Reynaert, P., Liu, G. and Niknejad, A.M. (2008). A 5.8GHz 1 V Linear Power Amplifier Using a Novel On-Chip Transformer Power Combiner in Standard 90 nm CMOS. *IEEE Journal of Solid-State Circuits*, Vol. 43, no. 5, May 2008, pp. 1054-1063.

Hanington, G., Chen, P., Asbeck, P., and Larson, L. (1999). High-Efficiency Power Amplifier using Dynamic Power-Supply Voltage for CDMA Applications. *IEEE Transactions on Microwave Theory and Techniques*, Vol. 47, no. 8, August 1999, pp. 1471-1476.

Jayaraman, A., Chen, P.F., Hanington, G., Larson, L. and Asbeck, P. (1998). Linear High-Efficiency Microwave Power Amplifiers Using Bandpass Delta-Sigma Modulators. *IEEE Microwave and Guided Wave Letters*, Vol. 8, no. 3, March 1998, pp. 121-123.

Jeon, Y.-S., Yang, H.-S. and Nam, S. (2005). A Novel High-Efficiency Linear Transmitter using Injection-Locked Pulsed Oscillator. *IEEE Microwave and Wireless Components Letters*, Vol. 15, no. 4, April 2005, pp. 214-216.

Jeong, J., Pornpromlikit, S., Asbeck, P.M. and Kelly, D. (2006). A 20 dBm Linear RF Power Amplifier Using Stacked Silicon-on-Sapphire MOSFETs. *IEEE Microwave and Wireless Components Letters*, Vol. 16, no. 12, December 2006, pp. 684-686.

Kavousian, A. Su, D:K., Hekmat, M., Shirvani, A. and Wooley, B.A. (2008). A Digitally Modulated Polar CMOS Power Amplifier with a 20-MHz Channel Bandwidth. *IEEE Journal of Solid-State Circuits*, Vol. 43, no. 10, October 2008, pp. 2251-2258.

Khan, L. (1952). Single Sideband transmission by Envelope Elimination and Restoration, in *Proceedings of the IRE*, pp. 803-806, July 1952, IRE.

Kitchen, J.N., Chu, C., Kiaei, S., and Bakkaloglu, B. (2009). Combined Linear and Δ-Modulated Switch-Mode PA Supply Modulator for Polar Transmitters. *IEEE Journal of Solid-State Circuits*, Vol. 44, no. 2, February 2009, pp. 404-413.

Kitchen, J.N., Deligoz, I., Kiaei, S., and Bakkaloglu, B. (2007). Polar SiGe Class E and F Amplifiers Using Switch-Mode Supply Modulation. *IEEE Transactions on Microwave Theory and Techniques*, Vol. 55, no. 5, May 2007, pp. 845-856.

Kobayashi, H., Hinrichs, J.M., and Asbeck, P.M. (2001). Current-Mode Class-D Power Amplifiers for High-Efficiency RF Applications. *IEEE Transactions on Microwave Theory and Techniques*, Vol. 49, no. 12, December 2001, pp. 2480-2485.

Kousai,S. and Hajimiri, A. (2009). An Octave-Range Watt-Level Fully Integrated CMOS Switching Power Mixer Array for Linearization and Back-Off Efficiency Improvement, in *The 2009 IEEE International Solid-State Circuits Conference (ISSCC) Digest of Technical Papers*, pp. 376-377, February 2009, IEEE, Piscataway, NJ, USA.

Kuo, T.C. and Lusignan, B.B. (2001). A 1.5W Class-F RF Power Amplifier in 0.2μm CMOS Technology, in *The 2001 IEEE International Solid-State Circuits Conference (ISSCC) Digest of Technical Papers*, pp. 154-155, February 2001, IEEE, Piscataway, NJ, USA.

Long, A., Yao, J., Long, S.I. (2002). A 13W Current Mode Class D High Efficiency 1 GHz Power Amplifier, in *Proceedings of The 2002 45th Midwest Symposium on Circuits and Systems (MWSCAS)*, vol. 1, pp. 33-36, August 2002, IEEE, Piscataway, NJ, USA.

McCune, E. (2005). High-Efficiency, Multi-Mode, Multi-Band Terminal Power Amplifiers. *IEEE Microwave Magazine*, March 2005, pp. 44-55.

Nielsen, M. and Larsen, T. (2008). A 2-Ghz GaAs HBT RF Pulsewidth Modulator. *IEEE Transactions on Microwave Theory and Techniques*, Vol. 56, no. 2, February 2008, pp. 300-304.

Niknejad, A.M., Emami, S., Heydari, B., Bohsali, M., and Adabi, E. (2007). Nanoscale CMOS for mm-Wave Applications, in *Proceedings of IEEE Compound Semiconductor Integrated Circuit Symposium (CSICS)*, pp. 1-4, October 2007, IEEE, Piscataway, NJ, USA.

Presti, C.D., Carrara, F., Scuderi, A., Asbeck, P.M., and Palmisano, G. (2009). A 25 dBm Digitally Modulated CMOS Power Amplifier for WCDMA/EDGE/OFDM with Adaptive Digital Predistortion and Efficient Power Control. *IEEE Journal of Solid-State Circuits*, Vol. 44, no. 7, July 2009, pp. 1883-1896.

Raab, F.H. (2001). Class-E, Class-C, and Class-F Power Amplifiers Based Upon a Finite Number of Harmonics. *IEEE Transactions on Microwave Theory and Techniques*, Vol. 49, no. 8, August 2001, pp. 1462-1468.

Reynaert, P. and Steyaert, M.S.J. (2005). A 1.75-GHz Polar Modulated CMOS RF Power Amplifier for GSM-EDGE. *IEEE Journal of Solid-State Circuits*, Vol. 40, no. 12, December 2005, pp. 2598-2608.

Shameli, A., Safarian, A., Rofougaran, A., Rofougaran, M. and De Flaviis, F. (2008). A Two-Point Modulation Technique for CMOS Power Amplifier in Polar Transmitter Architecture. *IEEE Transactions on Microwave Theory and Techniques*, Vol. 56, no. 1, January 2008, pp. 31-38.

Sjöland, H., Bryant, C., Bassoo, V., Faulkner, M. (2009). Switched Mode Transmitter Architectures, in *Proceedings of the 18th Workshop on Advances in Analog Circuit Design (AACD)*, pp. 315-333, April 2009, Ericsson AB, Lund, Sweden.

Smely, D., Ingruber, B., Wachutka, M. and Magerl, G. (1998). Improvement of Efficiency and Linearity of a Harmonic Control Amplifier by Envelope Controlled Bias Voltage, in The 1998 IEEE MTT-S Digest, pp. 1667-1670, 1998, IEEE, Piscataway, NJ, USA.

Sowlati,T. and Leenaerts, D.M.W. (2003). A 2.4-GHz 0.18-μm CMOS Self-Biased Cascode Power Amplifier. *IEEE Journal of Solid-State Circuits*, Vol. 38, no. 8, August 2003, pp. 1318-1324.

Stauth, J.T. and Sanders, S.R. (2008). A 2.4GHz, 20dBm Class-D PA with Single-Bit Digital Polar Modulation in 90nm CMOS, in *Proceedings of The IEEE 2008 Custom Integrated Circuits Conference (CICC)*, pp. 737-740, September 2008, IEEE, Piscataway, NJ, USA.

Su, D. and McFarland, W. (1997). A 2.5-V, 1-W Monolithic CMOS RF Power Amplifier, in *Proceedings of The IEEE 1997 Custom Integrated Circuits Conference (CICC)*, pp. 189-192, September 1997, IEEE, Piscataway, NJ, USA.

Su, D. and McFarland, W. (1998). An IC for Linearizing RF Power Amplifiers using Envelope Elimination and Restoration. *IEEE Journal of Solid-State Circuits*, Vol. 33, no. 12, December 1998, pp. 2252-2258.

Takahashi, K., Yamanouchi, S., Hirayama, T., and Kunihiro, K. (2008). An Envelope Tracking Power Amplifier using an Adaptive Biased Envelope Amplifier for WCDMA Handsets, in *Proceedings of the 2008 IEEE Radio Frequency Integrated Circuits Symposium (RFIC)*, pp. 405-408, June 2008, IEEE, Piscataway, NJ, USA.

Tsai, K-C. and Gray, P.R. (1999). A 1.9-GHz, 1-W CMOS Class-E Power Amplifier for Wireless Communications. *IEEE Journal of Solid-State Circuits*, Vol. 34, no. 7, July 1999, pp. 962-970.

Wagh, P. and Midya, P. (1999). High-Efficiency Switched Mode RF Power Amplifier, in *Proceedings of The 1999 42nd Midwest Symposium on Circuits and Systems (MWSCAS)*, vol. 2, pp. 1044-1047, August 1999, IEEE, Piscataway, NJ, USA.

Walling, J., Lakdawala, H., Palaskas, Y., Ravi, A., Degani, O., Soumyanath, K. and Allstot, D. (2008). A 28.6dBm 65nm Class-E PA with Envelope Restoration by Pulse-Width and Pulse-Position Modulation, in *The 2008 IEEE International Solid-State Circuits Conference (ISSCC) Digest of Technical Papers*, pp. 566-567, February 2008, IEEE, Piscataway, NJ, USA.

Wang, F., Kimball, D., Popp, J.D., Yang, A.H., Lie, D.Y., Asbeck, P.M., and Larson, L.E. (2006). An Improved Power-Added Efficiency 19-dBm Hybrid Envelope Elimination and Restoration Power Amplifier for 802.11g WLAN Applications. *IEEE Transactions on Microwave Theory and Techniques*, Vol. 54, no. 12, December 2006, pp. 4086-4098.

Yang, K., Haddad, G. and East, J.R: (1999). High-Efficiency Class-A Power Amplifiers with a Dual-Bias-Control Scheme. *IEEE Transactions on Microwave Theory and Techniques*, Vol. 47, no. 8, August 1999, pp. 1426-1432.

Yoo, C. and Huang, Q. (2001). A Common-Gate Switched 0.9W Class E Power Amplifier with 41% PAE in 0.25μm CMOS. *IEEE Journal of Solid-State Circuits*, Vol. 36, no. 5, May 2001, pp. 823-830.

Zhou, Y. and Yuan, J. (2003). A 10-bit Wide-Band CMOS Direct Digital RF Amplitude Modulator. *IEEE Journal of Solid-State Circuits*, Vol. 38, no. 7, July 2003, pp. 1182-1188.

15

Dimension Increase in Metal-Oxide-Semiconductor Memories and Transistors

Hideo Sunami
Hiroshima University
Japan

1. Introduction

This chapter describes a trend in dimension increase in structures of semiconductor memories and transistors focusing on metal-oxide-semiconductor, MOS devices. One, two, and three-dimensional (3-D) structures correspond to a legacy of grown-junction bipolar transistor, planar MOS transistor, and trench-capacitor dynamic-random-access memory, DRAM (Sunami et al., 1982-b & 1984), respectively. Flash memory has recently begun to employ 3-D stack of memory cells (Endoh et al., 2001).

To maintain the sufficient margin in DRAM operation, storage capacitance value should be kept as big as possible against scaling of memory cell area. In response to the requirement, 3-D capacitor has been introduced. The capacitor can be increased with the increase in the height of the capacitor without enlargement of planar area of the memory cell. First commercially available trench-capacitor DRAM appeared in mid 1980's at 1-M bit era together with stack-capacitor cell (Koyanagi et al., 1982). Recent stack-capacitor DRAM has begun to utilize a 3-D cylindrical capacitor same as trench capacitor cell.

While, MOS transistor has been shrunk continually from 12 μm to 45 nm with planar 2-D structure since early 1970's to date. An empirical fact that smaller MOS device leads to higher performance was theorized with the scaling theory (Dennard et al., 1968). However, hazardous short channel effects become obvious in sub-μm channel length regime. It is predicted that commercially usable minimum MOS device might be in the range of 5-10 nm. To cope with the short-channel effects vertical-channel transistors such as trench transistor (Richardson et al., 1981), surrounding gate transistor, SGT (Takato et al., 1988), and DELTA (Hisamoto et al., 1991) were proposed. Subsequently they have been extensively investigated these ten years. It is expected that the vertical transistor such as FINFET (Choi et al., 2001) will soon be applied to products to overcome the short-channel effects of 2-D transistor leading to a new era of 3-D LSI.

To summarize device trends in volume and size, increase in device count per chip and shrinkage of feature size are shown in Fig. 1. More than one-million fold increase in the device count has been achieved these 40 years leading to almost the same increase in processor performance. This has been driving enormous development of electronics and information technology.

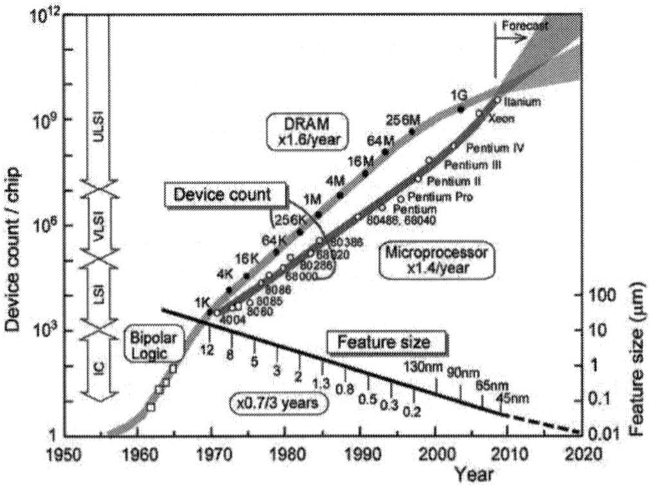

Fig. 1. Trends in device count/chip and feature size of MOS device. A DRAM cell consists of two devices of a cell transistor and a storage capacitor.

2. Grown-junction bipolar transistor as 1-D structure

It is well known that the first transistor invented in mid 1940's was a point-contact germanium bipolar transistor. Then, grown-junction type bipolar transistor became the first commercially successful semiconductor device (Teal et al., 1951). Although real devices are actually fabricated in 3-D structure, an operation mechanism of this bipolar transistor is based on 1-D current flow in principle.

Bipolar devices had been dominant in semiconductor market until early 1970's, and then MOS devices took over their position featuring less power consumption, denser packing, superior thermal stability, etc. Bipolar devices still survive in limited applications in fields of figh-frequency low-noise, inexpensive small scale IC, and high power. Besides, a kind of combination structure of bipolar and MOS transistors is insulated-gate bipolar transistor, IGBT. IGBT utilizes both an advantage of voltage-driven gate of MOS transistor and that of high current drivability of bipolar transistor. IGBT becomes a major device in power electronics such as electricity control in electric and hybrid cars.

No further description is made in this chapter since the major topic here is "scaling and higher integration of semiconductor device."

3. Invention of trench-capacitor DRAM cell as a quasi-3-D structure

3.1 Advent of DRAM

First DRAM was introduced to the market in 1970 by Intel with a 1-Kbit chip using three-transistor DRAM cells (Regitz & Karp, 1970). Subsequently, former 4-Kbit DRAM which was still employing 3-transistor cell began to be installed in IBM's mainframe computers. This was just the time when MOS devices were proven to deserve application as highly reliable main memory in mainframes. Until that time, MOS devices had been regarded as insufficiently stable.

A few years later 4-Kbit DRAM using the one-transistor cell (Dennard, 1968) was being widely manufactured. This memory cell trend is shown in Fig. 2 for equivalent circuit configuration. Since one-transistor cell was much smaller than two of others, very low-cost manufacturing was possible. Its low cost has been contributing to the development of personal computer. Then, the DRAM capacity has been increasing by a factor of four every three years until today. As modern computers are based on von Neumann's architecture, main memory is a key device together with processor. Along with the prosperity of computing, the demand for memory has increased to produce a world-wide 30-B$ market in 2009 for DRAM. Even if the main customer is still personal computer, various applications are extending DRAM's usage, e. g. cell phone, game machine, personal audio, and video machine, etc.

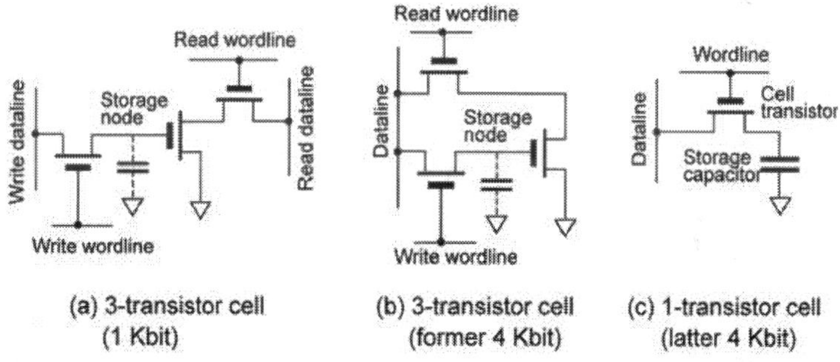

Fig. 2. Trend in DRAM cell configuration.

3.2 Key factor of cost

The strongest driving force for growing of DRAM market is undoubtedly "price". Therefore, various development efforts have been focusing on reduction of manufacturing cost. One of major effort is primarily devoted to finer patterning. The bit cost has decreased by a factor of 10^{-6} during 30 years since 1970, and 1-Gbit product has already been sold at the less price of 1-Kbit. Since chip cost is closely related to number of chips on a wafer, the wafer size has been continually increased to be such as 50, 75, 100, 125, 150, 200, and now 300 mm in diameter. Together with the diameter increase, memory cell size has been reduced to be 1/3 in each DRAM generation in volume production to absorb chip size increase. Consequently, the chip size has been enlarged at most up to 10 times despite the bit increase by a factor of 10^6 from 1 Kbit to 1 Gbit. Then, memory cell size decreases down to a factor of 10^{-5} as shown in Fig. 3.

3.3 Invention of trench-capacitor DRAM cell

In response to chip size reduction to cope with 4-times increase in memory capacity, the memory cell size has been reduced to almost one-third in each generation, previously shown in Fig. 3. The DRAM cell, so-called 1-transistor cell, consists of one cell transistor and one storage capacitor. Key specifications in DRAM operation, such as noise margin, soft-error durability, operational speed, power consumption, strongly depend on the capacitance of storage capacitor (Dennard, 1984). The capacitance value, C_S is expressed as

$$C_S = \epsilon_i A / T_i \quad (1)$$

where ϵ_i, A, T_i, are permittivity of storage insulator, area of capacitor electrode, and insulator thickness, respectively. Therefore, the cell size reduction through scaling leads to the area reduction and subsequent decrease in capacitance value.

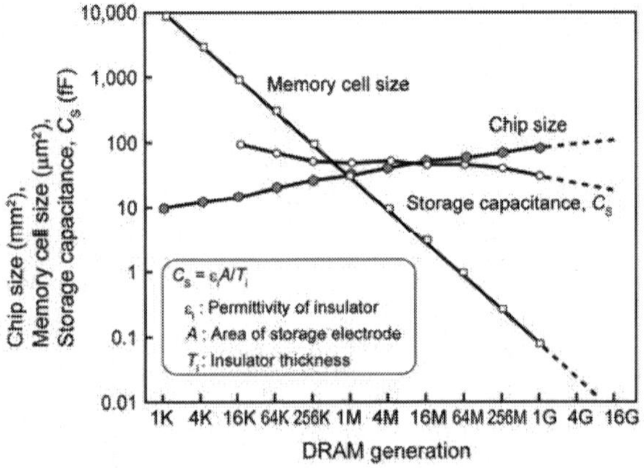

Fig. 3. Memory cell size shrinkage at DRAM in volume production.

To cope with the dilemma as to cell size vs. capacitance, insulator thickness was reduced by a factor of 10 from 100 nm in 1-Kbit to 10 nm in 1-Mbit chips, becoming adversely close to dielectric field breakdown. When the author took a glimpse at some conference presentation from Texas Instruments Inc. in 1974 introducing a highly efficient silicon solar cell with plural steep trenches, as shown in Fig.4 (a), forecasting the upcoming issue of cell size vs. capacitance, he got an idea of a trench capacitor DRAM cell. Even though his job at that time was to characterize the silicon surface with photoemission spectroscopy, his amateur-radio hobby connected the shape of trimmer condenser, which has two coaxial cylindrical opposite electrodes as illustrated in Fig. 4 (b), with the need of the 1-transistor cell. From that idea, he invented a trench capacitor cell and applied for a Japanese patent in 1975 (Sunami and Nishimatsu, 1975). Due to its low score of assessment, this was not applied to any overseas patent.

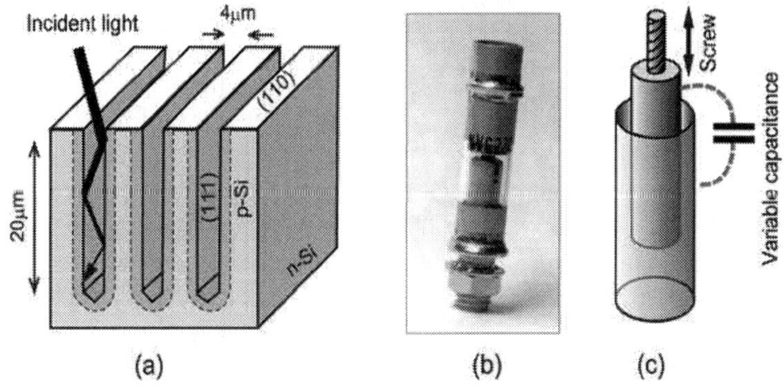

Fig. 4. Hints to create a trench-capacitor DRAM cell concept: proposed solar cell with steep trench, (a), a photograph of trimmer condenser, (b), and its equivalent model, (c).

As Hitachi had won a leader's position in 64-Kbit DRAM products with a 5-V single power supply (Itoh et al., 1980) and folded bit-line arrangement (Itoh, 1975), its research and development group could afford to challenge for novel cell development together with the development of integration processes at 1.3-μm technology node. After several years' development, the first 1-Mbit level trench cell in trial production was successfully implemented and then presented in IEDM (Sunami et al., 1982). The development story is described hereafter.

The memory cell obtained measured 4 μm by 8 μm with a 2.5-μm deep trench. The capacitor insulator is a triple layer of $SiO_2/Si_3N_4/SiO_2$ of which thickness was equivalent to that of 15-nm SiO_2. Resultant capacitance per unit area was 2.2 fF/μm^2. Then, obtained storage capacitance, C_S with a trench of 2.5 μm in depth and bit-line capacitance, C_B were 45 and 400 fF, respectively with 128-bit folded bit-line arrangement. Resultant C_S/C_B ratio is 0.11. This value is sufficiently large for the stable operation. A scanning-electron micrograph of a cross-section of the cell is shown in Fig. 5.

This first trial 1-Mbit cell array with trenches won a signal voltage of 200 mV at 5-V power supply, as shown in Fig. 6. Since it was empirically recognized that sufficient signal voltage was around 100 mV in those days, an obtained S/N ratio was large enough to obtain stable DRAM operation. Therefore, high immunity to alpha particle hit was being strongly expecxted for coming megabit DRAM products until the time when actual soft-error measurement was made.

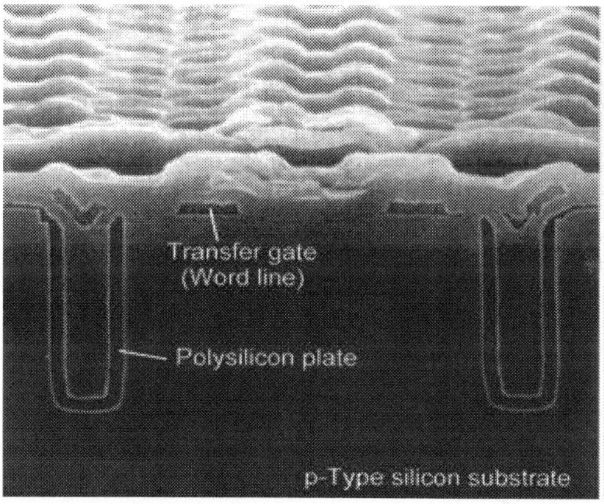

Fig. 5. An SEM cross section of memory-cell array of 1-Mbit DRAM in trial production. The memory cell measures 4 μm by 8 μm.

3.4 Changes of trench cell employment

In a R&D project, 1-Mbit DRAM was a prime vehicle to drive 1.3-μm node MOS technologies, such as lithography, dry etching, film deposition, gate material selection, metallization, etc. except packging. In the course of trench DRAM development, several issues were given birth to. Major ones were

a. degraded oxide uniformity on trench wall, which leads to degradation of oxide integrity,
b. trench to trench leakage which limits further denser packing of cells, and
c. increased soft error which is fatal in application to reliability-conscious computers.

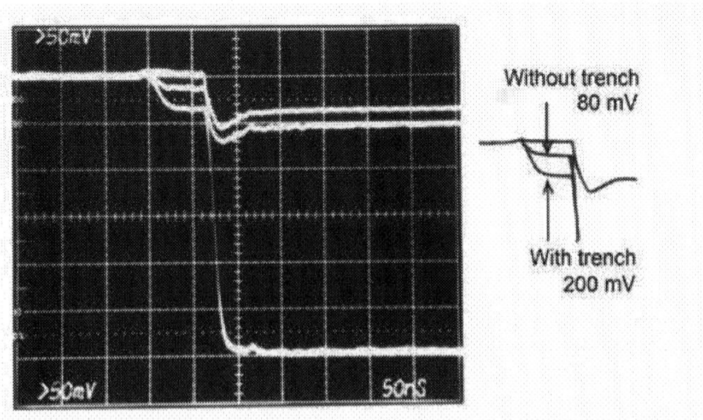

Fig. 6. Output signals of a sense amplifier for a 128-bit folded bit-line cell array with trench of 2.5 μm in depth and that without trench.

Beside these major issues, formation of neat trench shape, avoiding of dislocation formation at the bottom of the trench, high-energy boron implantation into deeper potion of the substrate, uniform capacitor film deposition, polysilicon filling into the trench with phosphorus doping to the polysilicon, etc. should have been solved in a limited period.

a. Oxide uniformity

Crystallographic orientation of trench surface varies resulting in different oxidation rate. It was observed that oxidation rate was higher in order of (110)>polysilicon>(111)>(100). Even though lower oxidation temperature gives rise to more enhanced oxidation rate (Sunami, 1978), this phenomenon still exists at relatively higher temperature range. Thus dielectric breakdown voltage was lowered at the thinnest portion on the trench wall. A transmission-electron micrograph of an experimental result is shown in Fig. 7 in case of 1000°C dry oxidation.

A drastic solution to overcome this problem, it is desirable to utilize chemical-vapor depotion, CVD. An $SiO_2/Si_3N_4/SiO_2$ film was made full use of in trial production. Thickness ratio should be carefully chosen in order to avoid non-volatile memory effect due to the existence of Si_3N_4/SiO_2 interface.

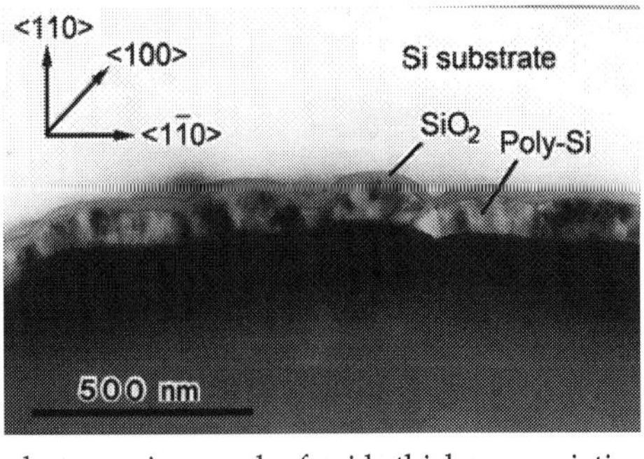

Fig. 7. A transmission-electron micrograph of oxide thickness variation on trench wall with 1000°C dry oxidation.

b. Trench to trench leakage

Another serious problem against further scaling was a leakage current flowing through the deeper portion of adjacent two trenches. The current gradually fills up an empty cell with charges turning "1" into "0". This is a fatal failure for DRAM. This is attributed to a parasitic MOS transistor formed spreading over ajacent two memory cells. Two trenches work as deep source and drain; capacitor plate is the gate; and thick field oxide is the gate oxide. This is regarded as a typical MOS transistor simply causing large punch through current at deeper portion between source and drain.

To outline the leakage current qualitatively, two-dimensional device simulation using CADDET (Toyabe, 1978) was carried out (Sunami et al., 1985). Resultant potential distribution with leakage current flow and a method of leakage current suppression are shown in Fig. 8 (a) and (b), respectively. In the simulation result, one notable fact is that the current flows in the deeper portion of the substrate and a potential mound is located at the substrate surface. These results may be attributable to a field implantation of which peak concentration exists at the surface.

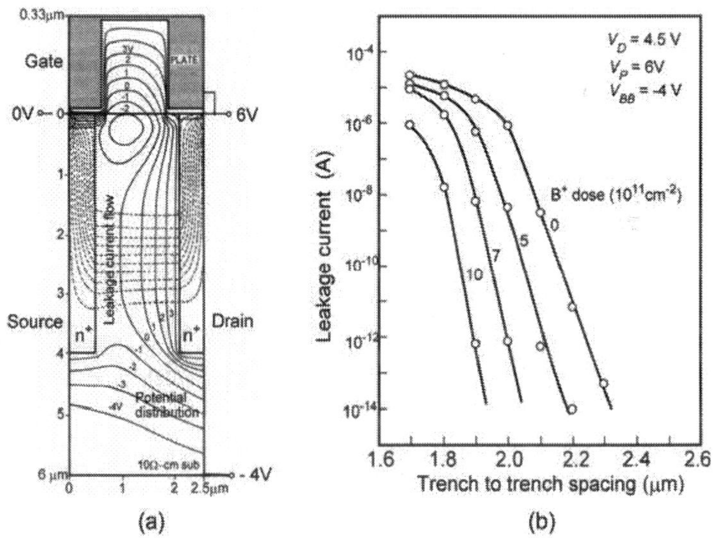

Fig. 8. Leakage current characteristics for two ajacent trenches. Resultant equipotential curves are denoted by solid lines and broken curves are leakage current paths, in (a). A method of leakage suppression with p-type well is shown in (b) using ion implantation with boron.

It is well known that punchthrough stopper with relatively higher dose of p-type dopant can suppress the punchthrough current. As is previously shown in Fig. 8 (b), the leakage current decreases inversely with the increase in a boron implantation dose. Since the impurity concentration of the substrate is 1.5×10^{15}cm^{-3}, boron implantation doses of 1, 5, 7, and 10×10^{11}cm^{-2} generate accepter concentrations of 1.5, 1.9, 2.1, and 2.5×10^{15} cm^{-3} at 3-μm deep portion between two adjacent trenches. It is noted that even small increase in the concentration can drastically reduce leakage current.

One of radical solutions is to provide a storage node being isolated from the current path in the substrate. Substrate plate or sheeth plate configurations are good candidates which will be referred to hereafter.

c. Soft-error

In the final stage of the development, most serious problem of soft-error was found caused by the alpha-particle hit as shown in Fig. 9. A difference of few orders of magnitude was observed between the planar and the trench cells at cell-failure mode. While, the same performance was observed for both of them in bit-line failure mode. This is because the bit-lines were formed in the same configuration. In this result, it was observed that the trench cell with 40% increase in signal charges provided the same soft-error rate as compared to the planar cell. Even though the trench cell provides stable DRAM operation due to larger signal charges, it loses the advantage of increased signal charges.

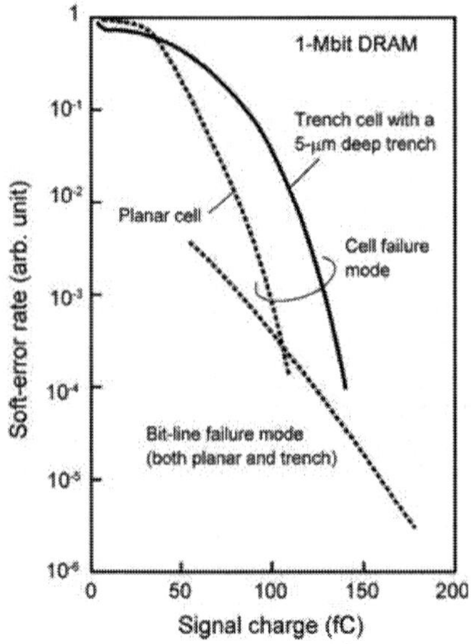

Fig. 9. Measured soft error rates of planar and trench cells.

One alpha particle at maximum 5-MeV energy generates almost one million electron-hole pairs. One million electrons is about 190 fC which is almost equivalent to signal charges stored in one storage capacitor of 1-Mbit DRAM cell. Due to extended depletion layer of the storage capacitor in the trench cell, it "effectively" collects generated electrons, as shown in Fig. 10.

In addition to the soft-error problem, it was predicted that punch-through current between any adjacent two capacitors would soon limit further shrinkage of the cell. That was a serious decision point about whether the trench cell should be improved or abandoned.

In those days, most DRAM manufacturers made efforts to supply their DRAM products to very limited leading mainframe makers. That was a kind of certificate that their products achieved first-grade reliability. The certificate surely made their business fruitful. Even with a half-year delay in product shipment, they might lose their business in the mainframe

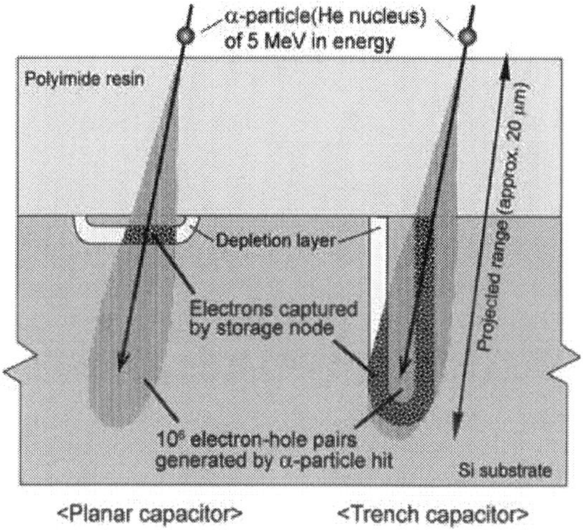

Fig. 10. A model of electron-hole pair generation by an alpha-particle hit.

market during one DRAM generation. There is a clear evidence that a leading maker has changed in each DRAM generation, Intel at 1 K, then, TI, MOSTEK, Hitachi, NEC, Toshiba, Samsung ...

Since Hitachi has been a DRAM manufacturer as well as mainframe supplier, it focused keenly on the mainframe application with highest-grade reliability compared to those of personal-use electronic appliances with relatively low reliability. Thus, Hitachi had abandoned the trench cell putting aside several ideas already proposed by the device development group for improved structures to reduce the soft-error problem (Sunami, 2008-a). Additional development was thought to need more than half a year. Since leading mainframe makers accept only a few DRAM suppliers, new product shipment with half-year delay would be fatal.

In the same period, a new configuration of array operation with half-V_{cc} plate (Kumanoya et al., 1985) was proposed as shown in Fig. 11. This has a strong potential of storage-capacitance doubling. As a result, it could prolong the use of conventional planar cell. This proposal had also influenced Hitachi's decision that conventional planar cell should be

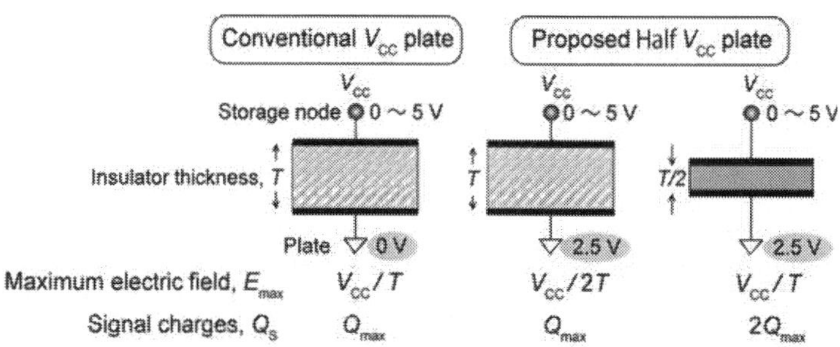

Fig. 11. Half-Vcc plate configuration has a possibility of doubling signal charges keeping maximum electric filed strength applied to capacitor insulator.

applied to 1-Mbit DRAM products. The conventional structure with conventional fabrication technologies are strongly desirable from manufacturability and cost points of view in general.

While, several innovative trench cells had been proposed and then developed to drastically improve soft-error problem in the same development project. A prime key fator was to avoid inflow of charges generated by alpha hit. Those cells were substrate-plate cell (Sunami et al., 1982-a) and sheath-plate cell (Kaga et al., 1988) as shown in Fig. 12. Storage nodes of these cells are surrounded by capacitor insulator being isolated from charges generated in the substrate by an alpha-particel hit. A portion of p-n juction exposed to generated charges is very small clearly illustrated in the figure. The substrate-plate trench cell amazingly improves soft-error tolerance due to its highly shrunk depletion layer.

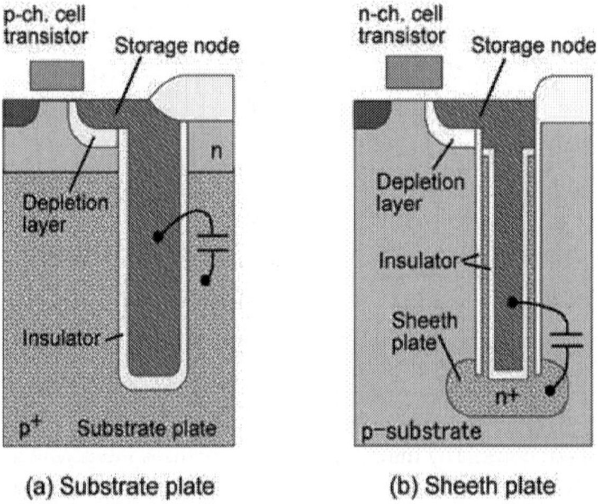

Fig. 12. Proposed DRAM cells to drastically improve soft-error caused by alpha-particle hit. Storage nodes are isolated from substrate by capacitor insulator

Despite alpha-immunity problem, several major manufacturers employed the trench and have been improving the structure until today. Together with the trench, the stacked capacitor cell was also applied in products. In addition to these cell structure innovations, the hemi-spherical grain (HSG) structure (Watanabe et al., 1992) was an inevitable technique to double the storage capacitance due to increased surface area.

3.5 DRAM cell trend

Major advancement in cell innovation is shown in Fig. 13. Cylinder-type stack and substrate-plate trench, both with HSG, are the major cells being manufactured today. These DRAM cell innodations are divided into three phases.

Phase I (1K→1M): Shrinkage of planar area of memory cell together with the decrease in capacitor insulator thickness. Thinning of the insulator finally brought about catastrophic dielectric breakdown of the insulator. Even with utilizing of half-Vcc configuration, planar cell could not survive at 4-Mbit era.

Phase II (1M→1G): 3-D capacitor structure with planar cell transistor. The capacitance does not suffer from planar area shrinkage in principle. Two categories of stack and trench capacitor cells were proposed. In the latter part of the phase II, high-k materials became inevitable to keep capacitance value against cell area shrinkage.

Phase III ((1G→1T): Three-dimensional stack of the capacitor and the cell transistor. This will be described later in section 4.5.

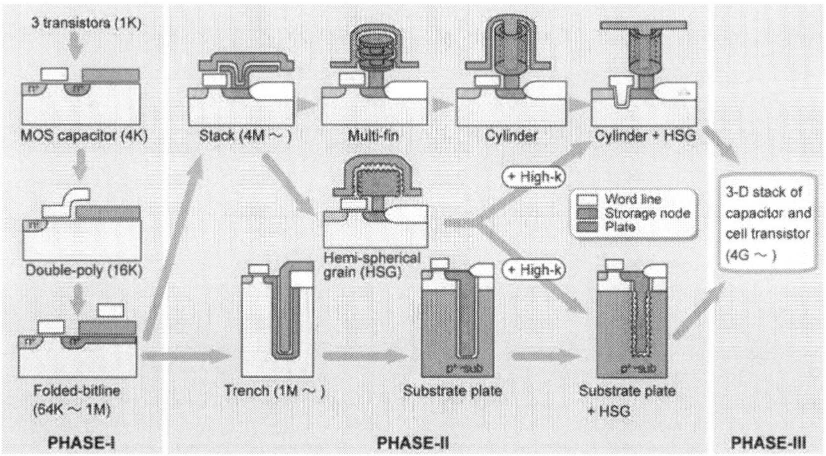

Fig. 13. DRAM cell trend. Phases I, II, and III correspond to planar area shrinkage, 3-D capacitor, and 3-D stack of cell transistor and storage capacitor, respectively.

A typical memory cell of commercially available 1-Gbit level DRAM is shown in Fig. 14 (Sunami, 2008-c). This shows one kind of combination to utilize various technologies. This virtual structure is not necessarily the exact one of commercially available real product.

Extended channel length with trench gate is aiming much less sub-threshold current to keep sufficient refresh time. Relatively low concentration of n-type dopant at junction also provides lower leakage current due to reduced electric filed across the junction. Since it is predicted that there will certainly exist an ultimate limit in size of hemi-spherical grain, diameter of the cylinder will also cease to shrink due to the grain size.

3.6 Material revolution

From 1 K to 1 M, size scaling was the key issue. The storage capacitance value was kept almost the same over several DRAM generations by reducing insulator thickness compensating memory cell shrinkage. Consequently, the reduced thickness made the electric field across the insulator close to 5 MV/cm which was recognized to be the upper limit for keeping insulator integrity and refresh time in DRAM operation. Thus, innovative techniques other than thickness reduction were strongly required.

In response, three-dimensional structures were proposed. From 1 M to 1 G, three-dimensional structure innovation has been achieved as previously shown in Fig. 13. However, as the aspect ratio of the storage capacitor exceeds more than 10, manufacturability becomes a much more serious issue. The final parameter to be handled in the relation expressed in Eq. (1) is permittivity, ϵ_i. Thus, various kinds of high-*k* materials

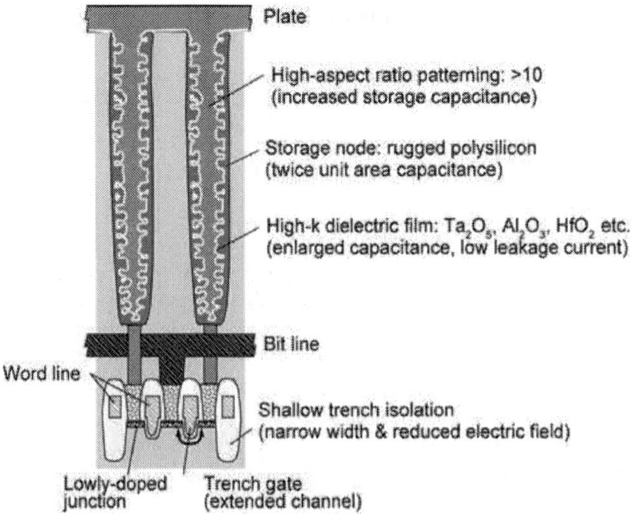

Fig. 14. A typical 1-Gbit level DRAM cell utilizing various kinds of proposed technologies. This may not necessarily be the exact memory cell in commercially available products.

have been developed as shown in Fig. 15. But there is a serious fact that the thinner the thickness is, the less its permittivity is. An empirical equation regarding the relation between leakage current, I_{leak} and barrier height in silicon-insulator system is expressed as

$$I_{leak} \propto \exp[-(m\phi)^{1/2}T], \tag{2}$$

where, m, ϕ, and T are effective mass, barrier height, and film thickness, respectively. Thus, high-k film may not be a unique ultimate solution at this moment. Material revolution with ultra high-k material is solicited to extend DRAM further toward terabit DRAM on a chip.

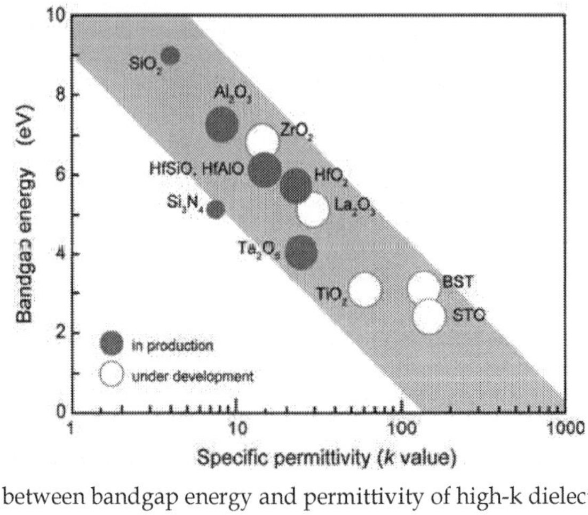

Fig. 15. Relation between bandgap energy and permittivity of high-k dielectric films.

To summarize innovation achieved in the past and requirements to the future, there are three eras for DRAM development.

- 1 K to 1 M ---- dimension improvement: smaller cell and reduced insulator thickness.
- 1 M to 1 G ---- structure or material innovation: stack or trench cell with high-k film.
- 1 G to 1 T ---- 3-D stack: cell transistor and storage capacitor with material revolution.

The final parameter which affects advanced shrinkage of the cell should be the insulator thickness itself. If the insulator is thick enough to fill the internal hole of the trench of the trench cell or cylinder of the stacked cell, the plate of the capacitor cannot penetrate inside the trench or the cylinder, resulting in no capacitor formation (Itoh et al., 1998), as shown in Fig. 16. In this sense, high-*k* films should be thin enough, simultaneously keeping their high-*k* value. This may be the deadlock for realizing smaller cells of the 1-transistor type DRAM cell. Even utilizing cutting-edge high-*k* films at present, 32 or 64-Gbit DRAM will be the biggest capacity on a chip without chip stack. We would like to expect novel main memory candidates in near future.

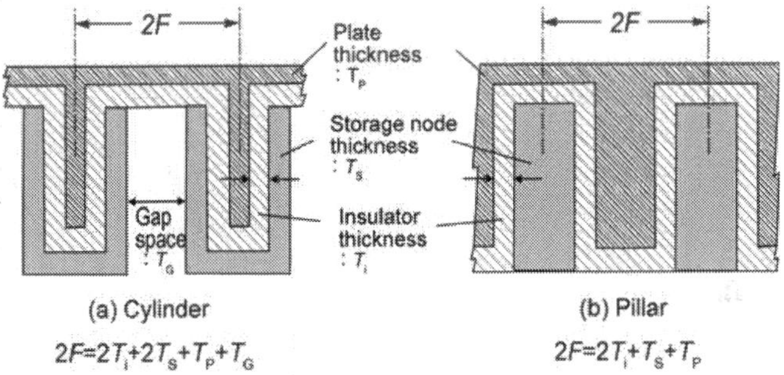

Fig. 16. Relations among several film elements constructing the storage capacitor.

4. Two- and three-dimensional MOS transistors

Since integrated circuits, particularly MOS memory and processor, were introduced to the market in early 1970's, almost four-fold increase in both memory's volume and processor's performance has been continually achieved every three years, as previously shown in Fig. 1. The strongest driving force for the increase is undoubtedly "cost" as previously described in section 3.2. The volume increase has been attained maily by shrinkage of all components on a chip. MOSFET (field-effect transistor) is particularly suitable to the shrinkage because the scaled transistor provides better performance. This transistor's behavior was theoretically analysed (Dennard et al., 1974) and named "scaling principle" later in semiconductor industry.

4.1 Innovation of 2-D transistors

Even though scaled transistor provides better performance, various kinds of problems become more serious in response to the scaling. They are so-called "short channel effects"; drain-to-source breakdown voltage is decreased; hot-carrier immunity gets worse; subthreshold current becomes more harmful against cut-off performance; gate leakage current increases with decreasing of gate oxide thickness; and mobility degradation sacrifices the scaling itself .

To cope with these short channel effects, 2-D transistor structure has been improved, as shown in Fig. 17. The structure has been improved so that electric field in the vicinity of drain is reduced. High electric field results increased leakage current and reduced breakdown voltage of source to drain. Thus DD was developed to reduce the electric filed with more graded impurity profile around n+ drain. However, the graded impurity profile increases punch-through current in deep portion between source and drain.

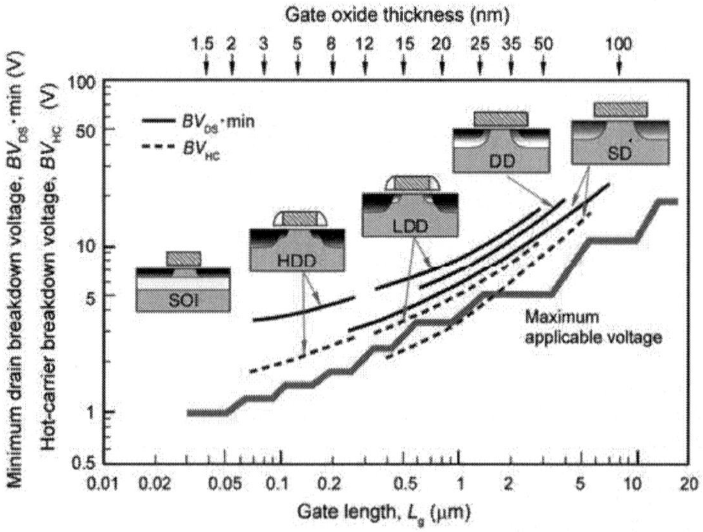

Fig. 17. Improvement of MOS transistor structure regarding source and drain regions. SD, DD, LDD, HDD, and SOI denote single drain, double drain, lightly-doped drain, highly-doped drain, and silicon-on-insulator, respectively.

Then, LDD was developed so as to suppress the punch-through current with graded impurity profile regions which were located only at edges of drain and source, as shown in Fig. 17. Due to relatively higher resistivity associated with the graded impurity profile, LDD's drivability was not satisfactory because of relatively higher series resistance between source and drain. Then HDD was developed to reduce the effect.

Even though these innovations were made, mobility degradation problem still remained. Based on a physical aspect that tensile and compressive strains enhance the electron and the hole mobilities respectively, a strained MOS transistor was proposed (Kesan et al., 1991; Ismail, 1995). Typical strained silicon MOS transistors are shown in Fig. 18.

The strained transistor (a) in Fig. 18 consists of SOI structure with a Si-Ge layer underneath source and drain. Since an overlayer silicon has to be epitaxially deposited on Si-Ge layer, complicated fabrication processes are likely to delay the practical use of it.

As a more practical structure, the usage of compressive and tensile chemical-vapor deposited (CVD) silicon-nitride films was proposed (Pidin et al., 2004), as shown (b) in Fig. 18 . Stresses of about -2 and +2 GPa were successfully introduced into p- and n-channel regions, respectively. Minus and plus signs of stress denote compressive and tensile, respectively. Even though real deposition methods of the films were not disclosed in the meeting, it is well presumed that the tensile strain may be introduced by thermally

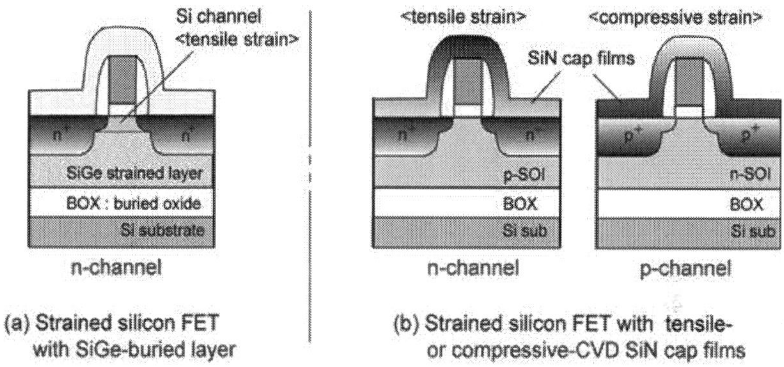

Fig. 18. Typical strained silicon MOSFET's: SiGe buried layer, (a) and CVD SiN cap films, (b).

decomposited CVD whereas the compressive strain may be given by plasma-enhanced CVD. Almost 50% increase in carrier mobilities of both n- and p-channel transistors were obtained.

4.2 Proposals of quasi 3-D transistors

To cope with short-channel effects which will be more and more serious in response to the scaling of conventional 2-D transistors, transistors of which channel was formed on both side walls of a silicon beam, named trench-isolated transistor using side-wall gates, TIS (Hieda et al., 1987) and fully depleted lean-channel transistor, DELTA (Hisamoto et al., 1989) were proposed as shown in Fig. 19 (a) and (b), respectively. Because of horizontal current flow of the transistor, this kind of transistors is called "quasi 3-D" in this article.

In TIS, full side walls were not used, while main channel was formed on side walls of the thin silicon beam in DELTA. The bottom of the silicon beam is fully oxidized with local-oxidation of silicon process (LOCOS), the beam is isolated from silicon substrate like SOI substrate. Advantages of the thin silicon channel were estimated.

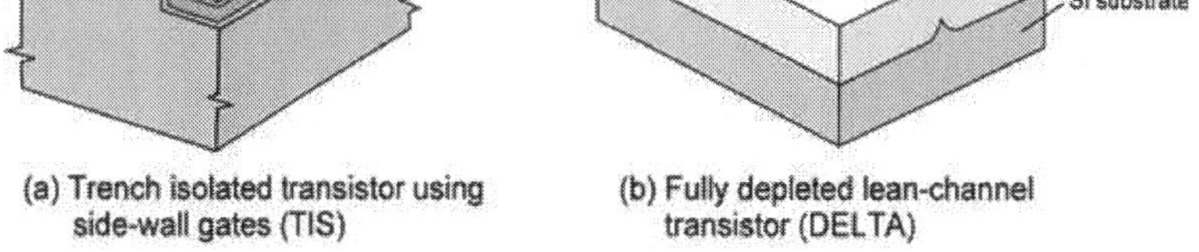

Fig. 19. Proposed quasi 3-D transistors of trench-isolated transistor using side-wall gates (TIS), (a) and fully depleted lean-channel transistor (DELTA), (b)

The author's group has proposed several devices with respect to quasi-3-D structures. One of them is corrugated channel transistor, CCT (Furukawa et al, 2003; Sunami et al, 2004) as shown in Fig. 20. Plural beam channels with {111} surface are formed by a crystallographically preferential etching with tetramethylammonium hydroxide, TMAH, atomically flat channel surface can be formed expecting less mobility degradation by avoiding rough surface of the channel.
The current drivability of CCT is proportional to the number of the beams as shown in Fig. 21. This is suitable for area-conscious applications such as power transistor and/or high-voltage transistor.

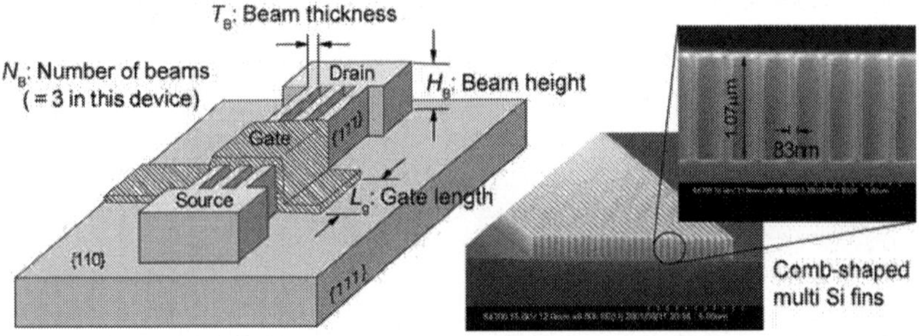

Fig. 20. A corrugated-channel transistor, CCT featuring.

Other proposal is super self-aligned triple gate transistor (Okuyama et al., 2007) as shown in Fig. 22. As two sidewall gates are delineated with an etching mask of a top gate, triple gates are selg-aligned each other leading to much smaller area occupation on a silicon die. One of device performance is shown in Fig. 23. Three gates operate three transistors independently with unified source and drain. At single-gate operation, subthreshold current can be controlled by other two side gates, namely, a variable threshold-voltage transistor can be realized in a certain voltage range.

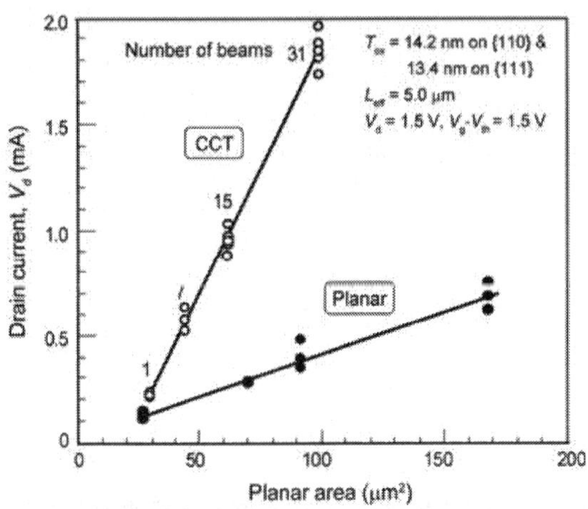

Fig. 21. Drivability of corrugated-channel transistor, CCT in terms of planaer area.

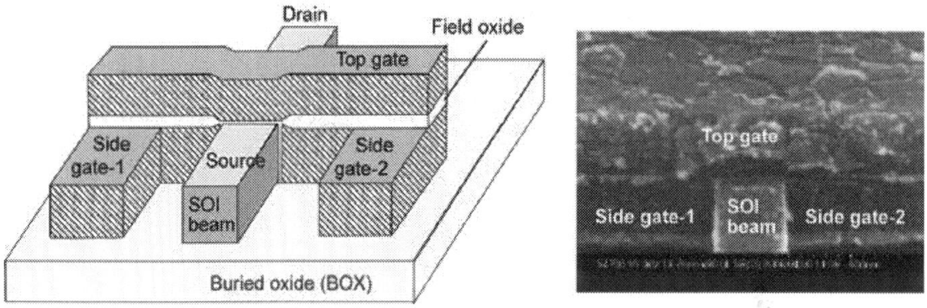

Fig. 22. Super self-aligned triple gate transistor featuring three gates of top gate, side gate-1, and side gate-2 formed in self-aligned manner.

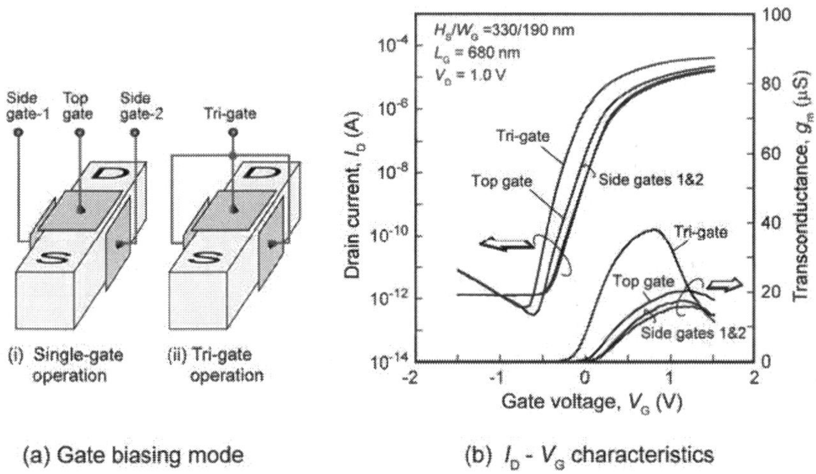

Fig. 23. Drain current characteristics of the triple gate transistor. Three gates provide independent three transistors with a unified drain and a unified source.

In these quasi-2-D transistors, there exist several serious issues caused by the formation of tall and thin steep silicon beam. They are (1) delineation of steep vertical silicon beam, (2) conformal gate material formation, (3) low-resistive source and drain, and (4) low resistive contacts to source and drain. The former two can be solved by advanced lithography with multi-level resist technique, CVD, and dry etching with high material selectivity. The latter two may be achieved by silicidation of silicon beam and wrapped metal contact as shown in Fig. 24.

In the figure, current paths of beam channel transistor are illustrated. It is obvious that longer current paths in relatively high resistivity area are illustrated in top contact as shown in Fig. 24 (a). On the other hand, relatively shorter current paths are formed in wrapped contact as shown in Fig. 24 (b)

Simulated drain currents and transconductances are described in Fig. 25 in case of typical impurity concentration and silicidation (Matsumura et al., 2007). Top contact transistor structure scrifices the advantage of beam-channel transistor to a considerable extent.

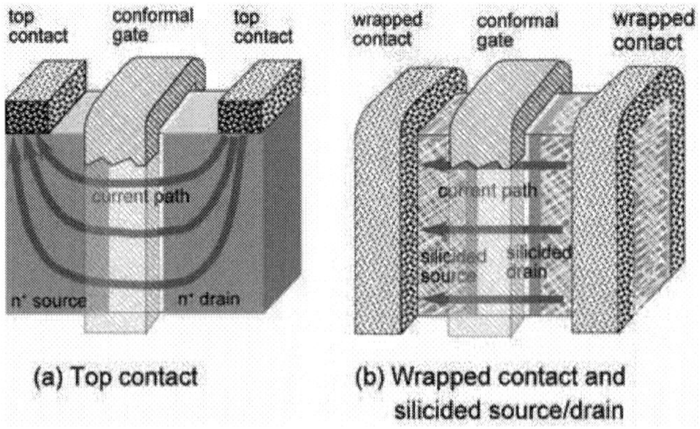

Fig. 24. One of drain current characteristics of the triple gate transistor at two modes of gate voltage application.

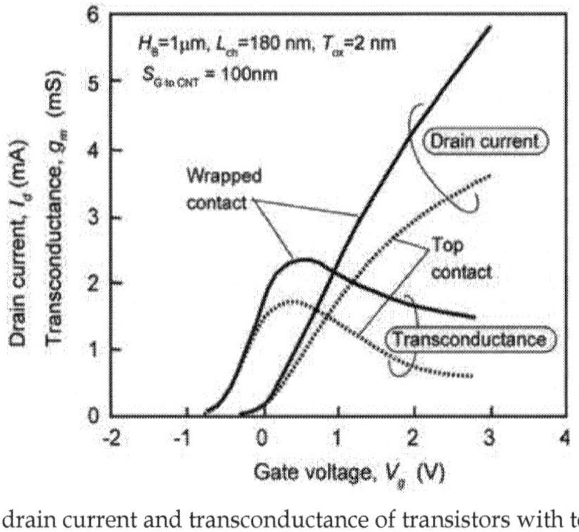

Fig. 25. Simulated drain current and transconductance of transistors with top contact and wrapped contact. Transistor structures are shown in Fig. 24.

4.3 Proposal of 3-D transistors

To summarize quasi-3-D transistors described above, a possible scenario of transistor structure innovation is illustrated in Fig. 26. Transistors with horizontal current flow inside a silicon beam are called FINFET today (Choi et al., 2001). Then, 3-D FET's with vertical current flow will be a next candidate for 3-D LSI.

With respect to the vertical transistor, a few DRAM cells utilizing vertical current flow structure have already been proposed in mid 1980's. They are trench-transistor cell, TTC (Richardson et al, 1985) and surrounding gate transistor, SGT (Takato et al., 1988). However, they are not manufactured in real products yet. One reason is probably that fabrication technologies do not become matured yet in general.

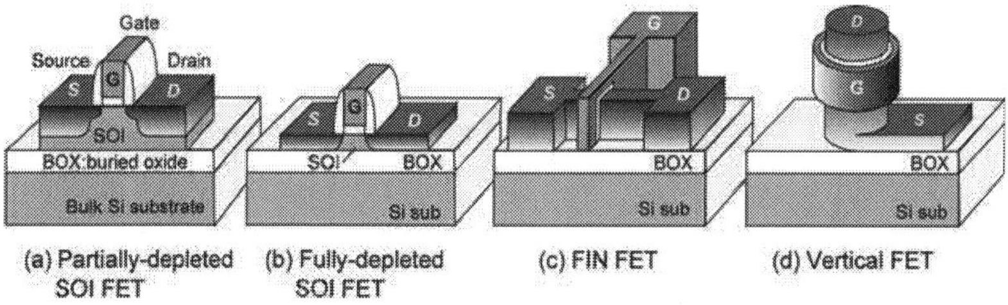

Fig. 26. Recent trend in transistor structure. It is not reported yet in 2009 that both FINFET or vertical FET is already shipped to the semiconductor market.

These structures may be almost the tiniest configuration in one-transistor DRAM cell. A theoretical area of these cells is $4F^2$. F is a feature size of device, in other word, technology node itself. In conventional array configurations, theoretical memory cell sizes of open bit-line and folded bit-line arrangements are $6F^2$ and $8F^2$, respectively. A vertical stack cell as shown in Fig. 27 (c) will be one of the most promising structures in near future.

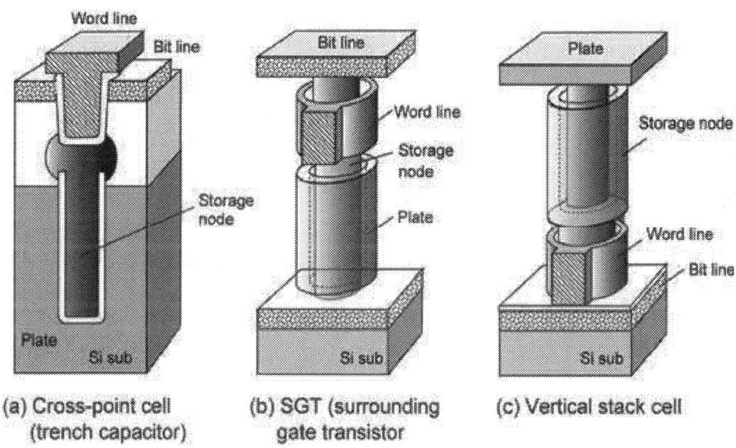

Fig. 27. Proposed vertical cell transistors applied to one-transistor DRAM cell.

4.4 A vertical transistor having a potential of $2F^2$ cell area

The author's group has proposed a super pillar transistor, SPT which has a potential of realizing $2F^2$ DRAM cell (Sugimura et al., 2008). This SPT can double the packing density of DRAM cell as compared to $4F^2$ cells previously shown in Fig. 27. A bird's eye view of SPT is shown in Fig. 28.

Fabrication process folw is as follows. Selected portions of a silicon beam are covered with CVD Si_3N_4 films. Then high temperature oxidation is performed at 1000°C to the extent that the beam is fully oxidized. Portions which are not covered with the Si_3N_4 films are converted into SiO_2 remaining physically and electrically separated silicon pillars. Subsequently, gate oxidation is processed and gate film is entirely deposited. Then, directional dry etching is performed entirely on a wafer remaining two gates located on both sides of the beam as residues associated with the dry etching. The resultant structure is already shown in Fig. 28.

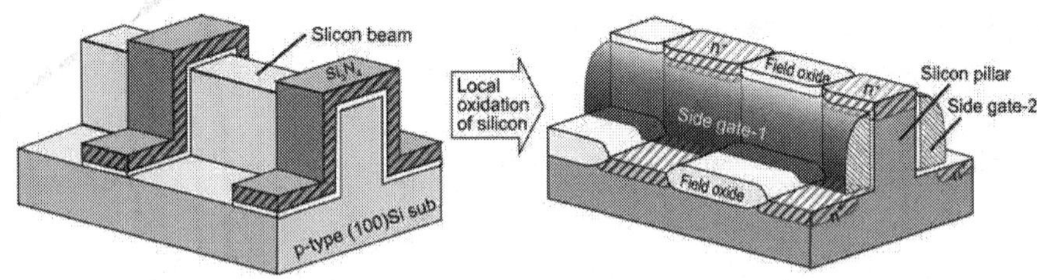

Fig. 28. A fundamental process sequence to fabricate super pillar transistor, SPT. The pillar is isolated with field oxide which is converted from silicon beam itself with well-known local oxidation of silicon, LOCOS technique. Side gate-1 and -2 are self-aligned to silicon and oxide beam.

An SEM plane view of SPT is shown in Fig. 29. Even though the thickness of field SiO_2 film is twice as much as that of silicon beam, removal of the Si_3N_4 film and scrificed oxidation reduce the thickness by a factor of 0.5. Thus the field oxide thickness shown in Fig. 29 is almost equivalent to that of silicon pillar.

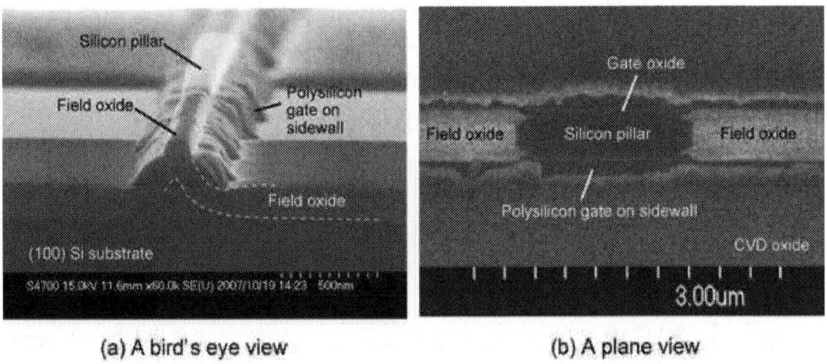

Fig. 29. SEM images of a bird's eye view, (a) and a plane view, (b) of super pillar transistor, SPT. Field oxide is thinned by a factor of 0.5 with a controlled wet etching.

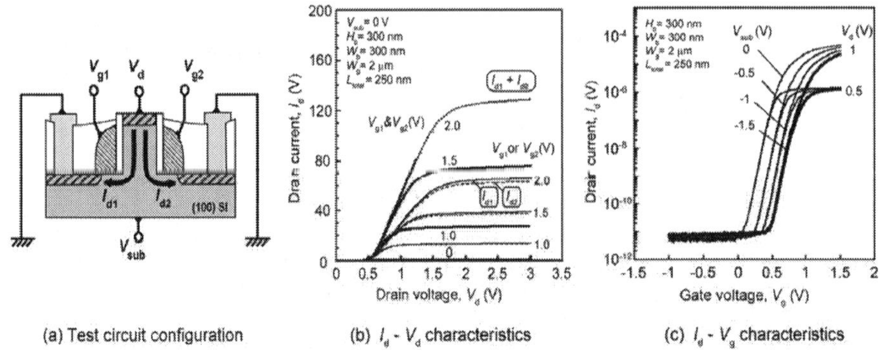

Fig. 30. A test circuit configuration, (a), characteristics of I_d-V_d, (b) and I_d-V_g, (c) for super pillar transistor, SPT.

Side-wall gates on both sides of the pillar make two transistors in one pillar. Typical I_d-V_g characteristics are shown in Fig. 30. Drain currents of I_{d1} and I_{d2} denote those of two sidewall gate transistors. As shown in the figure, two drain currents can be controlled separately.
With additional new technique of forming two capacitors on a pillar, two DRAM cells on a pillar can be obtained leading to $2F^2$ cell. Consequently doubled density of DRAM can be realized at the same technology node.

4.5 Prospect of vertical 3-D transistor

Even though a lot of advantages in vertical 3-D transistor are expected compared to 2-D transistor, there still exists a fundamental limit due to the vertical structure. Except the complexity in fabrication technologies, one of the biggest problems may be practically unchangeable gate length. As an LSI consists of various gate lengths to optimize the performance such as speed/power consumption, chip size, operational margin etc., vertical transistors with single gate length can not be applied to LSI's of processors and ASIC's in particular.
Under these circumstances, one of promising applications may be memory cell array. Cell transistors in a cell array should be identical in order to obtain compact array area and stable operation. Figure 31 proposes possible candidates of super pillar transistor, SPT to memory application. If a certain memory element is chosen, various kinds of memory will be possible. SPT can work as "a universal cell transistor" for almost all memories with one-transistor cell and also can be applied to static memory cell with plural transistors.

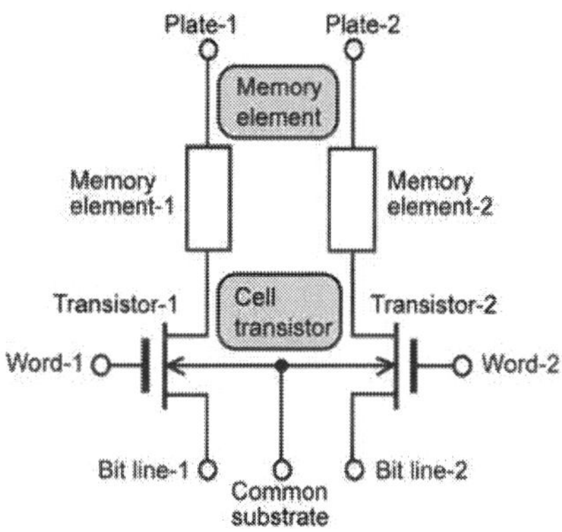

Memory element	Memory
(a) Capacitor	DRAM
(b) Phase change	PCM
(c) Resister	ReRAM
(d) Magnetic film	MRAM
(e) New material	New memory

Cell transistor	Memory
(f) Ferroelectric	FeRAM
(g) Floating gate	Flash memory
(h) SiON film	Flash memory
(i) New material	New memory

Fig. 31. Various applications of super pillar transistor, SPT which can be operated as a universal cell-transistor.

In addition to this kind of a cell transistor and a memory element stack, a transistor stack structure is proposed. At present, 16 stack layers of NAND flash memory, named pipe-shaped bit cost scalable (P-BiCS) flash memory, is proposed (Katsumata et al.; 2009), as shown in Fig. 32. As a silicon body of transistors is filled into a hole which is etched after 16 gate-layer stack formation, it is no need for the formation of thin and tall silicon pillar. In this sense, the manufacturability of P-BiCS is expected to be more stable than that of the pillar type in multi-stack memory, however, it is speculated that transistor performance problem exists due to the polycrystalline silicon body.

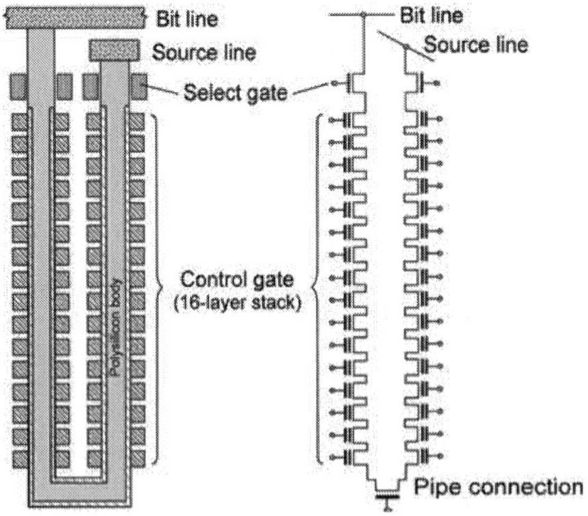

Fig. 32. Proposed 16 layer stack of NAND flash memory named as pipe-shaped bit cost scalable as P-BiCS.

5. Other approaches to 2.5-D stack LSI

A few kinds of 3-D stack of active transistors were extensively investigated in 1980's mainly using laser recrystallization. But they were almost abandoned in the next decade due to poor integrity of overlaid single crystal layer causing much poorer productivity. In place of this active transistor stack, two kinds of chip-stack techniques have been developed as shown in Fig. 33. Flash memory and DRAM are already utilizing bonding-wire connection and 6 to 8 chip stack are now available in flash and DRAM products. An example on a test chip is shown in Fig. 34.

Recently a through-silicon-via type connection has been extensively developed. This provides more flexibility of inter-chip connection and higher productivity due to the batch processing for via formation and inter-via contact. Nevertheless, this may not be a real 3-D stack, because the chip thickness measures tens of 10 μm which is much larger than the device arrangement pitch of tens of 100 nm. Therefore, the chip stack is called "2.5 dimensional" in this article.

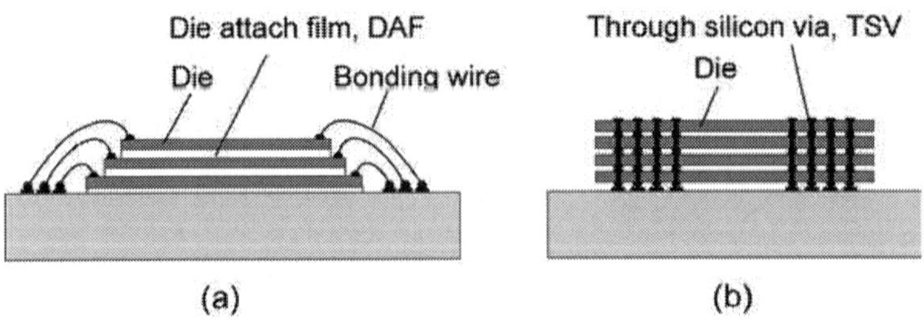

Fig. 33. Two kinds of chip-stack LSI's: bonding-wire connection type, (a) and through-silicon-via, TSV type (b).

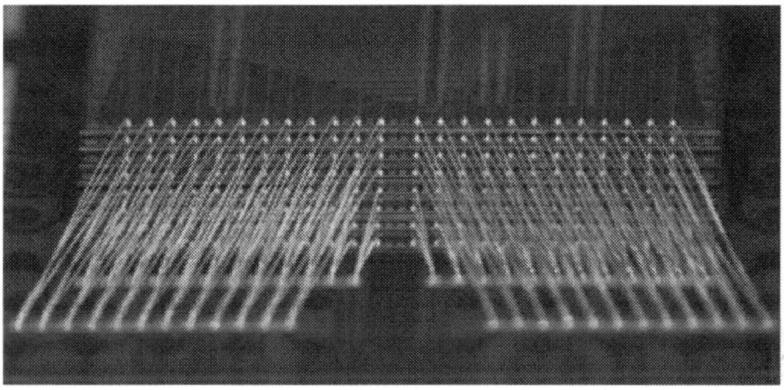

Fig. 34. Eight-layered bonding-wire connection on a test substrare.

6. Conclusion

In response to the ceaseless requirement for extended performance of transistor in LSI, continual scaling has been achieved since early 1970's. Sizes of transistors in products measured 12 μm in 1970 and around 45 nm in 2009. The scaling of device size has been brought about 4-fold increase in memory's volume and processor's performance every three years. Since there existed a limitation of amount of signal charges in DRAM against the cell size scaling, DRAM had first encountered the imitation of the volume size at 1 megabit in mid 1980's. To overcome the limitation, it began to employ a 3-D capacitor structure such as trench capacitor or stack capacitor.

Even with the 3-D structures, its maximum volume of DRAM in a chip is estimated to be 64 gigabit provided that the amount of signal charges stored in a cell must be kept constant against the cell scaling. To solve the deadlock, the employment of an extra high-*k* dielectrics, and a vertical stack of a cell transistor with a capacitor will be inevitable in near future.

Regarding NAND flash memory, multi-stacks of flash transistors have already been proposed. Since flash memory cell consists of one cell transistor in a memory cell and no contact is needed to source and drain in a string of cell transistors, the multi-stack is relatively easier than that of DRAM.

On the other hand, field-effect transistor itself will encounter the ultimate size limit of 5-10 nm. Only about several tens of silicon atoms exist in the channel region of 10-nm transistor. Normal filed-effect operation will be impossible due to fatal short-channel effects in that dimension range. Particularly a ratio of off current to on current becomes worse causing unacceptably large stand-by power consumption.

If the scaling pace is still kept constant, the ultimate limit will be encountered within 15 years. Forecasting the limitation, various kinds of 3-D transistors have been proposed, however, they will still suffer from the short-channel effects same as 2-D transistors. Due to a limitation of invariable channel length of vertical transistor, it will be practical in products that the vertical transistor is employed together with 2-D one in an LSI chip.

To cope with these fundamental limits in miniaturization of devices, various kinds of chip stack will be dominant in LSI products in response to the requirement for smaller package used in personal-use, hand-held products.

7. Acknowledgements

The author wishes to thank all of colleagues, who have done research and development with respect to trench capacitors and 3-D transistors together with him, M. Koyanagi, K. Itoh, T. Kure, Y. Kawamoto, S. Iijima, M. Ohkura, S. Kimura, T. Kaga, R. Hori, T. Toyabe, T. Furukawa, S. Matsumura, A. Sugimura, and K. Okumura for their cooperation. He is also thankful to N. Hashimoto, S. Asai, M. Kubo, and S. Harada for their continuous encouragement.

8. References

Choi, Y.-K.; Lindert, N.; Xuan, P.; Tang, S.; Ha, D.; Anderson, E.; King,T.-J.; Bokor, J. and Hu, C. (2001). Sub-20nm CMOS FinFET technologies, *IEDM Tech. Dig.*, pp. 421-424, December 2001, Washington, D. C.

Dennard, R. H. (1968). Field-effect transistor memory, US Patent 3,387,286

Dennard, R. H.; Gaensslen, F. H.; Yu, H. N.; Rideout, V. L.; Bassous, E. & LeBlanc, A. R. (1974). Design of ion-implanted MOSFETs with very small physical dimensions, *IEEE J. Solid-State Circuits*, Vol. SC-9, No. 5, pp. 256-268

Dennard, R. H. (1984). Evolution of the MOSFET Dynamic RAM - A Personal View, *IEEE Trans. Electron Devices*, Vol. ED-31, pp. 1549-1555

Endoh, T.; Kinoshita, K.; Tanigami, T.; Wada, Y.; Sato, K.; Yamada, K.; Yokoyama, T.; Takeuchi, N.; Tanaka, K.; Awaya, N.; Sakiyama, K. & Masuoka, F. (2001). Novel ultra high density flash memory with a stacked-surrounding gate transistor (S-SGT) structured cell, *IEDM Tech. Dig.*, pp. 33 - 36, December 2001, Washington, D. C.

Hieda, K.; Horiguchi, F.; Watanabe, H.; Sunouchi, K.; Inoue, I. & Hamamoto, T. (1987). New effects of trench isolated transistor using side-wall gates, *IEDM Tech. Dig.*, pp. 736 - 739, December 1987, Wahington, D. C.

Furukawa, T.; Yamashita, H. & H. Sunami, H. (2003). A Proposal of Corrugated-Channel Transistor (CCT) with Vertically-Formed Channels for Area-Conscious Applications, *Jpn. J. Appl. Phys.*, Vol. 42, Part 1, No. 4B, pp. 2067-2072

Hisamoto, D.; Kaga, T.; Kawamoto, Y. & Takeda, E. (1989). A fully depleted lean-channel transistor (DELTA) – A novel vertical ultra thin SOI MOSFET, *IEDM Tech. Dig.*, pp. 833 - 836, December 1989, Washington, D. C.

Itoh, K. (1975). Semiconductor memory, U. S. Patent-4044340, Dec. 29, 1985

Ismail, K. (1995). Si/SiGe high-speed field-effect transistors, *IEDM Tech. Dig.*, pp. 509-512, December 1995, Washington, D. C.

Itoh, K.; Hori, R.; Masuda, H.; Kamigaki, K.; Kawamoto, H. & H. Katto, H. (1980). A single 5V 64K dynamic RAM, *ISSCC Dig. Tech. Papers*, pp. 228 229, February 1980, San Francisco

Itoh, K.; Sunami, H.; Nakazato, K. & Horiguchi, M. (1998). Pathways to DRAM Design and Technology for the 21st Century, *Proc. the 8th Internat. Symp. Silicon Materials Science and Technology*, Vol. 98-1, pp. 350-369

Kaga, T.; Kawamoto, Y.; Kure, T. ; Nakagome, Y. ; Aoki, M. ; Makino, M. & H. Sunami, H. (1987). A 5.4 mm^2 Sheath- Plate-Capacitor DRAM Cell with Self-Aligned Storage Node Insulation, *Extended Abstracts of the 19th Conf. Solid-State Devices and Materials*, pp. 15-18, Tokyo, August 25-27, 1987.

Kaga, T.; Kawamoto, Y.; Kure, T. ; Nakagome, Y. ; Aoki, M. ; Sunami, H. ; Makino, M. ; Ohki, N. & Itoh, K. (1988). Half- V_{cc} Sheath-Plate Capacitor DRAM Cell with Self-Aligned Buried Plate Wiring, *IEEE Trans. Electron Devices*, Vol. 35, No. 8, pp. 1257-1263

Katsumata, R.; Kito, M.; Fukuzumi, Y.; Kido, M.; Tanaka, H.; Komori, Y.; Ishiduki, M.; Matsunami, J.; Fujiwara, T.; Nagata, Y.; Zhang, L.; Iwata, Y.; Kirisawa, R.; Aochi, H. & Nitayama, A. (2009). Pipe-shaped BiCS Flash Memory with 16 Srtacked Layers and Multi-Level-Cell Operation, *Symp. VLSI Tech. Dig.*, pp. 136-137, June 2009, Kyoyo

Kesan, V. P.; Subbanna, S.; Restle, P. J.; Tejwari, M. J.; Aitkin, J. M.; Iyer, S. S. & Ott, J. A. (1991). High Performance 0.25 mm p-MOSFETSs with silicon-germanium channels for 300 K and 77 K operation, *IEDM Tech. Dig.*, pp. 25-28, December 1991, Washington, D. C.

Koyanagi, M.; Sunami, H.; Hashimoto, N. & Ashikawa, M. (1978). Novel High Density, Stacked Capacitor MOS RAM, *IEDM Tech. Dig.*, pp. 348-358, December 1978, Washington, D. C.

Kumanoya, M.; Fujishima, K.; Miyatake, H.; Nishimura, Y.; Saito, K.; Matsukawa, T.; Yoshihara, T. & Nakano, T. (1985). A reliable 1-Mbit DRAM with a multi-bit-test mode, *IEEE J. Solid-State Circuits*, Vol. 20, Issue 5, pp. 909–913

Matsumura, S.; Sugimura, A.; Okuyama, K. & , Sunami, H. (2007), Anomalous Whisker Generation in Ni-Silicided Source and Drain for Three-Dimensional Beam-Channel MOS Transistor on SOI Substrate, *Proc. Advanced Metallization Conference 2006*, pp. 631-635, September 2006, Tokyo

Ohkura, M.; Kusukawa, K.; Sunami, H.; Hayashida, T. & Tokuyama, T. (1985). A Three-Dimensional DRAM Cell of Stacked Switching-Transistor in SOI (SSS), *IEDM Tech. Dig.*, pp. 718-721, Decem-ber 1985, San Francisco

Okuyama, K.; Yoshikawa, K. & Sunami, H. (2007). Control of Subthreshold-Characteristics of Narrow-channel SOI nMOS Transistor Utilized Additional Side Gate Electrodes, *Jpn. J. Appl. Phys.*, Vol. 46, No. 4B, pp. 2050-2053

Pidin, S.; Mori, T.; Inoue, K.; Fukuta, S.; Itoh, N.; Mutoh, E.; Ohkoshi, K.; Nakamura, R.; Kobayashi, K.; Kawamura, K.; Saiki, T.; Fukuyama, S.; Satoh, S.; Kase, M. & Hashimoto, K. (2004). A novel strain enhanced CMOS architecture using selectively deposited high tensile and high compressive silicon nitride films, *IEDM Tech. Dig.*, pp. 213-216, December 2004, San Francisco

Regitz, R & Karp, J. (1970). A three-transistor-cell, 1024-bit, 500 ns MOS RAM, *ISSCC Dig. Tech. Papers*, pp. 42-43, February 1970, San Francisco

Richardson, W. F; Bordelon, D. M.; Pollack, G. P.; Shah,A. H.; Malhi, S. D. S.; Shichijo, H.; Banerjee,S. K.; M. Elahy,M.; Womack, R. H.; Wang,C-P.; Gallia,J.; Davis,H. E. & Chatterjee, P. K. (1985). A trench transistor cross-point DRAM cell, *IEDM Tech. Dig.*, pp. 714-717, December 1985

Sugimura, A.; Okuyama, K. & Sunami, H. (2008). Proposal of a Vertical-Channel Metal-Oxide-Semiconductor Field- Effect Transistor with Entirely Oxidized Silicon Beam Isolation, *Jpn. J. Appl. Phys.*, Vol. 48, No. 4, pp. 04C049-1-4

Sunami, H. & Nishimatsu, S. (1975). *Semiconductor memory*, Japanese patent application: Tokugansho 50-53883

Sunami, H. (1978). Thermal Oxidation of Phosphorus Doped Polycrystalline Silicon in Wet Oxygen, *J. Electrochem. Soc.*, Vol. 125, No. 6, pp. 892-897

Sunami, H.; Kure, T. ; Kawamoto, Y. and Miyao, M. (1982-a). *Semiconductor memory*, Japanese patent application : Tokugansho 57-192478 ; Semiconductor memory, US Patent 4,901,128

Sunami, H.; Kure, T.; Hashimoto, N.; Itoh, K.; Toyabe, T. & Asai, S. (1982-b). A Corrugated Capacitor Cell (CCC) for Megabit Dynamic MOS Memories, *IEDM Tech. Dig.*, pp. 806-808, December 1982, San Francisco

Sunami, H.; Kure, T.; Hashimoto, N.; Itoh, K.; Toyabe, T. & Asai, S. (1984). A Corrugated Capacitor Cell (CCC). *IEEE Trans. Electron Devices*, Vol. ED-31, No. 6, pp. 746-753

Sunami, H.; Kure, T.; Yagi, K.; Yamaguchi, K. & Shimizu, S. (1985). Scaling Consideration and Dielectric Breakdown Improvement of Corrugated Capacitor Cell (CCC) for Future dRAM, *IEEE Trans. Electron Devices*, Vol. ED-32, No. 2, pp. 296-303

Sunami, H.; Furukawa, T. & and Masuda, T. (2004). A Three-Dimensional MOS Transistor Formation Technique with Crystallographic Orientation-Dependent TMAH Etchant, *SENSORS and ACTUATORS A: PHYSICAL*, A111, pp. 310-316

Sunami, H. (2008-a). The Role of the Trench Capacitor in DRAM Innovation, *IEEE Solid-State Circuits Society News*, Vol. 13, No.1, pp. 42-44

Sunami, H. (2008-b). Development of three-dimensional MOS structures from trench-capacitor DRAM cell to pillar-type transistor, *Proc. 9th Internat. Conf. Solid-State and Integrated-Circuit Technology*, pp. 853-856, Oct. 2008, Beijin

Sunami, H. (2008-c). *Semiconductor Memory (in Japanese)*, Corona Pub. Co., Ltd., ISBN 978-4-339-00798-5, Tokyo Takato, H.; Sunouchi, K.; Okabe, N.; Nitayama, A.; Hieda, K.; Horiguchi, F. & Masuoka, F. (1988). High performance CMOS surrounding gate transistor (SGT) for ultra high density LSIs, *IEDM Tech. Dig.*, pp. 222-225, December 1988

Teal, G. K.; Sparks, M. & Buehlor, E. (1951). Growth of Germanium Single Crystals Containing p-n Junctions, *Phys. Rev.*, Vol. 81, p. 657

Toyabe, T.; Yamaguchi, K.; Asai, S. & Mock, M. S. (1978). A numerical model of avalanche breakdown in MOSFET's, *IEEE Trans. Electron Devices*, Vol. ED-25, pp. 825-832

Watanabe, H.; Tatsumi, T.; Ohnishi, S.; Hamada, T.; Honma, I. & Kikkawa, T. (1992). A new cylindrical capacitor using hemispherical grained Si (HSG-Si) for 256Mb DRAMs, *IEDM Tech. Dig.*, pp. 259-262, December 1992, San Francisco

Hafnium-based High-k Gate Dielectrics

A. P. Huang[1,2], Z. C. Yang[1] and Paul K. Chu[2]
[1]*Department of Physics, Beijing University of Aeronautics and Astronautics, Beijing 100191,*
[2]*Department of Physics and Materials Science, City University of Hong Kong, Tat Chee Avenue, Kowloon, Hong Kong,*
[1]*Hong Kong, China*
[2]*China*

1. Introduction

Scaling of silicon dioxide dielectrics has once been viewed as an effective approach to enhance transistor performance in complementary metal-oxide semiconductor (C-MOS) technologies as predicted by Moore's law [1]. Thus, in the past few decades, reduction in the thickness of silicon dioxide gate dielectrics has enabled increased numbers of transistors per chip with enhanced circuit functionality and performance at low costs (Fig. 1). However, as devices approach the sub-45 nm scale, the effective oxide thickness (EOT) of the traditional silicon dioxide dielectrics are required to be smaller than 1 nm, which is approximately 3 monolayers and close to the physical limit (Fig. 2), thus resulting in high gate leakage currents due to the obvious quantum tunneling effect at this scale (Fig. 3). To continue the downward scaling, dielectrics with a higher dielectric constant (high-k) are being suggested as a solution to achieve the same transistor performance while maintaining a relatively thick physical thickness [2]. Many candidates of possible high-k gate dielectrics have been suggested to replace SiO_2 and they include nitrided SiO_2, Hf-based oxides, and Zr-based oxides. Hf-based oxides have been recently highlighted as the most suitable dielectric materials because of their comprehensive performance. One of the key issues concerning new gate dielectrics is the low crystallization temperature. Owing to this shortcoming, it is difficult to integrate them into traditional CMOS processes. To solve these problems, additional elements such as N, Si, Al, Ti, Ta and La have been incorporated into the high-k gate dielectrics, especially Hf-based oxides. In the following sections, the requirements of high-k oxides, brief history of high-k development, various candidates of high-k, and the latest hafnium-based high-k materials are discussed.

2. Requirements of high-k oxides

Among the various requirements of gate dielectric materials, the most important are good insulating properties and capacitance performance (Fig. 4). Because the gate dielectric materials constitute the interlayer in the gate stacks, they should also have the ability to prevent diffusion of dopants such as boron and phosphorus and have few electrical defects which often compromise the breakdown performance. Meanwhile, they must have good thermal stability, high recrystallization temperature, sound interface qualities, and so on.

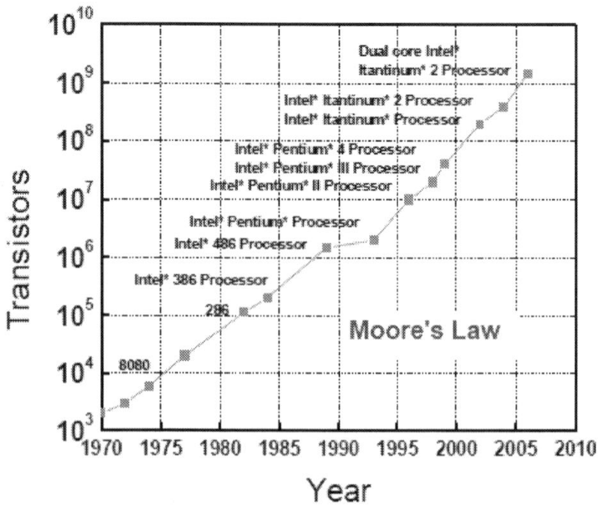

Fig. 1. Enhanced Performance Trend as Predicted by Moore's Law. Processing power has steadily risen as transistors become more complex [1].

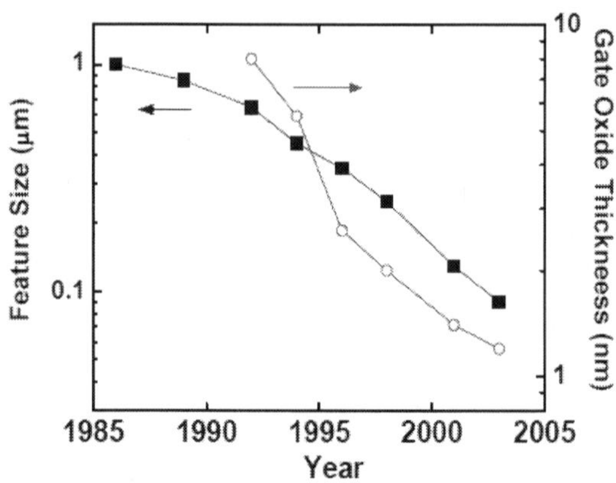

Fig. 2. Feature size of transistors downscales with time and the gate oxide thickness decreases accordingly [1].

2.1 K value, band gap and band offset

With regard to capacitance performance, the requirement is that the k value should be over 12, preferably 25-30. An appropriate k value means that the dielectrics will have a reasonable physical thickness which is enough to prevent gate leakage and not too thick to hamper physical scaling when achieving the target EOT. On the other hand, a very large k value is undesirable in CMOS design because they cause unfavorable large fringing fields at the source and drain regions [4]. Table 1 and Fig. 5 show that the k values of some oxides vary inversely with the band gap, so a relatively low k value is needed [5]. There are

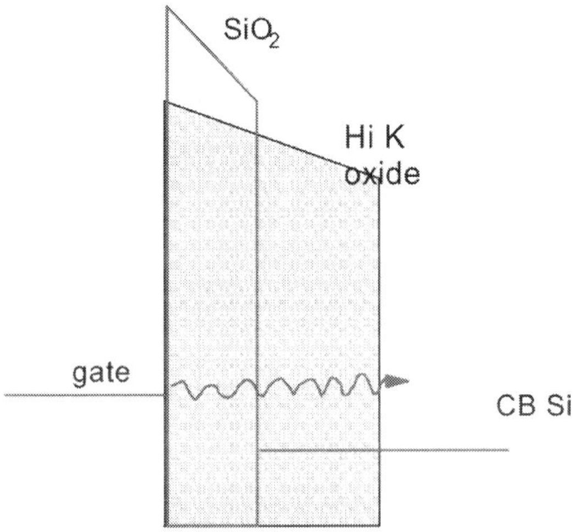

Fig. 3. Schematic of direct tunneling through SiO_2 [3].

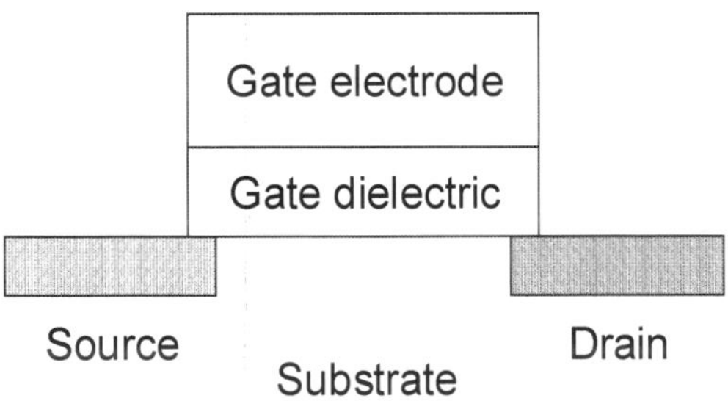

Fig. 4. Schematic drawing of a MOS stack.

numerous oxides with extremely large k values, such as $SrTiO_3$, which are candidates in DRAM capacitors [6], but their band gap is too small. According to the required insulating properties, the gate dielectrics must exhibit at least the band offset of 1 eV while in contact with the Si substrate in order to avoid serious gate leakage and breakdown. The band offset is required to be over 1 eV in order to inhibit conduction by the Schottky emission of electrons or holes into the oxide bands [5, 7], as schematically shown in Fig. 6. This means that the materials must have both the conduction band offset (CB) and valence band offset (VB) over 1 eV. In fact, the CB offset is less than the VB offset, which suggests oxides with band gaps wider than 5 eV may be excluded as gate dielectrics. For those oxides with narrow band gaps, either the CB offsets or the VB offsets may be smaller than 1 eV, also limiting the choice of these materials.

Dielectric	k	Band gap (eV)	CB offset (eV)
Si	-	1.1	-
SiO_2	3.9	9	3.2
Si_3N_4	7	5.3	2.4
Al_2O_3	9	8.8	2.8
Y_2O_3	15	6	2.3
Ta_2O_5	22	4.4	0.35
TiO_2	80	3.5	0
La_2O_3	30	6	2.3
a- $LaAlO_3$	30	5.6	1.8
$SrTiO_3$	2000	3.2	0
ZrO_2	25	5.8	1.5
HfO_2	25	5.8	1.4
$HfSiO_4$	11	6.5	1.8

Table I. Dielectric constant (k), band gap and CB offset on Si of the candidate gate dielectrics.

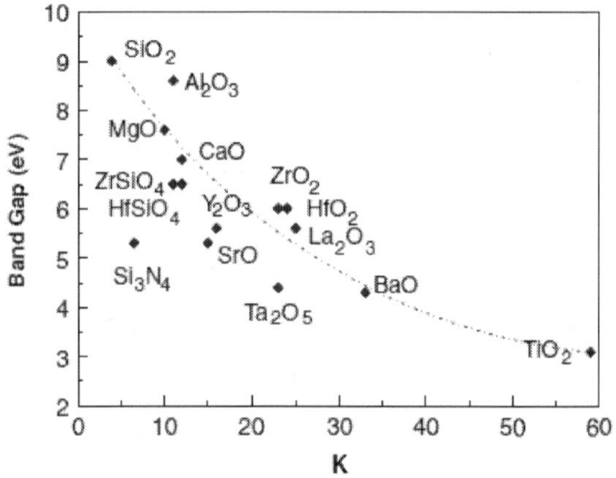

Fig. 5. Static dielectric constant versus band gap for candidate gate oxides [5].

2.2 Thermal stability

In present CMOS processes, the gate stacks must undergo rapid thermal annealing (RTA) of 1000 °C for 5s. This requires that the gate oxides must be thermally and chemically stable especially with the contacting materials. Thus, group II, III, IV oxides with a higher heat of formation than SiO_2 such as SrO, CaO, BaO, Al_2O_3, ZrO_2, HfO_2, Y_2O_3, La_2O_3 and lanthanides may be useful. Additionally, group II oxides which react with water are not favorable. Therefore, from the thermal stability point of view, only Al_2O_3, ZrO_2, HfO_2, Y_2O_3, La_2O_3, Sc_2O_3 and some lanthanides such as Pr_2O_3, Gd_2O_3 and Lu_2O_3 are left [3]. However, some materials with higher heat of formation than SiO_2 may also be slightly reactive with Si such as ZrO_2, forming the silicide, $ZrSi_2$ [8, 9]. Among these high k dielectrics, HfO_2 has both a high k value as well as chemical stability with water and Si.

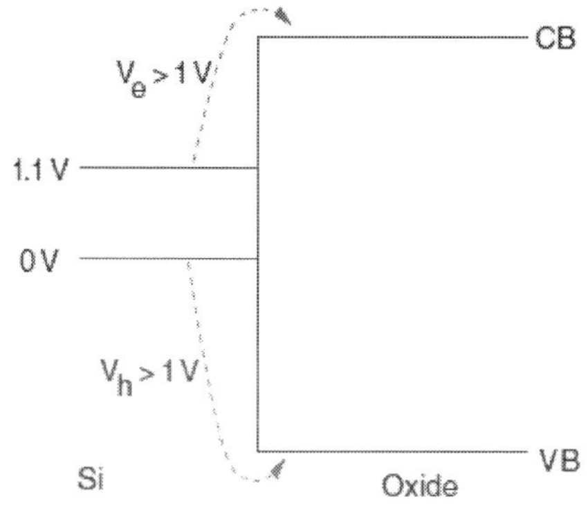

Fig. 6. Schematic of band offsets dependence of carrier injection in oxide band states.

2.3 Crystallization temperature

Owing to the absence of grains and good diffusion barrier properties, amorphous materials are preferred to crystalline ones. The grains which lie in the crystalline systems can often be the pathways for dopants diffusion and breakdown. Unlike SiO_2, high-k oxides usually have low crystalline temperature and can easily crystallize when subjected to RTA. In particular, HfO_2 and ZrO_2 crystallize at much lower temperatures at ~400 °C and ~300 °C, respectively (Fig. 7). According to the above factors, the approach to improve the crystallization temperature of HfO_2 and ZrO_2 should be considered. The crystallized HfO_2 has a much lower leakage current which has convinced many companies such as Intel and Freescale to adopt binary oxides because of their relative higher k values.

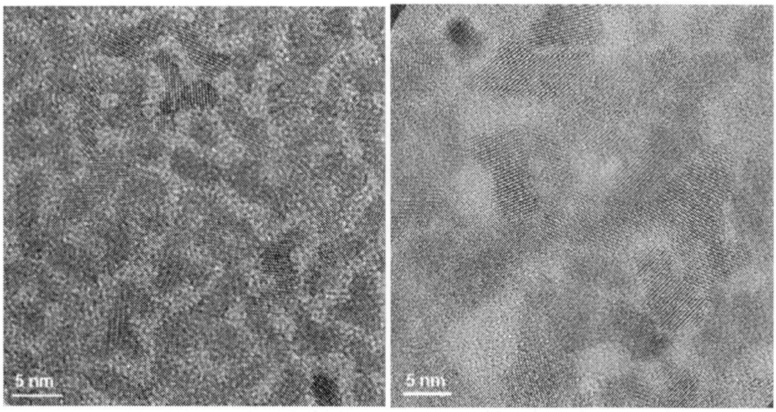

Fig. 7. TEM image of crystallization in HfO_2/SiO_2 dielectrics with (a) 40% HfO_2 and(b) 80% HfO_2 [10].

2.4 Interface quality

The interface between the high-k dielectrics and Si substrate must have the highest electrical quality and flatness, absence of interface defects, and low interface state density D_{it}. Bad interface quality can cause high fixed charge density, inducing a large shift in the flat band voltage (V_{fb}) which severely reduces the performance and reliability of the transistor. Most of the high-k materials reported in this chapter have D_{it} ~10^{11}–10^{12} eV/cm^2 and also exhibit a substantial flatband voltage shift larger than 300 mV [11]. Therefore, it is crucial to improve the quality of the interface. There are two ways to ensure a high quality interface, either using a crystalline oxide grown epitaxially on the Si or an amorphous oxide. An amorphous oxide has many advantages over a poly-crystalline oxide. Firstly, it is more economically and more compatible with existing processes. Secondly, the amorphous oxide can minimize the number of interface defects. Thirdly, it is possible to gradually vary the composition of an amorphous oxide without creating a new phase, as in silicate alloys, or when adding nitrogen or other metal elements. Fourthly, an amorphous oxide and its dielectric constant are isotropic, so that fluctuations in polarization from differently oriented oxide grains will not cause scattering of carriers. Finally, amorphous phases have no grain boundaries. The advantages of epitaxial oxides may come in the future, where their more abrupt interfaces allows us to reach lower EOTs. Besides the above consideration, the configuration of interface bonding is also significant. As the SiO_2/Si interface has high quality, the ideal gate dielectric stack may well turn out to have an interface comprising several monolayers of Si-O (and possibly N) containing materials, which can be a pseudobinary layer at the channel interface. This layer can serve to preserve the critical, high-quality nature of the SiO_2 interface (D_{it} ~2x10^{10} eV/cm^2) while providing a higher-k value for that thin layer. The same pseudobinary materials can also extend beyond the interface, or a different high-k material can be used on top of the interfacial layer.

2.5 Defects

Similar to interface defects, bulk defects formed in high-k oxides during deposition also causes degraded transistor performance due to the rising number of defect-related fixed charges. In addition, charges trapped in defects will cause a shift in the gate threshold voltage of the transistor, which is the key characteristic of performance. Furthermore, the trapped charges change with time and so the threshold voltage also shifts with time, leading to problems associated with negative bias temperature instability (NBTI) and positive bias temperature instability (PBTI). Meanwhile, trapped charges scatter carriers in the channel causing reduced carrier mobility. Lastly, they are the starting points for electrical failure and oxide breakdown. Typically, these defects are sites of excessive or deficient oxygen or

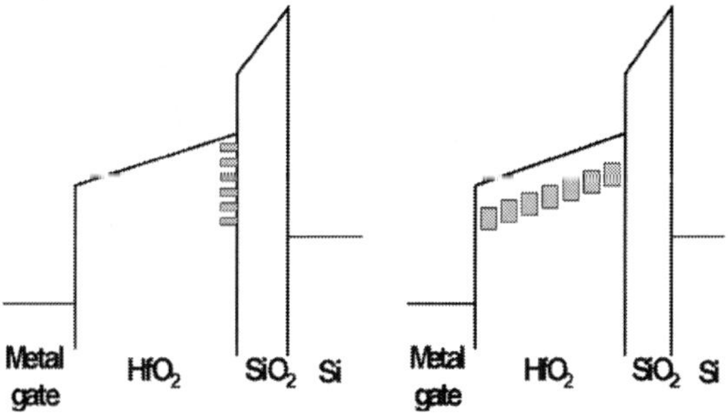

Fig. 8. Schematic diagram of two types of defects located (a) at the HfO_2/SiO_2 interface and (b) in the bulk of HfO_2 film.

impurities. Unfortunately, most of the high-k oxides inherently have more interface defects in contact with the Si substrate and bulk defects than SiO_2 because their bonding cannot relax as easily [12] (Fig. 8). Nowadays, many groups are endeavoring to reduce defect densities by either processing control or engineering of materials.

3. Brief history of high-k dielectric development

To overcome gate leakage problems and extend the usefulness of SiO_2-based dielectric, incorporation of nitrogen into SiO_2 has been adopted. There are several ways to introduce nitrogen into SiO_2, such as post deposition annealing in nitrogen ambient and forming a nitride/oxide stack structure. By incorporating nitrogen into SiO_2, it not only increases the dielectric constant but also acts as a better barrier against boron penetration. In addition, a nitride/oxide stack structure maintains the benefits of good interface quality between the oxide and substrate [13, 14], as schematically shown in Fig. 9.

Despite the immense development with SiO_2, these oxynitrides still have low k values and so a relatively thick layer is required to prevent direct tunneling current. Therefore, alternative materials with a higher k than SiO_2 (3.9) are needed to achieve the required capacitance without tunneling currents [15]. Oxides of group II, III, IV such as Al_2O_3, Y_2O_3, La_2O_3, Sc_2O_3 and some lanthanides such as Pr_2O_3, Gd_2O_3 and Lu_2O_3 have been proposed. Unfortunately, these dielectrics will only last a few generations due to limitations dictated by low power applications, scalability, or serious reactions with the Si substrate. Yet, these problems are much smaller for oxides and silicates of Hf and Zr. Thus, the choice of alternative gate dielectrics has been narrowed to HfO_2, ZrO_2 and their silicates due to their excellent electrical properties and high thermal stability in contact with Si [16]. However, another problem, namely low crystallization temperature, is associated with Hf-based and Zr-based oxides. They can easily crystallize during standard CMOS processes. These crystalline structures can increase the gate leakage by orders of magnitude and provide pathways for diffusion of dopants and dielectric breakdown. Up to date, many groups have focused on the improvement of the crystallization temperature of these oxides. Thus, elements such as N, Si, Al, Ta and La have been incorporated into these high-k oxides. Hf-based oxides are preferred over Zr-based oxides for its relative higher crystalline temperature.

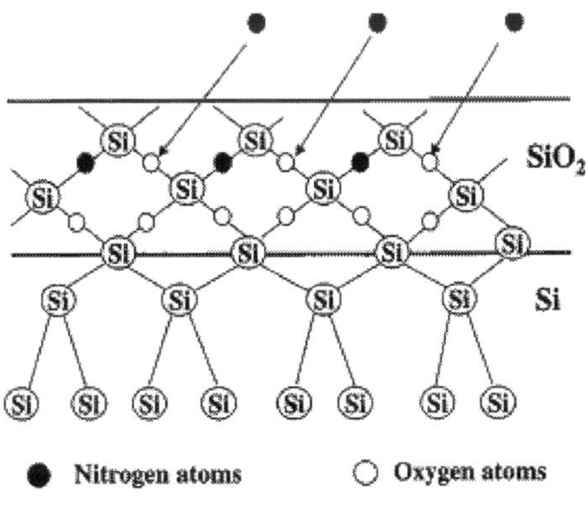

Fig. 9. Schematic showing incoming nitrogen radicals replace oxygen atoms to form Si–N bonds [17].

4. Latest development in Hf-based high-k oxides

4.1 Fabrication methods

Hf-based High-k dielectric oxides have replaced conventional SiO_2 as the gate dielectric in sub-0.1 μm complementary metal-oxide-semiconductor devices [18, 19]. The fabrication technology of Hf-based high-k ultrathin dielectrics has been developed very quickly. Overall, the techniques can be categorized into two major approaches based on the reaction mechanism during preparation, namely CVD (chemical vapor deposition) and PVD (Physical Vapor Deposition) processes. CVD-based approaches include metal-organic chemical vapor deposition (MOCVD) [20], plasma-enhanced chemical vapor deposition (PECVD) [21], atomic-layer chemical vapor deposition (ALCVD) [22], photo-assisted CVD synthesis [23] and so on. These growth methods provide more flexibility and have relatively low cost. Among them, ALCVD is considered particularly promising, since this is the only feasible method to control the thickness down to the nanometer range and layer-by-layer composition of the metal oxide ultrathin film [24].

4.2 Doping of Hf-based high-k oxides

Crystallization of pure HfO_2 occurs at only about 400–450 °C causing grain boundary leakage current and nonuniformity of the film thickness [25]. As a result, impurities such as O, B, and P can penetrate the grain boundaries during high temperature postprocessing. It causes equivalent oxide thickness (EOT) scaling and reliability concerns when Hf-based high-k ultrathin gate oxides are integrated into high temperature CMOS processes [26].

Recently, nitrogen incorporation has been extensively investigated in the field of high-k thin films [27, 28]. Nitrogen introduction into HfO_2 films has significantly improved the electric properties as well as crystallinity [29, 30]. On the contrary, nitrogen doping leads to decreased band gap. This is because it adds N 2p states which lie above the O 2p states in the free atoms and so the VB is raised and the CB is reduced due to the interaction between the nonbonding Hf 5d states and adjacent O and N states. The delocalized Hfd-Np bonding states contribute an indirect band gap E_g of 1.8 eV, which is smaller than the Op-Hfd band gap of larger than 5.8 eV [31, 32]. Despite the disadvantages, the introduced nitrogen can suppress the growth of microstructure and interfacial layer. When N is added to HfO_2, it is expected to distort the equilibrium of the lattice and produce disordered states. Choi et al. have demonstrated that adding nitrogen results in the reduction of the mobility of Hf and O atoms as well as increase in the nucleation temperature and consequently the crystalline temperature [33, 34]. All these indicate that nitrogen acts as a crystallization inhibitor and causes an increase in the crystallization temperature in Hf-based gate dielectrics (Fig. 10).

The interfacial layer between the high-k dielectrics and Si substrate is one of the key factors determining the performance and reliability of a MOS transistor. Hence, it is extremely crucial to fabricate a SiO_2/Si like interface. From this viewpoint, a SiO_2 interfacial layer is often grown between Hf-based oxide and Si by thermal oxidation. However, this HfO_2/SiO_2 gate dielectric stack usually introduces an additional EOT increase due to the low k SiO_x interfacial layer. In order to solve this problem, addition of Si into Hf-based oxide to form Hf silicate may be a plausible means. Besides improvement in the interface quality, incorporation of Si into Hf-based oxides can also foster the formation of amorphous or near-amorphous structures [36, 37]. A negative effect is the reduction in the k value. The k value decreases inversely with increasing Si concentration in Hf-based oxides. When the Si content approaches 100% (alternatively, Si-based oxide), the k value comes close to the lowest value of 3.9. Accordingly, the Si content must be selected to keep a balance between gains and defects.

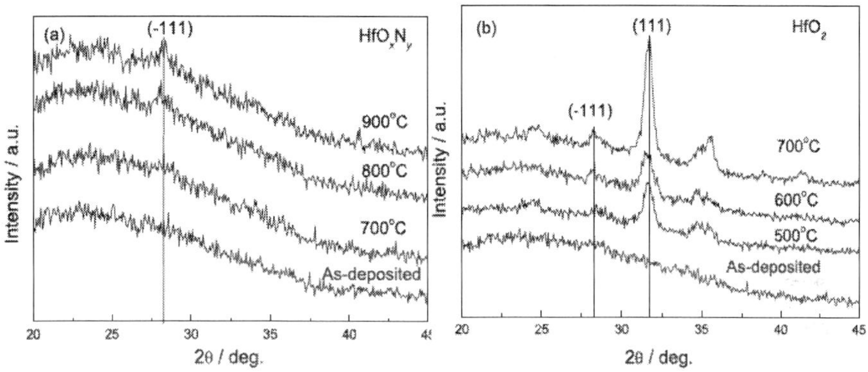

Fig. 10. XRD spectra for the HfO_2 and HfO_xN_y films: (a) as-deposited and HfO_xN_y films annealed at different temperatures and (b) as-deposited and HfO_2 films annealed at different temperatures [35].

HfSiON is thermally stable compared to HfO_2 due to the Si-N bonds that are created by the nitridation step, and thus HfSiON has the potential for implementation in a conventional gate-first process with high temperature activation annealing. By using nitrogen-incorporated HfSiO films, both the oxidation and reduction reactions can be suppressed in the annealing process at a proper partial pressure of N_2 gas. The N_2 gas suppresses only the reduction reaction, while nitrogen atoms incorporated in the dielectrics suppress both oxidation and reduction reactions, greatly improving the electrical characteristic of Hf-based high-k dielectrics [38]. Fig. 11 schematically shows the mechanism of the suppression of reaction and the results of suppression of interfacial layer growth can be seen in Fig. 12.

Many groups have reported that the crystallization temperature of HfO_2 (400–450 °C) can be increased by incorporation of Al_2O_3 forming an HfAlO alloy. Zhu et al. [39] have shown that Al inclusion in HfO_2 significantly increases the crystallization temperature. At an Al content of 31.7%, the crystallization temperature is about 400-500 °C higher than that without Al.

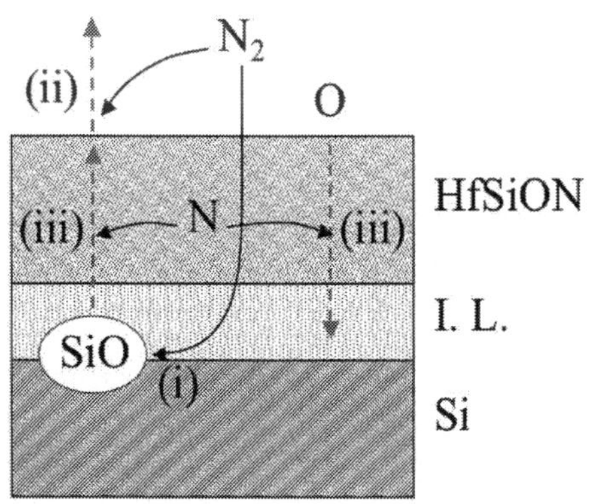

Fig. 11. Schematic of the mechanism for the suppression of reaction. N_2 ambient gas can suppress (i) SiO formation and (ii) SiO desorption. Nitrogen atoms in the dielectric film can suppress (iii) SiO and O diffusion [38].

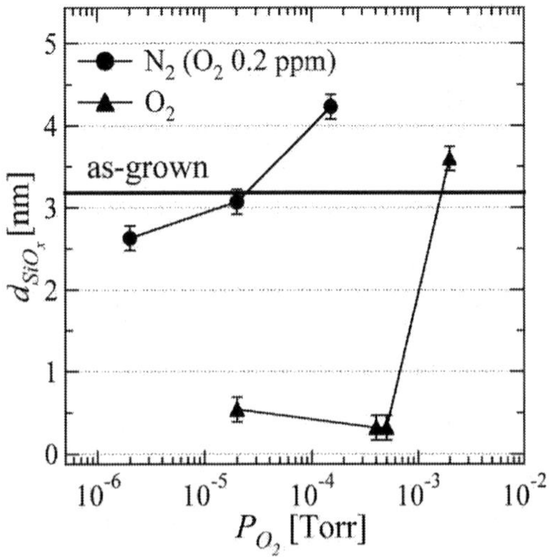

Fig. 12. SiO_2 equivalent thickness of dielectric films as a function of O_2 partial pressure (PO_2). These thicknesses were calculated from the peak-area ratio of the Si-oxide to the Si substrate, regarding the Si-oxide component as SiO_2 for simplicity. A straight line at around 3.2 nm denotes the thickness of the as-grown sample [38].

This additional Al increases the band gap of the dielectrics from 5.8 eV for HfO_2 without Al to 6.5 eV for HfAlO with 45.5% Al but reduced dielectric constant from 19.6 for HfO_2 without Al to 7.4 for Al_2O_3 without Hf. Considering the factors including the crystallization temperature, band gap, and dielectric constant, they conclude that the optimum Al concentration is about 30% for conventional CMOS gate processing technology. Moon et al. [40] have presented the similar trend in the change of the electrical and structural properties due to the Al incorporation. Their results suggest that the HfAlO film with 10% Al_2O_3 shows a great improvement in thermal stability and significant reduction of interfacial layer growth during subsequent thermal processes while maintaining a high k value (~19), leading to reduction in the leakage current by around 2 orders of magnitude compared to pure HfO_2. The HfAlO film also has good compatibility with the gate electrode in high temperature annealing process (Fig. 13). Bae et al. [41] have pointed out that while Al doping significantly increases the crystallization temperature in HfO_2 to up to 900 °C and improves its thermal stability, it also introduces negative fixed oxide charges due to Al accumulation at the HfAlO-Si interface, resulting in mobility degradation. The effects of Al concentration on the crystallization temperature, fixed oxide charge density, and mobility degradation in HfAlO have been characterized and correlated. In spite of these analyses, there are still a lot of issues to be settled in order to maximize the performance of the materials.

On account of the good thermal stability and electrical characteristics, HfTaO gate dielectrics have attracted attention. Incorporation of Ta into HfO_2 enhances the crystallization temperature dramatically while keeping a high k value of ~17 [42]. Compared to HfO_2 gate dielectrics, HfTaO also has the advantages of much lower charge trapping as well as BTI degradation and increased channel mobility [43]. Yu et al. [44] have confirmed that HfTaO

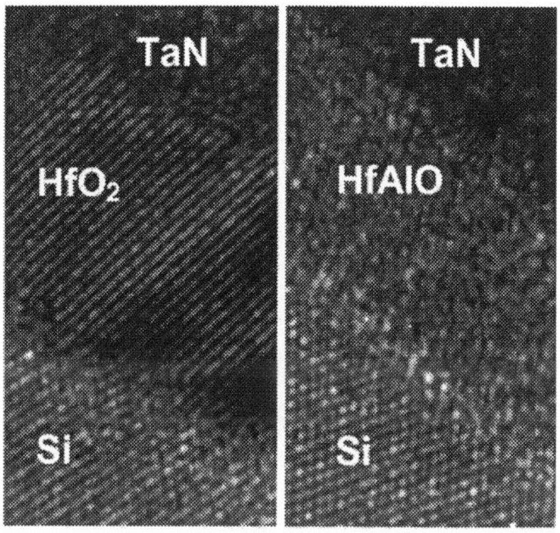

(a) (b)

Fig. 13. XTEM images of HfO_2 and HfAlO after 700 °C in-situ PDA treatment. HfAlO layer remains amorphous while HfO_2 is crystallized. Both films were deposited at 400 °C without surface nitridation [40].

with 43% Ta remains amorphous even after annealing at 950 °C for 30 s, and the formation of low-k interfacial layer is reduced (Fig. 14). The results indicate good interface properties between the HfTaO and Si substrate and sufficiently suppressed boron penetration behavior in the HfTaO film. The negligible flat-band voltage shift in HfTaO with 43% Ta film is observed and attributed to its amorphous structure after device fabrication. It also contributes to the improvement in performance and reliability of the devices. Zhang et al. [45] have found that HfTaO with 40% Ta exhibits the highest crystallization temperature of 900 °C, while 35% and 52% HfTaO films show crystallization temperature of 800 °C (Fig. 15). The results demonstrate that HfTaO N-MOSFETs possess higher electron mobility than controlled HfO_2 devices. Among them, the transistors with 40% Ta doped HfTaO as the gate dielectrics have the highest electron mobility (Fig. 16).

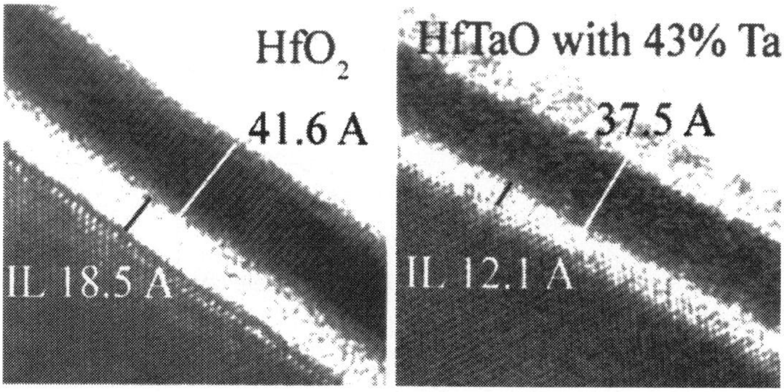

Fig. 14. TEM images of HfO_2 and HfTaO with 43% Ta after PDA at 700 °C for 40 s and activation annealing at 950 °C for 30 s. Pure HfO_2 film is fully crystallized whereas the HfTaO with 43% Ta film remains amorphous [44].

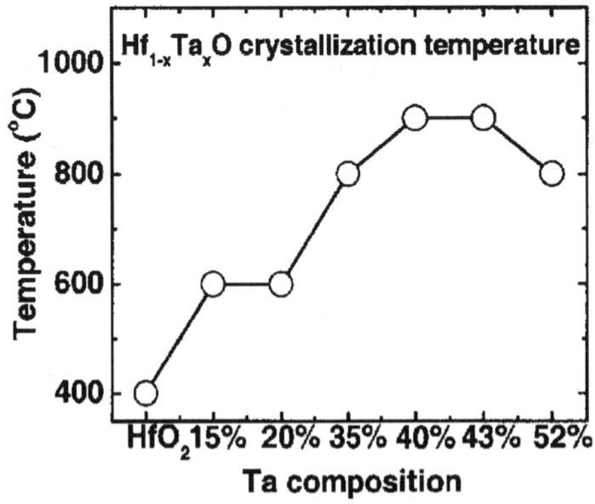

Fig. 15. Crystallization temperature of HfTaO with different Ta composition measured by XRD with incident angle of x ray: 3° [45].

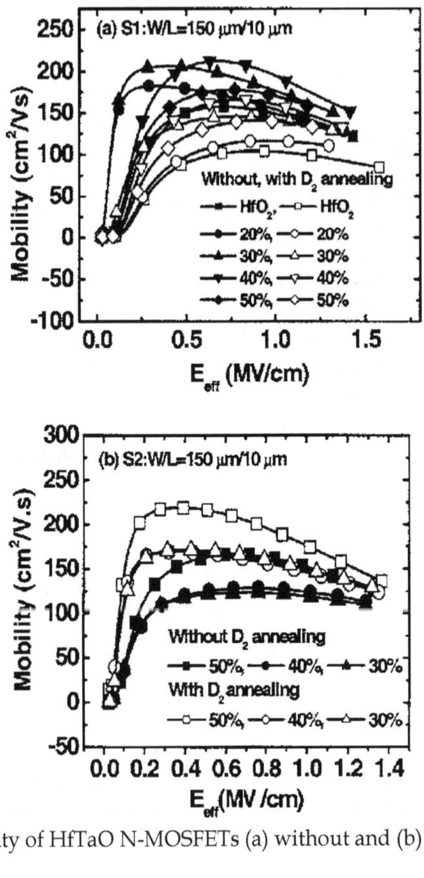

Fig. 16. Effective mobility of HfTaO N-MOSFETs (a) without and (b) with D_2 annealing [45].

In addition, some rare earth elements such as La can also improve the characteristics of Hf-based high-k dielectrics. Introduction of La_2O_3 into HfO_2 causes an increase in the crystallization temperature (Fig. 17). Furthermore, unlike other Hf-based amorphous materials such as $HfSiO_x$ or $HfAlO_x$, the permittivity of $HfLaO_x$ still yields a high k value (>20) [46] (Fig. 18). Besides, HfLaO also has the advantages of much lower charge trapping as well as BTI degradation and increased channel mobility. In addition, varying the La concentrations in the TaN/HfLaO or HfN/HfLaO gate stack can effectively tune the metal work function for N-MOSFETs [43]. In the capacitance-voltage curve of metal oxide semiconductor capacitor, Yamamoto et al. [46] have shown that the $HfLaO_x$ dielectric film exhibits very small degradation in both the interface and bulk properties, as shown in Fig. 19. A very low fixed charge density in $HfLaO_x$ films is demonstrated from a very small film thickness dependence on the flatband voltage in their study.

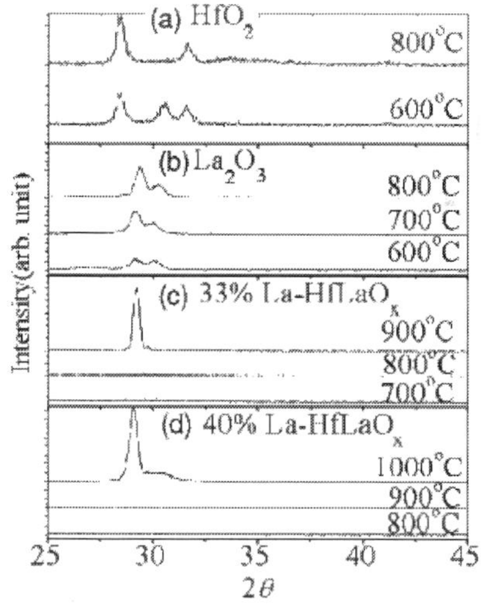

Fig. 17. XRD spectra of 30 nm films of (a) HfO_2, (b) La_2O_3, (c) 33% La-$HfLaO_x$, and (d) 40% La-HfLaOx annealed at various temperatures. HfO_2 and La_2O_3 films crystallize under 600 °C. On the other hand, 40% La-HfLaOx film remains amorphous after 900 °C annealing [46].

An et al. [47] have synthesized ultrathin HfO_2 and $HfLaO_x$ films with La/(Hf+La) ratios of 42%, 57%, and 64% by an atomic layer deposition process. By measuring the leakage current at different temperatures, they propose that the conduction mechanism of HfO_2 and $HfLaO_x$ films follow the Poole-Frenkel emission model under the gate injection condition. They have also demonstrated that the intrinsic trap energy levels are 1.42, 1.34, 1.03, and 0.98 eV in the $HfLaO_x$ samples with La/(Hf+La) ratios of 0%, 42%, 57%, and 64%, respectively, showing a decreasing behavior as the La content is increased (Fig. 20).

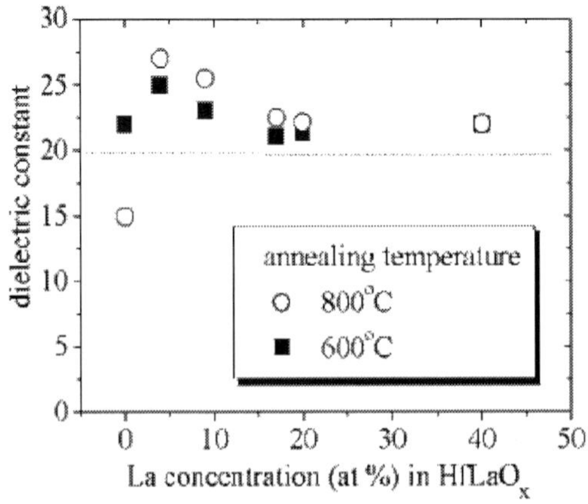

Fig. 18. Dielectric constants of HfLaO$_x$ film as a function of La concentrations. The dielectric constants are determined by MIM capacitors for the samples with La concentrations of 0%, 4%, 9%, and 17%. For 20% and 40%, CET vs physical thickness plots were used [46].

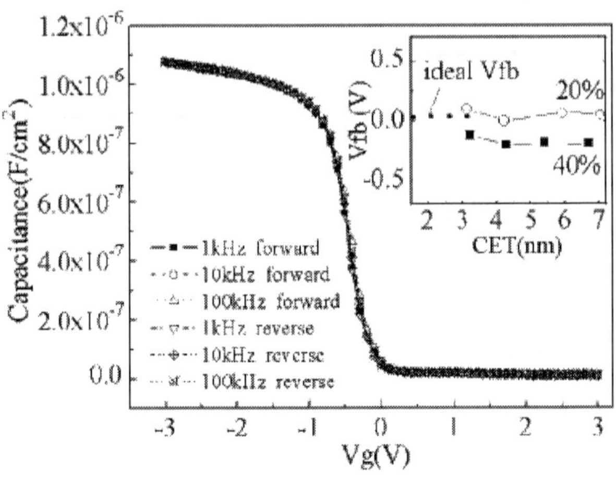

Fig. 19. C-V characteristics of Au/40% La-HfLaO$_x$/p-Si MOS capacitor annealed at 600 °C. The film thickness was 8.4 nm. It shows very small hysteresis and frequency dispersion. The inset in the upper right shows the flatband voltages of Au/20% La-HfLaO/p-Si or 40% La-HfLaO/p-Si MOS [46].

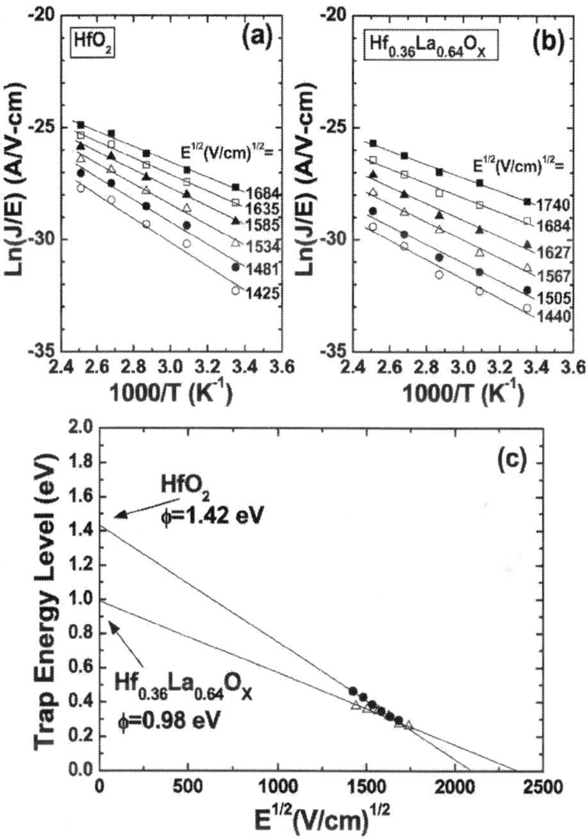

Fig. 20. ln(J/E) vs 1/T plots measured at various applied electric fields for (a) HfO_2 and (b) $Hf_{0.36}La_{0.64}O_x$ films, and (c) trap energy level as a function of $E^{1/2}$ for both samples [47].

From the above results, it can be easily inferred that $HfLaO_x$ is a potential dielectric material for amorphous high-k gate insulator in further advanced complementary metal oxide semiconductor (CMOS) devices.

5. Conclusion

This chapter succinctly reviews the motivation to replace traditional SiO_2 gate dielectrics, requirements of high-k dielectrics, brief history of high-k materials development, and latest development in Hf-based high-k dielectrics. In order to improve the performance of CMOS devices, Hf-based gate layers are being integrated into MOSFETs to achieve low leakage current. Excellent gate transistors with improved performance based on Hf-based gate dielectrics as the insulating layers are expected. Although much progress has been made in fabricating novel gate dielectrics, investigation of these Hf-based high-k gate dielectrics continues to be exciting and the final target has not yet been reached. There is still room for development and many issues need better understanding.

6. Acknowledgments

This work was jointly supported by National Natural Science Foundation of China (Grant No. 50802005), Program for New Century Excellent Talents in University (NCET-08-0035), Ph.D. Program Foundation of Ministry of Education of China (Grant No. 200800061055), and Hong Kong Research Grants Council (RGC) General Research Funds (GRF) No. CityU 112608.

7. References

[1] Moore, G.E. (1975). *Progress in digital integrated electronics*. in Electron Devices Meeting, 1975 International.

[2] Chau, R.; Datta, S.; Doczy, M.; Doyle, B.; Kavalieros J.; & Metz M. (2004). *High-k/Metal-Gate Stack and Its MOSFET Characteristics*, vol. 25, pp. 408-410, IEEE Electron Device Lett.

[3] Robertson, J. (2006). *High dielectric constant gate oxides for metal oxide Si transistors*, vol. 69, pp. 327-396, Rep. Prog. Phys.

[4] Plummer, J. D. & Griffin, P. B. (2001). *Material and process limits in silicon VLSI technology*, pp. 240-258, Proc. IEEE, 2001.

[5] Robertson, J. (2000). *Band offsets of wide-band-gap oxides and implications for future electronic devices*, vol. 18, pp. 1785-1791, J. Vac. Sci. Technol. B.

[6] Kingon, A. I.; Kingon, A. I.; Maria, J. P. & Streiffer, S. K. (2000). *Alternative dielectrics to silicon dioxide for memory and logic devices*, vol. 406, pp.1032-1038, Nature.

[7] Robertson, J. & Chen, C. W. (1999). *Schottky barrier heights of tantalum oxide, barium strontium titanate, lead titanate, and strontium bismuth tantalate*, vol. 74, pp. 1168-1170, Appl. Phys. Lett.

[8] Copel, M.; Gribelyuk, M. & Gusev, E. (2000). *Structure and stability of ultrathin zirconium oxide layers on Si(001)*, vol. 76, pp. 436-438, Appl. Phys. Lett.

[9] Schlom, D. G. & Haeni, H. J. (2002). *A thermodynamic approach to selecting alternative gate dielectrics*, vol. 27, pp. 198-204, MRS Bull.

[10] Stemmer, S.; Li. Y; Foran, B.; Lysaght, P. S.; Streiffer, S. K.; Fuoss, P. & Seifert, S. (2003). *Grazing-incidence small angle x-ray scattering studies of phase separation in hafnium silicate films*, vol. 83, pp. 3141-3143, Appl. Phys. Lett.

[11] Wilk, G. D.; Wallace, R.M. & Anthony, J.M. (2001). *High-kappa gate dielectrics: Current status and materials properties considerations*, vol 89, pp. 5243-5275, Journal of Applied Physics.

[12] Robertson, J. (2005). *Interfaces and defects of high-K oxides on silicon*, vol. 49, pp. 283-293, Solid State Electron.

[13] Buhrman, R. A. & Ellis, K. A. (1999). *Time-dependent diffusivity of boron in silicon oxide and oxynitrides*, vol. 74, pp. 967-969, Appl. Phys. Lett.

[14] Kumar, K.; Chou, A. L.; Lin, C.; Choudhury, P. & Lee, J. C. (1997). *Optimization of sub 3 nm gate dielectrics grown by rapid thermal oxidation in a nitric oxide ambient*, vol. 70, pp. 384-386, Appl. Phys. Lett.

[15] Kizilyalli, I.C.; Huang, R.Y.S. & Roy, R.K. (1998). *MOS transistors with stacked SiO2-Ta2O5 -SiO2 gate dielectrics for giga-scale integration of CMOS technologies*, vol. 19, pp. 423-425, IEEE Electron. Device Lett.

[16] Qi, W. J.; Nieh, R.; Lee, B. H.; Kang, L. G.; Jeon, Y.; Onishi, T.N. K.; Banerjee, S. & Lee, J.C. (1999). MOSCAP and MOSFET characteristics using ZrO2 gate dielectric deposited directly on Si, pp. 145-148, Tech. Dig. Int. Electron. Device. Meet.

[17] Alshareef, H. N.; Niimi, H.; Varghese, A.; Bevan, M.; Kuan, R.; Holt, J.; Tiner, P. & Khamankar, R. (2005). *Intrinsic reoxidation of microwave plasma-nitrided gate dielectrics*, vol. 86, pp. 132901-3, Appl. Phys. Lett.

[18] G.Pant, A. G.; Kim, M. J.; Wallace, R.M.; Quevedo-Lopez, M. A. & Kirsch, P. D. (2006). *Effect of thickness on the crystallization of ultrathin HfSiON gate dielectrics*, vol. 88, pp. 032901-3 Appl. Phys. Lett.

[19] Kita, K.; Kyuno, K. & Toriumi, A. (2005). *Permittivity increase of yttrium-doped HfO2 through structural phase transformation*, vol. 86, pp. 102906-3, Appl. Phys. Lett.

[20] Lu, X.-b.; Liu, Z.-G.; Wang, Y.-P.; Yang, Y.; Wang, X.-P.; Zhou, H.-W. & Nguyen, B.-Y. (2003). *Structure and dielectric properties of amorphous LaAlO[sub 3] and LaAlO[sub x]N[sub y] films as alternative gate dielectric materials*, vol. 94, pp. 1229-1234, J. Appl. Phys.

[21] Kim, H. & McIntyre, P. C. (2002). *Spinodal decomposition in amorphous metal–silicate thin films: Phase diagram analysis and interface effects on kinetics*, vol. 92, pp. 5094-5096, J. Appl. Phys.

[22] Biercuk, M.J.; Monsma, D. J.; Marcus, C. M.; Becker, J. S.; Gordon, R. G. (2003). *Low-temperature atomic-layer-deposition lift-off method for microelectronic and nanoelectronic applications*, vol. 83, pp. 2405-2407, Appl. Phys. Lett.

[23] Tokita, K & Okada, F. (1996). *Growth of metal oxide thin films by laser-induced metalorganic chemical vapor deposition*, vol. 80, pp. 7073-7075, J. Appl. Phys.

[24] Hausmann, D.; Becker, J.; Wang, S. & Gordon, R. G. (2002). *Rapid Vapor Deposition of Highly Conformal Silica Nanolaminates*, vol. 298, pp. 402-406, Science.

[25] Zhu, W.; Ma, T. P.; Tamagawa, T.; Di, Y.; Kim, J.; Carruthers, R.; Gibso, M. & Furukawa, T. (2001). pp. 4.1, Tech. Dig. Int. Electron Device Meet.

[26] Choi, C.H.; Rhee, S. J.; Jeon, T.S.; Lu, N.; Sim, J. H.; Clark, R.; Niwa, M. & Kwang, D. L. (2002). pp. 865, Tech. Dig. Int. Electron Device Meet.

[27] Chen, P.; Bhandari, H. B. & Klein, T. M. (2004). *Effect of nitrogen containing plasmas on interface stability of hafnium oxide ultrathin films on Si(100)*, vol. 85, pp. 1574-1576, Appl. Phys. Lett.

[28] Choi, C.H.; Jeon, T.S.; Clark, R. & Kwong, D.L. (2003). *Electrical properties and thermal stability of CVD HfOxNy gate dielectric with poly-Si gate electrode*, vol. 24, pp. 215-217, IEEE Electron Device Lett.

[29] Choi, K. J.; Kim, J. H. & Yoon, S. G. (2004) *Characterization of HfO2 and HfOxNy Gate Dielectrics Grown by PE Metallorganic CVD with a TaN Gate Electrode*, vol. 151, pp. G262-G265, J. Electrochem. Soc.

[30] Kang, C. S.; Cho, H. J.; Onishi, K.; Nieh, R.; Choi, R.; S.Gopanlan, S.K.; Han, J.H. & Lee, J. C. (2002). *Bonding states and electrical properties of ultrathin HfOxNy gate dielectrics*, vol. 81, pp. 2539-2541, Appl. Phys. Lett.

[31] Kroll, P. (2003). *Hafnium Nitride with Thorium Phosphide Structure: Physical Properties and an Assessment of the Hf-N, Zr-N, and Ti-N Phase Diagrams at High Pressures and Temperatures*, vol. 90, pp. 125501, Phys. Rev. Lett.

[32] Lim, S. G.; Kriventsov, S.; Jackson, T. N.; Haeni, J. H.; Schlom, D. G.; & Balbashov, A. M. (2002). *Dielectric functions and optical bandgaps of high-K dielectrics for metal-oxide-semiconductor field-effect transistors by far ultraviolet spectroscopic ellipsometry*, vol. 91, pp. 4500, J. Appl. Phys.

[33] Cho, M. H.; Roh, Y. S.; Whang, C. N.; Jeong, K.; Choi, H. J.; Nam, S. W.; Ko. D. H.; Lee, J. H.; Lee, N. I. & Fujihara, K. (2002). *Dielectric characteristics of Al2O3–HfO2 nanolaminates on Si(100)*, vol. 81, pp. 1071-1073, Appl. Phys. Lett.

[34] K.J.Choi, J.H.Kim, J. H.; Yoon, S.G. & Shin, W.C. (2004) *Structural and electrical properties of HfOxNy and HfO2 gate dielectrics in TaN gated nMOSCAP and nMOSFET devices*, vol. 22, pp. 1755-1758, J. Vac. Sci. Technol. B.

[35] He, G.; Fang, Q. & Zhang, L. D. (2006). vol. 9, pp. 870, Mater. Sci. Semicon. Proc.

[36] Hu, J.C.; Yang, H.; Kraft, R.; Rotondaro, A.L.P.; Hattangady, S.; Lee, W.W.; Chapman, R.A.; Chao, C.-P.; Chatterjee, A.; Hanratty, M.; Rodder, M. & Chen, I.-C. (1997). *Feasibility of using W/TiN as metal gate for conventional 0.13 μm CMOS technology and beyond*, pp. 825-828, Tech. Dig. Int. Electron Devices Meet.

[37] ITRS. (2003). ITRS Roadmap.

[38] Kamada, H.; Tanimura, T.; Toyoda, S.; Kumigashira, H.; Oshima, M.; Liu, G. L.; Liu, Z. & Ikeda K. (2008). *Control of oxidation and reduction reactions at HfSiO/Si interfaces through N exposure or incorporation*, vol. 93, pp. 212903-3, Appl. Phys. Lett.

[39] Zhu, W. J.; Tamagawa, T.; Gibson, M.; Furukawa, T. & Ma, T. P. (2002). *Effect of Al inclusion in HfO₂ on the physical and electrical properties of the dielectrics*, vol. 23, pp. 649-651, IEEE Electron Device Lett.

[40] Joo, M. S.; Cho, B. J.; Yeo, C. C.; Chan, S. H.; Whoang, S. J.; Mathew, S.; Kanta, B. L. & Balasubramanian, N. (2003). Dim-Lee, Kwong *Formation of hafnium-aluminum-oxide gate dielectric using single cocktail liquid source in MOCVD process*, vol. 50, pp. 2088-2094, IEEE Transactions on Electron Devices.

[41] Bae, S. H.; Lee, C. H.; Clark, R. & Kwong, D. L. (2003). *MOS characteristics of ultrathin CVD HfAlO gate dielectrics*, vol. 24, pp. 556-558, IEEE Electron Device Lett.

[42] Lu, X.-b.; Maruyama, K. & Ishiwara, H. (2008). *Characterization of HfTaO films for gate oxide and metal-ferroelectric-insulator-silicon device applications*, vol. 103, pp. 044105-5, J. Appl. Phys.

[43] Li, M.-F.; Zhu, C.-X; Shen, C. Y.; Xiong, F. F.; Yuan, P.; Yeo, Y. C.; Chin, A.; Kwong, D.-L.; Wang, S. H.; Du, A. Y. & Samudra, G. (2006). *New Insights in Hf Based High-k Gate Deielectrics in MOSFETs*, vol. 1, pp. 717-730, *ECS Transactions.*

[44] Yu, X.-F.; Zhu, C.-X.; Li, M. F.; Chin, A.; Du, A. Y.; Wang, W. D. & Kwong, D.-L. (2004). *Electrical characteristics and suppressed boron penetration behavior of thermally stable HfTaO gate dielectrics with polycrystalline-silicon gate*, vol. 85, pp. 2893-2895, Appl. Phys. Lett.

[45] Zhang, M.H.; Rhee, S. J.; Kang, C. Y.; Choi, C. H.; Akbar, M. S.; Krishnan, S. A.; Lee, T.; Ok, I. J.; Zhu, F.; Kim, H. S. & Lee, J. C. (2005). *Improved electrical and material characteristics of HfTaO gate dielectrics with high crystallization temperature*, vol. 87, pp. 232901-3, Appl. Phys. Lett.

[46] Yamamoto, Y.; Kita, K.; Kyuno, K. & Toriumi, A. (2006). *Structural and electrical properties of HfLaO[sub x] films for an amorphous high-k gate insulator*, vol. 89, pp. 032903-3, Appl. Phys. Lett.

[47] An, C.-H.; Lee, M. S.; Choi, J.-Y. & Kim, H. (2009). *Change of the trap energy levels of the atomic layer deposited HfLaO[sub x] films with different La concentration*, vol. 94, pp. 262901-3, Appl. Phys. Lett.

17

Liquid Phase Oxidation on InGaP and Its Applications

Yeong-Her Wang[1] and Kuan-Wei Lee[2]
[1]*Institute of Microelectronics, Department of Electrical Engineering, Advanced Optoelectronic Technology Center, National Cheng-Kung University, Tainan 701,*
[2]*Department of Electronic Engineering, I-Shou University, Kaohsiung County 840, Taiwan*

1. Introduction

High-electron-mobility transistors (HEMTs) and heterojunction bipolar transistors (HBTs) have attracted many attentions in high speed and power applications due to the superior transport properties. As compared to AlGaAs pseudomorphic HEMTs (PHEMTs), InGaP-related devices have advantages, such as higher band gaps, higher valence-band discontinuity [1], negligible deep-complex (DX) centers [2], excellent etching selectivity between InGaP and GaAs, good thermal stabilities [3-5], higher Schottky barrier heights [3], and so on. Particularly, the use of an undoped InGaP insulator takes the advantages of its low DX centers and low reactivity with oxygen [6-10], which may still suffer from the high gate leakage issue. In order to inhibit the gate leakage issue, increase the power handling capabilities, and improve the breakdown voltages, a metal-oxide-semiconductor (MOS) structure has been widely investigated. However, it is still lacks a reliable native oxide film growing on InGaP, and very few papers have reported on InGaP/InGaAs MOS-PHEMTs. In addition, the MOS-PHEMT not only has the advantages of the MOS structure (e.g., lower leakage current and higher breakdown voltage) but also has the high-density, high-mobility 2DEG channel.

Over the past years, a study on the liquid phase oxidation (LPO) of InGaP near room temperature has been done [11-14]. The application of surface passivation to improve the InGaP/GaAs HBTs' performance has also been first demonstrated [13]. The InGaP/GaAs HBTs with surface passivation by LPO exhibit significant improvement in current gain at low collector current regimes due to the reduction of surface recombination current, as compared to those without surface passivation. Moreover, a larger breakdown voltage and a lower base recombination current are also obtained. In this chapter, the oxide film composition and some issues are addressed. Then a thin InGaP native oxide film prepared by the LPO as the gate dielectric for InGaP/InGaAs MOS-PHEMTs application are discussed, and the comparisons between devices with and without LPO passivation on the InGaP/GaAs HBTs are also reviewed.

2. Characterization of the oxide film

The root mean square (rms) value of surface roughness for the $In_{0.49}Ga_{0.51}P$ sample is estimated to be 1.1 nm before oxidation (i.e., as received) by AFM measurement, and can be improved to 0.95 nm after oxidation (i.e., as grown), as shown in Fig. 1. Fig. 2 shows the SIMS depth profiles before and after liquid phase oxidation on $In_{0.49}Ga_{0.51}P$. Although LPO on InGaP material has a much slower oxidation rate which is less than 10 nm/h, as comparing to that of the GaAs material, however, it is still feasible to grow a thin oxide film without pH control [15, 16]. The oxidation rate becomes significantly saturated when the oxidation time is longer than an hour, which is measured using a Veeco Instrument DEKTAK and confirmed by SEM.

The XPS depth profiles of the LPO-grown oxide for $In_{0.49}Ga_{0.51}P$ are shown in Fig. 3(a). Fig. 3(b)-(d) show the XPS surface spectra of the Ga-3d, In-3d, and P-2p core levels, respectively. The binding energies for all spectra are calibrated with the reference (as-received) signal. The as-received sample was dipped into a solution of $HF:H_2O$ = 1:200 for 30 s before measurement. From Fig. 3(c)-(d), in comparison with the previous paper [17], the spectrum is rather similar to that of $InPO_4$. This is also confirmed by the values of the O-1s peak energy and energy separations between the main core levels (i.e., Ga-3d, In-3d, and P-2p) in the oxide phases [18]. This clearly suggests that the oxide film is mostly composed of $InPO_4$-like and Ga oxide. In addition, the oxide film may appear to be etched back in the growth solution after 2 h of oxidation. The thermal stability of the oxide layer is also important in

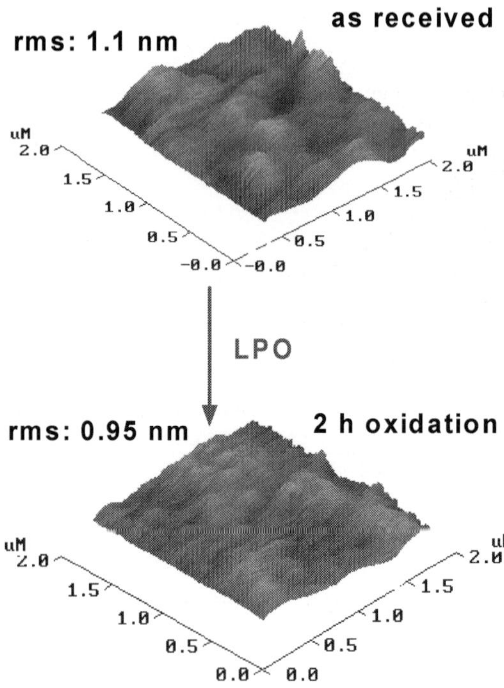

Fig. 1. AFM images of the $In_{0.49}Ga_{0.51}P$ sample before (i.e., as received) and after (i.e., as grown) liquid phase oxidation.

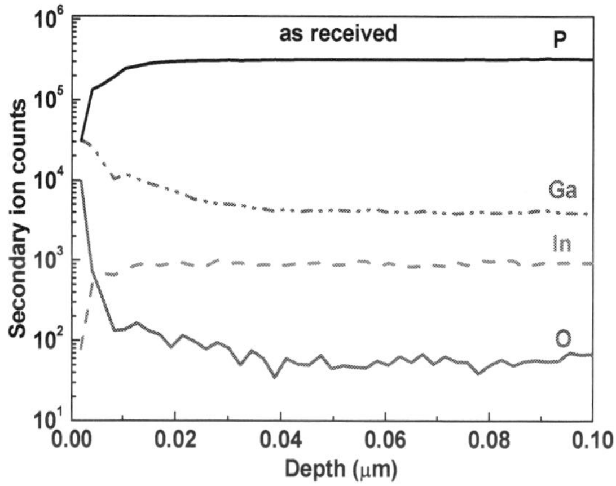

(a)

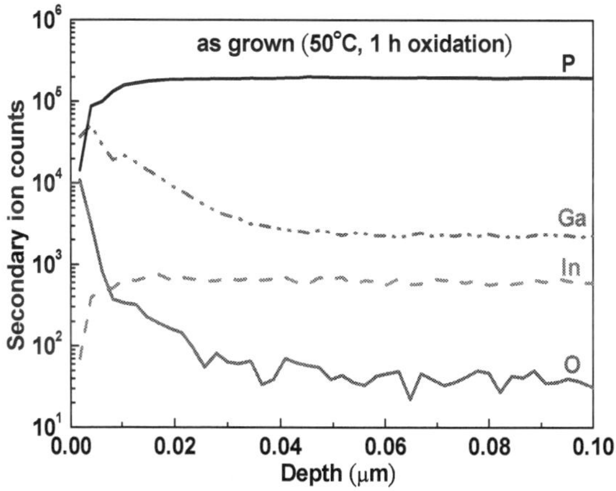

(b)

Fig. 2. SIMS depth profiles of the $In_{0.49}Ga_{0.51}P$ sample before (i.e., as received) and after (i.e., as grown) liquid phase oxidation.

device fabrications because high-temperature processes are usually required. Again, XPS is utilized to also important in device fabrications because high-temperature processes are usually required. Again, XPS is utilized to analyze the surface chemistry of the oxide films, as shown in Fig. 3. After 2 h of oxidation, the RTA processes were performed in a furnace with N_2 flowing at 300-700 °C for 1 min [13]; however, a peak of $InPO_4$-like is still observed. $InPO_4$ (bandgap energy = 4.5 eV) is chemically stable and has rather good dielectric properties [19]. As a result, the $InPO_4$ probably acts as a capping layer for the entire oxide film to enhance the thermal stability. However, the experimental results show that high-temperature treatments (700 °C) will change the properties of Ga_2O_3, since the XPS energy peak of Ga_2O_3 shifts to a lower binding energy, and the binding energy is inferred to form the GaO_x or Ga_2O_x.

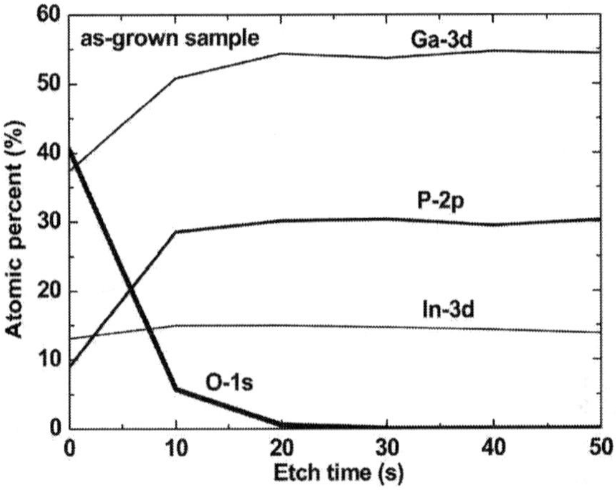

(a)

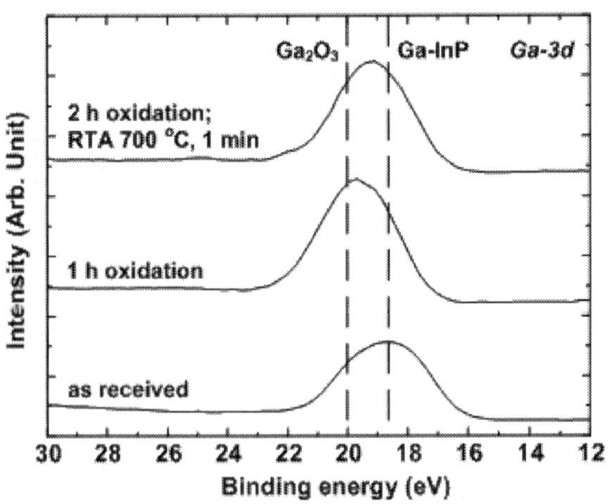

(b)

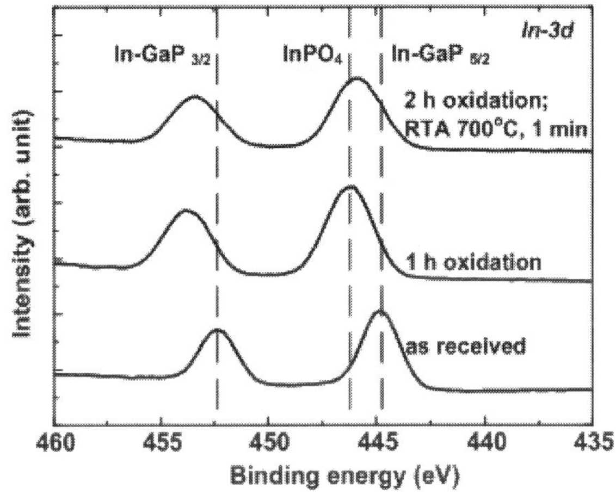

(c)

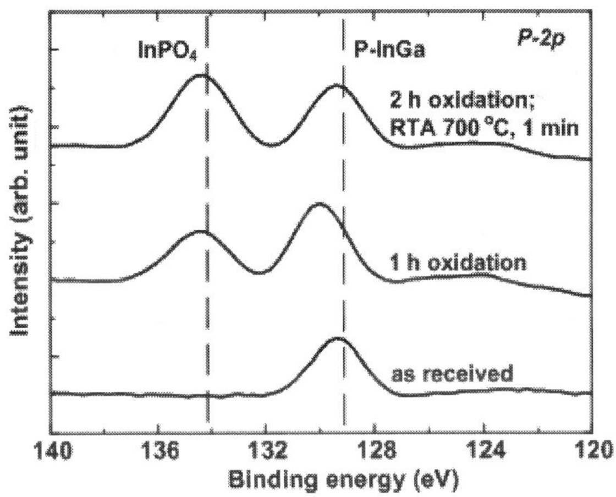

(d)

Fig. 3. (a) The XPS depth profiles of the as-grown oxide film on $In_{0.49}Ga_{0.51}P$.
The (b)-(d) show the XPS surface spectra for the Ga-3d, In-3d, and P-2p core levels, respectively.

3. InGaP/InGaAs MOS-PHEMT

3.1 Experimental

Figure 4 schematically shows the PHEMT structure grown by the metallorganic chemical vapor deposition (MOCVD) on a semi-insulating GaAs substrate. Hall measurement indicates that the electron mobility is 4000 $cm^2/V \cdot s$, and the electron sheet density is 2.2×10^{12} cm^{-2} at room temperature [11]. The device isolation was accomplished by mesa wet etching down to the buffer layer. The ohmic contacts of the Au/Ge/Ni metal were deposited by evaporation and then were patterned by lift-off processes, followed by RTA. The depth of gate recess is 110 nm for reference PHEMT and 100 nm for MOS-PHEMT. After etching the capping layer and the partial Schottky layer, an LPO growth solution was used to generate the gate oxide for the MOS-PHEMT at 50 °C for 30 min. Finally, the gate electrode was formed with Au. Moreover, the oxide layer, as illustrated in the figure, also selectively and simultaneously passivated the isolated surface sidewall. The gate dimension is 2×100 μm² with a drain-to-source spacing of 5 μm.

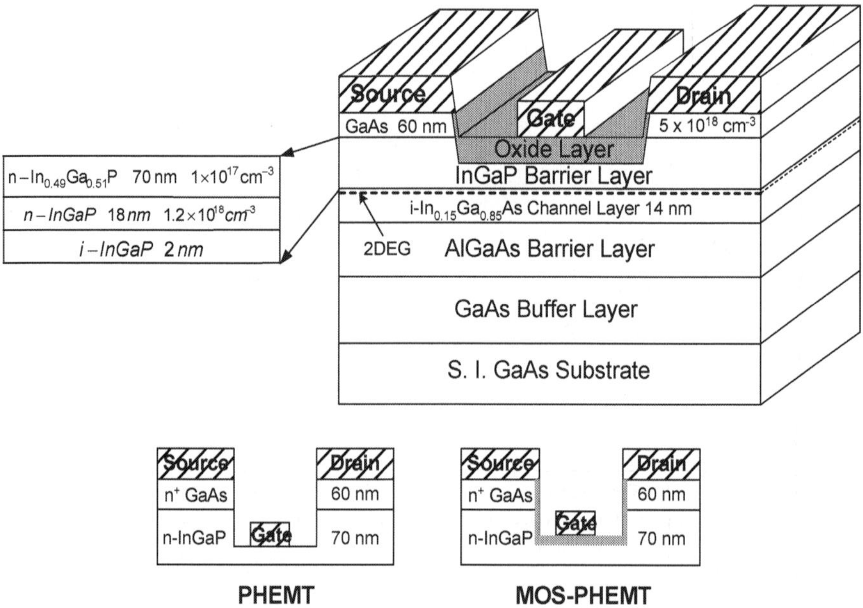

Fig. 4. The schematic drawing of the InGaP/InGaAs MOS-PHEMT.

3.2 Results and discussion

Figure 5(a) compares the measured I-V characteristics of the MOS-PHEMT with those of the reference PHEMT fabricated under identical conditions. Clearly, good pinch-off and saturation current characteristics are obtained. Due to the higher energy barriers between the metal gate and the Schottky layer, the MOS PHEMT can be operated at higher gate to source voltage (V_{GS}) and drain-to-source voltage (V_{DS}) than those of the conventional Schottky gate PHEMT, which can enhance the current driving capability. Fig. 5(b) compares the transconductance g_m and the drain current density I_D as a function of V_{GS} at V_{DS} = 4 V of the MOS-PHEMTs with those of the reference PHEMT. For MOS-PHEMT, the 1.8 V-wide gate voltage swing (defined by 10% reduction from the maximum g_m) is higher than that of the PHEMT. The threshold voltage V_{th} of MOS-PHEMT shifts to the left, which is similar to the result of the one with oxide deposited on the Schottky layer [20, 21]. However, the separation region between the oxide-InGaP interface and the InGaAs channel for MOS-

PHEMT is still larger than that of the reference PHEMT in this study, so the drain current density of the PHEMT is smaller than that of the MOS-PHEMT at the same bias V_{GS} due to the decrease of the carrier concentration within the InGaAs 2DEG channel.

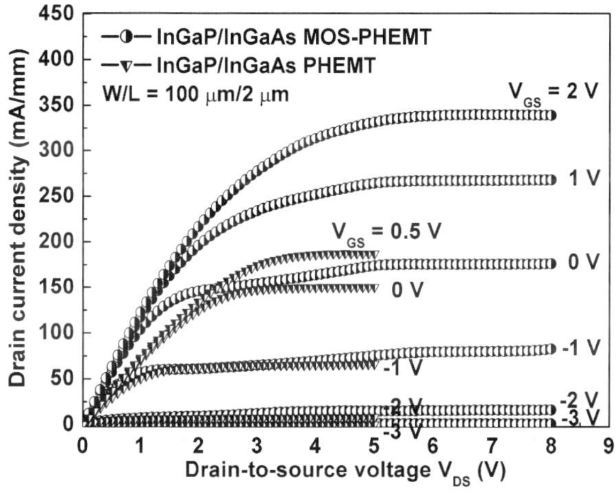

(a)

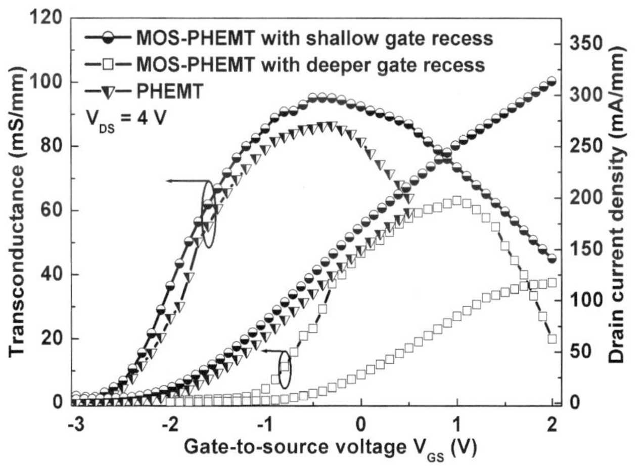

(b)

Fig. 5. (a) Measured I-V characteristics of MOS-PHEMT and PHEMT. (b) The transconductance and the drain current density versus V_{GS} at V_{DS} = 4 V for the MOS-PHEMT and the reference PHEMT.

In addition, if the depth of gate recess is etched to be 120 nm, the V_{th} becomes more positive, -0.5 V, for MOS-PHEMT with the identical processing conditions including initial pH value (5.0), temperature (50 °C), and oxidation time (30 min). For V_{th} shifts to the right, the separation between the oxide-InGaP layer interface and the InGaAs channel layer is decreased due to the consumption of the InGaP during the processes of gate recess and the unique properties of the LPO with the reaction of InGaP, leading to the increase of the total effect of the gate bias on the control of V_{th}. However, a decrease in the maximum g_m, 63 mS/mm, accompanies the degradation in the saturation current, 84 mA/mm at V_{GS} = 1 V. The result is also confirmed by a longer oxidation time, i.e., a thicker oxide layer. This drawback can be overcome by suitable device structures, such as inserting a Si-planar doping layer under the InGaAs channel to increase the carrier density.

The oxide film provides an improvement in the breakdown voltage in terms of the gate leakage current of the MOS structure, supported by the typical gate-to-drain I-V characteristics, as shown in Fig. 6(a). For InGaP MOS-PHEMT, the turn-on voltage, 2.2 V, is obviously higher than that of InGaP PHEMT, 0.8 V, and the corresponding reverse gate-to-drain breakdown voltages, BV_{GD}, are -14.1 V and -6.5 V, respectively. The turn-on voltage and the BV_{GD} are defined as the voltage at which the gate current reaches 1 mA/mm. The gate leakage current can be suppressed at least by more than two orders of magnitude with an oxide film at V_{GD} = -4 V. The smaller gate leakage current of MOS-PHEMT is due to the MOS structure and the elimination of sidewall leakage paths that are directly passivated during the oxidation, which is consistent with the result of Fig. 5. In addition, the gate leakage current observed in MOS-PHEMT comes from a gate leakage path at the edge of the mesa [22] that is not present in the MOS capacitor, which may contribute to the Schottky-like I-V characteristics for forward biases. Fig. 6(b) shows the gate current density as a function of reverse V_{GS} at different V_{DS}. Due to the high electric field existing in the gate-to-drain region, hot electron phenomena occur in the narrow band-gap InGaAs channel. Electrons can obtain higher energy to generate electron-hole pairs through the enhanced impact ionization, resulting in easy injection of the holes into the gate terminal [23]. However, in InGaP-related devices, it is more difficult for the holes generated by the impact ionization to overcome the valence band discontinuity and to reach the gate [4], so the bell shaped behavior of the impact ionization does not appear in Fig. 6. Moreover, the gate current density of MOS-PHEMT is significantly improved, which is less than 0.5 μA/mm, as compared to that of PHEMT. In other words, the electrons and holes generated by the impact ionization are decreased to further reduce the drain and gate currents owing to the oxide layer with a high barrier height.

In order to have a better insight into the transient behavior of the studied devices, the gate pulse measurements were performed using a Tektronix 370A curve tracer [24]. V_{GS} was pulsed from the V_{th} to 0 V with a pulsewidth of 80 μs, while V_{DS} was swept from 0 to 4 V. The comparisons between the static and pulsed I-V characteristics for PHEMT and MOS-PHEMT are shown in Fig. 7. The drain current of PHEMT decreased by 9.8%, while the MOS-PHEMT decreased by only 0.63%. To the best of our knowledge, if the pulsewidth is too short, electrons captured by the traps do not have enough time to be fully emitted. However, if the pulsewidth is long enough, all the trapped electrons are de-trapped and will contribute to the drain current. We believe that the differences between dc and pulsed I-V become evident by applying shorter voltage pulses to the gate such as less than 10-μs pulses for PHEMT and MOS-PHEMT. Therefore, it is clear that the oxide passivation on the

Schottky layer can minimize the effect of surface traps, which is consistent with the lower gate leakage current in Fig. 6.

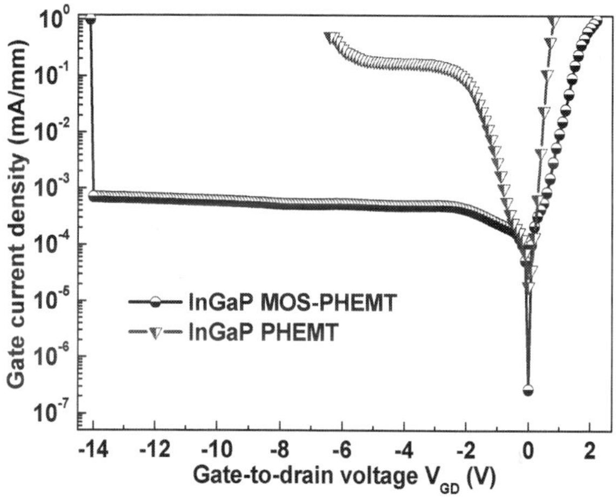

(a)

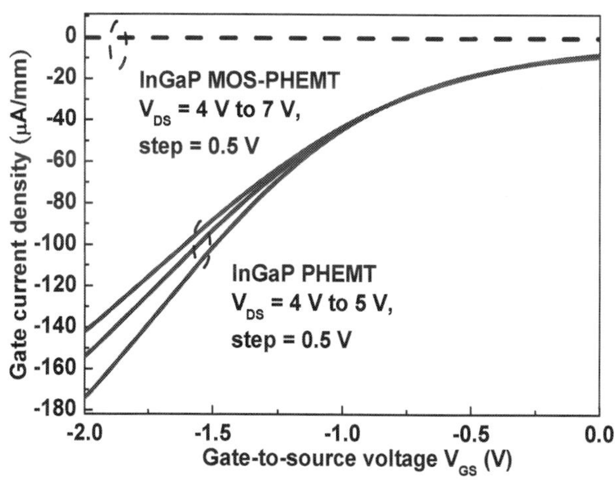

(b)

Fig. 6. (a) The typical I_G-V_{GD} characteristics of PHEMT with and without an oxide film. (b) The gate current density versus reverse V_{GS} at different V_{DS}.

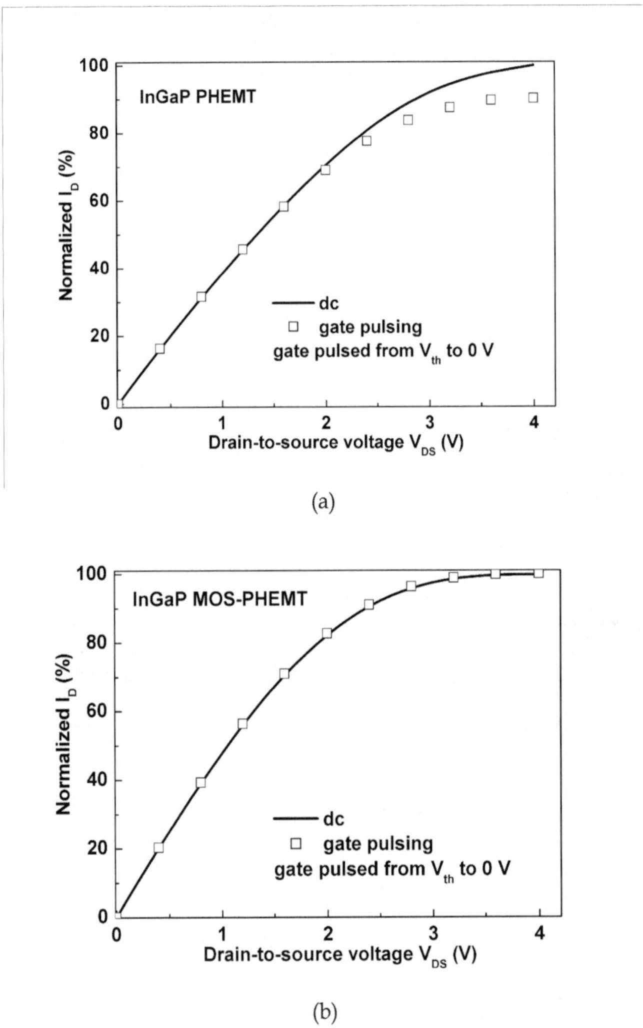

Fig. 7. Gate pulse measurements for (a) reference PHEMT and (b) MOS-PHEMT with V_{GS} pulsed from V_{th} to 0 V with a pulsewidth of 80 μs, while V_{DS} was swept from 0 to 4 V.

4. InGaP/GaAs HBT with LPO passivation

4.1 Experimental

The structure used for HBT is given in Table 1. The epilayers were grown by a low-pressure MOCVD system on an (100)-oriented semi-insulating (S.I.) GaAs substrate. For InGaP/GaAs HBTs, device fabrication began with emitter definition. The emitter cap layer was removed and stopped at the InGaP active layer. After removing the InGaP layer, a growth solution was used to form the base oxide (passivation) on the exposed extrinsic surface of base and the base contact was then deposited. Finally, the mesa of base was defined and etched to sub-collector before the collector contact deposition. H_3PO_4-based etchant was used for GaAs and InGaP. The Au/Ge and Au/Be metals were deposited by evaporation and patterned by lift-off processing to form emitter, base and collector regions, respectively.

Layer	Material	Thickness (nm)	Dopant (cm^{-3})
Cap	InGaAs	45	1×10^{19}
Graded	InGaAs	45	1×10^{19}
Sub-emitter	GaAs	130	5×10^{18}
Emitter	InGaP	40	3×10^{17}
Base	GaAs	100	4×10^{19}
Collector	GaAs	750	1×10^{16}
Etching-stop	InGaP	20	5×10^{18}
Sub-collector	GaAs	600	5×10^{18}
S.I. GaAs substrate			

Table 1. The epitaxial structure of InGaP/GaAs HBT.

4.2 Results and discussion

Figure 8 shows the common-emitter I-V characteristics of the HBT with and without surface passivation by LPO. Clearly, the dc current gain (β) of HBTs with passivation is improved (increased) 15% when comparing to HBTs without passivation. The higher β with surface passivation is due to the reduction of the surface recombination current in the exposed extrinsic base regions by LPO method. The common-emitter I-V characteristics of the devices with and without surface passivation at low collector current regimes are shown

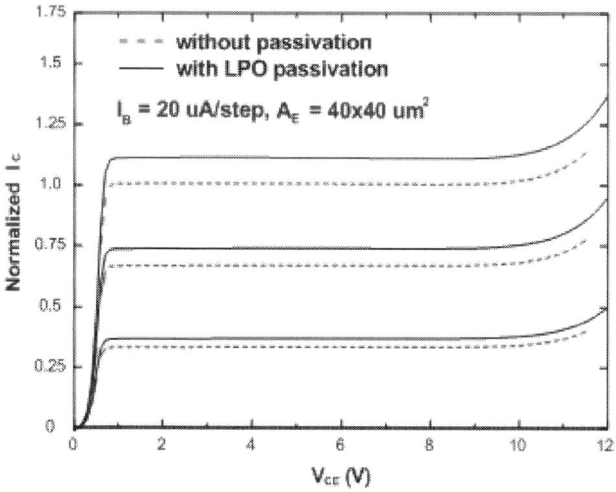

Fig. 8. Common-emitter I-V characteristics of the HBTs with and without LPO passivation.

in Fig. 9. The devices with surface passivation have higher common-emitter β than those devices without passivation, due to the reduction of the surface combination velocity by using an oxide layer on the base surface. In addition, the β values with and without passivation are 13.3 and 2 at I_B = 900 pA, respectively. The maximum increase of 7 fold in the current gain at collector current down to nA level.

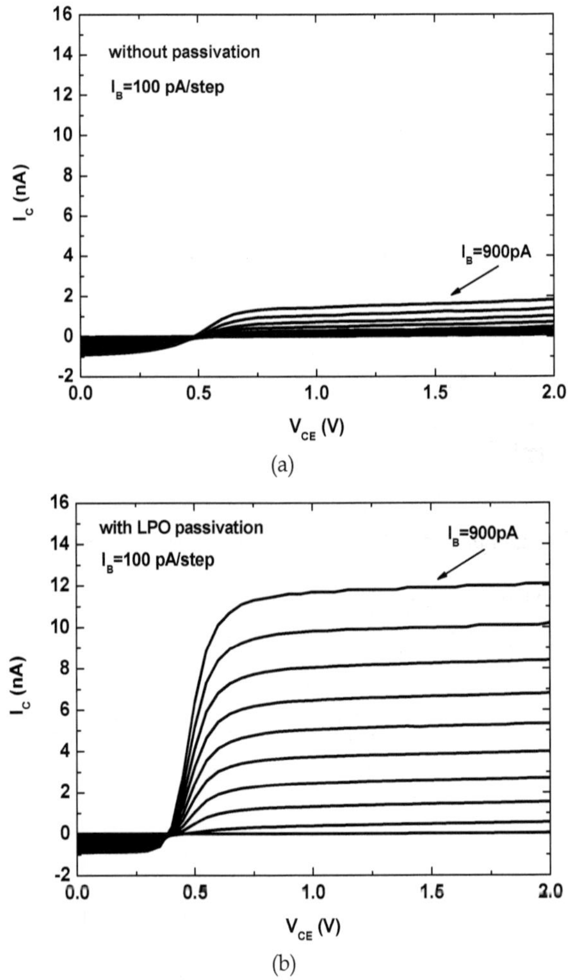

Fig. 9. Common-emitter I-V characteristics of the HBTs (a) without and (b) with LPO passivation at low collector current regimes.

Figure 10 illustrates the measured Gummel plots of the devices with and without LPO passivation. The collector currents are almost identical without being affected by the passivation treatment. However, a decrease of the base leakage current at low collector current levels is obviously observed after oxidation. Moreover, it is found that the recombination current at the extrinsic base region and the base-emitter perimeter are competed against one another, resulting in current reduction at lower base-emitter bias V_{BE} = 0.4 V. The increasing β is owing to the reduction of the surface recombination current. It can also be indicated that the device with pasivation exhibits higher β than that without passivation at lower V_{BE} bias. The comparison of β versus the collector current is shown in Fig. 11. The collector-base bias is maintained at 0 V. Clearly, the device with LPO passivation shows wider collector regimes from 10^{-10} A to 0.1 A. And the maximum shift of

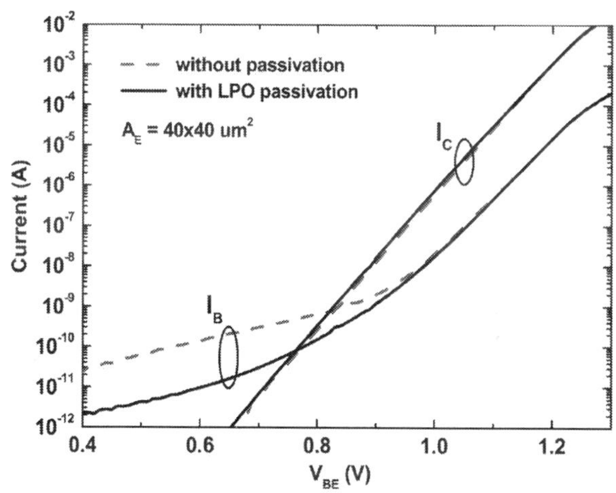

Fig. 10. Typical Gummel plots of InGaP/GaAs HBTs with and without LPO passivation.

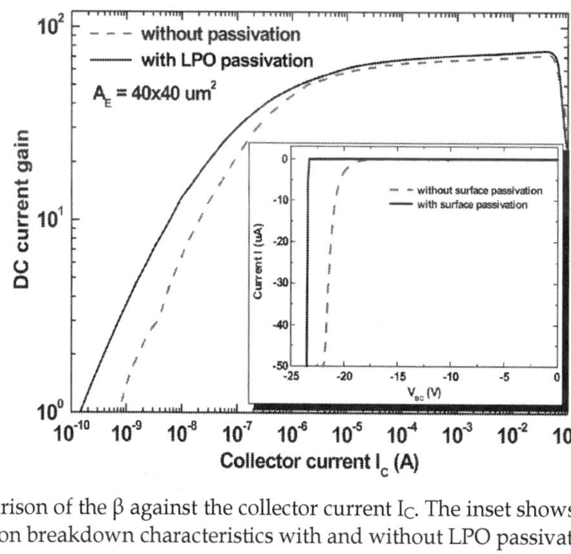

Fig. 11. Comparison of the β against the collector current I_C. The inset shows the base-collector junction breakdown characteristics with and without LPO passivation.

5 fold in the current gain from collector current of 8.1×10^{-10} A to 1.6×10^{-10} A can be achieved. This is attributed to the surface state density are suppressed, i.e., the surface recombination current is effectively reduced. The inset shows the base-collector junction current against bias voltage for the devices with and without passivation. For the device with passivation, the breakdown voltage (23.5 V) is higher than that (21.9 V) without passivation at I = 50 μA. The smaller leakage current is owing to the reduction of the surface recombination by the native oxide passivation in the base region. Above results clearly indicate that the β at low (medium) collector current regimes and the breakdown voltage will be increased. Additionally, the base current is decreased for the devices with passivation when comparing to those without passivation, which will be beneficial to low-power electronics and communication applications.

5. Conclusion

The InGaP/InGaAs/GaAs MOS-PHEMT with the $In_{0.49}Ga_{0.51}P$ oxide as the gate insulator prepared by LPO has been demonstrated. As compared to the counterpart of the conventional InGaP PHEMT, the proposed InGaP MOS-PHEMT can further reduce the gate leakage current at least by two orders of magnitude, increase the breakdown voltage by 200%, and enhance the gate voltage swing. Also, the pulse transient measurement shows much less impact of the surface trap effects for the InGaP MOS-PHEMT. In addition, as compared to the conventional InGaP/GaAs HBTs without surface passivation, the HBTs with LPO passivation possess the characteristics of lower surface recombination currents, higher breakdown voltage and improved higher dc current gain. The HBTs with LPO passivation exhibit 700% improvement in current gain at low collector current regimes by the reduction of surface recombination current, as compared to those without passivation. Therefore, the proposed low-temperature and low-cost LPO can easily be implemented and can provide new opportunities in device applications.

6. Acknowledgements

The authors wish to thank Nan-Ying Yang whose research made this work possible. This work was supported in part by the National Science Council of Taiwan under contract number NSC95-2221-E-006-428-MY3, NSC97-2221-E-214-063, and the MOE Program for Promoting Academic Excellence of Universities.

7. References

[1] M. A. Rao, E. J. Caine, H. Kroemer, S. I. Long, and D. I. Babic, "Determination of valence and conduction-band discontinuities at the (Ga,In) P/GaAs heterojunction by C-V profiling," *J. Appl. Phys.*, vol. 61, pp. 643-649, 1987.

[2] R. Menozzi, P. Cova, C. Canali, and F. Fantini, "Breakdown walkout in pseudomorphic HEMT's," *IEEE Trans. Electron Dev.*, vol. 43, pp. 543-546, 1996.

[3] S. Fujita, T. Noda, A. Wagai, C. Nozaki, and Y. Ashizawa, "Novel HEMT structures using a strained InGaP Schottky layer," *in Proceedings of the 5th Indium Phosphide and Related Materials (IPRM)*, Paris, France, April 19-22, 1993, pp. 497-500.

[4] H. K. Huang, C. S. Wang, Y. H. Wang, C. L. Wu, and C. S. Chang, "Temperature effects of low noise InGaP/InGaAs/GaAs PHEMTs," *Solid-State Electron*, vol. 47, pp. 1989-1994, 2003.

[5] H. K. Huang, C. S. Wang, C. P. Chang, Y. H. Wang, C. L. Wu, and C. S. Chang, "Noise characteristics of InGaP-gated PHEMTs under high current and thermal accelerated stresses," *IEEE Trans. Electron Dev.*, vol. 52, pp. 1706-1712, 2005.

[6] Y. S. Lin, S. S. Lu and Y. J. Wang, "High-performance $Ga_{0.51}In_{0.49}P$/GaAs airbridge gate MISFET's grown by gas-source MBE," *IEEE Trans. Electron Dev.*, vol. 44, pp. 921-929, 1997.

[7] L. W. Laih, S. Y. Cheng, W. C. Wang, P. H. Lin, J. Y. Chen, W. C. Liu, and W. Lin, "High-performance InGaP/InGaAs/GaAs step-compositioned doped- channel field-effect transistor (SCDCFET)," *Electron. Lett.*, vol. 33, pp. 98-99, 1997.

[8] W. C. Liu, W. L. Chang, W. S. Lour, H. J. Pan, W. C. Wang, J. Y. Chen, K. H. Yu and S. C. Feng, "High-performance InGaP/$In_xGa_{1-x}As$ HEMT with an inverted delta-doped V-shaped channel structure," *IEEE Electron Dev. Lett.*, vol. 20, pp. 548-550, 1999.

[9] K. K. Yu, H. M. Chuang, K. W. Lin, S. Y. Cheng, C. C. Cheng, J. Y. Chen, and W. C. Liu, "Improved temperature-dependent performances of a novel InGaP-InGaAs-GaAs double channel pseudomorphic high electron mobility transistor (DC-PHEMT)," *IEEE Trans. Electron Dev.*, vol. 49, pp. 1687-1693, 2002.

[10] H. M. Chuang, S. Y. Cheng, C. Y. Chen, X. D. Liao, R. C. Liu, and W. C. Liu, "Investigation of a new InGaP-InGaAs pseudomorphic double doped-channel heterostructure field-effect transistor (PDDCHFET)," *IEEE Trans. Electron Dev.*, vol. 50, pp. 1717-1723, 2003.

[11] K. W. Lee, P. W. Sze, Y. J. Lin, N. Y. Yang, M. P. Houng, and Y. H. Wang, "InGaP/InGaAs metal-oxide-semiconductor pseudomorphic high-electron-mobility transistor with a liquid-phase-oxidized InGaP as gate dielectric," *IEEE Electron Device Lett.*, vol. 26, pp. 864-866, 2005.

[12] K. W. Lee, Y. J. Lin, N. Y. Yang, Y. C. Lee, P. W. Sze, Y. H. Wang, and M. P. Houng, "InGaP/InGaAs/GaAs metal-oxide-semiconductor pseudomorphic high electron mobility transistor with a liquid phase oxidized InGaP gate," *in Proceedings of the 7th IEEE International Conference on Solid-State and Integrated Circuits Technology (ICSICT)*, Beijing, China, Oct. 18-21, 2004, pp. 2301-2304.

[13] K. W. Lee, N. Y. Yang, K. L. Lee, P. W. Sze, M. P. Houng, and Y. H. Wang, "Liquid phase oxidation on InGaP and its application to InGaP/GaAs HBTs surface passivation," *in Proceedings of the 17th Indium Phosphide and Related Materials (IPRM)*, Glasgow, Scotland, UK, May 8-12, 2005, pp. 516-519.

[14] K. W. Lee, P. W. Sze, K. L. Lee, M. P. Houng, and Y. H. Wang, "InGaP PHEMT with a liquid phase oxidized InGaP as gate dielectric," *in Proceedings of IEEE International Conference on Electron Devices and Solid-State Circuits (EDSSC)*, Hong Kong, China, Dec. 19-21, 2005, pp. 609-612.

[15] H. H. Wang, J. Y. Wu, Y. H. Wang, and M. P. Houng, "Effects of pH values on the kinetics of liquid phase chemical enhanced oxidation of GaAs," *J. Electrochem. Soc.*, vol. 146, pp. 2328-2332, 1999.

[16] H. H. Wang, "Investigation of Liquid Phase Chemical-Enhanced Oxidation Technique for GaAs and Its Application," Ph.D. dissertation, National Cheng-Kung University, Taiwan, Republic of China, 2000.

[17] G. Hollinger, E. Bergignat, J. Joseph, and Y. Robach, "On the nature of oxides on InP surfaces," *J. Vac. Sci. Technol. A*, vol. 3, pp. 2082-2088, 1985.

[18] T. Hashizume and T. Saitoh, "Natural oxides on air-exposed and chemically treated InGaP surfaces grown by metalorganic vapor phase epitaxy," *Appl. Phys. Lett.*, vol. 78, pp. 2318-2320, 2001.

[19] G. Hollinger, J. Joseph, Y. Robach, E. Bergignat, B. Commere, P. Viktorovitch, and M. Froment, "On the chemistry of passivated oxide-InP interfaces," *J. Vac. Sci. Technol. B*, vol. 5, pp. 1108-1112, 1987.

[20] M. A. Khan, X. Hu, G. Sumin, A. Lunev, J. Yang, R. Gaska, and M. S. Shur, "AlGaN/GaN metal oxide semiconductor heterostructure field transistor," *IEEE Electron Device Lett.*, vol. 21, pp. 63-65, 2000.

[21] D. W. Chou, K. W. Lee, J. J. Huang, P. W. Sze, H. R. Wu, Y. H. Wang, M. P. Houng, S. J. Chang, and Y. K. Su, "AlGaN/GaN metal oxide semiconductor heterostructure field-effect transistor based on a liquid phase deposited oxide," *Jpn. J. Appl. Phys.*, vol. 41, pp. L748-L750, 2002.

[22] S. R. Bahl, M. H. Leary and J. A. del Alamo, "Mesa-sidewall gate leakage in InAlAs/InGaAs heterostructure field-effect transistors," *IEEE Trans. Electron Devices*, vol. 39, pp. 2037-2043, 1992.

[23] T. Suemitsu, T. Enoki, N. Sano, M. Tomizawa, and Y. Ishii, "An analysis of the kink phenomena in InAlAs/InGaAs HEMT's using two-dimensional device simulation," *IEEE Trans. Electron Devices*, vol. 45, pp. 2390-2399, 1998.

[24] K. Balachander, S. Arulkumaran, T. Egawa, Y. Sano, and K. Baskar, "Demonstration of AlGaN/GaN metal-oxide-semiconductor high-electron-mobility transistors with silicon-oxy-nitride as the gate insulator," *Materials Science and Engineering: B*, vol. 119, pp. 36-40, 2005.

18

Germanium Doped Czochralski Silicon

Jiahe Chen and Deren Yang
State Key Lab of Silicon Materials, Zhejiang University
People's Republic of China

1. Introduction

Due to the development of higher integrity electronic devices, it is required to improve the quality of Czochralski (CZ) silicon. On one hand, voids at the near-surface of wafers degrade gate oxide integration (GOI) of MOS devices and therefore reduce the yield of devices. On the other hand, it is a trend for the oxygen concentration of CZ silicon used for ultra large scale integrated circuits (ULSI) to become lower, so it will be difficult to form oxygen precipitates and create gettering sites in the bulk for undesirable metallic contaminants on silicon wafers. In addition, with increasing the diameter of wafers, the dislocations due to higher thermal stress and gravitational stress will generate easily, therefore it is desirable to enhance the mechanical property of wafers.

As an important consist for the novel "impurity engineering" for CZ silicon materials (Chen et al., 2010; Chen & Yang, 2009; Yang et al., 2009), the behaviors of germanium in CZ silicon have attracted considerable attention in recent years, which was invented by our group. Compared to normal dopant elements, germanium doping will not induce electrical centers such as shallow thermal donors due to its equivalent electrons with silicon. Furthermore, the solubility of germanium in silicon is so large that germanium doping will not have influence on the growth of CZ silicon, if germanium concentration is lower than 10^{19} cm^{-3}. And, it is believed that germanium doping in CZ silicon could be much easier to control, so that the influences of germanium doping to the properties of CZ silicon wafers could be adjusted".

Recently, we have investigated the effect of germanium with concentration of 10^{15}-10^{19} cm^{-3} on the mechanical stress, the formation of oxygen-related donors, oxygen precipitation and void defects in CZ silicon materials. It has been established that the mechanical strength of silicon wafers could be improved by germanium doping, which benefits the improved production yield of wafers (Chen et al., 2008). It is also found that germanium suppresses thermal donors (TDs) and new donors (NDs), which benefits the stable electrical property of wafers (Cui et al., 2006; Li et al., 2004b). More importantly, germanium has been found to suppress the formation of crystal originated particles (COPs) related to void defects, which can be annihilated easily during high temperature treatments (Chen et al., 2007a; Yang et al., 2002). Meanwhile, the enhancement of oxygen precipitation can be obtained by germanium doping (Chen et al., 2009; Chen et al., 2006a; Chen et al., 2006b; Li et al., 2004a), and therefore internal gettering (IG) capability could be improved (Chen et al., 2007b; Chen et al., 2007c). Up to now, ascribing to the novel properties induced from germanium atoms, it

is considered that germanium doped CZ (GCZ) silicon could probably become one of the new type silicon materials met requirements of higher performance ULSI.
In this chapter, the behaviors of germanium doping CZ silicon will be reviewed mainly based on our recent work, and two preliminary applications of GCZ silicon wafers will be shown as examples.

2. Mechanical strength

By alloying with oxygen and some dopants, such as nitrogen, the mechanical strength of silicon single crystals could be increased. The strengthening is believed to be associated with impurity concentrations and dislocation densities. Like that for nitrogen-doped float zone (FZ) silicon, it shows a much higher yield strength than usual FZ silicon because nitrogen atoms bring about the hardening of silicon crystals through locking of dislocations upon congregating on the latter one (Kishino et al., 1982; Yonenaga, 2005). As a similar story, the mechanical strength improvement of silicon crystals doped with germanium is considered to be effective at immobilizing and retarding the velocity of dislocations while germanium doping level excessed 6×10^{19} cm^{-3} (Fukuda & Ohsawa, 1992). Furthermore, dislocation-free CZ silicon crystal could be obtained using a heavily germanium doped seed without Dash necking(Huang et al., 2003). Recently, we emphasize that the lightly germanium doping benefits the mechanical stress improvements for CZ silicon wafers.
Table 1 lists the statistical Total Thickness Variation (TTV), Warp and Bow data from 100 pieces of the as-processed wafer during a mass production for both the CZ and GCZ silicon (with the germanium level of 10^{18} cm^{-3}) (Chen et al., 2008). Normally, Warp represents the total amount of maximum variations between the medium and reference surfaces of wafers, while Bow is defined as a half dispersion of concave and convexo maximum between the medium and reference wafer surfaces, both of which are believed to characterize the extent of warpage for silicon wafers and are controlled in production lines extensively: the smaller they are, the slighter the warpage would be. As can be seen in Table 1, both the Warp and Bow merits were relatively smaller in percentage for the GCZ silicon wafers than that for the CZ silicon wafers, indicating that germanium doping in silicon inclines hardly to cause warpage during the wafer making from monocrystalline ingots. Moreover, the fact of the slightly smaller data for the GCZ wafers than the CZ wafers shows that the mechanical strengths of CZ wafers might be improved slightly by germanium doping, which is coincident with the fact that a higher yield of polished wafer could be obtained for GCZ wafer during the assemble wafer making: the yields of polished CZ and GCZ silicon wafers were 89.9% and 92.8%, respectively. It is therefore concluded that a slight suppression on the warpage of CZ silicon wafer could be presented by light germanium doping. It is considered that, compared with normal CZ silicon, grown-in oxygen precipitation could be enhanced in GCZ silicon, which will be discussed below. Then, the enhanced grown-in oxygen precipitates could pin up dislocations and retard their movements, so that the macroscopical mechanical strength of GCZ silicon wafers could be increased.
Herein, it is believed that the novel concept of "mechanical strength improvement by germanium doping" is of great merit, not limiting to the application field of IC used silicon wafers. Especially, it is worthwhile to point out that this novel concept could be adopted in improving the wafer production yield and producing super thin wafer support for solar cells.

Statistical Index	TTV (μm)		Warp (μm)		Bow (μm)	
	CZ	GCZ	CZ	GCZ	CZ	GCZ
Mean	2.7	2.9	15.9	11.6	-0.31	0.26
Range	4.7	4.9	20.1	16.8	25.6	16.9
Quartile	2.0	2.3	4.7	4.5	6.2	3.4
Variance	1.6	1.6	28.5	14.1	16.2	14.4
Standard Deviation	1.3	1.3	5.4	3.3	4.2	3.5

Table 1. Data dispersion degrees of TTV, Warp and Bow for polished CZ and GCZ silicon wafers during the making processing. (from 300 pieces) (Chen et al., 2008)

As a detailed clarification for the mechanical strength of as-processed silicon wafers, the indention tests performed at room temperature followed by a high temperature annealing, which is considered to be one of the popular approaches for investigating the behaviors of dislocations in silicon wafers (Akatsuka et al., 1997; Fukuda & Ohsawa, 1992), were also adopted in our investigation. Fig. 1 shows the classical optical images of the indentation (as indented) and the rosette pattern of punched out dislocations (PODs) introduced by indentations (subjected to 1100°C/2h anneal) in GCZ silicon wafer. Herein, the POD diffusion length stands for the capability of mechanical strength of silicon wafers. From the rosette sizes shown in the GCZ silicon wafers with germanium doping (from the concentrations from 10^{16} to 10^{19} cm^{-3}) subjected to 1100°C/2h anneal (Chen et al., 2008), it could be found that the mechanical strength was improved by germanium doping. With the increase of germanium doping level, the POD diffusion length decreases, which should be ascribed to the intensive dislocation pin up effects by the micro-defects (such as small-sized oxygen precipitates).

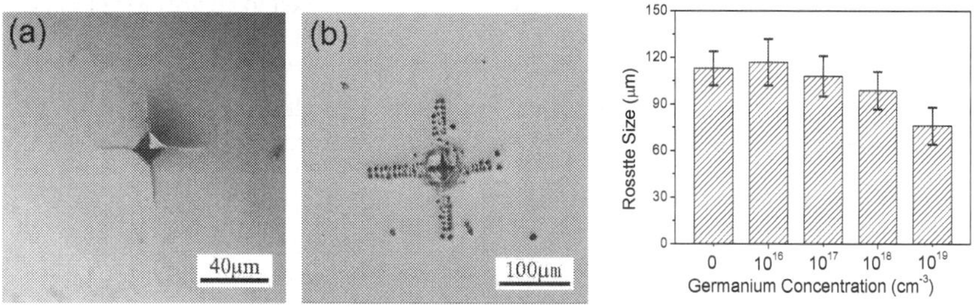

Fig. 1. Optical images of (a) the indentation and (b) the rosette pattern of PODs introduced by indentations in the GCZ silicon wafer subjected to 1100°C/ 2h anneal; (c) Rosette size in the GCZ silicon wafers with different germanium doping subjected to 1100°C/ 2h anneal (Chen et al., 2008).

During ULSI device fabrication, the mechanical strength during thermal processing affects the cracked-wafer breakdown yield and even the lithography accuracy. Considering this, the indentation tests on thermal treated silicon wafers have been studied via varied pre-annealing. Fig. 2 shows the optical images of PODs for the CZ and GCZ silicon, which were annealed at 800°C for 16 h or plus re-annealed at 1000°C for 4 h. Actually, the amorphous silicon and dislocations could be formed around the indentation positions at room temperature and then high stress could occur under a highly localized stress (Minowa &

Sumino, 1992). The amorphous silicon inclined to transform to the heavy dislocated crystalline silicon and the dislocations began to move so as to release the stresses when high temperature annealing was adopted. Herein, the travel distance of PODs in the GCZ silicon samples after 800°C/16h annealing was calculated to be somewhat shorter than that of the CZ silicon samples. Whereas, for 800°C/16h + 1000°C/4h annealing, the moving distance of PODs in the GCZ silicon sample seemed to be unambiguously shorter than that in the CZ silicon samples. And, these phenomena were consistent with the conclusions drawn from the fracture strength measurement (Chen et al., 2008).

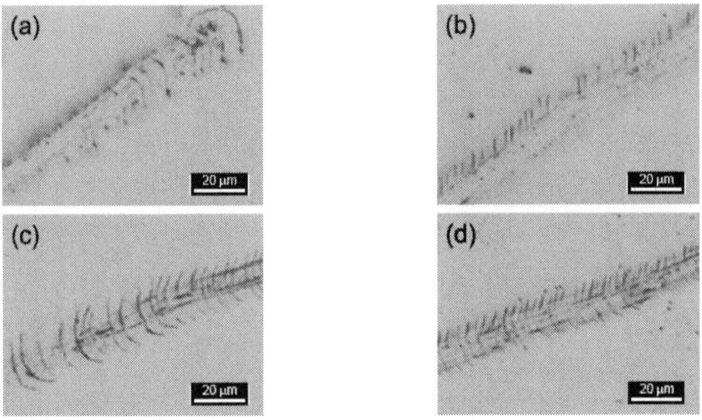

Fig. 2. Typical optical micrographs for the scratch-introduced CZ and GCZ silicon samples annealed at 1000°C/2.5 h. (a) CZ, 800°C/16h, (b) GCZ, 800°C/16h; (c) CZ, 800°C/16h + 1000°C/4h; (d) GCZ, 800°C/16h + 1000°C/4h.

Additionally, the influence of germanium doping levels in CZ silicon on the mechanical strength during device fabrication processing has been clarified by strain-stress checking. The rectangular-parallelepiped-shape samples of both the 2000 μm thickness normal CZ and GCZ silicon (GCZ2 and GCZ3, with the germanium doping level of 10^{17} and 10^{18} cm^{-3}, respectively) wafers were investigated after a pre-treated low-high temperature two-step thermal anneal (800°C for 16h + 1000°C for 4h). As can be seen from the typical stress-strain curves for both the CZ and GCZ silicon samples shown in Fig. 3, it is indicated that the higher content of germanium benefits the improvement of the critical fracture stress (Chen et al., 2008). It is considered that the strain field introduced by germanium doping might not directly lead to the suppression of dislocations, whereas, the germanium-doping-related small-sized but higher-density oxygen precipitates within the GCZ silicon can contribute to the excess of mechanical strengths compared to the normal CZ silicon wafers.

It is considered that, the light germanium doping with the concentration of 10^{16}-10^{19} cm^{-3} is expected to introduce the compressive strain field into silicon matrix due to the larger atom size of germanium. The strain fields would generally give rise to the retardation of dislocation movements due to the potential barrier related with the interaction between the dislocations and matrix. However, the geometrical influences induced by light germanium doping are too slight to retard dislocation mobilization. Instead, it is considered that germanium could combine with some point defects in CZ silicon, such as vacancy and /or interstitial oxygen, and seeds for oxygen precipitates of smaller sizes but higher density.

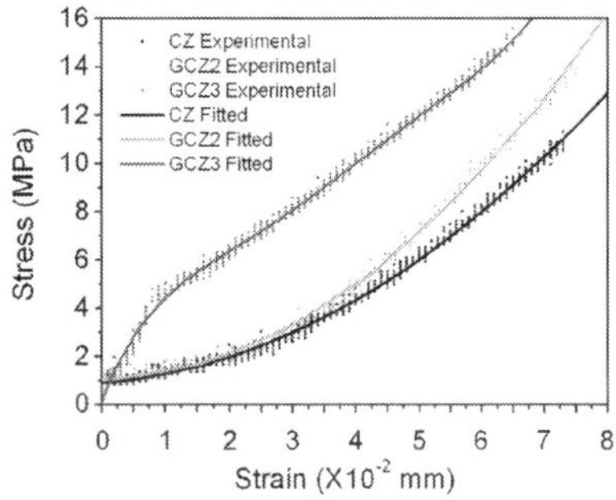

Fig. 3. Typical stress-strain curves for the CZ (CZ) and GCZ silicon samples (GCZ2 and GCZ3, with the germanium concentrations of 10^{17} and 10^{18} cm^{-3}, respectively) annealed at 800°C/16h + 1000°C/4h. (Chen et al., 2008)

Therefore, for both the grown-in case and the thermal treatment case, the oxygen precipitate nucleation at the sites of the dislocation cores could be enhanced by light germanium doping and the precipitates could act as the strong pinning complexes for the dislocation mobilization. In this viewpoint, it is reasonable to understand that the higher concentration of germanium atoms in CZ silicon could reduce the dislocation velocity and then decrease their moving distance.

3. Oxygen-related donors

Oxygen-related donors, including thermal donors (TDs) and new donors (NDs), which are believed to generate normally in the temperature ranges of 350-550°C (Fuller & Logan, 1957) and 600-700°C (Capper et al., 1977), respectively, can deteriorate the electrical properties of wafers. Impurities like germanium and nitrogen have been reported to retard TD formation (Hild et al., 1998). Based on the experimental facts, it is considered that germanium doping suppress the formation of TDs, but does not affect the microscopic structure of TDs, which suggested to be the result of the reaction of germanium with point defects (like silicon interstitial, boron, vacancy and interstitial oxygen dimer) in CZ silicon; whereas, the germanium doping could enhance the formation of NDs in CZ silicon, which is proposed as a process associated with the nucleation enhancement of oxygen precipitation by germanium doping.

In this section, a conventional CZ silicon and two GCZ silicon (GCZ1 and GCZ2, with the germanium concentrations of 10^{16} and 10^{18} cm^{-3} at the seed-ends, respectively) ingots were grown under almost the same conditions. Samples from different position of CZ and GCZ2 silicon ingots were annealed at 650°C for 30min to annihilate as-grown TDs. The resistivity of the annealed samples was measured by means of four-point probe, and the TD concentration ([TD]s) was converted from resistivity according to ASTM F723-88. Fig. 4 shows the distribution of the as-grown TD concentrations along the axial orientation in CZ and GCZ2 silicon crystals (Yang et al., 2004). Compared with the CZ silicon, the TD concentrations in the middle and the tail part of the GCZ2 silicon are much lower. The segregation coefficient of germanium in crystal is about 0.33, indicating that the germanium concentration would increase from the seed-end to the tang end of the crystal ingot. It is therefore believed that germanium suppresses the formation of TDs during crystal growth so that the TD concentration is lower in the tail. Furhtermore, the TD concentration variation in the GCZ1 ingot was similar to that in the GCZ2, which is further inferred that TDs are inhibited in GCZ silicon when germanium concentration is above $10^{16}cm^{-3}$.

In fact, it was also found that the TD concentrations in the GCZ samples are always lower than those in the CZ wafers during low temperature annealing. In our experiments, the samples were annealed at different temperatures from 350°C to 500°C for different time to investigate the suppression effect of germanium on TD formation. The TD concentrations of the CZ and GCZ2 samples were plotted as a function of annealing time, as shown in Fig. 5 (Yang et al., 2004). When annealed at 350°C or 500°C, there is nearly no change of the [TD]s in both the CZ and GCZ2 samples, meaning that almost no donors have been generated at these temperatures. When annealed at 400°C, [TD]s increased with the annealing time, however, the increase speed in the thermal donors in the GCZ2 is lower than that of the CZ samples. When annealed at 450°C, the [TD] variation speed is the most rapid one among all the anneal temperatures, while the [TD]s of the GCZ2 increases still lower than that of the CZ silicon. That is, germanium doping could suppress the formation of TDs.

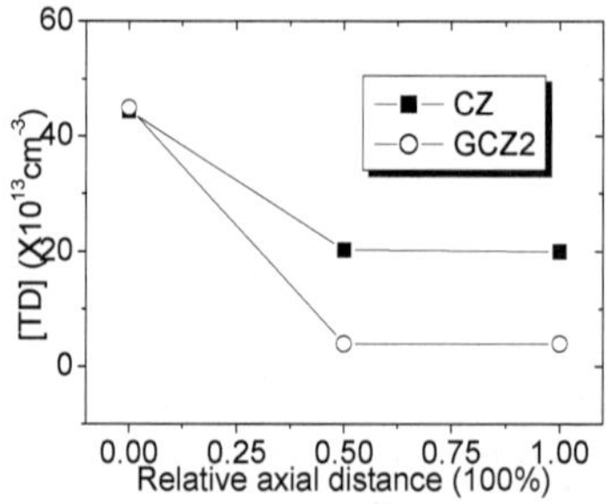

Fig. 4. Distribution of the as-grown [TD]s along the axial orientation in the CZ and GCZ2 silicon. (Yang et al., 2004)

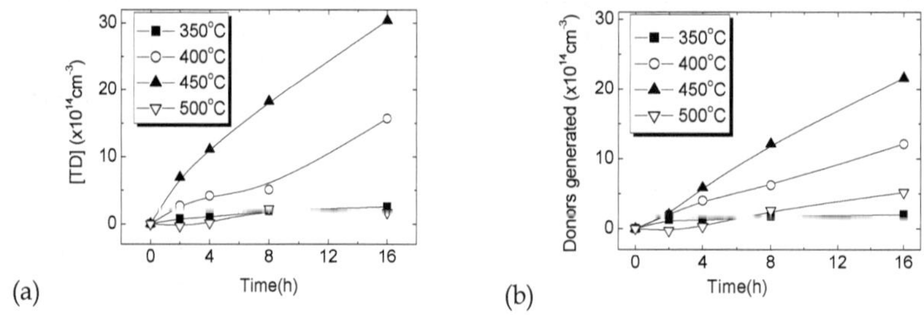

Fig. 5. TD concentrations of the CZ (a) and GCZ2 (b) samples as a function of annealing time (Yang et al., 2004).

The low temperature Fourie Tansmission Infrared (FTIR) absorption spectra of thermal donors (TDs) in GCZ silicon were found to be similar to the one in CZ silicon, but their density is different. Therefore, it is considered that light germanium doping suppresses the formation of TDs but does not affect the microscopic structure of TDs. Fig. 6 shows the low-temperature FTIR spectra of the CZ and GCZ samples in far- (350-650 cm^{-1}) and mid-IR (650-1200 cm^{-1}) range, respectively(Cui et al., 2006). As can be seen in Fig. 6(a), a series of individual FTIR absorption lines related to TDs in silicon are observed in both the CZ and GCZ silicon. These absorption lines were caused by the transitions of neutral TDDs into the conduction band at low temperature of 10K and different absorption lines correspond with different donor energy levels (Wagner & Hage, 1989), and the neutral donors in the GCZ sample have the same energy levels as those in the CZ sample. Meanwhile, from the low-temperature FTIR absorption spectra of the CZ and GCZ silicon in the range 650-1200 cm^{-1} illustrated in Fig. 5(b), the similar situation could be found. These series of FTIR absorption lines are reported to correspond with the singly ionized TDs (Wagner & Hage, 1989). It is obviously that the FTIR absorption spectrum of the singly ionized donors in the GCZ agrees quite well with that in the CZ silicon, but its density is much stronger. These results further confirm that the TDs in both the silicon samples are the thermal double donors (TDDs) with the same energy levels and microstructures. Therefore, it is considered that germanium doping in silicon suppresses the generation of TDs, but has little influence on their structures, which different from the results in heavily germanium content silicon, GeSi. In GeSi, the TDs were found to be broadbands in the FTIR spectra measured at low-temperature (Hild et al., 1998).

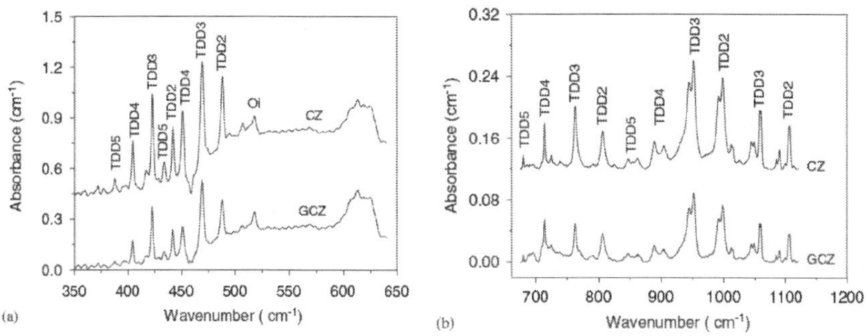

Fig. 6. (a) Low-temperature far-IR spectra of the CZ and GCZ silicon samples subjected to 650°C/30min + 450°C/4h annealing, (b) low temperature mid-IR spectra of the CZ and silicon GCZ samples subjected to 450°C/30 min + 450°C/4h annealing. The TDD_n refers to the *n*th (*n*=1-5) neutral donor in Fig. 6(a) and singly ionized donor in Fig. 6(b). (Cui et al., 2006)

When iso-electrical germanium atoms are incorporated into silicon lattice, they locate at substitutional sites and usually cause the increase of internal stress. During crystal growth, point defects could interact with germanium atoms. Vacancies incline to combine with germanium atoms to form Ge-V_n complexes, which have been identified by DLTS measurements in GCZ silicon crystals (Budtz-Jorgensen et al., 1998). We have clarified that germanium can enhance the nucleation of oxygen precipitation in the wide temperature range of 650-1200°C, which is based on the assumed Ge-O and Ge-O-V complexes. Normally, the TDs generated around 450°C is due to the aggregation of oxygen atoms (Kaiser et al., 1958). The molar volume of TDs is larger than that of silicon, thus, during the TD formation the lattice strain must be released by attracting free vacancies whose concentration is greatly decreased by the formation of Ge-V complexes. Meanwhile, the generation of TDs is a process of oxygen clustering, so that the interactions between the germanium and oxygen atoms together with the complexes of Ge-V with the fast diffusion O_{2i} dimmer will reduce the oxygen flux to form the smaller oxygen clusters during lower temperatures and therefore suppress the TD formation.

Considering the effect of germanium doping on NDs, it has ever been reported that germanium could suppress the formation of NDs (Babitskii et al., 1985) in heavily germanium doping cases, and it was also suggested that the generate rate of oxygen precipitates and NDs was lowered down by the lattice deformation caused by germanium doping in silicon (Babich et al., 1995; Babitskii et al., 1988). However, our investigation showed an opposite results in light germanium doping silicon materials.

Both the CZ and GCZ2 silicon were annealed at 650°C/128h and the ND concentrations ([ND]s) in the wafers as a function of the annealing time was drawn in Fig. 6 (Li et al., 2004b). The [ND]s for both the silicon increased with the annealing time at 650°C due to the formation of NDs. However, the ND formation rate in the GCZ2 sample is dramatically higher than that in the CZ one, so that the conductivity type reversed from p-type (all the original CZ silicon ingots are boron doped) to n-type after anneal for 128h in the GCZ2 silicon, meaning that large number of NDs have generated due to the enhancement of germanium on the ND formation. Besides, from the oxygen concentration variation of the annealed samples, it is found that more oxygen atoms have precipitated in the GCZ2 samples than in the CZ samples after 650°C/128h annealing. Generally, NDs are considered to be the bigger oxygen clusters compared to TDs, which are generally nuclei of oxygen precipitates during lower temperature anneal (Pensl et al., 1989). It is considered that, the enhanced ND formation by the germanium doping, is believed to be relative to the enhancement of oxygen precipitation. As germanium can enhance the nucleation of oxygen precipitates based on Ge-O complexes, some precipitate nuclei might become NDs. Thus, it is reasonable to suggest that most of these denser small oxygen precipitate nuclei become NDs with electrical activity at 650°C anneal. However, when germanium concentration is much larger than oxygen concentration, most of oxygen will be trapped by germanium to form Ge-O complexes, resulting in the reduction of oxygen flux to form NDs. Therefore, the formation of NDs will be suppressed, which was reported by Babitskii's work (Babitskii et al., 1985).

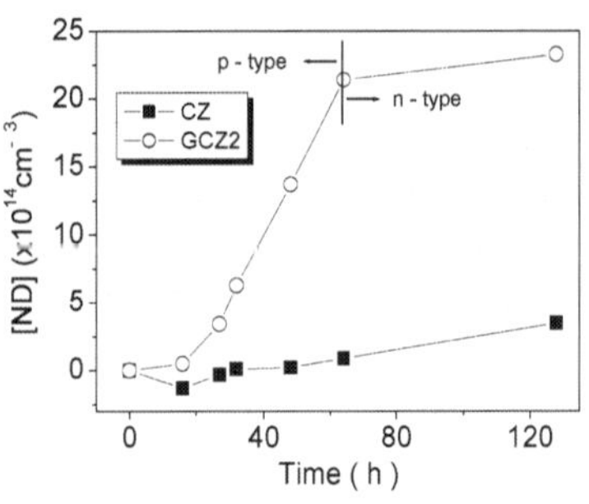

Fig. 7. ND concentration in the CZ and GCZ2 wafers annealed at 650°C as a function of the annealing time (Li et al., 2004b).

4. Oxygen precipitation

Oxygen precipitates, the main micro-defects in CZ silicon especially used for the bulk isolated devices in early years, could not only deteriorate the electrical properties itself but also induce the secondary defects such as stacking faults and dislocations which increase the breakdown current of devices. However, oxygen precipitates with suitable density in bulk benefit both for the improvement of mechanical properties and for the enhancement of internal gettering capacitance for wafers. The super-saturated interstitial oxygen atoms in CZ silicon will accumulate to form grown-in oxygen precipitates due to post-anneal in crystal pullers, resulting in so called as-grown oxygen precipitates. It is also widely accepted that the supersaturated oxygen atoms in silicon matrix can precipitate and further induce secondary defects, so-called bulk micro-defects (BMDs), within device fabrication processes. Oxygen precipitates as well as BMDs are believed to be the gettering sites for metallic contamination. Thus, normally, it is required to control the concentration and distribution of oxygen precipitates in silicon bulk so that the optimum comprehensive effects benefit the quality of CZ silicon material.

The germanium doping in CZ silicon is found to enhance not only as-grown oxygen precipitation but also oxygen precipitation during successive thermal anneals within a large temperature range; and it could also vary both the distribution situations of BMDs and microscopic morphology of oxygen precipitates, resulting in poor thermal stability of oxygen precipitates at elevated high temperatures. We consider that a certain complexes, the so called germanium-related complexes, could be generated in the GCZ silicon and thus change the behavior of oxygen precipitates in GCZ silicon.

A CZ and two GCZ (GCZ1 and GCZ2 with [Ge] ~10^{16} and 10^{17} cm^{-3}, respectively) silicon ingots with the comparable initial oxygen concentration have been selected to investigate the formation of grown-in oxygen precipitation: after annealing at 1270°C/2h to annihilate the thermal history, both the CZ and GCZ silicon were cooled down by a controlled rate of 0.5°C/min and were taken out at 1150-850°C separately. The reductions of [O_i] (Δ[O_i]s) in the CZ and GCZ samples as a function of the taking out temperatures is shown in Fig. 8(a) (Chen et al., 2006b). Generally, the thermal history of wafers can well influence the oxygen precipitation of CZ silicon during the successive annealing, while grown-in precipitates can be dissolved when annealed at considerably high temperatures above 1250°C (Kishino et al., 1982). The Δ[O_i] variation of the CZ and GCZ silicon annealed at 1270°C/2h is shown in Fig. 8(b)(Chen et al., 2006b). It can be seen that the ratio of increased [O_i] and as-received [O_i] in the CZ silicon before and after annealing is a bit smaller than that of the GCZ silicon, which indicates the grown-in precipitates in the GCZ silicon is more than those in the CZ silicon. It is considered that germanium enhances the formation of grown-in oxygen precipitation during crystal growth. From Fig. 8(a), it can be also found that the [O_i]s of the GCZ2 silicon decreased much more dramatically than that of the CZ silicon in the whole temperature range and that the GCZ1 silicon decreased more slightly than that of the CZ silicon below 1050°C, indicating that oxygen can precipitate more easily in the GCZ silicon crystals, even at the temperatures higher than 1150°C.

Another ramping-up processing was also performed to investigate the effect of germanium on as-grown oxygen precipitation in GCZ silicon. Samples were annealed at a heating rate of 1°C/min starting at 750°C, 850°C, 950°C or 1050°C, and ending at 1050°C with a isothermal anneal for 16 h,. The Δ[O_i]s as a function of the starting ramping temperature is shown in Fig. 9 (Chen et al., 2006b). It is believed that 1°C/min is a suitable heating rate to grow up oxygen precipitate nuclei, if their radius is larger than the critical nucleation radius

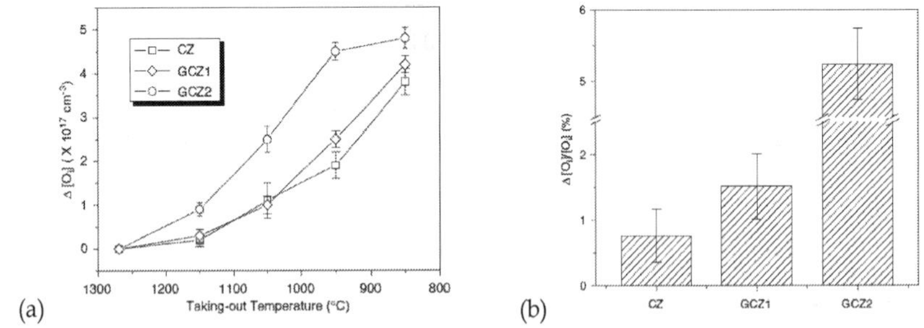

Fig. 8. (a) Evolution of the Δ[O_i]s in the CZ and GCZ silicon after 1270°C/2h pre-anneal as a function of the taking out ramping temperatures during the cooling-down process, (b) Δ[O_i] of the CZ and GCZ silicon before and after annealing at 1270°C/2h. (Chen et al., 2006b)

of oxygen precipitates (r_c) at the starting temperatures of the ramping process, while the formation of new oxygen precipitate nuclei during the ramping is suppressed(Kissinger et al., 1998). Accordingly, the precipitated oxygen after 1050°C/16h anneal ramped from different starting temperatures are considered to be roughly related to the grown-in oxygen precipitates whose radius is larger than r_c at the starting ramping temperature in the corresponding grown-in crystals. Thus, with the increase of starting temperatures, the amount of grown-in precipitates larger than the r_c decreases, which results in the reduction of oxygen precipitates. As can be seen, the decreased Δ[O_i]s of the GCZ wafers was much more than that of the CZ wafers at every starting temperatures, which indicates that more grown-in oxygen precipitates have been generated in the GCZ wafers than that in the CZ wafers in the investigated temperature range (between 850 and 1050°C). Meanwhile, the [O_i] curve of the GCZ2 wafer moved rightward relative to that of the CZ one as shown in Fig. 9. In this case, there is a sharp decrease of [O_i] when the temperature below 950°C, which means the radius of majority of as-grown oxygen precipitates in the GCZ wafers was smaller than r_c at 950°C, while that of most as-grown oxygen precipitates in the CZ wafer was smaller than r_c at 850°C. It is therefore believed that the germanium incorporation increases the forming temperatures of as-grown oxygen precipitation during the cooling-down process of crystal growth, and thus larger as-grown oxygen precipitates could be presented in GCZ silicon when the cooling-down processing completed.

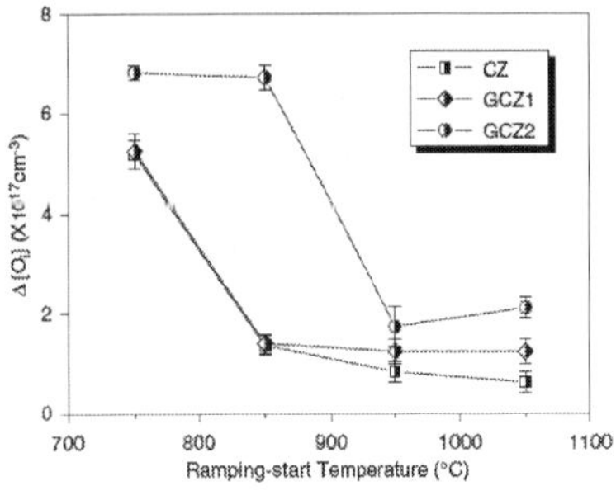

Fig. 9. Evolution of Δ[O_i]s in the CZ and GCZ silicon as a function of the starting ramping temperatures in the ramping process with 1°C/min ramping-up rate.(Chen et al., 2006b)

It is also suggested that germanium doping could enhance the oxygen precipitation in CZ silicon wafer in a large temperature range (from 550 to 1050°C) during successive annealing. The as-received samples of both the CZ and GCZ silicon were put into a diffusion furnace at 550-950°C at every 100°C and then isothermally annealed for 2-64h, following by annealing at 1050°C for 16h. All the thermal treatments were preformed in an argon atmosphere. Fig. 10 shows the $\Delta[O_i]$s in the CZ and GCZ silicon annealed for 64 h as a function of the annealing temperatures (Chen et al., 2006a). As can be seen, the $\Delta[O_i]$s in both the CZ and GCZ silicon samples annealed at above 850 °C were larger. Moreover, the amount of $\Delta[O_i]$ was always larger in the GCZ silicon than in the CZ silicon. It is therefore suggested that germanium doping could enhance oxygen precipitation in CZ silicon in a wide temperature range.

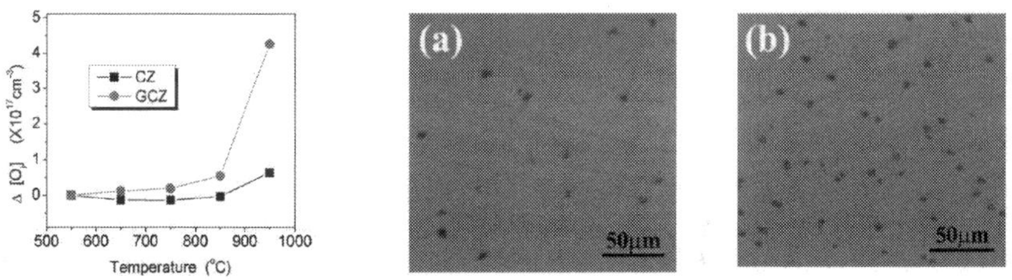

Fig. 10. Left: $\Delta[O_i]$s in the CZ and GCZ silicon annealed for 64 h as a function of annealing temperatures. Right: Optical micrographs of the BMDs in the CZ (a) and GCZ (b) silicon subjected to 950°C/64 h anneal. (Chen et al., 2006a)

Normally, oxygen precipitate growth is limited by oxygen diffusion especially at low temperatures. (Joly & Robert, 1994) When annealed at low temperatures (such as 750 °C or below), the diffusivity of oxygen is considerably small, thus the growth of oxygen precipitates is not remarkable. However, a part of oxygen atoms can still aggregate into precipitate nuclei and embryos, so that the $\Delta[O_i]$s in the GCZ silicon subjected to 64h anneal at 750 °C is somewhat larger than that in the CZ silicon. It is thus believed that the formation of precipitate nuclei is enhanced by germanium doping. When the silicon wafers were annealed at higher temperatures (such as 950°C and above), the oxygen diffusion coefficient greatly increased, while the supersaturation of interstitial oxygen in silicon crystal decreased. In this case, oxygen precipitation in silicon was primarily based on the as-grown precipitate nuclei. (Borghesi et al., 1995) The typical optical micrographs of BMDs induced by oxygen precipitates in the CZ and GCZ silicon samples subjected to 950°C/64 h anneal are also shown in Fig. 10. It can be clearly seen that the BMD density was much higher in the GCZ silicon than in the CZ silicon. It is generally believed that the grown-in oxygen precipitates have a size distribution following the *Boltzmann*'s statistics. Only the oxygen precipitates with radius larger than r_c at the annealing temperatures can survive and further grow up. Accordingly, the density of the grown-in oxygen precipitates with r_c at 950 °C is much higher in the GCZ silicon than that in the CZ silicon. That is, again, germanium doping can enhance the formation of larger grown-in oxygen precipitates during crystal growth.

Furthermore, if as-grown oxygen precipitates were eliminated by high temperature annealing, oxygen precipitation in GCZ silicon wafers during successive thermal cycles could still be enhanced by germanium doping. Fig. 11 shows the $\Delta[O_i]$s of the thermal-history-eliminated CZ and GCZ silicon wafers subjected to the 1050-1150°C/2h anneal. As can be seen, a bit larger oxygen precipitates could be generated in the GCZ silicon wafer than in the CZ silicon, which should be ascribed to the presentation as nucleation embryos of the germanium-related complexes formed in the GCZ silicon.

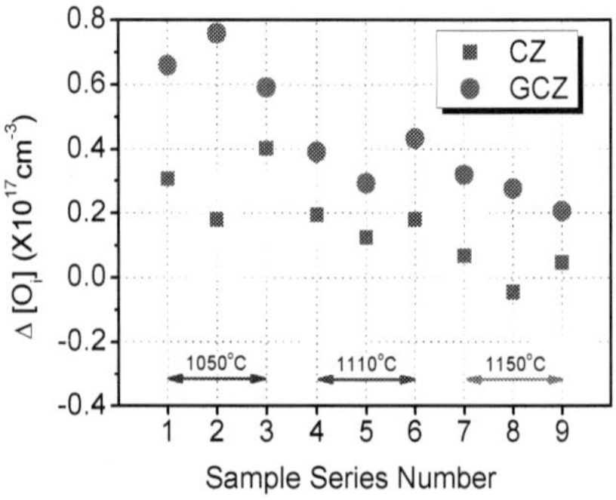

Fig. 11. $\Delta[O_i]$s of the thermal-history-eliminated CZ and GCZ silicon wafers subjected to the 1050-1150°C/2h anneal.

Ascribed to the enhancement of oxygen precipitation nucleation at low temperatures by germanium doping, oxygen precipitation during successive annealing processing will be no doubt enhanced. Fig. 12 shows the comparison of $[O_i]$s and the optical micrographs of BMDs, which corresponds to oxygen precipitates and induced defects, in the CZ and GCZ silicon subjected to the two-step anneals consisted of different low temperature pre-annealing plus the same high temperature anneal at 1050°C (Chen et al., 2006a). In the experiments, the precipitate nuclei subsisted after the prior annealing would coarsen during the subsequent high temperature annealing since the oxygen precipitate is characteristics of growth at high temperatures. As can be seen in the left of Fig.12, with the pre-anneal at 650 and 750 °C, the $[O_i]$s in the CZ silicon decreased almost to the oxygen solubility at 1050°C, while the $[O_i]$s remained at a bit higher levels in the GCZ silicon. Correspondingly, the dense BMDs in larger sizes were formed in the CZ silicon while denser BMDs in smaller sizes were generated in the GCZ silicon, as shown in Figs. 12R(a) and 12R(b). With the 850°C pre-anneal, the $[O_i]$ in the CZ silicon remained at the level much higher than the oxygen solubility at 1050 °C, while the $[O_i]$ in the GCZ silicon was much lower. Moreover, low density BMDs in smaller sizes were formed in the CZ silicon, while high density BMDs in larger sizes were formed in the GCZ silicon, which are illustrated in Figs. 12R(c) and 12R(d). Consequently, it is illuminated that oxygen precipitation is greatly enhanced by germanium-doping during low-high two step annealing.

Generally, the nuclei of oxygen precipitates formed at lower temperatures have a size distribution and not all of them can survive in subsequent thermal cycles. That is, the nuclei with smaller size will dissolve while those with larger size will grow up. As shown in Figs. 12R(a) and 12R(b), denser BMDs in smaller sizes were generated in the GCZ silicon in comparison with those in the CZ silicon, which is probably due to the much more nuclei formed at the lower temperatures by germanium-doping. Actually, high density of nuclei in the GCZ silicon was in a competition to attract interstitial oxygen atoms. Therefore, oxygen precipitation was to a certain extent retarded in the 1050°C anneal for the GCZ silicon and

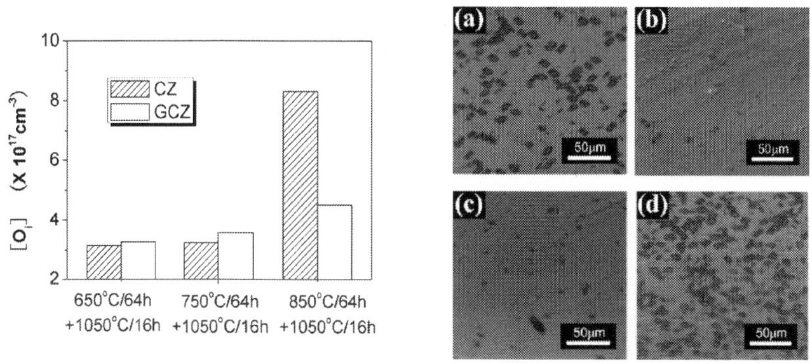

Fig. 12. Left: $[O_i]$s in the CZ and GCZ silicon subjected to 1050°C/16h anneal following 64h pre-anneals at different temperatures of 650-850°C. Right: Optical micrographs of BMDs in the CZ and GCZ silicon subjected to two-step anneals: (a) CZ, 650°C/64h+1050°C/16h, (b) GCZ, 650°C/64h+1050°C/16h, (c) CZ, 850°C/64h+1050°C/16h,, and (d) GCZ, 850°C/64h+ 1050°C/16h. (Chen et al., 2006a)

the $[O_i]$ reduced while the BMD density increased in the GCZ silicon when the annealing duration was prolonged. That is, germanium-doping could greatly enhance the nucleation for oxygen precipitate at low temperatures, especially below 750 °C. For 850°C/64h pre-anneal case, only the oxygen precipitates whose sizes are larger than the r_c at 850 °C could survive and further grow up in the subsequent 1050°C anneal. Most of the grown-in oxygen precipitates in the CZ silicon is smaller than r_c at 850 °C, thus the oxygen precipitation in 1050°C/16h anneal is slight. Whereas, from Figs. 12R(c) and 12R(d), the oxygen precipitate nucleation is enhanced in the GCZ silicon during the 850°C/64h anneal. Consequently, germanium-doping can increase the onset temperature up to 850°C for precipitate nucleation in the GCZ silicon, while, it is usually below 750 °C in CZ silicon. Furthermore, it is considered that the critical radius r_c at 850 °C is reduced by germanium-doping, and the oxygen precipitates with smaller radius could generate and survive in the GCZ silicon.

The morphology of oxygen precipitates in GCZ silicon is different from the ones in CZ silicon after different thermal treatments. Fig. 13 shows the transmission electron microscopy images of the oxygen precipitates and induced defects in CZ and GCZ samples subjected to 800°C/225h and 1000°C/225h anneal respectively (Yang et al., 2006b). After prolonged anneal at 800°C, platelet precipitates were typical in the CZ silicon, while particle precipitates besides platelet ones were also generated in the GCZ silicon [Figs. 13(a) and 13(b)]; however, after 1000°C/ 225h annealing, the oxygen precipitates generated in the CZ silicon are mainly in polyhedral morphology, while entangled and mixed morphologies consisting of polyhedral and platelet were formed in the GCZ silicon [Figs. 13(c) and 13(d)]. It is reported that the platelet oxygen precipitates could be dissolved easier than the polyhedral ones (Shimura, 1994), thus, the microscopic morphology's variation of oxygen precipitates in GCZ silicon could decline their thermal stability at high temperatures.

Fig. 14(a) shows the $[O_i]$s for both the CZ and GCZ silicon wafers in the statuses of as-grown, after the 1270°C/1h conventional furnace annealing (CFA) and after the 1280°C/60s rapid thermal annealing (RTA) treatment, respectively. The $[O_i]$s after CFA or RTA treatments for both the CZ and GCZ silicon were higher than that for the as-grown ones, ascribing to the dissolution of grown-in oxygen precipitates. The germanium doping in CZ silicon could decline the thermal stability of grown-in oxygen precipitates by generating the platelet shape precipitates. Furthermore, the concentration of oxygen in both the CZ and GCZ silicon as a function of duration at RTA preformed at 1260°C with pre-annealing at 800-1000 °C for 225h has been shown in Fig. 14(b). As can be seen, the $[O_i]$s ia recovered and is slightly higher in the GCZ than in the CZ silicon, indicating the easier dissolution of oxygen precipitates in the GCZ silicon.

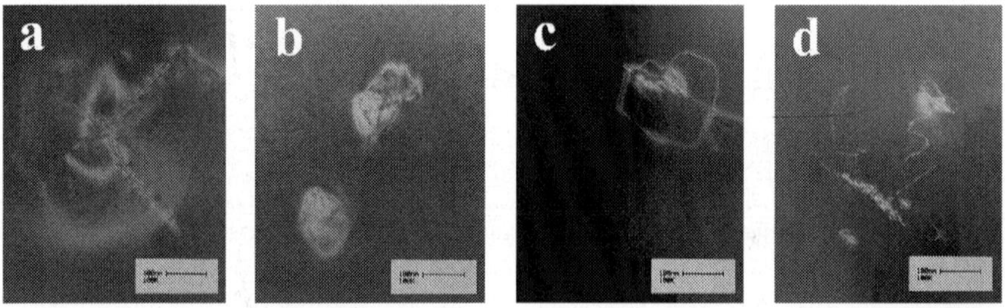

Fig. 13. Transmission electron microscopy images of the oxygen precipitates in the annealed CZ and GCZ silicon. (a) CZ, 800°C/225h, (b) GCZ, 800°C/225h, (c) CZ, 1000°C/225h, (d) GCZ, 1000°C/225h. (Yang et al., 2006b)

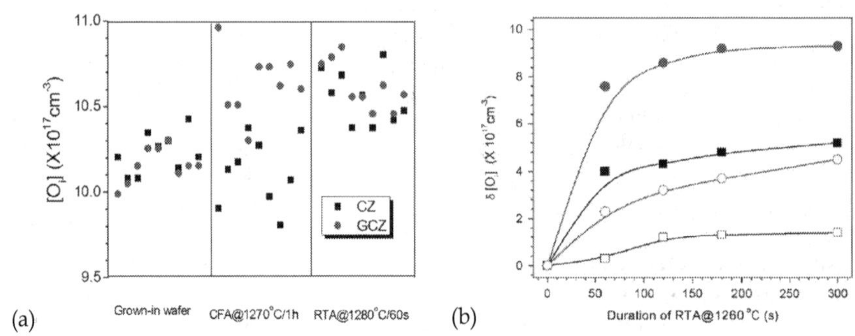

Fig. 14. (a) $[O_i]$s in both the CZ and GCZ silicon wafers with as-grown status, conventional furnace annealing (CFA) at 1270 °C/1h and rapid thermal annealing (RTA) at 1280°C/60s. (b) Variation of oxygen recoveries ($\delta[O_i]$s) in both the CZ (square points) and GCZ (circle points) silicon as a function of duration at RTA preformed at 1260°C with pre-annealing at 800°C (full points) or 1000 °C (open points) for 225h (Chen et al., 2007d)

Germanium atoms locate at the substitutional sites in CZ silicon, and induce distortion and local stresses in silicon lattice due to their larger atom radius. So, the lattice sites where germanium atoms locate are provided with potential activities and inclined to interact with other structural defects and / or impurities. Ge-V_m or Ge-V_m-O_n (m, n≥1) complexes, in the great amounts, are supposed to form for relieving the lattice stresses, and they could further act as heterogeneous precipitate nuclei to accumulate interstitial oxygen atoms in GCZ silicon. Due to the limit of oxygen content, the oxygen precipitates in GCZ silicon inclines to present with much smaller size than that in CZ silicon. It is said that vacancies in CZ silicon could be gathered by germanium atoms to generate germanium-vacancy-related complexes and thus benefit the generation of polyhedral precipitates, so that the oxygen precipitates could be presented as mixed morphologies in GCZ silicon. Normally, when subjected to the high temperature treatments, the inner Si-O and Si-Si bonding in the oxygen precipitates can be easily cracked and the oxygen atoms situated in the precipitate originally could revert to interstitial oxygen atoms and finally diffuse out the precipitates. Ascribed to the distribution of smaller-sized and higher-density precipitates, the total surface area of oxygen precipitates in GCZ silicon can be dramatically heightened. The net oxygen flux out of precipitates is enhanced and the precipitates can be therefore dissolved easier in GCZ silicon.

5. Void defects

Voids, the main micro-defects in modern large diameter silicon crystal, play more important roles in the reliability and yield of ULSI devices. It is well established that voids, especially those locate in the near-surface region of wafers, can deteriorate gate oxide integration (GOI) and enhance the leakage current of metal-oxide-semiconductor devices (Huth et al., 2000; Park et al., 2000). As a result of the agglomerations of excess vacancies during crystal growth, it is believed that voids are normally of an octahedral structure, about 100-300 nm in size and with a thin oxide film of about 2nm on their {111} surfaces (Itsumi et al., 1995; Yamagishi et al., 1992). It has been reported that during cooling-down process of silicon crystal from the melting point to room temperature, grown-in voids are formed with densities between 10^5-10^7cm^{-3} (Yamagishi et al., 1992).

The techniques to control voids have been studied extensively over years, and three different ways to achieve this have been widely accepted: 1) thermally controlled CZ silicon crystal growth (Voronkov, 1982), 2) high-temperature annealing (Wijaranakula, 1994) and 3) nitroge doping (Yu et al., 2002). It is believed that the GOI failure of devices can be improved by germanium doping. The characteristics of the grown-in voids in GCZ wafers, including flow pattern defects (FPDs) and crystal originated particles (COPs) [two main formations of void defects], suggested that germanium can suppress larger voids, resulting in denser and smaller voids. Meanwhile, it has been found that the density of voids can be decreased by germanium doping and then can be eliminated easily in GCZ silicon crystals through high temperature annealing.

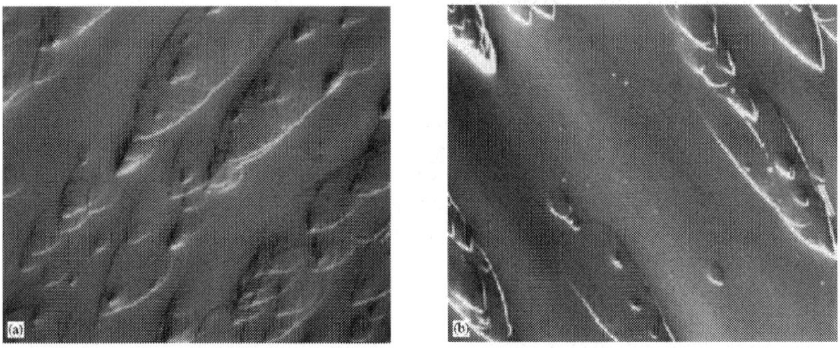

Fig. 15. Optical microscopic photographs of FPDs in the head samples of (a) CZ and (b) GCZ silicon crystal. (Yang et al., 2002)

Three *p*-type GCZ silicon crystal ingots with different germanium concentrations ([Ge]s) ($10^{15}cm^{-3}$, $10^{16}cm^{-3}$ and $10^{17}cm^{-3}$ in the head portions while/and $10^{16}cm^{-3}$, $10^{17}cm^{-3}$ and $10^{18}cm^{-3}$ in the tail portions and were named as GCZ1, GCZ2, and GCZ3 silicon, respectively) and a conventional CZ Silicon crystal were pulled under almost the same growth conditions. Typical optical microscopic photographs of FPDs in the head portion samples of the CZ and GCZ3 silicon crystals are shown in Fig. 15 (Yang et al., 2002). The FPD density in the GCZ3 silicon wafer was much less than that in the CZ silicon crystal. Similar results were also found in the tail samples. It can accordingly be concluded that germanium doping could significantly suppress the voids in GCZ silicon crystals. The FPD densities in the as-grown silicon wafers sliced from different portions of the four ingots are shown in Fig. 16 (Yang et al., 2002). As can be seen, the FPD densities in the head samples of the CZ, GCZ1 and GCZ2 silicon wafers were almost the same, while that of the head sample of the GCZ3 with a relatively higher [Ge] of $10^{17}cm^{-3}$ was much lower. For the CZ silicon crystal, the FPD density of the tail sample was almost the same as that of the head sample. However, for the GCZ1, GCZ2 and GCZ3 silicon crystals, the FPD densities of the tail samples were less than those of the head. Due to the segregation coefficient of germanium in silicon crystal is 0.33, [Ge] in the tail portion of the GCZ silicon is believed to be higher than that in the head portion. It is therefore clear that the FPD densities in the GCZ silicon wafer decreased with the increase of [Ge], and the FPD density in the grown-in GCZ silicon wafer is much less than that in the conventional CZ wafer. Germanium doping in CZ silicon could significantly suppress voids during crystal growth.

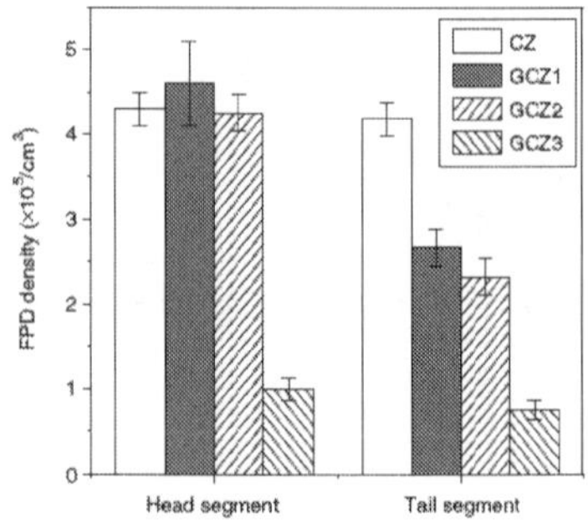

Fig. 16. FPD densities in the head and tail portions of the as-grown CZ and GCZ silicon crystals samples with different germanium concentrations. (Yang et al., 2002)

Furthermore, it is suggested that the thermal stability of FPDs in GCZ silicon is much poorer than that in CZ silicon. Fig. 17 indicates the FPD densities in both the CZ and GCZ silicon samples before and after different annealing. As can be seen, after the 1050°C/2h annealing, the FPD density in the GCZ silicon is significantly reduced, while that in the CZ silicon crystals remains almost constant. Although the FPD density in the CZ silicon wafer decreased to a considerable extent after 1150°C/2h annealing, it was still much higher than that in the GCZ1 wafer. However, after 1200°C/2h annealing, the FPD densities in both the CZ and GCZ1 silicon wafers decreased to nearly the same level. The prolonged annealing at high temperatures has no notable effect on the annihilation of FPDs. That is, the FPDs in the GCZ silicon crystals can be annihilated at lower temperatures than those in the CZ crystal, implying the thermal stability of voids in the GCZ silicon crystals is much poorer, i.e., the voids in the GCZ silicon crystals can be eliminated by high temperature anneals with a low-cost heat budget.

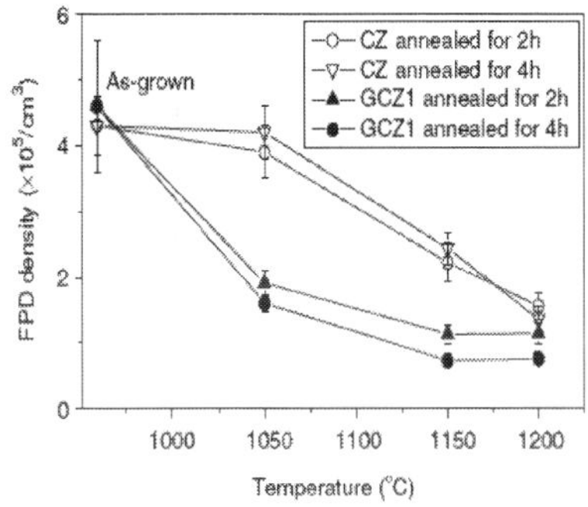

Fig. 17. FPD densities in both the CZ and GCZ silicon samples before and after different high temperature annealing. (Yang et al., 2002)

Fig. 18 shows the size profiles of grown-in COPs in both the CZ and GCZ silicon wafers (Yang et al., 2006a). As can be seen, an increase in the percentage of COPs which are smaller (0.11-0.12 μm), and a decrease in the percentage of COPs which are larger (over 0.12μm) in the GCZ silicon wafers compared to those in the CZ silicon wafer has been suggested. The total amount of grown-in COPs on the GCZ silicon wafers was actually more than that on the CZ wafers, meaning germanium doping could induce a higher density of COPs generated with smaller sizes. As noted, the evolution of COPs in as-grown GCZ silicon seems not to coincide with the result given by FPDs detection. It is worthwhile to point out that the FPDs are believed to be deduced by larger voids, *i.e.*, only those whose radius is larger than the critical radius r_c can bring enough hydrogen bubbles to etch wafer surface and leave flow patterns. Suggested by the results of COPs detection, the quantity of larger voids in GCZ silicon crystals is less than that in CZ silicon. Therefore, it is reasonable to conclude that the fewer FPDs in the GCZ silicon samples is associated with the lack of larger voids while the higher density COPs on the GCZ silicon wafers is mainly contributed by smaller size voids.

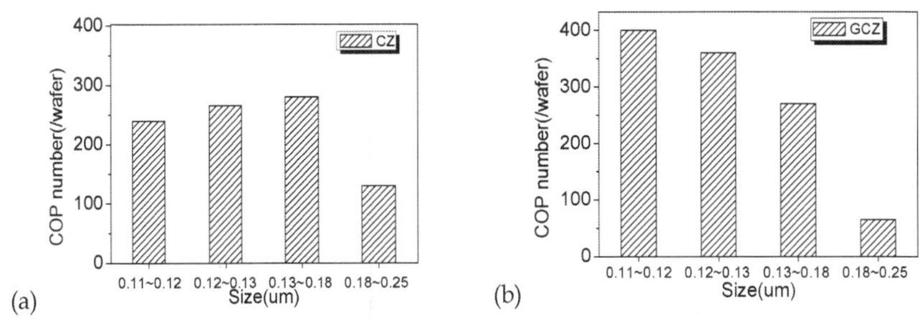

Fig. 18. Density and size profiles of the COPs on (a) CZand (b) GCZ S ilicon wafers. (Yang et al., 2006a)

Similar with the FPDs, poorer thermal stability of COPs could be also detected. Fig. 19 shows the COP maps for both the CZ and GCZ silicon wafers sampled from the tail portions of the crystals before and after annealing in hydrogen at 1200°C (Yang et al., 2006a). COP density on the GCZ silicon was much lower than that on the CZ silicon after the annealing, indicating that the COPs on CZ silicon wafer can be annihilated more easily by germanium doping. Actually, at the subsurface (such as at the depth of 30μm) in the annealed wafers, it was also found that more grown-in COPs were annihilated on the GCZ silicon wafers than on the CZ ones. Also, from the comparison of COP densities of the CZ and GCZ silicon annealed in Ar or H_2 atmosphere shown in Fig. 20 (Chen et al., 2007a), it could be found that germanium doping could reduce the thermal stability of grown-in COPs not only on the surface but also in the bulk of the GCZ silicon wafers. Consequently, it is suggested that germanium doping could effectively deteriorate the thermal stability of grown-in COPs on wafers.

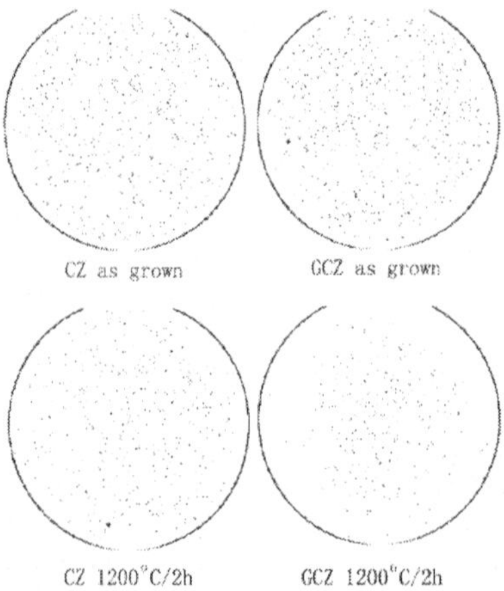

Fig. 19. COP maps of the CZand GCZ silicon wafers before and after annealing in hydrogen at 1200°C. (Yang et al., 2006a)

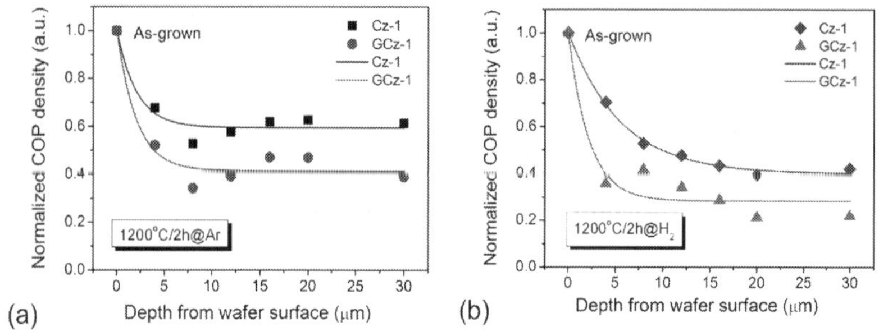

Fig. 20. Normalized COP densities of the CZand GCZ silicon wafers annealed in (a) Ar or (b) H_2 atmosphere as a function of the depth from the wafer surface. Notice that the curves were fitted following exponential growth method. (Chen et al., 2007a)

Herein, we discuss on the mechanism of germanium doping on void defects by forming germanium-related complexes. It is considered that, germanium atoms can react with the intrinsic point defects in CZ silicon crystals, so that the formation of vacancy-based micro-defects, such as P-band and voids, will be influenced by germanium doping. Meanwhile, the germanium atoms located at substitutional sites of silicon lattice cause lattice distortion and lattice stress. To relieve the lattice stress, germanium inclines to react with vacancy and/or oxygen to form Ge-V_m or Ge-V_n-O_m (m, n≥1) complexes when GCZ wafers are annealed at high temperatures, and that the complexes would survive at low temperatures and become the nuclei of oxygen precipitates. Thus, prior to the nucleation of voids, the nuclei of oxygen precipitates can grow by the rapid diffusion of oxygen and absorption of a considerable number of vacancies at high temperatures. Accordingly, the number of surviving vacancies contributing to the formation of voids during the subsequent cooling is reduced.

The driving force for void formation is the gain in volume free energy per vacancy associated with vacancy super-saturation, *i.e.*, the vacancy chemical potential f (Voronkov & Falster, 1998):

$$f = k_B T \log\left(\frac{C_0}{C_e}\right) \tag{1}$$

where k_B is *Bolztman*'s constant, T is the void nucleation temperature, C_e is the equilibrium vacancy concentration, and C_0 is the initial vacancy concentration (the actual vacancy concentration in as-grown silicon). From equation (1), it can be found that the void nucleation temperature T will be lower when the initial vacancy concentration C_0 is reduced by germanium doping in CZ silicon crystal. Therefore, voids, especially for those with large volume voids which are believed to be the origin of FPDs, are suppressed in as-grown GCZ silicon crystal. This can also explain the fact that the FPD density decreases with the increase of germanium concentration shown in Fig. 16. Additionally, the voids could be formed during lower temperature annealing because of the plentiful vacancy consumption caused by the formation of the germanium-related complexes, which is illustrated in Fig. 18. In fact, when binding temperature of germanium and vacancies T_b is higher than nucleation temperature of voids T_n, the void formation will be strongly or completely suppressed, due to a lack of free vacancies (Voronkov & Falster, 2002). Because T_b is probably higher than T_n, the void formation will be suppressed due to the decrease in free vacancies which results in the decrease of C_0. According to *Voronkov*'s results, the density N and size R (assuming the voids to be spheres in silicon lattice and the radius R standing for their size) of voids in CZ silicon crystals accord with the relational expression as follows:

$$N = \left(\frac{1.72}{4\pi m^*}\right)\left(\frac{qE^*}{Dk_B T^2}\right)^{\frac{3}{2}}\left(\frac{2C_0}{\rho}\right)^{-\frac{1}{2}} \tag{2}$$

$$R = 1.35(m^*)^{\frac{1}{3}}\left(\frac{C_0 D k_B T^2}{qE^*}\right)^{\frac{1}{2}} \tag{3}$$

From which, one could conclude that the N and R of voids is direct proportional to the initial vacancy concentration C_0. Therefore, the formation of lower density FPDs and denser

COPs with smaller size were believed to be enhanced in GCZ silicon crystals, due to the decrease of the initial vacancy concentration C_0, as well as the decrease of the formation temperature T of voids. Furthermore, higher germanium concentration in CZ silicon benefits the higher COP density, thus the COP density in the tail portion is higher than that of the head and middle portion of the GCZ silicon crystals, which is shown in Fig. 16.

Moreover, voids in CZ silicon usually form in a narrow temperature range about 30°C below 1100°C during crystal growth. They could be annihilated especially in hydrogen gas during elevated temperatures annealing due to dissolving the inner oxide films surrounding voids. The removal of oxide films on the inner walls of grown-in void defects is believed to be the first step in the reduction process, which is an oxygen diffusion-determined process (Adachi et al., 2000). Then the second step is the shrinkage of voids through the diffusion of vacancies, which is a diffusion-determined process. For GCZ silicon crystal, due to the decrease of void formation temperature T and the increase of void density N, the thickness of inner oxide film of voids in GCZ silicon crystals might be thinner than that in CZ silicon; additionally, the volume of voids in GCZ silicon crystals is considered to be smaller than that in CZ silicon. Therefore, the voids in GCZ silicon could be dissolved by thermal cycles easier comparable to those in CZ silicon.

6. Application of germanium doped Czochralski silicon: two examples

6.1 Thick epitaxial layers on germanium doped CZ silicon substrate

Misfit dislocations (MDs) would lead significant junction leakage into transistors, while the generation of MDs is still a serious issue in the volume fabrication of p/p+ epi-wafer to date. It has been suggested that germanium doping can suppress the epi-layer MDs on high boron doped CZ silicon substrates (Jiang et al., 2006). A 50μm thick p/p^+ epi-wafers were grown on the conventional heavily boron-doped (B-doped) substrate and germanium boron co-doping (Ge-B-co-doped) silicon substrates. The germanium content in the CZ silicon is calculated aiming to balance the stress induced by boron doping. However, in principle, the co-doping of germanium and boron in CZ silicon substrate can be tailored to achieve misfit dislocation-free epi-layer with required thickness. It is therefore expected that this solution to elimination of MDs in p/p^+ silicon wafers can be applied in volume production.

Fig. 21 shows the optical images of the etched interface of the p/p+ epi-wafers with 11 μm thick epi-layer grown on the conventional heavily boron doped and Ge–B-codoped substrates, respectively. As can be seen, in the p/p+ epi-wafer grown on the conventional heavily boron-doped substrate, there were three sets of MDs on the etched interface, which can even be distinguished by naked eye under a spotlight. While, there were no MDs in the p/p+ epi-wafer using the Ge-B-codoped substrate wafer. It is definite that the MDs in the p/p+ epi-wafers can be avoided by using the Ge-B-codoped substrates. Furthermore, a much thicker epi-layer could be fabricated on the Ge-B-copdoped substrate wafer without misfit dislocations. Fig. 22 shows both the classical cross-view and top-view optical images of the etched silicon samples. Fig. 22(a) reveals that, in the p/p^+ epi-wafer grown on the conventional heavily B-doped substrate, the MDs penetrated into the epi-layer. Whereas, in the top-view optical images of the etched interface of the p/p^+ epi-wafers, the triangularly intersected MDs are clearly demonstrated [Fig. 22(c)]. On the contrary, for the p/p^+ epi-wafers using the Ge-B-co-doped silicon substrate, MDs could hardly be observed [Figs. 22(b) and 22(d)].

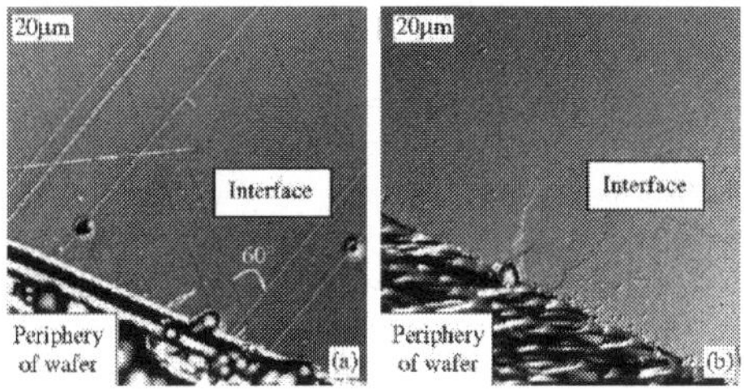

Fig. 21. Plan-view optical images of the etched interface in the 11 μm thick p/p+ epi-wafers using the conventional (a) heavily boron-doped substrate and (b) Ge-B-codoped substrate. (Jiang et al., 2006).

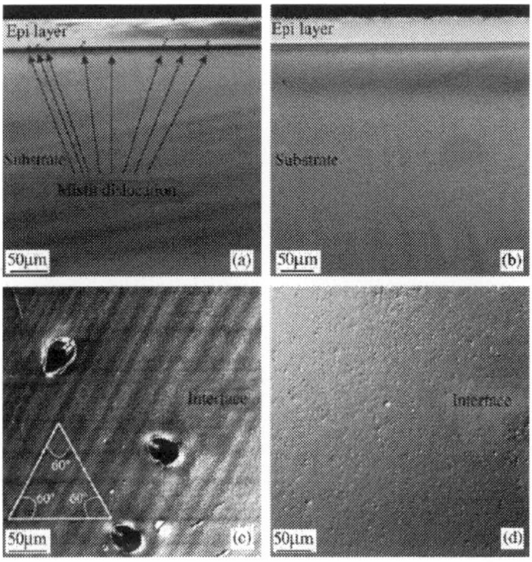

Fig. 22. Cross-sectional-view optical images of the 50μm thick p/p+ epi-wafers grown using conventional heavily boron-doped substrate (a) and Ge-B-co-doped substrate (b). And plan-view optical images of the 50 μm thick p/p+ epi-wafers grown using conventional heavily boron-doped substrate (c) and Ge-B-co- doped substrate (d) (Jiang et al., 2006).

6.2 Improved internal gettering capability

Double-side mirror polished wafers will be adopted for industrial manufacturing processes of large diameter CZ silicon, such as 300mm diameter silicon, ascribed to the higher requirements of wafer surface flatness. Therefore, the external gettering processes (such as sand sputtering processes and polycrystalline silicon depositing processes) on backside of CZ silicon wafers will be out of date and replaced by internal gettering (IG) processes based on the formation of high density BMDs in bulk and the thin defect-free denuded zone (DZ) in sub-surface of wafers simultaneously, which can be illustrated in Fig. 23(c) (Chen & Yang, 2009). However, with the ever-decreasing feature size of integrated circuits, the thermal budget for advanced devices is reduced to improve the characteristics; meanwhile, the application of magnetic-filed CZ-grown method to large diameter crystal growth leads to the reduction of oxygen concentration in silicon. Both trends led to the density reduction of BMDs which are related to gettering sites for metallic contamination.

Fig. 23 illustrates the model of the influence of germanium on generation of IG structure for CZ silicon wafer. Generally, for IG effect, both the high density BMDs and the suitable width of DZ could be generated in the CZ silicon doped with some types of impurities, so as to improve the IG capability of the metal contamination and improve the quality of IC devices. Compared to the CZ silicon, germanium atoms could generally induce germanium-related complexes and then seed for oxygen precipitation in bulk silicon during IG denudation processing based on either CFA or RTA processing. Both the good-quality defect-free DZ in sub-surface region and the BMD region with higher density in bulk silicon could be obtained simultaneously in the GCZ silicon. Generally, the DZ shrinks and is

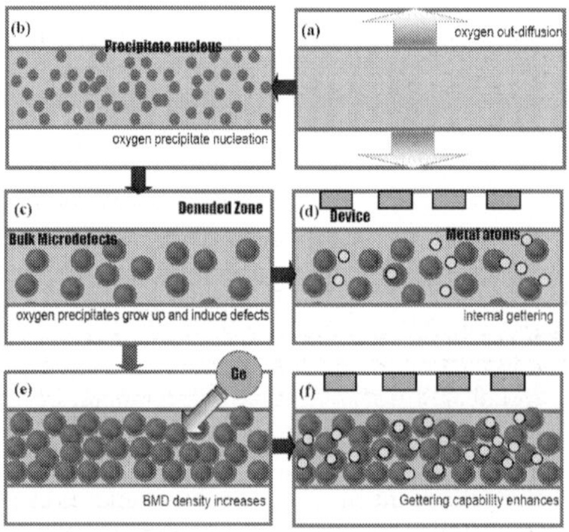

Fig. 23. Schematic illustrations for internal gettering (IG) structure' in GCZ silicon wafers. (a)-(d) shows the normal steps generating IG structure for silicon wafer and the gettering capability. As an example, (e)-(f) shows the germanium effects upon IG structure and capability. (Chen & Yang, 2009)

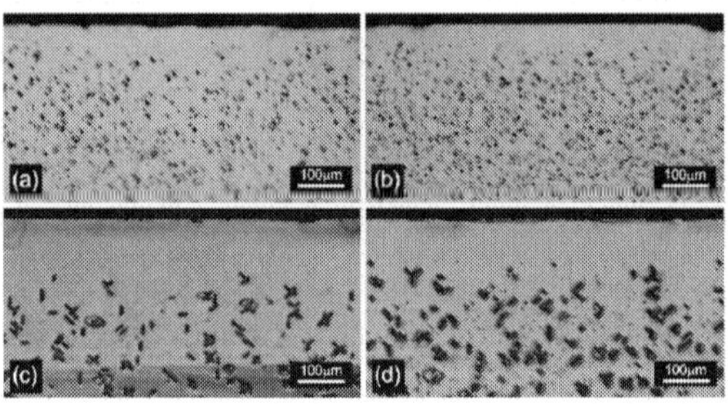

Fig. 24. Representative cross-sectional etched optical microphotographs in both the normal CZ and GCZ silicon wafers. (a) CZ, before Cu in-diffusion; (b) GCZ, before Cu in-diffusion; (c) CZ, after Cu in-diffusion; and (d) GCZ, after Cu in-diffusion. (Chen et al., 2007c)

slightly smaller than that of the CZ silicon wafer, which might be ascribed to the denser small precipitates located at the boundary of DZ and BMD region. Nevertheless, it has been also indicated that the DZs could present in the GCZ silicon wafers after a certain critical anneals despite the width shrinkage (Chen et al., 2007c).

IG capability for metallic contamination could be therefore enhanced by intentional germanium doping in CZ silicon wafers. Taking copper contamination as an example (Chen et al., 2007c). Fig. 24 shows the cross-sectional etching optical photographs of both the normal CZ and GCZ silicon wafers before and after Cu diffusion in 1100°C/1h. As can be seen, denser BMDs of smaller size with denser Cu precipitates were presented in bulk of the GCZ silicon wafers in comparison with the CZ silicon, indicating a stronger IG capability in the GCZ silicon. The explanation could be, the denser gettering sites (even with smaller size) can lower down the total interstitial Cu concentration in wafer bulk, therefore more Cu atoms could be gettered in the GCZ silicon due to the denser but smaller BMDs. It is noted that the fairly clean DZs near surfaces remained in both the silicon wafers, which ensures the integrity of wafer sub-surface for device fabrication.

7. Summary

We have illustrated the effect of germanium doping in CZ silicon on mechanical strength, oxygen-related donors, oxygen precipitation and void defects. It has been established that the mechanical strength of silicon wafers could be improved by intended germanium doping, which benefits the improved production yield of wafers. It is also found that germanium suppresses the generation of TDs, which benefits the stable electrical property of wafers. More importantly, germanium has been found to suppress the formation of void defects, which can be annihilated easily during high temperature treatments. Moreover, oxygen precipitation can be enhanced by germanium doping, and therefore IG capability could be improved. Additionally, compared to nitrogen doped CZ silicon, germanium doping level in CZ silicon could be much easier to control, and no electrical Centers such as shallow thermal donors will be introduced. Ascribing to the novel properties, it is considered that GCZ silicon could satisfy the higher requirements of ULSI.

8. References

Adachi, N., Hisatomi, T., Sano, M., & Tsuya, H. (2000). Reduction of Grown-In Defects by High Temperature Annealing. *Journal of The Electrochemical Society*, 147, 350.

Akatsuka, M., Sueoka, K., Katahama, H., Morimoto, N., & Adachi, N. (1997). Pinning effect on punched-out dislocations in silicon wafers investigated using indentation method. *Japanese Journal of Applied Physics Part 2-Letters & Express Letters*, 36, 11A, L1422-L1425.

Babich, V.M., Baran, N.P., Zotov, K.I., Kiritsa, V.L., & Kovalchuk, V.B. (1995). Low-temperature diffusion of oxygen and formation of thermal donors in silicon doped with an isovalent germanium impurity. *Fizika i Tekhnika Poluprovodnikov*, 29, 0015-3222.

Babitskii, Y.M., Gorbacheva, N.I., Grinshtein, P.M., Il'in, M.A., Kuznetsov, V.P., Mil'vidskii, M.G., & Turovskii, B.M. (1988). Kinetics of generation of low-temperature oxygen donors in silicon containing isovalent impurities. *Fizika i Tekhnika Poluprovodnikov*, 22, 2, 307-312, 0015-3222

Babitskii, Y.M., Grinshtein, P.M., Il'in, M.A., Kuznetsov, V.P., & Mil'vidskii, M. (1985). Behavior of oxygen in silicon doped with isovalent impurities. *Fizika i Tekhnika Poluprovodnikov*, 19, 11, 1982-1985, 0015-3222

Borghesi, A., Pivac, B., Sassella, A., & Stella, A. (1995). Oxygen precipitation in silicon. *Journal of Applied Physics*, 77, 4169.

Budtz-Jorgensen, C.V., Kringhoj, P., Larsen, A.N., & Abrosimov, N.V. (1998). Deep-level transient spectroscopy of the Ge-vacancy pair in Ge-doped n-type silicon. *Physical Review B*, 58, 3, 1110-1113, 0163-1829.

Capper, P., Jones, A.W., Wallhouse, E.J., & Wilkes, J.G. (1977). The effects of heat treatment on dislocation-free oxygen-containing silicon crystals. *Journal of Applied Physics*, 48, 1646.

Chen, J., Ma, X., & Yang, D. (2010). Impurity Engineering of Czochralski Silicon. *Solid State Phenomena*, 156-158, 261-267.

Chen, J., & Yang, D. (2009). Impurity engineering for germanium-doped Czochralski silicon wafer used for ultra large scale integrated circuit. *Physica Status Solidi C - Current Topics in Solid State Physics, Vol 6, No 3*, 6, 3, 625-632, 1610-1634.

Chen, J., Yang, D., Li, H., Ma, X., & Que, D. (2006a). Enhancement effect of germanium on oxygen precipitation in Czochralski silicon. *Journal of Applied Physics*, 99, 7, 0021-8979.

Chen, J., Yang, D., Li, H., Ma, X., & Que, D. (2006b). Germanium effect on as-grown oxygen precipitation in Czochralski silicon. *Journal of Crystal Growth*, 291, 1, 66-71, 0022-0248.

Chen, J., Yang, D., Li, H., Ma, X., Tian, D., Li, L., & Que, D. (2007a). Crystal-originated particles in germanium-doped czochralski silicon crystal. *Journal of Crystal Growth*, 306, 2, 262-268, 0022-0248.

Chen, J., Yang, D., Ma, X., Li, H., & Que, D. (2007b). Intrinsic gettering based on rapid thermal annealing in germanium-doped Czochralski silicon. *Journal of Applied Physics*, 101, 3, 0021-8979.

Chen, J., Yang, D., Ma, X., & Que, D. (2009). Rapid-thermal-anneal-based internal gettering for germanium-doped Czochralski silicon. *APPLIED PHYSICS A-MATERIALS SCIENCE & PROCESSING*, 94, 4, 905-910, 0947-8396.

Chen, J., Yang, D., Ma, X., Wang, W., Zeng, Y., & Que, D. (2007c). Investigation of intrinsic gettering for germanium doped Czochralski silicon wafer. *Journal of Applied Physics*, 101, 11, 0021-8979.

Chen, J., Yang, D., Ma, X., Zeng, Z., Tian, D., Li, L., Que, D., & Gong, L. (2008). Influence of germanium doping on the mechanical strength of Czochralski silicon wafers. *Journal of Applied Physics*, 103, 12, 0021-8979.

Chen, J.H., Yang, D.R., Ma, X.Y., Li, H., Fu, L.M., Li, M., & Que, D.L. (2007d). Dissolution of oxygen precipitates in germanium-doped Czochralski silicon during rapid thermal annealing. *Journal of Crystal Growth*, 308, 2, 247-251, 0022-0248.

Cui, C., Yang, D., Ma, X., Li, M., & Que, D. (2006). Effect of light germanium doping on thermal donors in Czochralski silicon wafers. *Materials Science in Semiconductor Processing*, 9, 1-3, 110-113, 1369-8001.

Fukuda, T., & Ohsawa, A. (1992). Mechanical strength of silicon crystals with oxygen and/or germanium impurities. *Applied Physics Letters*, 60, 1184.

Fuller, C.S., & Logan, R.A. (1957). Effect of Heat Treatment upon the Electrical Properties of Silicon Crystals. *Journal of Applied Physics*, 28, 1427.

Hild, E., Gaworzewski, P., Franz, M., & Pressel, K. (1998). Thermal donors in silicon-rich SiGe. *Applied Physics Letters*, 72, 11, 1362-1364, 0003-6951.

Huang, X., Sato, T., Nakanishi, M., Taishi, T., & Hoshikawa, K. (2003). High strength Si wafers with heavy B and Ge codoping. *Japanese Journal of Applied Physics Part 2-Letters*, 42, 12B, L1489-L1491, 0021-4922.

Huth, S., Breitenstein, O., Huber, A., & Lambert, U. (2000). Localization of gate oxide integrity defects in silicon metal-oxide-semiconductor structures with lock-in IR thermography. *Journal of Applied Physics*, 88, 7, 4000-4003, 0021-8979.

Itsumi, M., Akiya, H., Ueki, T., Tomita, M., & Yamawaki, M. (1995). The composition of octahedron structures that act as an origin of defects in thermal SiO on Czochralski silicon. *Journal of Applied Physics*, 78, 5984.

Jiang, H., Yang, D., Ma, X., Tian, D., Li, L., & Que, D. (2006). Growth of misfit dislocation-free p/p(+) thick epitaxial silicon wafers on Ge-B-codoped substrates. *Physica B-Condensed Matter*, 376, 841-844, 0921-4526.

Joly, J.P., & Robert, V. (1994). silicon wafers: a detailed analysis. *Semicond. Sci. Technol*, 9, 105-111.

Kaiser, W., Frisch, H.L., & Reiss, H. (1958). Mechanism of the formation of donor states in heat-treated silicon. *Physical Review*, 112, 5, 1546-1554, 0031-899X.

Kishino, S., Matsushita, Y., Kanamori, M., & Iizuka, T. (1982). Thermally Induced Microdefects in Czochralski-Grown Silicon - Nucleation and Growth-Behavior. *Japanese Journal of Applied Physics Part 1-Regular Papers Short Notes & Review Papers*, 21, 1, 1-12, 0021-4922.

Kissinger, G., Vanhellemont, J., Lambert, U., Graf, D., Dornberger, E., & Richter, H. (1998). Influence of Residual Point Defect Supersaturation on the Formation of Grown-In Oxide Precipitate Nuclei in CZ-Si. *Journal of The Electrochemical Society*, 145, 5, L75-L78.

Li, H., Yang, D., Ma, X., Yu, X., & Que, D. (2004a). Germanium effect on oxygen precipitation in Czochralski silicon. *Journal of Applied Physics*, 96, 8, 4161-4165, 0021-8979.

Li, H., Yang, D., Yu, X., Ma, X., Tian, D., Li, L., & Que, D. (2004b). The effect of germanium doping on oxygen donors in Czochralski-grown silicon. *Journal of Physics-Condensed Matter*, 16, 32, 5745-5750, 0953-8984.

Minowa, K., & Sumino, K. (1992). Stress-induced amorphization of silicon crystal by mechanical scratching. *Physical Review Letters*, 69, 2, 320-322, 0031-9007.

Park, J.G., Lee, G.S., Kwack, K.D., & Park, J.M. (2000). Crystal Originated Particle Induced Isolation Failure in Czochralski Silicon Wafers. *Japanese Journal of Applied Physics*, 39, 1, 197.

Pensl, G., Schulz, M., Holzlein, K., Bergholz, W., & Hutchison, J.L. (1989). New oxygen donors in silicon *Applied Physics a-Materials Science & Processing*, 48, 1, 49-57, 0947-8396.

Shimura, F. (1994). Oxygen in silicon. In R.K. Willardson, E.R. Weber, & A.C. Beer, *Semiconductors and semimetals*. New York: Academic Press, p. 434.

Voronkov, V.V. (1982). The mechanism of swirl defects formation in silicon *Journal of Crystal Growth*, 59, 3, 625-643, 0022-0248.

Voronkov, V.V., & Falster, R. (1998). Vacancy-type microdefect formation in Czochralski silicon. *Journal of Crystal Growth*, 194, 1, 76-88, 0022-0248.

Voronkov, V.V., & Falster, R. (2002). Intrinsic point defects and impurities in silicon crystal growth. *Journal of the Electrochemical Society*, 149, 3, G167-G174, 0013-4651.

Wagner, P., & Hage, J. (1989). Thermal double donors in silicon. *Applied Physics a-Materials Science & Processing*, 49, 2, 123-138, 0947-8396.

Wijaranakula, W. (1994). Dissolution kinetics of D defects in Czochralski silicon. *Journal of Applied Physics*, 75, 7, 3678-3680, 0021-8979.

Yamagishi, H., Fusegawa, I., Fujimaki, N., & Katayama, M. (1992). Recognition of D defects in silicon single crystals by preferential etching and effect on gate oxide integrity. *Semicondutor Science and Technology*, 7, 1, A135-138.

Yang, D., Chen, J., Ma, X., & Que, D. (2009). Impurity engineering of Czochralski silicon used for ultra large-scaled-integrated circuits. *Journal of Crystal Growth*, 311, 3, 837-841, 0022-0248.

Yang, D., Li, H., Yu, X., Ma, X., Tian, D., Li, L., & Que, D. (2004). Oxygen-related donors in germanium doped Czochralski silcion. *High Purity Silicon VIII: the 206th Fall Meeting of The Electrochemical Society in Honolulu*. Hawaii, USA.

Yang, D., Yu, X., Ma, X., Xu, J., Li, L., & Que, D. (2002). Germanium effect on void defects in Czochralski silicon. *Journal of Crystal Growth*, 243, 3-4, 371-374, 0022-0248.

Yang, D., Chen, J., Li, H., Ma, X., Tian, D., Li, L., & Que, D. (2006a). Micro-defects in Ge doped Czochralski grown Si crystals. *Journal of Crystal Growth*, 292, 2, 266-271, 0022-0248.

Yang, D., Chen, J., Li, H., Ma, X., Tian, D., Li, L., & Que, D. (2006b). Germanium effect on oxygen-related defects in Czochralski silicon. *Physica Status Solidi a-Applications and Materials Science*, 203, 4, 685-695, 0031-8965.

Yonenaga, I. (2005). Nitrogen effects on generation and velocity of dislocations in Czochralski-grown silicon. *Journal of Applied Physics*, 98, 2, 023517, 0021-8979.

Yu, X., Yang, D., Ma, X., Yang, J., Li L., Que D.(2002). Grown-in defects in nitrogen-doped Czochralski Silicon. *Journal of Applied Physics*,92, 1, 188-194

19

Miniature Dual Axes Confocal Microscope for Real Time *In Vivo* Imaging

Wibool Piyawattanametha and Thomas D. Wang
[1]*Departmentsof Applied Physics, Biology, Electrical Engineering Microbiology & Immunology, Radiology, and Pediatrics James H. Clark Center (Bio-X), Stanford,*
[2]*NECTEC, Pathumthani,*
[3]*Faculty of Medicine, Chulalongkorn University, Prathumwan,*
[1]*CA*
[2,3]*Thailand*

1. Introduction

Today, disease interpretation of excised tissue is performed by analyzing biopsy specimens with a tabletop microscope [1]. While this method is effective, the process can be limited by sampling error, processing costs, and preparation time. In addition, the interpretive accuracy of the specimens can be affected by artefacts associated with tissue sectioning, paraffin embedding, and histochemical staining. Thus, a lot of effort has gone into the development of new methods that perform real time *in vivo* imaging with sub-cellular resolution.

Confocal microscopy is a powerful optical imaging method that can achieve sub-cellular resolution in real time. The technique of optical sectioning provides clear images from "optically thick" biological tissues that have previously been collected with large, tabletop instruments that occupy the size of a table [2, 3]. They can be used to collect either reflectance or fluorescence images to identify morphological or

molecular features of cells and tissues, respectively. Moreover, images in both modalities can be captured simultaneously with complete spatial registration. This approach uses a "pinhole" placed in between the objective lens and the detector to allow only the light that originates from within a tiny focal volume below the tissue surface to be collected. For miniature instruments, the core of an optical fiber is used as the "pinhole."

Recently, significant progress has been made in the development of endoscope-compatible confocal imaging instruments for visualizing inside the human body. This direction has been accelerated by the availability, variety and low cost of optical fibers, scanners, and light sources, in particular, semiconductor lasers. These methods are being developed for use in the clinic as well as in small animal imaging facilities. The addition of a miniature real-time, high resolution imaging instrument can help guide tissue biopsy and reduce pathology costs. However, these efforts are technically challenging because of the demanding performance requirements for small instrument size, high image resolution, deep tissue penetration depths, and fast frame rates.

The performance parameters for miniature *in vivo* confocal imaging instruments are governed by the specific application. An important goal is the early detection and image guided therapy of disease in hollow organs, including colon, esophagus, lung, oropharynx, and cervix. Applications can also be found for better understanding of the molecular mechanisms of disease in small animals. In particular, localization of pre-malignant (dysplastic) lesions in the digestive tract can guide tissue biopsy for early detection and prevention of cancer. In addition, visualization of over expressed molecular targets in small animal models can lead to the discovery of new drugs.

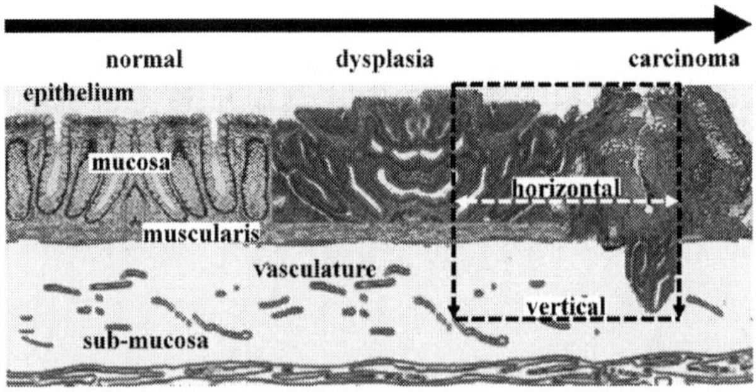

Fig. 1. Dysplasia represents a pre-malignant condition in the epithelium of hollow organs, such as the colon and esophagus. The dual axes confocal architecture has high dynamic range that is suitable for imaging in the vertical cross-sectional plane to visualize disease processes with greater tissue penetration depths.

As shown in Fig. 1, dysplasia originates in the epithelium and represents an important step in the transformation of normal mucosa to carcinoma. Dysplasia has a latency period of approximately 7 to 14 years before progressing onto cancer and offers a window of opportunity for evaluating patients by endoscopy who are at increased risk for developing cancer. The early detection and localization of dysplastic lesions can guide tissue resection and prevent future cancer progression. Dysplastic glands can be present from the mucosal surface down to the muscularis. Thus, an imaging depth of ~500 μm is sufficient to evaluate most early epithelial disease processes.

On reflectance imaging, sub-cellular resolution (typically <5 μm) is needed to identify nuclear features, such as nuclear-to-cytoplasm ratio. On fluorescence imaging, high contrast is needed to distinguish between the target and background. With both modalities, a fast imaging frame rate (>4 Hz) is necessary to avoid motion artefact.

2. Single axls confocal architecture

A. Configuration of optics

Recent advances in the development of microlenses and miniature scanners have resulted in the development of fiber optic coupled instruments that are endoscope compatible with high resolution, including single [4-8], and multiple fiber [7, 9] strategies. Different methods of scanning are also being explored [10-14].

All of these endoscope compatible designs use a single axis design, where the pinhole (fiber) and objective are located along one main optical axis. A high NA objective is used to achieve sub-cellular resolution and maximum light collection, and the same objective is used for both the illumination and collection of light. In order to scale down the dimension of these instruments for endoscope compatibility, the diameter of the objective must be reduced to ~5 mm or less. As a consequence, the working distance (WD) as well as the field-of-view (FOV) is also decreased, as shown by the progression of the 3 different objectives in Fig. 2. The tissue penetration depth also decreases, and is typically inadequate to assess the tissue down to the muscularis, which is located at a depth of ~500 μm and is an important landmark for defining the early presence of epithelial cancers.

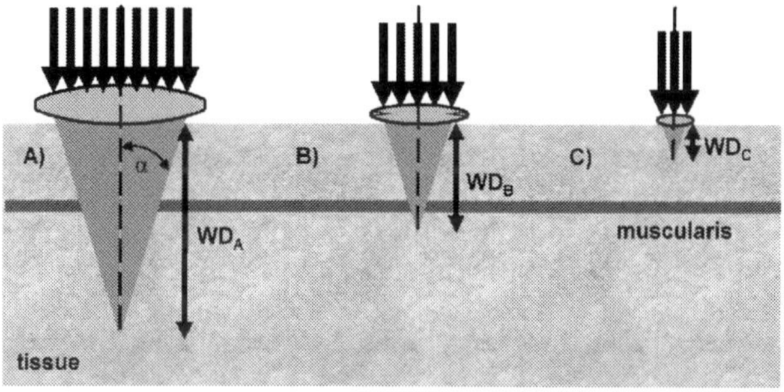

Fig. 2. For endoscope compatibility, the diameter of a single axis confocal microscope must be scaled down in size (A→B→C), resulting in a reduced working distance and limited tissue penetration depth.

B. Resolutions

For the conventional single axis architecture, the transverse, Δr_s, and axial, Δz_s, resolution between full-width-half-power (FWHP) points for uniform illumination of the lenses are defined by the following equations [3]:

$$\Delta r_s = \frac{0.37\lambda}{n\sin\alpha} \approx \frac{0.37\lambda}{n\alpha}; \quad \Delta z_s = \frac{0.89\lambda}{n(1-\cos\alpha)} \approx \frac{1.78\lambda}{n\alpha^2} \qquad (1)$$

where λ is the wavelength, n is the refractive index of the medium, α is the maximum convergence half-angle of the beam, $NA = n\sin\alpha$ is the numerical aperture, and $\sin\alpha \approx \alpha$ for low NA lenses. Eq. (1) implies that the transverse and axial resolution varies as 1/NA and $1/NA^2$, respectively. A resolution of less than 5 μm is adequate to identify sub-cellular structures that are important for medical and biological applications. To achieve this resolution in the axial dimension, the objective lens used requires a relatively large NA (>0.4). The optics can be reduced to the millimeter scale for *in vivo* imaging, but requires a sacrifice of resolution, FOV, or WD. Also, a high NA objective limits the available WD, and requires that the scanning mechanism be located in the pre-objective position, restricting the FOV and further increasing sensitivity to off-axis aberrations.

C. Commercial systems

Two endoscope compatible confocal imaging systems are commercially available for clinical use. The EC-3870K (Pentax Precision Instruments, Tokyo, Japan) has an integrated design where a confocal module (Optiscan Pty Ltd, Victoria, Australia) is built into the insertion tube of the endoscope, and results in an overall diameter of 12.8 mm, as shown in Fig. 3a [15]. This module uses the single axis optical configuration where a single mode optical fiber is aligned on-axis with an objective that has an NA ≈ 0.6. Scanning of the distal tip of the optical fiber is performed mechanically by coupling the fiber to a tuning fork that vibrates at resonance. Axial scanning is performed with a shape memory alloy (nitinol) actuator that can translate the focal volume over a distance of 0 to 250 μm below the tissue surface. Excitation is provide at 488 nm (peak absorption of fluorescein) by a semi-conductor laser, and a transverse and axial resolution of 0.7 and 7 μm, respectively, has been achieved. The images are collected at a frame rate of either 0.8 or 1.6 Hz to achieve a FOV of either 1024x1024 or 1024x512 pixels, respectively. The dimension of the confocal instrument by itself is ~5 mm. When a suspicious lesion is identified, the confocal window located on the distial tip is placed into contact with the tissue to collect images. A separate instrument channel can be used to obtain pinch biopsies of tissue.

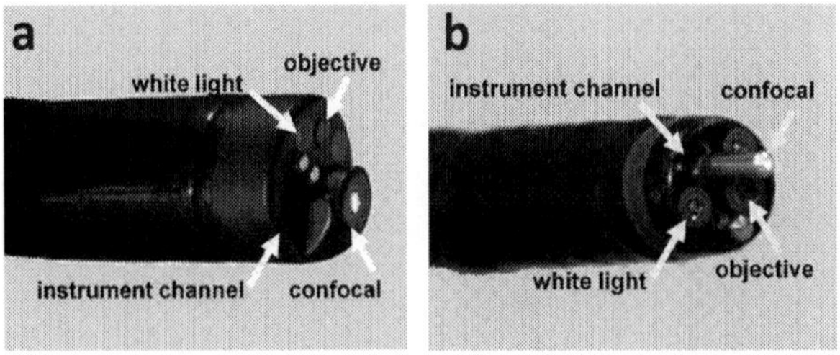

Fig. 3. a) The EC-3870K (Pentax) has a confocal module (Optiscan) integrated into the endoscope insertion tube. b) The Cellvizio® GI is a confocal miniprobe that passes through the instrument channel of the endoscope.

The Cellvizio® GI (Mauna Kea Technologies, Paris, France) uses a set of miniprobes that range in diameter from 1.5 to 2.5 mm, and passes through the standard instrument channel of medical endoscopes, as shown in Fig. 3b. This instrument moves independently of the endoscope, and its placement onto the tissue surface can be guided by the conventional white light image [8, 15]. This miniprobe consists of a fiber bundle with ~30,000 individual fibers that is aligned on-axis with an objective that has an NA ≈ 0.6. The core of each individual fiber acts as a collection pinhole to reject out-of-focus light. Scanning is performed at the proximal end of the bundle in the instrument control unit with a 4 kHz oscillating mirror for horizontal lines and a 12 Hz galvo mirror for frames. In this design, axial scanning cannot be performed. Instead, separate miniprobes that have different working distances are needed to optically section at different depths. Excitation is provided at 488 nm, and the transverse and axial resolution of these instruments ranges from 2.5 to 5 μm and 15 to 20 μm, respectively. Images are collected at a frame rate of 12 Hz with a FOV of either 600x500 μm^2 or 240x200 μm^2.

3. Dual axes confocal architecture

D. Configuration of optics

So far, the aforementioned miniaturization techniques in the previous section deploy a conventional single-axis confocal architecture that has the objective and optical fiber aligned along the same optical axis. In order to overcome some of these limitations for endoscope compatibility and *in vivo* imaging, we have developed the novel dual axes confocal configuration, shown in Fig. 4. We use two fibers oriented along separate optical axis of different low NA objectives to spatially separate the light paths for illumination and collection [16, 17]. The region of overlap between the two beams (crossed at a half angle θ from the midline) defines the focal volume, hence the resolution, and can achieve sub-cellular dimensions. A very low probability exists for light scattered by tissue along the illumination path (blue cone) to enter the low NA collection objective (green cone), thus significant improvement in the dynamic range of detection can be achieved.

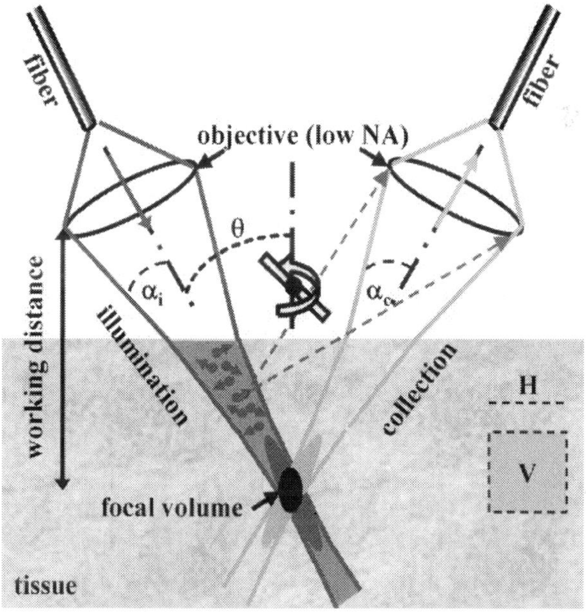

Fig. 4. Novel dual axes confocal architecture uses separate optical fibers and low NA lenses for off-axis light collection, achieving long working distance, high dynamic range, and scalability while preserving resolution.

Furthermore, the low NA objectives enable an increased working distance so that the scanning mirror can be placed on the distal (tissue) side of the lens (post-objective position), resulting in less sensitivity to off-axis aberrations [17]. In this configuration, the beams always pass through the low NA objectives on axis, resulting in a diffraction-limited focal volume that can then be scanned over a large FOV, limited by the performance of the scanner rather than by the optics. This design feature allows for the instrument to be scaled down in size to millimeter dimensions for compatibility with medical endoscopes without loss of performance.

We first develop the theory to explain the unique performance features of the dual axes confocal architecture by characterizing the point-spread function (PSF) and dynamic range. Then, we demonstrate the scaled down implementation of this configuration in miniature prototypes. Because of the challenges of packaging in such a small form factor, we first demonstrate a handheld (10 mm diameter) instrument and then an endoscope-compatible (5.5 mm diameter) prototype, using the same MEMS mirror and scanhead optics.

E. Definition of coordinates

The coordinates for the dual axes confocal configuration are shown in Fig. 5. The illumination (IO) and collection (CO) objectives represent separate low NA lenses. The maximum convergence half-angles of the illumination and collection beams are represented by α_i and α_c, respectively. The separate optical axes are defined to cross the z-axis (z_d) at an angle θ. The main lobe of the PSF of the illumination objective is represented by the light gray oval. This lobe has a narrow transverse but a wide axial dimension.

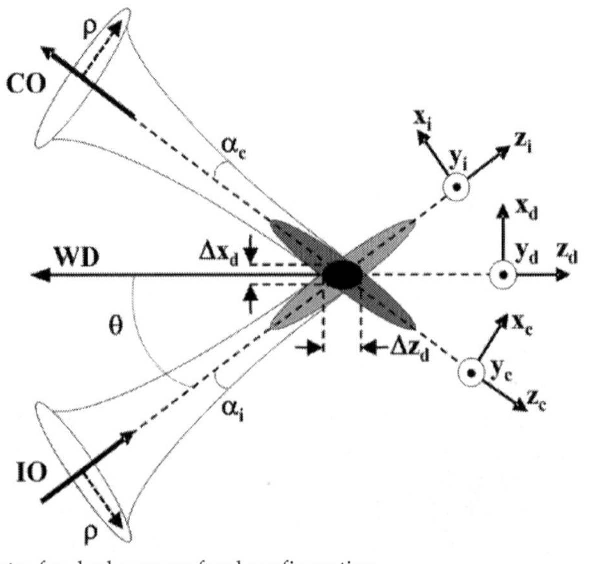

Fig. 5. Coordinates for dual axes confocal configuration

Similarly, the main lobe of the PSF of the collection objective is similar in shape but symmetrically reflected about z_d, as represented by the dark gray oval. For dual axes, the combined PSF is represented by the overlap of the two individual PSF's, represented by the black oval. This region is characterized by narrow transverse dimensions, Δx_d and Δy_d (out of the page), and by a significantly reduced axial dimension, Δz_d, which depends on the transverse rather than the axial dimension of the individual beams where they intersect.

F. Point spread function

The dual axes PSF can be derived using diffraction theory with paraxial approximations [18]. The coordinates for the illumination (x_i,y_i,z_i) and collection (x_c,y_c,z_c) beams are defined in terms of the coordinates of the main optical axis (x_d,y_d,z_d), and may be expressed as follows:

$$\begin{aligned} x_i &= x_d \cos\theta - z_d \sin\theta & x_c &= x_d \cos\theta + z_d \sin\theta \\ y_i &= y_d & y_c &= y_d \\ z_i &= x_d \sin\theta + z_d \cos\theta & z_c &= -x_d \sin\theta + z_d \cos\theta \end{aligned} \qquad (2)$$

The maximum convergence half-angles of the focused illumination and collection beams in the sample media are represented as α_i and α_c, respectively. The angle at which the two beams intersect the main optical axis is denoted as θ. A set of general dimensionless coordinates may be defined along the illumination and collection axes, as follows [19]:

$$\begin{aligned} u_i &= k_i n z_i \sin^2 \alpha_i & u_c &= k_c n z_c \sin^2 \alpha_c \\ v_i &= k_i n \sqrt{x_i^2 + y_i^2} \sin \alpha_i & v_c &= k_c n \sqrt{x_c^2 + y_c^2} \sin \alpha_c \end{aligned} \tag{3}$$

The wavenumbers for illumination and collection are defined as $k_i = 2\pi/\lambda_i$ and $k_c = 2\pi/\lambda_c$, respectively, where λ_i and λ_c are the wavelengths, and n is the index of refraction of the media.

The amplitude PSF describes the spatial distribution of the electric field of the focused beams. Diffraction theory may be used to show that the PSF of the illumination and collection beams is proportional to the Huygens-Fresnel integrals below [18]:

$$U_i(v_i, u_i) \propto \int_0^1 W_i(\rho) J_0(\rho v_i) e^{-j u_i \rho^2/2} \rho d\rho \tag{4}$$

$$U_c(v_c, u_c) \propto \int_0^1 W_c(\rho) J_0(\rho v_c) e^{-j u_c \rho^2/2} \rho d\rho \tag{5}$$

where J_0 is the Bessel function of order zero, and ρ is a normalized radial distance variable at the objective aperture. The weighting function, W(ρ), describes the truncation (apodization) of the beams. For uniform illumination, W(ρ) = 1. For Gaussian illumination, the objectives truncate the beams at the $1/e^2$ intensity, resulting in a weighting function of $W(\rho) = e^{-\rho^2}$. In practice, the beams are typically truncated so that 99% of the power is transmitted. For a Gaussian beam with a radius ($1/e^2$ intensity) given by w, an aperture with diameter πw passes ~99% of the power. In this case, the weighting function is given as follows:

$$W(\rho) = e^{-(\pi\rho/2)^2} \tag{6}$$

For the single axis configuration, the illumination and collection PSF's at the focal plane ($u_i = u_c = 0$) are identical functions of the radial distance ρ, and can both be given by U_s using the substitution $v = knr \sin\alpha$, as follows:

$$U_s(v) \propto \int_0^1 W_i(\rho) J_0(\rho v) \rho d\rho \tag{7}$$

The resulting signal at the detector V from a point source reflector in the media is proportional to the power received, and is given by the square of the product of the overlapping PSF's as follows:

$$V = A \left| U_i U_c \right|^2 \tag{8}$$

where *A* is a constant.

Similarly, since the depth of focus for each individual beam, described within the exponential term in the integral product of Eqs. (3) and (4), is much larger than that of the transverse width, the exponential term may be neglected. As a result, the detector output V_d for the dual axes configuration for uniform illumination ($W = 1$), is given as follows:

$$V_d \propto \left(\frac{2J_1(v_i)}{v_i}\right)^2 \left(\frac{2J_1(v_c)}{v_c}\right)^2 \tag{9}$$

This expression can be combined with Eqs. (2) and (3) to derive the result for transverse and axial resolution with uniform illumination as follows [16]:

$$\Delta x_d = \frac{0.37\lambda}{n\alpha \cos\theta}; \quad \Delta y_d = \frac{0.37\lambda}{n\alpha}; \quad \Delta z_d = \frac{0.37\lambda}{n\alpha \sin\theta} \tag{10}$$

Note that for the dual axes configuration, the axial resolution is proportional to $1/NA$, where $NA = n \sin\alpha \approx n\alpha$, rather than $1/NA^2$, as is the case for the single axis design [3]. For example, with uniform illumination and the following parameters: $\alpha = 0.21$ radians, $\theta = 30$ degrees, $\lambda = 0.785$ μm and $n = 1.4$ for tissue, Eq. (10) reveals a result for the dual axes configuration of $\Delta x_d = 1.1$ μm, $\Delta y_d = 1.0$ μm, and $\Delta z_d = 2$ μm for the transverse and axial resolutions, respectively. Thus, sub-cellular resolution can be achieved in both the transverse and axial dimensions with the dual axes configuration using low NA optics but not with the single axis architecture.

For an endoscope-compatible instrument, delivery of the illumination and collection light is performed with use of optical fibers and is more appropriately modeled by a Gaussian rather than a uniform beam. With this apodization, the detector response for the dual axes configuration from a point source reflector in the media, given by Eq. (9), may be solved numerically as a function of transverse (x_d and y_d) and axial (z_d) dimensions. The integrals are calculated in Matlab, and use the weighting function with 99% transmission. In comparison, this model reveals a result of $\Delta x_d = 2.4$ μm, $\Delta y_d = 2.1$ μm, and $\Delta z_d = 4.2$ μm for the transverse and axial resolutions, respectively. Thus, the use of optical fibers, modeled by a Gaussian beam, produces results that are slightly worse but still comparable to that of uniform illumination [19].

G. Dynamic range

Differences in the dynamic range between the single and dual axes confocal configurations can also be illustrated with this model [18]. The calculated axial response for the single axis design with Gaussian illumination is shown by the dashed line in Fig. 6a, where optical parameters are used that achieve the same axial resolution (FWHM) of 4.2 μm. The result reveals that the main lobe falls off in the axial (z-axis) direction as $1/z^2$, and reaches a value of approximately -25 dB at a distance of 10 μm from the focal plane (z = 0). In addition, a number of side lobes can be appreciated.

In comparison, the response for the dual axes configuration, shown by the solid line in Fig. 6a, reveals that the main lobe rolls off in the axial (z-axis) direction as $\exp(-kz^2)$, and reaches a value of -60 dB at a distance of 10 μm from the focal plane ($z = 0$). Thus, off-axis illumination and collection of light in the dual axes architecture results in a significant improvement in dynamic range and in an exponential rejection of out-of-focus scattered light in comparison to that for single axis. This advantage allows for the dual axes configuration to collect images with deeper tissue penetration and with a vertical cross-section orientation. The transverse response with Gaussian illumination is shown in Fig. 6b.

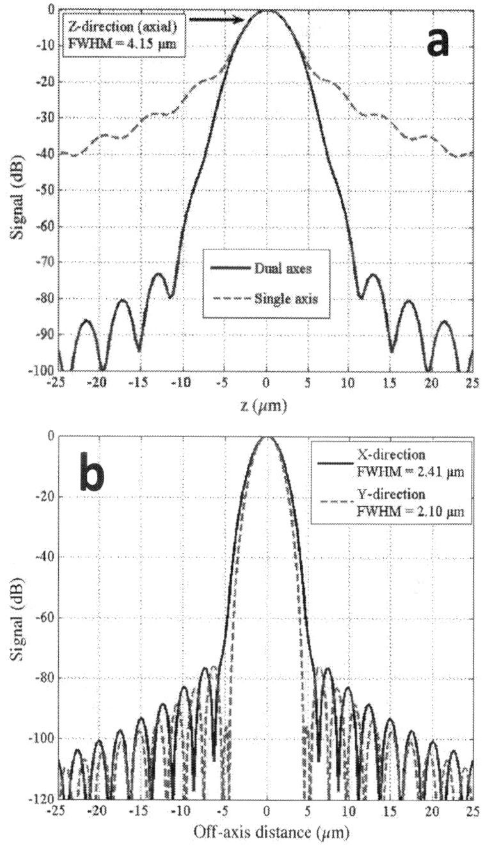

Fig. 6. Dynamic range of novel dual axes confocal architecture. a) The axial response of the single axis (dashed line) configuration falls off as $1/z^2$ and that for the dual axes (solid line) design falls off as $\exp(-kz^2)$, resulting in a significant improvement in dynamic range, allowing for vertical cross-sectional imaging to be performed. b) Transverse (X-Y direction) response.

H. Post-objective scanning

In confocal microscopes, scanning of the focal volume is necessary to create an image. In the single axis architecture, the high NA objectives used limit the working distance, thus the scan mirror is by convention placed on the pinhole (fiber) side of the objective, or in the pre-objective position, as shown in Fig. 7a. Scanning orients the beam at various angles to the optical axis and introduces off-axis aberrations that expand the focal volume. In addition, the FOV of pre-objective scanning systems is proportional to the scan angle and the focal length of the objective. The diameter of the objective limits the maximum scan angle, and as this dimension is reduced for endoscope compatibility, the focal length and FOV are also diminished.

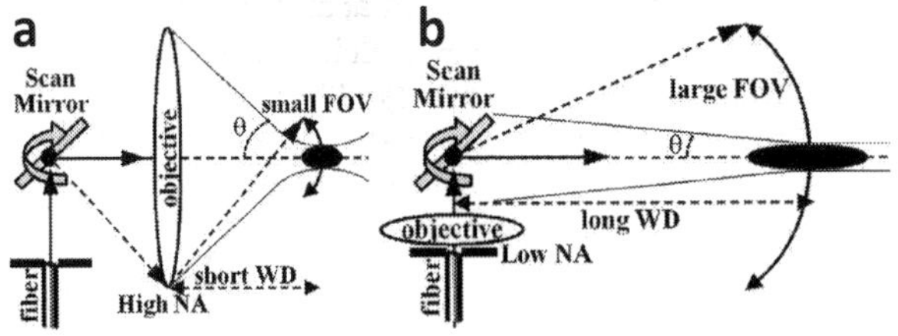

Fig. 7. a) For pre-objective scanning, illumination light is incident on the objective off-axis, resulting in more sensitivity to aberrations and limited FOV. b) With post-objective scanning, the incident light is on-axis, less sensitivity to aberrations, and large FOV. Post-objective scanning is made possible by the long WD produced by the low NA objectives used in the dual axes architecture.

In the dual axes configuration, the low NA objectives used creates a long working distance that allows for the scanner to be placed on the tissue side of the objective, or in the post-objective position [17]. This design feature is critical for scaling the size of the instrument down to millimeter dimensions for *in vivo* imaging applications without losing performance. As shown in Fig. 7b, the illumination light is always incident on-axis to the objective. In the post-objective location, the scan mirror can sweep a diffraction-limited focal volume over an arbitrarily large FOV, limited only by the maximum deflection angle of the mirror. Moreover, the scanner steers the illumination and collection beams together with the intersection of the two beams oriented at a constant angle θ and with the overlapping focal volume moving without changing shape along an arc-line. This property can be conceptualized by regarding the dual axes geometry as being equivalent to two separate beams produced from two circles in the outer annulus of a high NA lens containing a central obstruction (or a large central hole). A flat scan mirror deflects both beams equally, and thereby preserves the overlapping region without introducing aberrations to the beams.

I. Improved rejection of scattering

In the dual axes confocal architecture, the off-axis collection of light significantly reduces the deleterious effects of tissue scattering on the dynamic range of detection and allows for deeper ballistic photons to be resolved [20]. These features provide the unique capacity to collect vertical cross-sectional images in the plane perpendicular to the tissue surface. This is the preferred view of pathologists because differences from the normal patterns of tissue differentiation are revealed in the direction from the lumen to the sub-mucosa.

1. Optical configurations

The improvement in rejection of light scattered by tissue can be illustrated by comparing the dynamic range of detection between the single and dual axes optical configurations with equivalent axial resolution, as shown in Fig. 8a and 8b. The incident beams are modeled with a Gaussian profile because this is representative of light delivered through an optical fiber. For the single axis configuration, this beam is focused into the tissue by an ideal lens (L1). A mirror (M) is embedded in the tissue at the focal plane (parallel to the x-y plane) of the objective lens. In this scheme, the rays that reflect from the mirror pass back through the lens L1, deflect at an angle off the beam splitter, and are focused by an ideal lens (L2) on to a pinhole detector. For the dual axes set-up, the incident Gaussian beam is focused into the tissue by an ideal lens (L3) with its axis oriented at an angle θ = 30° to the z-axis, and an ideal lens (L4) focuses the backscattered beam, with its axis z′ at an angle -30° to the z-axis, onto the pinhole detector. As before, a mirror (M) with its plane perpendicular to the z-axis and passing through the coincident focuses of the lenses is embedded in the tissue to reflect the incident light to the detector. In both configurations, the lens system has a magnification of 1 from the focal plane to the pinhole detector.

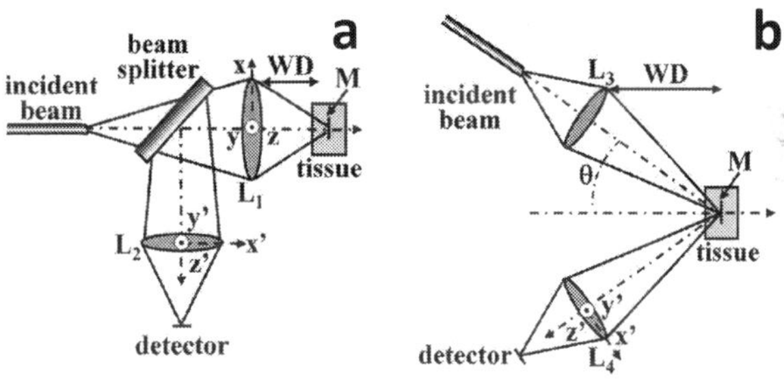

Fig. 8. a) single axis and b) dual axes optical configurations are used to evaluate the axial response at the detector.

In order to achieve an equivalent -3 dB axial resolution (FWHM), the NA's for the single and dual axes configurations are defined to be 0.58 and 0.21, respectively. From diffraction theory, discussed above, the theoretical transverse and axial resolutions for the PSF for dual axes at a wavelength λ = 633 nm with an average tissue refractive index of 1.4 and NA = 0.21 are found to be Δx = 1.16 μm, Δy = 1.00 μm, and Δz = 2.00 μm.3 The mirror is placed at a distance of 200 μm below the tissue surface in the focal plane of the objective lenses for both the single and dual axes configurations. This depth is representative of the imaging distance of interest in the epithelium of hollow organs. The calculations performed to analyze the effects of tissue scattering on light are based on Monte Carlo simulations using a non-sequential ray tracing program (ASAP® 2006 Breault Research Organization, Tucson, AZ). Three assumptions are made in this simulation study: 1) multiple scattering of an incoherent beam dominates over diffraction effects, 2) the non-scattering optical medium surrounding the lenses and the tissue (the scattering medium) is index matched to eliminate aberrations, and 3) absorption is not included to simplify this model and because there is much larger attenuation due to the scattering of ballistic photons.

2. *Mie scattering analysis*

We use Mie theory with the Henyey-Greenstein phase function *p(θ)* to model the angular dependence of tissue scattering, as follows [21, 22]:

$$p(\theta) = \frac{1}{4\pi}\frac{1-g^2}{(1+g^2-2g\cos\theta)^{3/2}} \tag{11}$$

where *g*, the anisotropy factor, is defined as

$$g = \langle\cos\theta\rangle = \int_0^{2\pi}\int_0^{\pi} \cos\theta\cdot p(\theta)\sin\theta d\theta d\varphi. \tag{12}$$

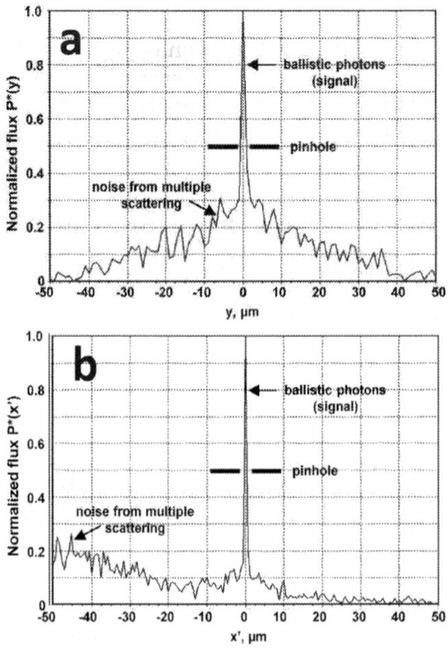

Fig. 9. Distributions of photon flux in tissue scattering model. The peak value of multiple scattered photons for A) single axis is co-located with the confocal pinhole while that for B) dual axes is separated by ~50 μm. As a consequence, the ballistic photons for dual axes result in a greater signal-to-noise ratio.

Given the average scatterer size, refractive index, and concentration, the attenuation coefficient μ_s and anisotropy g are determined and provided to the ASAP program as simulation parameters. For a tissue phantom composed of polystyrene spheres with a diameter of 0.48 μm, refractive index 1.59, and a concentration of 0.0394 spheres/μm^3 in water, the values $g = 0.81$ and $\mu_s = 5.0$ mm^{-1} at $\lambda = 633$ nm are calculated from Mie theory [23]. For single axis, P(y′) is defined as the photon flux distribution along the y′-axis at the detector. The photon flux can be normalized by defining $P^*(y') = P(y')/P_{max}$, where P_{max} is the maximum flux. The normalized flux $P^*(y')$ consists of ballistic (signal) and multiple scattered (noise) photons, as shown in Fig. 9a [24]. The maximum flux for both the signal and noise components arrive at center of the detector. A confocal pinhole placed in front of the detector can filter out some but not all of this "noise," resulting in a reduced signal-to-noise ratio (SNR). For dual axes, the detector is angled off the optical axis by 30 deg. P(x′) is defined as the photon flux distribution along the x′-axis at the detector. The photon flux can be normalized by defining $P^*(x') = P(x')/P_{max}$, where P_{max} is the maximum flux at the detector. Fig. 9b shows that normalized photon flux distribution for dual axes also exhibits a ballistic and multiple scattered components. However, for dual axes, the peak flux of multiple scattered photons arrives ~50 μm lateral to the center of the detector where the ballistic photons arrive, a consequence of off-axis collection. Thus, there is much less "noise" for the confocal pinhole (diameter ~1 μm) to filter out, resulting in a higher SNR and dynamic range.

3. Improvement in dynamic range

An implication of this result is that the dual axes configuration has improved dynamic range compared to that of single axis. This difference can be quantified by determining the axial response at the detector. This can be done by calculating the photon flux f(Δz) as the mirror is displaced along the *z*-axis in the tissue. The flux is calculated using Monte-Carlo simulations in ASAP with the mirror at positioned in the range -10 μm < Δ*z* < 10 μm with respect to the focal plane at *z* = 0, which is located at 200 μm below the tissue surface. The flux is then normalized according to F(Δz) = f(Δz)/f(0). The axial response is shown in Fig. 10a for various pinhole diameters D, including 1, 2 and 3 μm, which correspond to typical fiber core dimensions. Note that for each pinhole diameter, the dual axes (DA) configuration has significantly better dynamic range than that of single axis (SA). Note that the introduction of tissue scattering results in a reduction of the dynamic range compared to that found in free space, as shown by Fig. 6a.

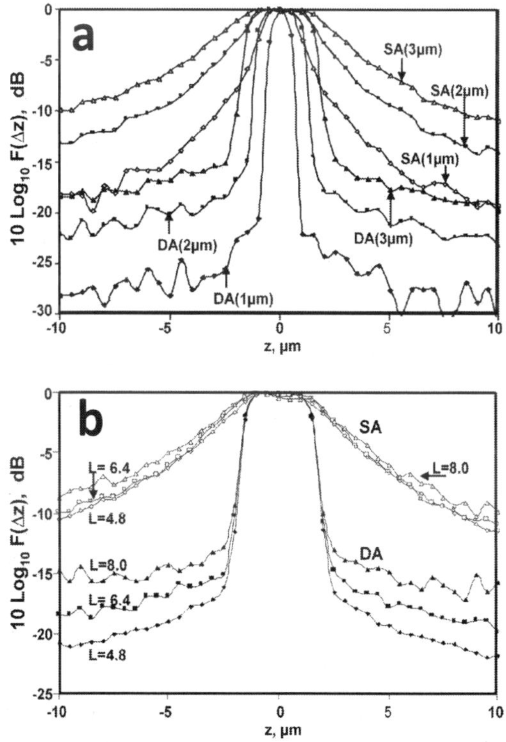

Fig. 10. Axial response for single and dual axes geometries. The dual axes (DA) configuration has a much greater dynamic range than that for single axis (SA) given different a) pinhole diameters (1, 2, and 3 μm) and b) optical lengths L (4.8, 6.4 and 8.0).

We can also determine the axial response of the detector for various optical lengths in tissue. This analysis reveals differences in the dynamic range between the single and dual axes configuration for tissues with various scattering properties. The total optical length L is defined as twice the product of the scattering coefficient μ_s and the tissue depth t, or $L = 2\mu_s t$. The factor of two originates from the fact that the total path length is twice the tissue depth. The axial response is shown in Fig. 10b for various optical lengths L, including 4.8, 6.4, and 8.0. Note that for each optical length L, the dual axes (DA) configuration has significantly better dynamic range than that of single axis (SA). These values of L are typical parameters of gastrointestinal epithelium. At λ = 633 nm, μ_s is about 7 mm^{-1} for esophagus tissue [24] and about 20 mm^{-1} for normal colon mucosa [25]. The range of tissue depths spanned by L = 4.8 to 8 for esophagus and colon is 340 μm to 570 μm and 120 μm to 200 μm, respectively. In addition, these results shows that for single axis only minimal changes occur in the dynamic range with approximately a factor of 2 difference in optical thickness L, while for dual axes significant changes occur over this thickness range. Furthermore, scattering does not appear to alter the FWHM of the axial response for either single or dual axes over this range of lengths.

4. Geometric differences produced by off-axis detection

The superior axial response of the dual axes confocal architecture has a simple geometric explanation. When the mirror moves away from the focal plane by $\pm\Delta$, the centroid of the beam is steered away from the optical axis by $\pm 2\Delta\sin\theta$ from where the center of the pinhole is located [20]. Even taking into consideration diffraction and the broadening of the out-of-focus beam, the beam intensity decreases exponentially when $\Delta > D/2$ (for $\theta = 30°$). But in the single axis case, many of the photons scattered near the vicinity of the focal plane ($\pm\Delta$) are collected by the detector through the pinhole. Thus, the spatial filtering effect by a pinhole for the single axis configuration is not as effective as that for dual axes. The implication of this effect for imaging deep in tissue is evident. In the single axis case, many of the multiple scattered photons that arrive from the same direction as that of the ballistic photons, starting from the surface to deep within the tissue, are collected by the detector despite the presence of a pinhole to filter the out-of-focus light. This explains why in Fig. 9a the single axis configuration has a large noise component alongside the ballistic component. Thus, the dual axes confocal architecture provides optical sectioning capability that is superior to that of the conventional single axis design in terms of SNR and dynamic range, and this result can be generalized to a range of relevant pinhole sizes. As a result, the dual axes architecture allows for imaging with greater tissue penetration depth, thus is capable of providing images in the vertical cross-section with high contrast. The implementation of the dual axes confocal configuration to an endoscope compatible instrument for collection of both reflectance and fluorescence has significant implications for *in vivo* imaging by providing both functional and structural information deep below the tissue surface.

4. Tabletop dual axes confocal imaging instruments

The dual axes confocal architecture was first implemented as a tabletop instrument using readily available optical components to demonstrate the proof of concept of off-axis illumination and collection with post-objective scanning. In particular, the primary advantages of the dual axes configuration including high dynamic range and deep tissue penetration are revealed by vertical cross-sectional images with either reflectance or fluorescence. The combination of these two imaging modes forms a powerful strategy for integrating structural with functional information.

The dual axes optical design incorporates a solid immersion lens (SIL) made from a fused-silica hemisphere at the interface where the two off-axis beams meet the tissue. This refractive element minimizes spherical aberrations that occur when light undergoes a step change in refractive index between two media. The curved surface of the SIL provides a normal interface for the two beams to cross the air-glass boundary. Fused silica is used because its index of refraction of $n = 1.45$ is closely matched to that of tissue. Note that as the beams are scanned away from their neutral positions, they will no longer be incident to the surface of the SIL and small aberrations will occur. Another feature of the SIL is that its curved surface increases the effective NA of the beams in the tissue by a factor of n, the index of refraction, and produces higher resolution and light collection efficiency. On the other hand, the SIL acts to reduce the scanning displacement of the beams in the tissue by a factor of $1/n$ so that larger deflections are needed to achieve the desired scan range.

J. Horizontal cross-sectional imaging instrument

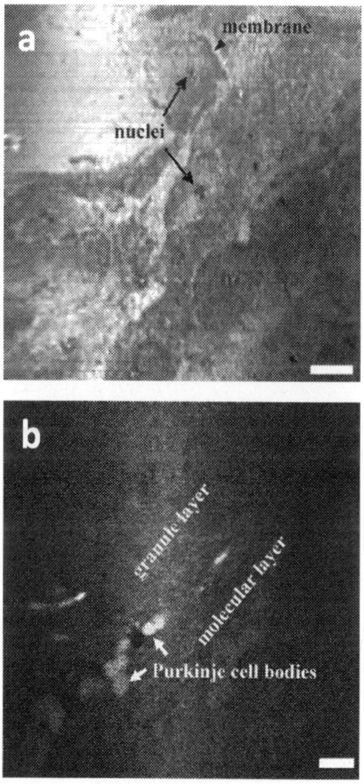

Fig. 11. Horizontal cross-sectional dual axes images ex vivo. a) Squamous esophageal mucosa collected at z = 0 μm with λ = 488 nm reveals sub-cellular features, including cell nuclei (arrows) and membrane (arrowhead), scale bar 20 μm. b) Normal colonic mucosa at z = 150 μm with λ = 1.3 μm illumination reveals circular crypts with colonocytes surround the lumen and lamina propria (LP) filling space in between the crypts, scale bar 50 μm.

Reflectance imaging takes advantage of subtle differences in the refractive indices of tissue micro-structures to generate contrast. The backscattered photons can provide plenty of signal to overcome the low NA objectives used for light collection in the dual axes configuration. The first reflectance images were collected with a tabletop system that used a 488 nm semiconductor laser that delivered illumination into a single mode optical fiber that was focused by a set of collimating lenses with NA = 0.16 to a spot size with ~400 μW of power [16]. These parameters produced a transverse and axial resolution of 1.1 and 2.1 μm, respectively. The reflected light was collected by a complementary set of optics. The off-axis illumination and collection was performed at $\theta = 30°$ to the main optical axis. Reflectance images were collected in horizontal cross-sections of freshly excised specimens of esophagus *ex vivo*. As shown in Fig. 11a, the cell membrane and individual nuclei of squamous (normal) esophageal mucosa can be appreciated in the image collected at z = 0 μm, scale bar 20 μm.

Much greater image contrast can be achieved with fluorescence imaging where the use of optical reporters, such as GFP, and exogenous probes can reveal over expression of molecular targets. The same tabletop dual axes microscope was also used to collect fluorescence images with a long pass filter to block the excitation light and photomultiplier tube (PMT) for detection [26]. In Fig. 11b, a fluorescence image of the cerebellum of a transgenic mouse that constituitively expresses GFP under the control of a β-actin promoter at a depth of z = 30 μm is shown, scale bar 50 μm. Purkinje cell bodies (arrows) can be seen as large round structures aligned side by side in a row, separating the granule layer and the molecular layer.

K. Vertical cross-sectional imaging instrument

1. Reflection imaging mode

In order to collect vertical cross-sectional images, heterodyning can be used to provide a coherence gate that filters out illumination photons that are multiply-scattered and travel over longer optical paths within the tissue [17]. This approach is demonstrated with a fiber optic Mach-Zehnder interferometer, shown in Fig. 12a. A broadband near-infrared source produces light centered at $\lambda = 1345$ nm with a 3 dB bandwidth of 35 nm and a coherence length in tissue of ~50 μm. A fiber coupler directs ~99% of the power to the illumination path, which consists of a single mode optical fiber (SMF_1) with a collimating (CL_1) and focusing lens (FL_1) with NA = 0.186. The axes of illumination and collection are oriented at $\theta = 30°$ to the midline. Light reflected from the tissue is collected by the second set of focusing (FL_2) and collimating (CL_2) lenses into another single mode fiber (SMF_2). The lens and fiber parameters are the same for both the illumination and collection beams. The fiber optic coupler directs ~1% of the source into a reference beam which is frequency shifted by an acousto-optic modulator at 55 MHz for heterodyne detection. An adjustable optical delay is used to increase the signal by matching the optical path length of the reference beam to that of the ballistic photons. An adjustable optical delay is used to increase the signal. In addition, a polarization controller consisting of two half-wave plates and a single quarter-wave plate is used to maximize the signal. The reference and collection beams are combined by a 50/50 coupler and the resulting heterodyne signal is detected by a balanced InGaAs detector (D_1, D_2) with a bandwidth of 80 MHz. The resulting electronic signal is then processed with a band pass filter (BPF) with a 3 MHz bandwidth centered at 55 MHz, then demodulated (DM), digitized by a frame grabber (FG), and displayed (D).

In this heterodyne detection scheme, the reference beam essentially provides amplification of the weak collection beam via coherent optical mixing, and enables the measurement of reflected light with a dynamic range larger than 70 dB. Post-objective scanning is performed with the scan mirror (SM) placed distal to the objective lenses. Reflectance images were collected from fresh biopsy specimens taken from the squamo-columnar junction of subjects with Barrett's esophagus. Specimens with dimensions of ~3 mm were resected with jumbo biopsy forceps, and the mucosal surface was oriented normal to the z-axis. Vertical cross-sectional images were collected with depth of 1 mm. From Fig. 12b, squamous (normal) mucosa is present over the left half of the image with an intact epithelium (EP). The other structures of normal esophageal mucosa, including the muscularis mucosa (MM), sub-mucosa (SM), and muscularis propria (MP), can also be identified. Columnar mucosa consistent with intestinal metaplasia is seen over the right half of the image, and reveals the presence of pit epithelium (PE) [17]. These findings correlate with the tissue micro-structures seen on histology.

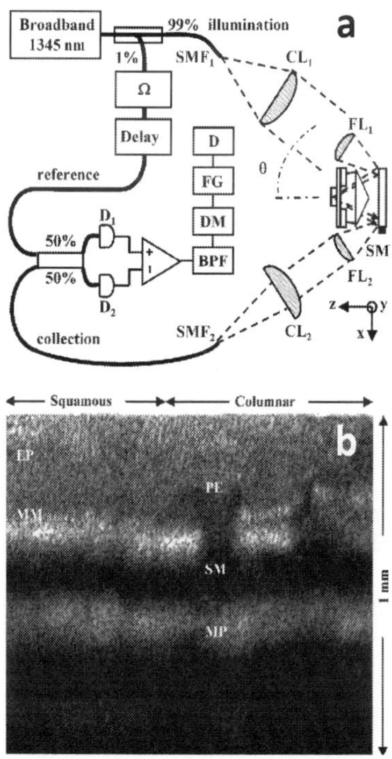

Fig. 12. Vertical cross-sectional dual axes confocal reflectance images ex vivo. a) Schematic of optical circuit for heterodyne detection, details discussed in the text. b) Reflectance image of squamo-columnar junction in esophagus with vertical depth of 1 mm. Squamous (normal) mucosa reveals epithelium (EP) and muscularis mucosa (MM) over left half. Columnar (intestinal metaplasia) mucosa shows pit epithelium (PE) over right half. Submucosa (SM) and musclaris propria (MP) are seen on both sides.

2. *Fluorescence imaging mode*

Fluorescence detection adds an entirely new dimension to the imaging capabilities of the dual axes architecture. Detection in this mode offers an opportunity to achieve much higher image contrast compared to that of reflectance and is sensitive to labeled molecular probes that can identify specific tissue and cellular targets. These features provide a method to perform functional as well as structural imaging, and allows for the study of a wide variety of molecular mechanisms. Although the use of low NA objectives in the dual axes configuration reduces the collection efficiency of the optics by a factor of ~NA^2, this deficit can be overcome by the use of bright fluorophores. In order to achieve the deep tissue penetration depths possible with the off-axis collection of light, a near-infrared laser at 785 nm is used as the source and a PMT with a long pass filter to block the excitation light is used as the detector [27]. The large dynamic range (>40 dB) of the dual axes confocal architecture encountered with collection of vertical cross-sectional images requires modulation of the PMT gain to compensate for the rapid decrease in fluorescence signal with axial depth due to tissue absorption and scattering.

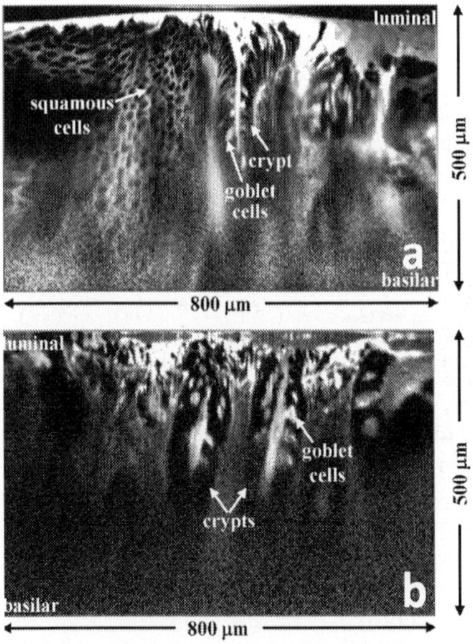

Fig. 13. Vertical cross-sectional dual axes confocal fluorescence images ex vivo. a) Squamo-columnar junction in esophagus with vertical depth of 500 µm. Squamous mucosa present over left half. Columnar (intestinal metaplasia) mucosa over right half shows crypts with goblet cells. b) Colon. Many goblet cells can be seen in dysplastic crypts from a flat colonic adenoma.

In Fig. 13, vertical cross-sectional fluorescence images of a) esophagus and b) colon collected with a tabletop dual axes confocal microscope are shown [27]. These specimens were incubated with a near-infrared dye, LI-COR IRDye® 800 CW NHS Ester (LI-COR Biosciences, Inc) prior to imaging after being freshly excised during endoscopy. These images were collected at 2 frames per second with a transverse and axial and resolution of 2 and 3 µm, respectively. With use of post-objective scanning, a very large FOV of 800×500 μm^2 deep was achieved. In Fig. 13a, the specimen was collected from the squamo-columnar junction of a patient with Barrett's esophagus. Over the left half of the image, the individual squamous cells from normal esophageal mucosa can be seen in the luminal to the basilar direction over a depth of 500 μm. Over the right half of the image, vertically oriented crypts with individual mucin-secreting goblet cells associated with intestinal metaplasia can be appreciated as brightly stained vacuoles. This diseased condition is associated with greater than 100 fold relative risk of developing cancer in the esophagus. In Fig. 13b, the specimen was collected from a flat colonic adenoma, and the image reveals vertically oriented dysplastic crypts with individual goblet cells.

Volume rendering can also be performed with the dual axes confocal microscope to illustrate three-dimensional (3D) imaging capabilities. These views are important for tracking cell movements, observing protein-protein interactions, and monitoring angiogenic development. A xenograft mouse model of glioblastoma multiforme has been developed by subcutaneously implanting ~10^7 human U87MG glioblastoma cells in the flank of a nude mouse. Horizontal cross-sectional fluorescence images were collected with a tabletop instrument when the tumors reached ~1 cm in size. The mice were anesthetized for the *in vivo* imaging session, and indocyanine green (ICG) at a concentration of 0.5 mg/ml was injected intravenously to produce contrast. A skin flap overlying the tumor was exposed, and horizontal cross-sectional images were collected with a FOV of 400 x 500 μm^2. A fluorescence image collected at 50 μm below the tissue surface, shown in Fig. 14a, reveals that the glioblastoma has developed a dense, complex network of tortuous vasculature. A total of 400 horizontal cross-sectional images acquired at 1 μm increments were used to

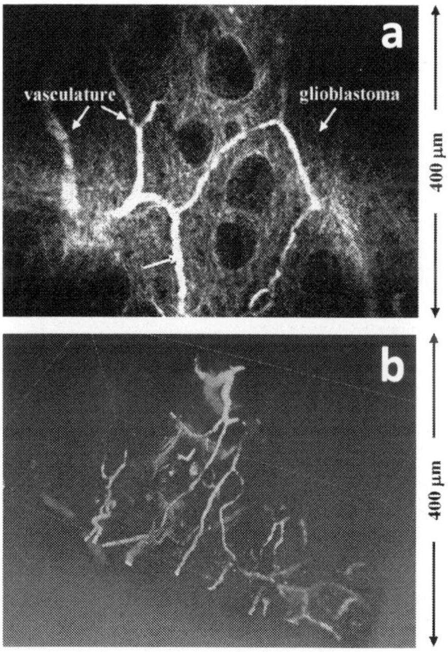

Fig. 14. Dual axes confocal fluorescence images in small animal models. a) A horizontal cross-sectional fluorescence image of a human U87MG xenograft glioblastoma tumor implanted subcutaneously in the flank of a nude mouse was collected in vivo at 50 μm depth using i.v. indocyanine green (ICG). b) A 3D volumetric image is generated from a z-stack of 400 sections collected at 1 μm intervals.

generate the 3D volumetric image, shown in Fig. 14b. Volume rendering was performed using Amira™ modeling software.

5. Mems-scanner based dual axes confocal imaging instruments

The long working distance created by the low NA objectives in the dual axes architecture provides space for a scanning mechanism to be placed in the post-objective position. This location is a key feature of the design that allows for scaling down of the optics to millimeter dimensions. Moreover, for *in vivo* imaging, a fast scan rate is needed to overcome motion artifacts introduced by organ peristalsis, heart beating, and respiratory activity, typically requiring a frame rate of >4 per second. As a result, we have developed a MEMS-scanner based miniature and endoscope compatible imaging instruments without loss of performance. This strategy is much more complex than other approaches being developed, but is well suited to meet the size and speed requirements for *in vivo* imaging in a compact package [28, 29].

L. MEMS Scanner Structure

The schematic of the 2-D MEMS scanner is shown in Fig. 15a. It has a gimbal structure, and is electrostatically actuated by self-aligned, vertical combdrives to give large deflection. The mirror can be actuated with respect to the frame by rotating around the springs that define the inner axis. The frame supporting the mirror can be actuated with respect to the substrate by rotating around the springs that define the outer axis. Fig. 2b shows the cross-sections of various structures of the device, which is made in double-stacked silicon-on-insulator (SOI). The two device layers are each 30 μm thick with an oxide layer of 0.38 μm in between. The substrate thickness is 530 μm, while the oxide layer between the lower device layer and substrate has a thickness of 1 μm. The thick device layers increase the tilt range of the mirror by deeper comb engagement, and lead to a larger FOV. The mirror, movable combteeth, and inner torsional springs are fabricated in the upper device layer. The fixed combteeth, outer torsional spring, and frame consist of double-stacked layers. A backside window is located below the gimbal structure to release the device and allow large-range motion. Four actuation voltages are supplied to the lower layer to actuate both sides of each axis (outer: V1 and V2, inner: V3 and V4, in Fig. 15a). The upper layer and substrate are both grounded. Electrical isolation between the device layers and the substrate is provided by buried oxide layers, as seen in Fig. 15b. The double-stacked layers of the outer torsional spring and frame deliver actuation voltages and ground to the inner combdrives.

Alignment of the off-axis illumination and collection are achieved with two mirrors connected together by a strut. The size of the mirrors (600 μm x 650 μm) and the distance between them (1.51 mm) enable an off-axis half-angle, θ, of 24.3°. The inner combdrives are placed on the connecting beam between the two mirrors and the inner springs are recessed into the mirror sides, to allow the die size to be reduced to 3.4 mm x 2.9 mm to fit inside the scanhead package. The frame width is designed to be 150 μm to prevent stress-induced curvature of the gimbal. In order to increase the torque with the same number of combs, the moment arm is lengthened by placing the outer combdrives further away from the outer torsional spring.

M. Theoretical Analysis

The outer spring consists of two silicon device layers (each 30 μm thick) with an oxide layer (0.38 μm) in between, and delivers three different voltages to the inner frame. Its dimensions

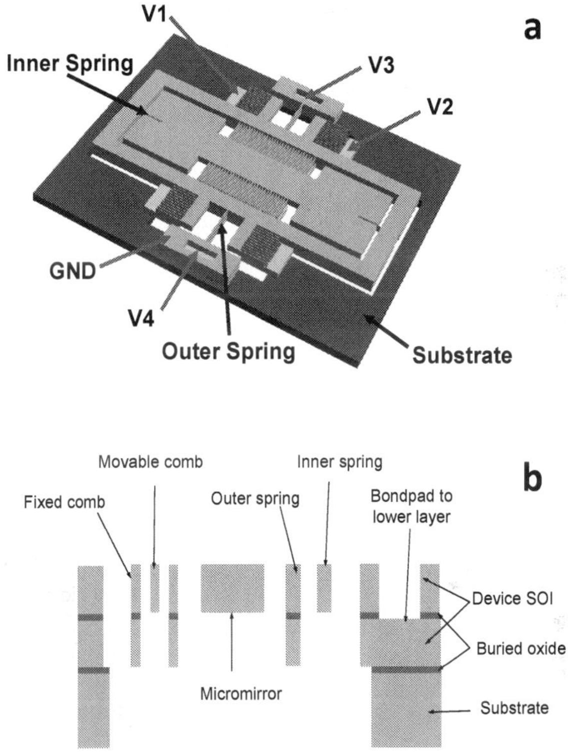

Fig. 15. a) Schematic drawing of the 2-D MEMS scanner. b) Cross-sectional view of various structures of the scanner.

are 60.38 μm thick and 350 μm long. The inner spring consists of one silicon layer and is therefore 30 μm thick. The mechanical torque of the torsional spring can be expressed as

$$T_{\phi}(\phi) = k_{\phi}(\phi)\,, \tag{13}$$

where k_{ϕ} is the torsional spring constant, and ϕ is the mechanical deflection angle. Both torsional springs are rectangular, so k_{ϕ} is given as follows [30].

$$k_{\phi} = \frac{2G}{3}\frac{tw^3}{l}\left[1 - \frac{192}{\pi^5}\frac{w}{t}\tanh\left(\frac{\pi}{2}\frac{t}{w}\right)\right], \text{ for } w < t \tag{14}$$

Here G is the shear modulus given by $G = E / 2(1+\nu)$, where E is the Young's modulus, and ν is Poisson's ratio. The parameters l, w, t represent the length, width, and thickness of the spring, respectively.

When an electrical voltage, V, is applied between the fixed and movable combs, the electrical torque, T_e, is given as follows.

$$T_e(\phi, V) = NV^2 \frac{\partial C_{unit}(\phi)}{\partial \phi} \tag{15}$$

Here N is the number of comb pairs, and C_{unit} is the capacitance of a comb pair. In steady-state, the mechanical torque of the spring is balanced by the electrostatic torque of the combdrives, and torques expressed by equations (13) and (15) are equal. A hybrid program which combines a 2-D finite element method (FEM) with analytical calculation [31] is used for generating the simulated DC transfer curves of Fig. 16.

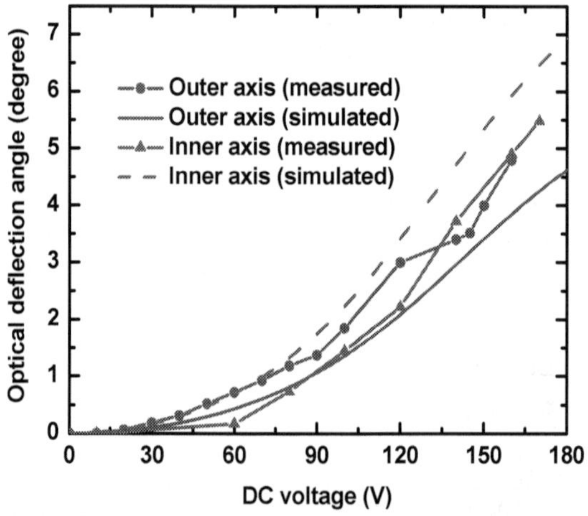

Fig. 16. Static optical deflection curve of MEMS scanner.

N. Device Fabrication

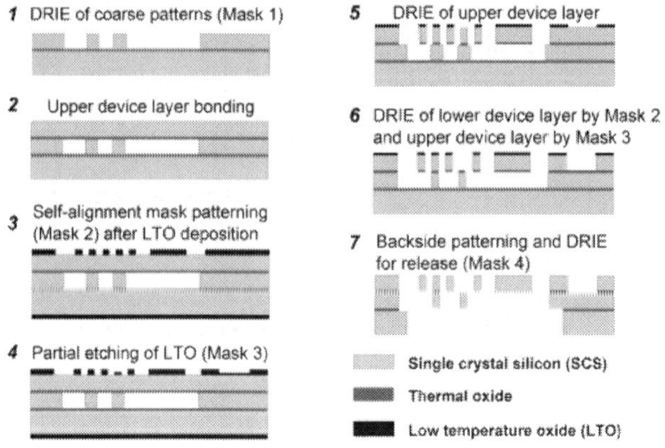

Fig. 17. MEMS scanner fabrication process flow.

The design of this mirror uses a gimbal geometry to perform scanning in the horizontal (X-Y) plane, and rotation around an inner and outer axes defined by the location of the respective springs. The overall structure has a barbell shape with two individual mirrors that have active surface dimensions of 600x650 μm². A 1.51 mm long strut connects these two mirrors so that the illumination and collection beams preserve the overlapping focal volume in the tissue. The fabrication process flow, shown in Fig. 17, starts with a SOI wafer composed of a silicon substrate, buried oxide, and silicon lower device layers that are 530, 0.38, and 30 μm thick, respectively [32]. A deep-reactive-ion-etch (DRIE) of coarse patterns, including the combdrives and trenches, is performed on the SOI wafer with Mask 1 (step 1). Next, an oxide layer is grown on a plain silicon wafer using a wet oxidation process. This wafer is then fusion bonded onto the etched surface of the SOI wafer (step 2). The yield is increased by bonding in vacuum, and the bonded plain wafer is ground and polished down to 30 μm thickness, forming the upper device layer.

The two oxide layers between the silicon layers provide electrical isolation, and act as etch stops, allowing for precise thickness control. The front side of the double-stacked SOI wafer is patterned and DRIE etched to expose the underlying alignment marks in the lower device layer. Then, a low temperature oxide (LTO) layer is deposited on both sides of the wafer. The front side layer is patterned by two masks. The first mask (Mask 2) is the self-alignment mask (step 3), and is etched into the full thickness of the upper LTO layer. The second mask (Mask 3) is mainly for patterning the electrodes for voltage supplied to the lower device layer (step 4). It goes through a partial etch leaving a thin layer of LTO. The alignment accuracy of each step needs to be $> g/2$, where g is the comb gap. Since most devices have 6 μm comb gaps, this leads to a required alignment accuracy of better than 3 μm.

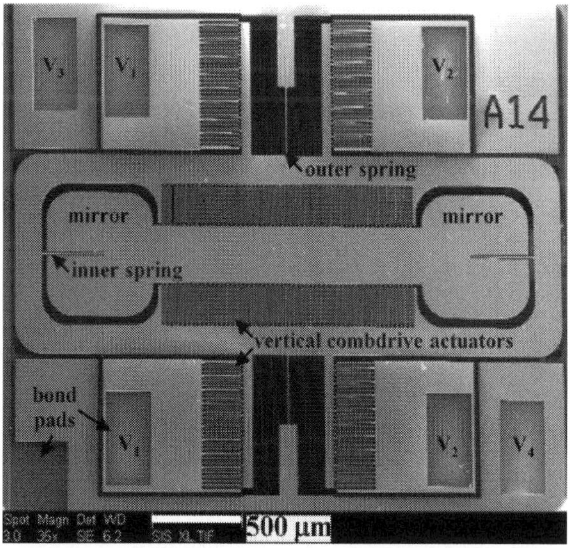

Fig. 18. SEM of 2D gimbaled MEMS scanner, scale bar 500 μm.

Good alignment accuracy is important to minimize failures due to electrostatic instability during actuation. These three masks eventually define the structures in the upper, lower, and double-stacked layers. After the front side patterning is done, the LTO layer on the wafer back side is stripped (step 5). The wafer is cleaned and photoresist is deposited on the back side. Then, front side alignment marks are patterned. Next, the upper silicon layer is etched with the features of Mask 2 in DRIE. Then, a thin LTO and buried oxide layer is anisotropically dry-etched. Finally, the lower and upper silicon layers are etched (DRIE) simultaneously with features patterned by Mask 2 and 3, respectively (step 6).

For backside processing, the wafer is bonded to an oxidized handle wafer with photoresist. The back side trenches are patterned with Mask 4 on photoresist (step 7). The back side trench should etch through the substrate to release the gimbal structure, so handle wafer bonding and thick resist is required for DRIE. Alignment to the front side features are accomplished by aligning to the previously etched patterns. After the substrate (530 μm) is etched by DRIE, the process wafer is separated from the handle wafer with acetone. After wafer cleaning, the exposed oxide layer is directionally dry-etched from the back side. Finally, the remaining masking LTO and exposed buried oxide layer is directionally etched from the front side.

O. Device Characterization

The two-dimensional (2D) MEMS scanner is actuated by electrostatic vertical combdrive actuators [33]. Electrostatic actuation in each direction is provided by two sets of vertical comb actuators that generate a large force to produce sizable deflection angles. The scanning electron micrograph (SEM) of the scanner is shown in Fig. 18. There are 4 actuation voltages (V_1, V_2, V_3, and V_4) that power the device. The parameters of the scanner are characterized for quality control purposes prior to use in the miniature dual axes confocal microscope. First, the flatness of the mirror is measured with an interferometric surface profiler to identify micro-mirrors that have a peak-to-valley surface deformation <0.1 μm. The scanner is metalized with 10 nm thick aluminum (reflectivity = 67% at 785 nm wavelength) to increase reflectivity.

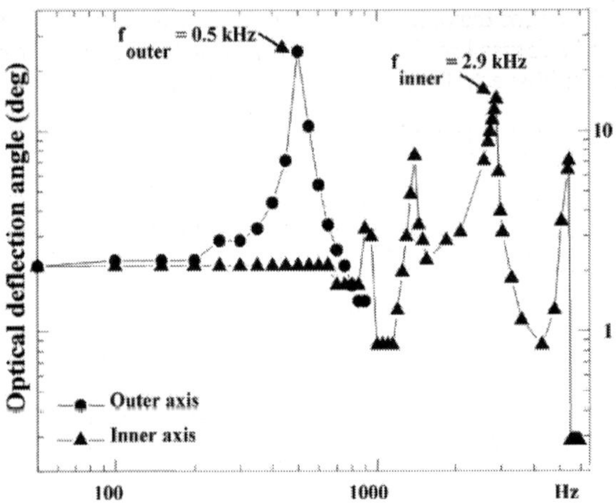

Fig. 19. Frequency response of MEMS mirror shows resonant peaks at 0.5 kHz (outer axis) and 2.9 kHz (inner axis) to achieve real time operation.

The radius of curvature of the mirror is greater than 60 cm with an average surface roughness of 7 nm. Static optical deflections of ±1.5 deg at 180 volts and ±4.25 deg at 150 volts are achieved for the outer and inner axis, respectively. The resonant frequencies are 0.5

kHz with ±10 deg optical angle at resonance for the outer axis and 2.9 kHz with ±17 deg optical angle at resonance for the inner axis. The frequency response of the device is shown in Fig. 19. The parametric resonances can sometimes be observed in the inner axis near frequencies of $2f_0/n$, where n is an integer ≥1 [34]. This phenomenon is caused by the non-linear response of the torsional combdrives, which leads to sub-harmonic oscillations.

6. Dual axes scanhead

P. Scanhead design

The design and integration of the miniature dual axes scanhead is a very challenging part of the development of this novel imaging technique because of the small size required for compatibility with medical endoscopes. This process requires a package that allows for precise mounting of the following optical elements: 1) two fiber-coupled collimators, 2) 2D MEMS scanner, 3) parabolic focusing mirror, and 4) hemispherical index-matching solid-immersion-lens (SIL) [35]. The basic design of the miniature scanhead is shown in Fig. 20. Two collimated beams are focused at an inclination angle θ to the z-axis by a parabolic mirror with a maximum cone half-angle α to an overlapping focal volume below the tissue surface after being deflected by the 2D MEMS scanner. The flat side of the SIL is placed against the tissue, and the curve surface accommodates the incident beams at normal incidence to minimize aberrations. The parabolic mirror is fabricated using a replicated molding process that provides a surface profile and smoothness needed for diffraction-limited focusing of the two collimated beams. Once the beams are aligned parallel to each other, the parabolic mirror then provides a "self-aligning" property that forces the focused beams to intersect at a common focal point below the tissue surface. Focusing is performed primarily by the parabolic mirror which is a non-refractive optical element with an NA of 0.12. This feature allows for the optical design to be achromatic. That is, light over a broad spectral regime can be focused to the same point below the tissue surface simultaneously, allowing for future multi-spectral imaging to be performed.

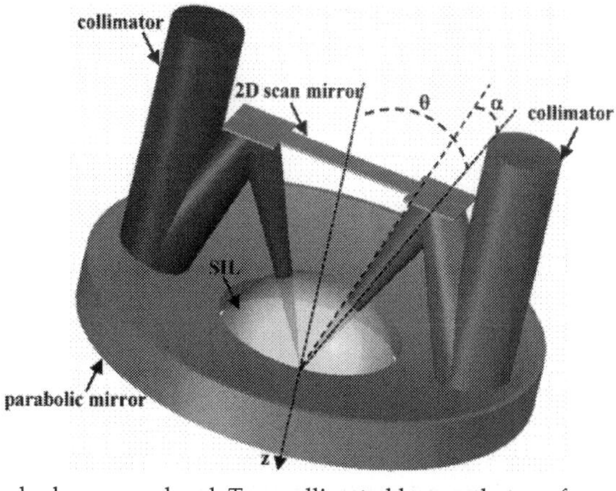

Fig. 20. Miniature dual axes scanhead. Two collimated beams that are focused by a parabolic mirror at angle θ to the z-axis for en face scanning by the 2D MEMS mirror. The solid-immersion lens (SIL) minimizes aberrations to the incident beams.

Q. Scanhead alignment and packaging

Alignment of the two beams in this configuration is a key step to maximizing imaging performance. This step is accomplished by locating the two fiber-pigtailed collimators in a pair of v-grooves that are precision machined into the housing, as shown in Fig. 21a [35]. An accuracy of 0.05 deg can be achieved in aligning the two beams parallel to one another using the v-grooves with pre-assembled fiber collimators. Additional precision in alignment can be attained with use of Risley prisms (optical wedges) introduced into the light paths to provide fine steering of the collimated beams to bring the system into final alignment. These prisms are angled at 0.1 deg, and can be rotated to steer the collimated beam in an arbitrary direction over a maximum range of ~0.05 deg. This feature maximizes the overlap of the two beams after they are focused by the parabolic mirror. Two wedges are used in each beam so that complete cancellation of the deflection by each can be achieved, if needed, to provide maximum flexibility.

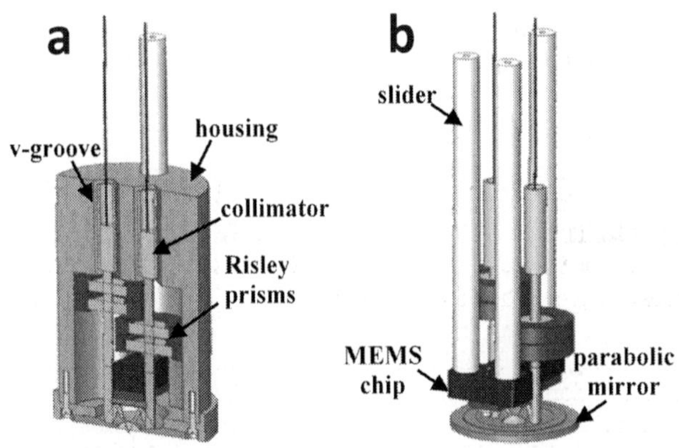

Fig. 21. Alignment and assembly of dual axes scanhead. a) Precision machined v-grooves and Risley prisms provide coarse and fine alignment, respectively, of the two beams. b) Axial (z-axis) displacement of the MEMS chip is made with a slider mechanism.

Axial (z-axis) displacement of the MEMS chip is performed with a computer-controlled piezoelectric actuator (Physik Instrumente GmbH, P-783.ZL actuator, and E-662.LR controller) that moves a slider along 3 mechanical supports, shown in Fig. 21b. This feature adjusts the imaging depths to collect a stack of *en face* images to produce the 3D volume rendered images. The distal end of the slider has a mounting surface to attach the printed circuit board (PCB), which supports the MEMS chip, wire bonding surfaces, and soldering terminals.

A mixture of conductive epoxy (adhesive resin ECCOBOND Solder 56 C and Catalyst 9, Emerson & Cuming, Inc.) is used to attach aluminum-1% silicon bonding wire (Semiconductor Packaging Materials, Inc.) from the bond pads on the MEMS die to that on the PCB. Electrical power is delivered to the mirror via wires that run through the middle of the housing, and are soldered onto the PCB terminals. The z-axis translational stage consists of a closed-loop piezoelectric linear actuator. Finally, the scanhead assembly is covered and sealed from the environment using UV-curing glue to prevent inward leakage of bodily fluids.

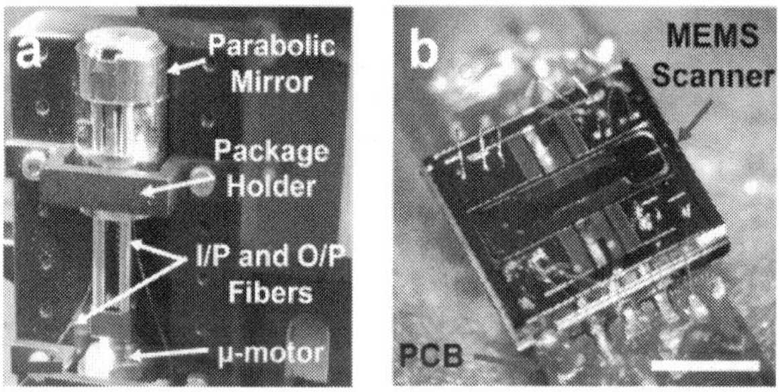

Fig. 22. a) Assembly of the dual axes confocal scanhead mounted on a V-block. b) Gimbaled 2D MEMS scanner wire bonded onto the PCB, scale bar 2 mm.

Packaging of the 10 mm diameter scanhead mounted on a V-block stage is shown in Fig. 22a. A piezoelectric (micro) μ-motor is used to perform vertical depth translation (z-axis). The MEMS scanner (die size is 3.2 (w) × 2.9 (h) mm^2) mounted on the PCB is shown in Fig. 22b.

R. Instrument control and data acquisition

Both the data acquisition and MEMS actuation systems are controlled using LabVIEW™ with Vision Acquisition software package and two National Instruments data acquisition (DAQ) boards (PXI-6711 and PXI-6115). The frequency and amplitude of the actuation signals control the frame rate and FOV of the MEMS scanner. There are 4 live and 1 ground wires that provide voltage to the device and are connected to the wirebond pads on the PCB via an ultrasonic wedge bonding technique.

For each 2D *en face* image, the MEMS scanner is resonantly driven 180 deg out of phase to maximize the linear region of the angular deflection [14, 29] around the outer axis (V_1 and V_2) at 1.22 kHz with a unipolar sine wave at a maximum of 70V, while rotation around the MEMS scanner inner axis (V_3 and V_4) is driven 180 deg out of phase in the DC mode (5 Hz) with a unipolar sawtooth waveform at a maximum of 200 V (AgilOptics, Inc). The unipolar sawtooth waveform is smoothed at the transition edges to mitigate higher frequency ringing from the inner axis. The step size and depth scan range of the piezoelectric actuator (vertical translation) can be adjusted to optimize the acquisition of the 3D datasets.

The PMT gain is synchronously adjusted to compensate for reduced light collected at increased tissue depths. Automated frame averaging and display can be performed to reduce noise and improve image quality during imaging. 2D *en face* images from the analog input channel are acquired and displayed in real time to enable continuous monitoring or visualization of the sample. All images are acquired in 16-bit data format. 3D volumetic data can be rendered by post-processing using Amira®software (Visage Imaging, Inc).

7. Prototype systems

S. Handheld instrument

We first developed a 10 mm diameter handheld instrument, schematic shown in Fig. 23a [14, 29]. A semiconductor laser delivers 25 mW of light (λ = 785 nm) into a single mode fiber (SMF, Fibercore Limited, SM750). The fiber terminates in a 1.8 mm diameter gradient index (GRIN) collimator (GRINTECH, GmbH). The output beam diameter (1/e^2) from the collimator is 0.9 mm. The half angle θ between the input (illumination) and output (collection) beams is 24.3°. The input beam is focused by an aluminum coated parabolic mirror (PM) with the focal length of 4.6 mm (Anteryon BV) and reflects off the first mirror surface of the MEMS scanner. The center-to-center distance between the two collimators is 3.7 mm. The focused beam continues through a fused silica SIL (hemispheric lens) until it reaches the focal plane below the tissue surface. The SIL has a refractive index (1.47) that is similar to that of the tissue and this material was chosen for index matching. The beams enter the air-silica interface at normal incidence to minimize aberrations as the focal point is scanned. Scattered light from the overlapping focal volume is collected through the optical window provided by the SIL and reflected off the MEMS mirror to the opposite surface of the parabolic mirror. The collected light is then focused onto the output fiber collimator for delivery to the PMT.

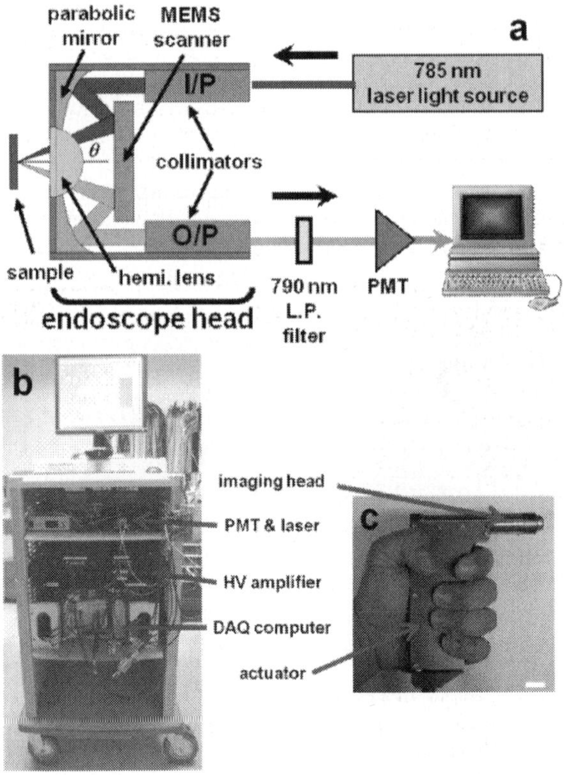

Fig. 23. Handheld prototype system. a) Schematic of complete instrument. b) Portable system demonstrated. c) Packaged handheld (10 mm diameter) dual axes confocal microscope with piezoelectric actuator in the handle, scale bar 10 mm.

As the MEMS mirror raster scans the overlapping beams, the 2D *en face* image is continuously displayed on a computer monitor using a frame grabber and image acquisition software. Intensity mapping of each 2D *en face* image is performed by reading the MEMS scanner driving voltages and estimating focal beam trajectory. A 3D volumetric stack is created with post-processing and rendering a series of 2D *en face* images. Each 3D volumetric image is obtained by translating the MEMS scanner with the piezoelectric micro-motor in the *z*-direction under computer-control. Imaging can be performed in reflectance or fluorescence mode by inserting a 790 nm long pass optical filter (LP-02-785RU-25, Semrock, Inc.) in the collection path for the latter case. The maximum output laser power on the sample is 2 mW. A photograph of a fully-packaged miniature dual axes confocal microscope is shown in Fig. 23c.

T. Endoscope-compatible instrument

We scale down the basic design of the 10 mm diameter handheld instrument to develop the 5.5 mm endoscope-compatible version, shown in Fig. 24 [36]. This prototype uses the same replicated parabolic focusing and MEMS mirrors as that employed in the larger prototype. A pair of smaller (1 mm) diameter fiber-coupled GRIN collimator lenses is used in the smaller version. Alignment is provided by a pair of 1 mm diameter rotating wedges (Risley prisms), which are inserted into the path of one of the collimated beams. The collimators and Risley prisms are both located by precision wire-EDM machined v-grooves and epoxied into place with UV curing glue. As with the larger prototype, the combined precision of the v-grooves and the pointing accuracy of the pre-assembled fiber collimators allow for the collimated beams to become parallel to each other to within ~0.05 deg accuracy. The alignment wedges have a small (0.1 deg) angle, which allows for steering of a collimated beam over a range of about 0.05 deg in any direction as each wedge is rotated.

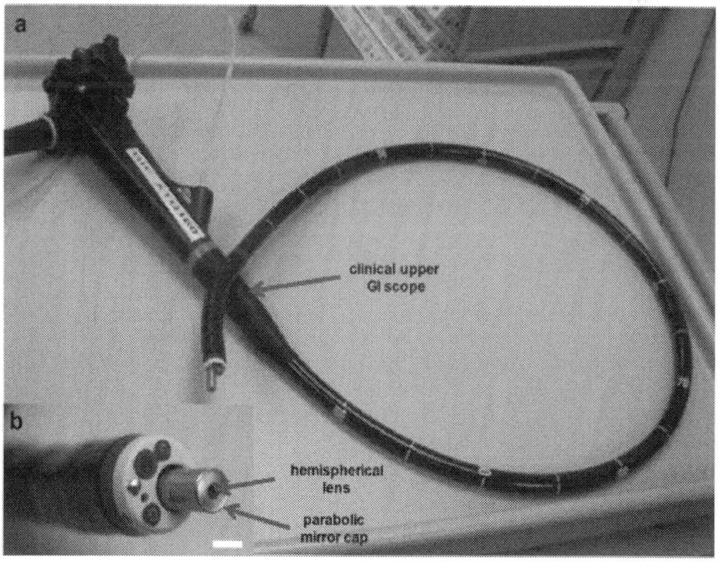

Fig. 24. Endoscope-compatible dual axes confocal microscope. a) Microscope passes through the instrument channel of an Olympus XTQ 160 therapeutic upper endoscope that has a 6 mm diameter instrument channel. b) Distal end of endoscope shows the protruding dual axes microendoscope.

This smaller package design also accommodates a slider mechanism, which is used for axial (z-axis) scanning of the MEMS chip to provide a variable imaging depths within the tissue and for generating 3D volumetric images. This smaller slider mechanism comprises a single rod, which moves within a precision hole drilled through the housing. The MEMS chip is mounted by an adhesive to a PCB, which is in-turn mounted onto the slider. The PCB provides bondpads to accommodate wire bonding to the MEMS chip and also to provide soldering terminals for the external control wires that power the scan mirror. In Fig. 24a, the endoscope-compatible dual axes confocal microscope is shown inserted through the 6 mm diameter instrument channel of a therapeutic upper endoscope (Olympus GIF XTQ160). A magnified view of the distal tip is shown in Fig. 24b.

8. Imaging results

U. Reflectance imaging

Instrument characterization is performed in reflectance mode by imaging a chrome surface of a standard (USAF) resolution target. It is also used as a sample to measure the image resolution and FOV. The transverse resolution was measured by the knife-edge method, defined by 10% to 90% of maximum intensity points, and found to be 5 μm [36]. The axial resolution, defined by FWHM, is measured by translating a plane mirror in the z-direction and was found to be 7 μm.

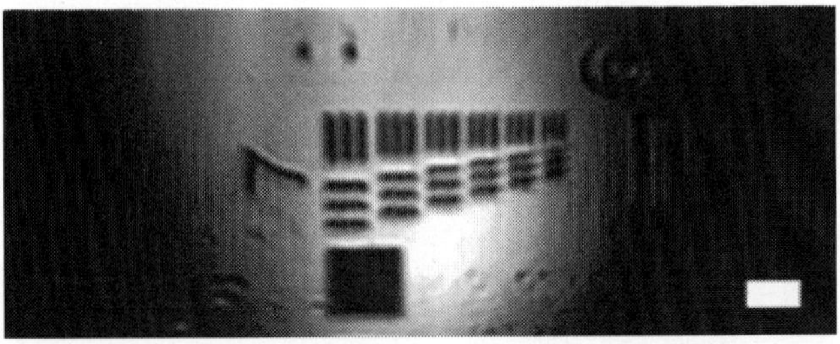

Fig. 25. Reflectance image of standard (USAF) resolution target collected with handheld dual axes confocal microscope, scale bar 20 μm.

Fig. 25 shows a reflectance image collected with the handheld confocal microscope that reveals clear visualization of group 7 of the USAF resolution target. The measured values are slightly larger than the theoretical resolutions of 4.5 μm for the transverse dimensions, and 6.0 μm for the axial dimension. This is mainly due to the decrease in effective NA of the imaging system from the truncation of both input and output collimated beams by the width dimension of the MEMS scanner die. All acquired images are captured at 5 frames/second with the largest FOV of 800×450 μm^2 (900×506 pixels^2). This FOV is much larger than that most other miniature confocal instruments, and is achieved with use of post-objective scanning.

V. Ex vivo fluorescence imaging

1. Handheld dual axes confocal instrument

The 3D fluorescence imaging capability of the handheld dual axes confocal instrument is shown in Fig. 26. Excised tissue specimens of normal and dysplastic colonic mucosa are soaked in 0.5 mg of LI-COR IRDye® 800 CW NHS Ester (LI-COR Biosciences, Inc) diluted in 10 ml of phosphate-buffered saline (PBS) at neutral pH for 5 minutes and then rinsed with water to remove excess dye. After imaging, the specimens are fixed in 10% buffered formalin, cut into 5 μm sections, and processed for histology with hematoxylin and eosin (H&E). All *ex vivo* images are obtained from freshly excised human tissues (obtained with informed consent at the VA Palo Alto Health Care System).

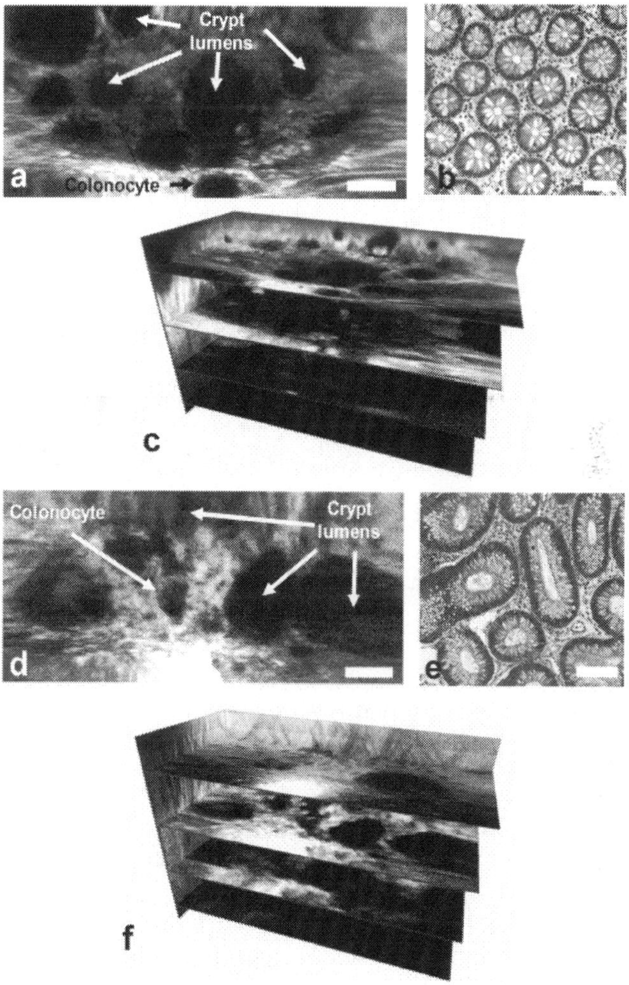

Fig. 26. Ex vivo images. En face dual axes confocal images of a) normal, d) dysplastic colonic mucosa. Corresponding histology (H&E) of b) normal and e) dysplasia. 3D volumetric images of c) normal and f) dysplasia, scale bar 100 μm.

Fig 26a, b, and c and Fig 26d, e, and f show the *en face* image, histology (H&E), and 3D volumetric images of normal and adenomatous (dysplastic) colonic mucosa, respectively, scale bar 100 µm[36]. Features of colonic crypts, including colonocytes and crypt lumens, are clearly resolved. Fig. 26c and 24f show three extracted *en face* planes at 50, 170, and 230 µm below the tissue surface. The gain is increased with depth to compensate for the lower signal levels.

2. Endoscope-compatible dual axes confocal instrument

A 2D *en face* fluorescence image of normal colonic mucosa collected *ex vivo* with the 5.5 mm diameter endoscope-compatible dual axes confocal prototype is shown in Fig. 27, scale bar 100 µm. ICG was topically applied to enhance contrast, and the pseudocolor image shows dye enhancement in the lamina propria surrounding the circular shaped crypts.

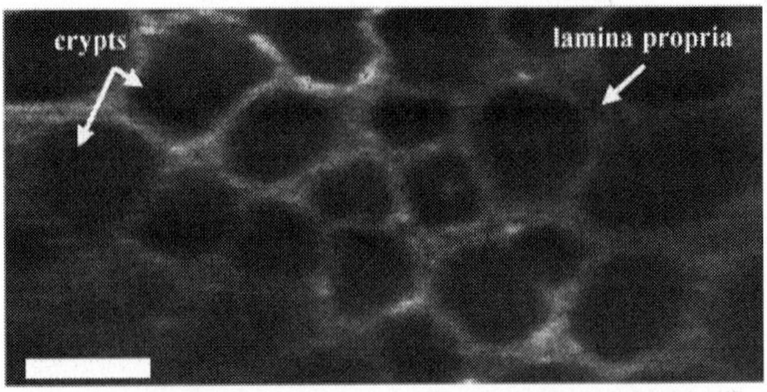

Fig. 27. En face fluorescence image of normal colonic mucosa collected with endoscope-compatible dual axes confocal prototype ex vivo using topically applied ICG to enhance contrast shows regular crypt pattern, scale bar 100 µm.

W. In vivo fluorescence imaging

In vivo imaging with the handheld dual axes confocal microscope has also been demonstrated. A mouse was anesthetized, and 10 mg of indocyanine green (Sigma-Aldrich, Corp) diluted in 10 ml of PBS was injected into the retro-orbital plexus of the mouse. Imaging was performed by resting the mouse on a translational stage and placing its ear intact on the SIL window of the microscope. Fig. 28a shows an *in vivo* image of blood vessels *en face* with a maximum intensity projection. Fig. 28b shows a 3D volumetric rendering of the image stack obtained by scanning from the surface to 150 µm deep into the intact ear. The images were collected in 3 µm intervals along the z-axis by using the piezoelectric actuator. All images were taken at 5 Hz with 5 frames averaging (1 second per image). The full 3D volume rendered image was acquired in 50 seconds, scale bar 100 µm.

In addition, *in vivo* images of human skin collected with the handheld dual axes confocal microscope are shown in Fig. 29. A sequence of approximately 300 individual *en face* images of human skin were collected at a fixed depth of 60 µm below the tissue surface (stratum corneum) with a speed of 5 Hz µm [36]. Topically applied indocyanine green was used for contrast. Image stitching or mosaicing was performed to enlarge the FOV and to increase the signal-to-noise ratio in real time with custom mosaicing software, shown in Fig. 29a. The

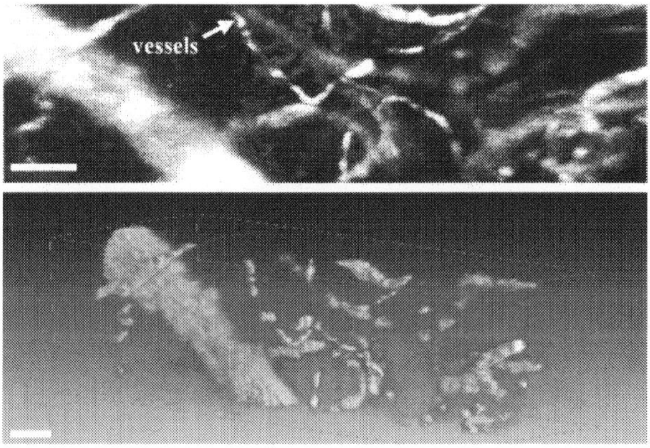

Fig. 28. a) A maximum intensity projected in vivo image of blood vessels in an intact mouse ear collected with handheld prototype. b) A 3D volume rendered image of blood vessels, scale bar 100 μm.

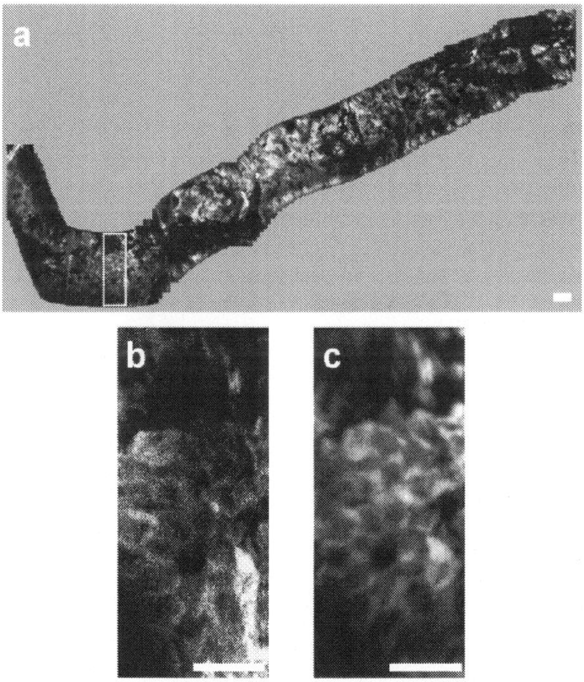

Fig. 29. a) Image mosaic of human skin acquired in vivo at a depth of 60 μm composed of roughly 300 images. The white box shows the corresponding location of individual images, b) A single input image; c) the corresponding area of the mosaic with improved signal-to-noise ratio, scale bars 50 μm.

white rectangle box in Fig. 29a represents an individual *en face* image (100×400 μm^2) obtained with the dual axes confocal microscope. The images were mosaiced by first correcting the image borders for scanning distortions. Then, each new image was registered and blended before proceeding to the next. Fig. 29b and 29c demonstrate how image mosaicing can increase the signal to noise ratio and dramatically improve image quality by tuning the amount of image overlap. The maximum input frame rate that our computer acquisition can process with the real-time mosaicing algorithm is 15 frames/second.

9. Conclusions and future directions

In this chapter, we present the theory, design and implementation of a novel dual axes confocal microscope in both tabletop and miniature form factors. Separate illumination and collection of light using the region of overlap between the two beams (focal volume) provide a number of advantages for purposes of miniaturization and *in vivo* imaging. The instruments were developed with 785 nm illumination to take advantage of the "optical window" in tissue where the high dynamic range and deep tissue penetration of this novel architecture can be demonstrated. This instrument is able to achieve sub-cellular resolution (~5 μm), sufficient for *in vivo* histopathological evaluation. Performance of the dual axes confocal microscope is demonstrated by collecting both *en face* images in real time and 3D volumetric images with post-processing at a maximum interrogating depth of 300 μm for both *ex vivo* and *in vivo* samples. Furthermore, we used this instrument as a test bed to further scale down the dimensions of this architecture to a 5.5 mm diameter package for endoscope compatibility. The size of the instrument has been reduced with a more compact aligning mechanism.

We have demonstrated a tissue penetration depth with the dual axes confocal microscope that is unmatched by any other endoscope-compatible instrument. From our *in vivo* experiments, fluorescence images can be collected up to a depth of 300 μm, limited by the maximum travel of the piezoelectric actuator. Greater depths have been achieved with our tabletop instruments (>500 μm). These results demonstrate the large working distance and high dynamic range of the dual axes confocal architecture to enable deep subsurface tissue imaging. Further improvements in performance can be achieved by increasing light throughput. The relatively low output power of 2 mW can be significantly increased with use of either silver or gold coatings as the reflective surfaces of the MEMS scanner and parabolic mirror, rather than aluminum. In addition, a higher power fiber-coupled laser source can be used.

Future development of dual axes confocal architecture will focus on achieving the theoretical levels of performance in a miniature instrument package. In addition, repeatability and reliability will be addressed. We will take advantage of the high dynamic range of the system by developing new z-axis actuators that rapidly scan the focal volume perpendicular to the tissue surface to achieve deep penetration in vertical cross-sections. This orientation provides a powerful view for studying the epithelium and presents a comprehensive picture of the biological differentiation patterns in this thin layer of metabolically active tissue. The epithelium forms the inner lining of all hollow organs, and is accessible by medical endoscopes. In addition, we will extend this approach to multi-spectral imaging capabilities by developing achromatic optics using the same basic optical design. Finally, smaller form factors will be developed to achieve compatibility with standard medical endoscopes.

As this novel approach matures, we will be able to use this high resolution imaging instrument to perform clinical investigation in human subjects and longitudinal studies in small animal models. Molecular specificity can be achieved by combining this microendoscope with use of affinity probes that bind to over expressed cell surface targets. This integrated imaging methodology will provide the ability to visualize molecular features of tissue micro-structures in the vertical plane with sub-mucosal axial depths. This powerful capability has tremendous potential to unravel previously unknown molecular mechanisms about important disease processes, such as cancer and inflammation.

10. Acknowledements

We thank Christopher H Contag, Shai Friedland, Gordon S Kino, Jonathon TC Liu, Michael J Mandella, Hyejun Ra, Roy M Soetikno, Olav Solgaard, and Larry K Wong for their technical support.

11. References

[1] R. S. Cotran, V. Kumar, and T. Collins, "Robbins Pathologic Basis of Disease, 6th ed.," Philadelphia: W. B. Saunders Company, 1999, pp. 781-87.

[2] J. Pawley, *Handbook of biological confocal microscopy,*, 3rd ed. New York: Plenum Press, 1996.

[3] T. R. C. a. G. S. Kino, "Confocal Scanning Optical Microscopy and Related Imaigng Systems, Academic Press, Boston," 1996.

[4] E. Laemmel, M. Genet, G. Le Goualher, A. Perchant, J. F. Le Gargasson, and E. Vicaut, "Fibered confocal fluorescence microscopy (Cell-viZio) facilitates extended imaging in the field of microcirculation. A comparison with intravital microscopy," *Journal of Vascular Research,* vol. 41, pp. 400-411, Sep-Oct 2004.

[5] K. Carlson, M. Chidley, K.-B. Sung, M. Descour, A. Gillenwater, M. Follen, and R. Richards-Kortum, "In vivo fiber-optic confocal reflectance microscope with an injection-molded miniature objective lens," *Applied Optics,* vol. 44, pp. 1792-1796, April 1, 2005.

[6] N. Thekkek and R. Richards-Kortum, "Optical imaging for cervical cancer detection: solutions for a continuing global problem," *Nat Rev Cancer,* vol. 8, pp. 725-731, 2008.

[7] P. M. Delaney, M. R. Harris, and R. G. King, "Fiber-optic laser scanning confocal microscope suitable for fluorescence imaging," *Applied Optics,* vol. 33, pp. 573-577, February 1, 1994.

[8] T. D. Wang, S. Friedland, P. Sahbaie, R. Soetikno, P.-L. Hsiung, J. T. C. Liu, J. M. Crawford, and C. H. Contag, "Functional Imaging of Colonic Mucosa With a Fibered Confocal Microscope for Real-Time In Vivo Pathology," *Clinical Gastroenterology and Hepatology,* vol. 5, pp. 1300-1305, 2007.

[9] D. L. Dickensheets and G. S. Kino, "A scanned optical fiber confocal microscope," in *SPIE Symposium on Electronic Imaging Science & Technology,* San Jose, CA, 1994, pp. 39-47.

[10] D. L. Dickensheets and G. S. Kino, "Silicon-micromachined scanning confocal optical microscope," *Journal of Microelectromechanical Systems,* vol. 7, pp. 38-47, March, 1998.

[11] W. Piyawattanametha, H. Toshiyoshi, J. LaCosse, and M. C. Wu, "Surface-micromachined confocal scanning optical microscope," Conference on Lasers and Electro-Optics (CLEO 2000), May 7-12 2000, San Francisco, CA, USA, pp. 447-448, 2000.

[12] K. Murakami, A. Murata, T. Suga, H. Kitagawa, Y. Kamiya, M. Kubo, K. Matsumoto, H. Miyajima, and M. Katashiro, "A miniature confocal optical microscope with MEMS gimbal scanner.," in IEEE International Solid-State Sensors and Actuators Conference, June 8-12, 2003, Boston, MA, USA, 2003, pp. 587-590, vol.1.

[13] W. Piyawattanametha, L. Fan, S. Hsu, M. Fujino, M. C. Wu, P. R. Herz, A. D. Aguirre, Y. Chen, and J. G. Fujimoto, "Two-dimensional endoscopic MEMS scanner for high resolution optical coherence tomography.," in *on Lasers and Electro-Optics (CLEO), May 16-21, 2004, San Francisco, CA, USA,* 2004, pp. 3, vol.1.

[14] W. Piyawattanametha, P. R. Patterson, D. Hah, H. Toshiyoshi, and M. C. Wu, "Surface- and bulk- micromachined two-dimensional scanner driven by angular vertical comb actuators.," *Journal of Microelectromechanical Systems,* vol. 14, pp. 1329-1338, 2005.

[15] L. Thiberville, S. Moreno-Swirc, T. Vercauteren, E. Peltier, C. Cave, and G. Bourg Heckly, "In Vivo Imaging of the Bronchial Wall Microstructure Using Fibered Confocal Fluorescence Microscopy," *Am. J. Respir. Crit. Care Med.,* vol. 175, pp. 22-31, January 1, 2007 2007.

[16] T. D. Wang, M. J. Mandella, C. H. Contag, and G. S. Kino, "Dual-axis confocal microscope for high-resolution in vivo imaging," *Optics Letters,* vol. 28, pp. 414-416, March 15, 2003.

[17] T. D. Wang, C. H. Contag, M. J. Mandella, N. Y. Chan, and G. S. Kino, "Dual-axes confocal microscopy with post-objective scanning and low-coherence heterodyne detection," *Optics Letters,* vol. 28, pp. 1915-1917, October 15, 2003.

[18] M. Born and E. Wolf, *Principles of Optics,* 7th ed. ed.: Cambridge University Press, 1999.

[19] J. T. C. Liu, M. J. Mandella, S. Friedland, R. Soetikno, J. M. Crawford, C. H. Contag, G. S. Kino, and T. D. Wang, "Dual-axes confocal reflectance microscope for distinguishing colonic neoplasia," *Journal of Biomedical Optics,* vol. 11, pp. -, SEP-OCT 2006.

[20] L. K. Wong, M. J. Mandella, G. S. Kino, and T. D. Wang, "Improved rejection of multiply scattered photons in confocal microscopy using dual-axes architecture," *Optics Letters,* vol. 32, pp. 1674-1676, June 15, 2007.

[21] C. Henyey, *Diffuse radiation in the Galaxy,* 2000.

[22] K. T. Mehta and H. S. Shah, "Correlating Parameters Of The Henyey-Greenstein Phase Function Equation With Size And Refractive-Index Of Colorants," *Applied optics-OT,* vol. 24, pp. 892-896, 1985.

[23] K. Pavlos, "Scattering, Absorption, and Emission of Light by Small Particles," *Bulletin of the American Meteorological Society,* vol. 84, p. 494, 2003.

[24] M. K. S. A. Prahl, S. L. Jacques, A. J. Welch, "Dosimetry of Laser Radiation in Medicine and Biology," Proc. SPIE, vol. 5, pp. 102-111 1989.

[25] T. D. Wang, C. H. Contag, M. J. Mandella, N. Y. Chan, and G. S. Kino, "Confocal fluorescence microscope with dual-axis architecture and biaxial postobjective scanning," *J Biomed Opt,* vol. 9, pp. 735-42, Jul-Aug 2004.

[26] T. D. Wang, C. H. Contag, M. J. Mandella, N. Y. Chan, and G. S. Kino, "Confocal fluorescence microscope with dual-axis architecture and biaxial postobjective scanning," *Journal of Biomedical Optics,* vol. 9, pp. 735-742, JUL-AUG 2004.

[27] J. T. C. Liu, M. Mandella, J. Crawford, C. Contag, T. Wang, and G. Kino, "Efficient rejection of scattered light enables deep optical sectioning in turbid media with low-numerical-aperture optics in a dual-axis confocal architecture," *Journal of Biomedical Optics,* vol. 13, p. 34020, 2008.

[28] D. L. Dickensheets and G. S. Kino, "Micromachined scanning confocal optical microscope," *Optics Letters,* vol. 21, pp. 764-766, May 15, 1996.

[29] W. Piyawattanametha, R. P. J. Barretto, T. H. Ko, B. A. Flusberg, E. D. Cocker, H. J. Ra, D. S. Lee, O. Solgaard, and M. J. Schnitzer, "Fast-scanning two-photon fluorescence imaging based on a microelectromechanical systems two-dimensional scanning mirror," *Optics Letters,* vol. 31, pp. 2018-2020, July 1, 2006.

[30] S. P. Timoshenko and J. N. Goodier, *Theory of Elasticity,* 3rd ed. New York: McGraw-Hill, 1970.

[31] W. Piyawattanametha, P. R. Patterson, D. Hah, H. Toshiyoshi, and M. C. Wu, "Surface- and bulk-micromachined two-dimensional scanner driven by angular vertical comb actuators," *Journal of Microelectromechanical Systems,* vol. 14, pp. 1329-1338, DEC 2005.

[32] H. Ra, Y. Taguchi, D. Lee, W. Piyawattanametha, and O. Solgaard, "Two-dimensional MEMS scanner for dual-axes confocal in vivo microscopy," *Proceedings of the 19th IEEE International Conference on Micro Electro Mechanical Systems (MEMS), Istanbul, Turkey,* vol. 2006, pp. 862-865, 2006.

[33] H. Ra, W. Piyawattanametha, Y. Taguchi, D. Lee, M. J. Mandella, and O. Solgaard, "Two-dimensional MEMS scanner for dual-axes confocal microscopy," *Journal of Microelectromechanical Systems,* vol. 16, pp. 969-976, AUG 2007.

[34] C. Ataman and H. Urey, "Nonlinear Frequency Response of Comb-Driven Microscanners," Proceedings of SPIE - The International Society for Optical EngineeringMOEMS Display and Imaging Systems II; Jan 26-27 2004; San Jose, CA., United States, vol. 5348, pp. 166-174, 2004.

[35] J. T. C. Liu, M. J. Mandella, H. Ra, L. K. Wong, O. Solgaard, G. S. Kino, W. Piyawattanametha, C. H. Contag, and T. D. Wang, "Miniature near-infrared dual-axes confocal microscope utilizing a two-dimensional microelectromechanical systems scanner," Optics Letters, vol. 32, pp. 256-258, FEB 1 2007.

[36] A. E. Siegman, M. W. Sasnett, and T. F. Johnston, "Choice of Clip Levels for Beam Width Measurements Using Knife-Edge Techniques," *IEEE Journal of Quantum Electronics,* vol. 27, pp. 1098-1104, April 1991.

20

Scanning Near-field Raman Spectroscopic Microscope

Sumio Hosaka
Gunma University
Japan

1. Introduction

Recently, a need to develop a high-speed semiconductor device increases remarkably. In order to achieve such a device characteristic, we have paid attentions to a device with localized stress such as Si-Ge device (Welser et al. 1994, Rim et al. 2000, Tezuka at al. 2001). Electron and hole mobility can be influenced by the stress distribution of active layer in very fine device. Thus, we have to control the strength of the stress as 2- or 3-dimensional distribution for development of advanced devices. It is very important to measure and analyze the stress of very fine semiconductor devices. We have expected to analyze the stress with Raman spectroscopy. In Raman spectrum, strong stokes and anti-stokes scattering peak shifts occur when compressing and stretching the local area of the device.

On the other hand, a minimum size of semiconductor device pattern will be miniaturized to reach to a region of 25 nm in the future. We have to analyze the stress distribution for such a structure with very fine probe in Raman spectroscopy. In optical microscopic Raman spectroscopy, there exists a diffraction limit so that we cannot obtain the spectrum with a resolution of less than submicron. In order to overcome the diffraction limit, scanning near-field optical microscopy (SNOM) with a resolution of less than wavelength of optical probe by small aperture has been proposed by E. H. Synge in 1932. Then, SNOM technology with a few 10s nm resolution has reported by D. Pohl in 1986 (Durig et al., 1986, Fischer et al. 1989). The technology has been based on near-field optical probe with small aperture on the metal-coated probe or fiber probe (Trautman et al. 1994). On the other hand, another approach, using apertureless near-field optics, has been proposed by J. Wessel in 1985. The concept is based on surface enhanced Raman spectroscopy (SERS). Since 1985, SERS and SNOM technologies have been advanced by using scanning tunneling microscopy (STM) or atomic force microscopy (AFM) technique. So far, many researchers have reported lots of works using various type near field optical microscopes.

In Raman spectroscopy, the significant technical issuer is how to detect too weak Raman scattering signal to increase spatial resolution. For improving the signal, a tip-enhanced Raman spectroscopy (TERS) has been proposed by A. Hartschuh, L. Novotny et al. in 2003. Then, many researchers proposed various type TERSes; bottom illumination (Stockle et al., 2000, Hayazawa et al.,2003), side-illumination (Nieman et al., 2001, Hayazawa et al. 2002, Mehtani et al. 2005), and modified top illumination (Poborchii et al.. 2005). The concepts are based on SERS, in which enhancement of the signal is obtained in the vicinity of the metal particle or tip. It can use the metal particle or metal tip as Raman scattering near-field probe with larger amplitude of Raman scattering. Although TERS technology has high spatial resolution based on Raman scattering signal enhancement in the vicinity of metal particle, the metal particle or metal tip has serious problem that they are one of contamination sources in Si device fabrication process.

In SNOM or NSOM (near-field scanning optical microscopy), many researchers have also reported these technologies such as illumination mode, collection mode, illumination-collection mode, etc (Pohl, 1986, Trautman et al. 1994). They are based on formation of fine electromagnetic field probe in the vicinity of small aperture on the metal tip as near-field optical probe. Although the aperture size controls the size of near-field optical probe, microscopic image is determined by a signal and noise ratio. Both probe size and optical power throughput are very important. So far, good spatial resolution was not demonstrated using aperture type SNOM (Hosaka et al., 1996, Ono et al., 2005).

In near-field Raman spectroscopy, M. Yoshikawa et al. have reported to develop near-field optical Raman spectroscopy with illumination-correction mode and fine aperture pyramidal probe (Yoshikawa et al., 2006). Using resonant Raman scattering, they have 2-dimensional stress distribution of the VLSI standard sample made by Si and SiO_2 for checking AFM The resolution is, however, not so high as about 250 nm even though they employed near-field light in Raman spectroscopy, though the Raman peak shift image is improved rather than optical microscopic Raman spectroscopy. In S. Hosaka group SNOM research, thus resolution could be also a little improved with near-field light in a case of using metal aperture probe in illumination-correction mode SNOM. The group have already pointed out optical aperture probe has limited to improve a spatial resolution because the optical probe has two components of near-field and far-field optical probes (Hosaka et al., 1999). The far-field optical probe power is gigantic larger than that of near-field probe to eliminate the near-field optical probe. Recently, 10 nm-less spatial resolution using surface plasmon effect near-field optical probe (Fischer et al., 1989) in an illumination-collection and depolarization mode SNOM (Hosaka et al., 2007).

The metal aperture on outside at the top of the pyramidal probe causes metal contamination on the device surface as described above. We have an idea to improve to get fine near-field optical probe and to protect the contamination in order to solve these problems. The idea is to utilize such a structure that the optical probe is made by plasmon resonance without the outside metal and its aperture. The probe is based on the commercial pyramidal probe. The outside and inside layers are made of insulator of Si_3N_4 or SiO_2 and metal, respectively. The structure can protect from the contamination. However, we are anxious if we get fine near-field optical probe without the aperture. It is very interested to study the aperture-less pyramidal probe for high resolution image (Hosaka et al., 2007).

In this section, I describe recent research states of TERS and SNOM technologies. Then, I describe SNOM technology in regard to plasmon resonace optical probe using apertureless cantilever, depolarization optical system, Raman spectroscopy, etc. I describe estimation of near-field light propagation from the aperture-less pyramidal probe using FDTD method. And I describe some features of prototyped atomic force cantilevered SNOM with the aperture-less pyramidal probe and its combination system with Raman spectroscopy, and discuss on the possibility to detect Raman spectra for measuring fine stress distribution of semiconductor devices.

2. Raman scattering, TERS and SNOM technologies

2.1 Raman scattering

When light incidents the sample, the reflected light is modulated by lattice vibration of the sample. Assuming that polarizability of the sample is given by $a = a_0 + a_1\cos2\pi v_s t$ and the electric field is given by $E = E_0 \cos2\pi vt$ the induced dipole moment P is given by $P = aE$ where v_1 and v are vibrations of the sample and incident light, respectively. The P is given by Eq. (1).

$$\begin{aligned} P &= aE = (a_0 + a_1 \cos 2\pi v_s t)E_o 2\pi vt \\ &= a_0 E_0 \cos 2\pi vt + \frac{1}{2} a_1 E_0 \cos 2\pi (v - v_s)t + \frac{1}{2} a_1 E_0 \cos 2\pi (v + v_s)t \end{aligned} \quad (1)$$

The reflected dipole moment has 3 components as presented in above equation. The 1st, the 2nd and the 3rd terms correspond to Rayleigh scattering light, stokes scattering light and anti-stokes scattering light, respectively, as shown in Fig. 1(a). Therefore, stokes and anti-stokes scattering lights have the sample's information of lattice vibration, polarizability, atomic binding, stress, etc.

Furthermore, when compression or tensile stress is applied to the sample, the frequencies of stokes and anti-stokes scattering peaks are shifted. The peak-frequency shift (peak shift) occurs as shown in Fig. 1(b). The peak shifts to low and high frequencies mean tensile stress and compression stress, respectively. The stress σ can be estimated from the shift Δv as represented by Eq. (2).

$$\sigma = 2.3 \times 10^4 \Delta v \ [Ncm^{-2}] \quad (2)$$

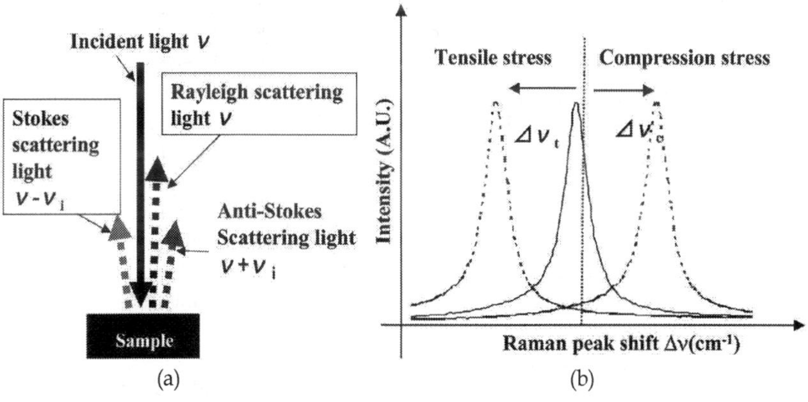

Fig. 1. Raman scattering phenomenon (a) and Raman peak shifts due to tensile andcompression stress (b).

2.2 TERS

The TERS has typically 3 types of bottom illumination, side illumination and modified top illumination as shown in Fig. 2. A. Hartschuh et al. have reported that near-field Raman spectroscopy with a spatial resolution of 20 nm has been demonstrated using a bottom

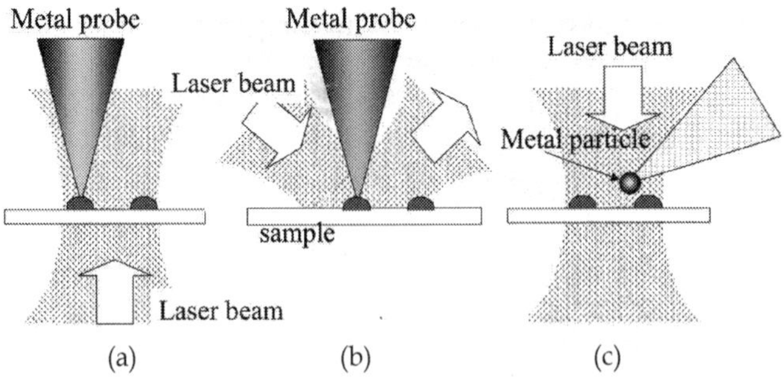

Fig. 2. Schematic diagrams of some TERSes; (a) bottom illumination, (b) side illuminationand (c) modified top illumination.

illumination mode TERS with vibration mode of single-wall carbon nanotubes (SWNTs) (Hartschuh et al., 2003). The tip material was gold. The tip was controlled within 10-50 pN using tuning fork detection in AFM. They demonstrated an enhancement of TERS signals of G and G′ bands by a comparison of near-field and far-field as shown in Fig. 3. They showed Raman scattering image of SWNT using G′ band peak, and topographic and Raman scattering signal profiles (Fig. 4). The G′ band Raman signal profile indicated that the spatial resolution was less than 30 nm because a FWHM of SWNT with G′ band peak was about 26 nm.

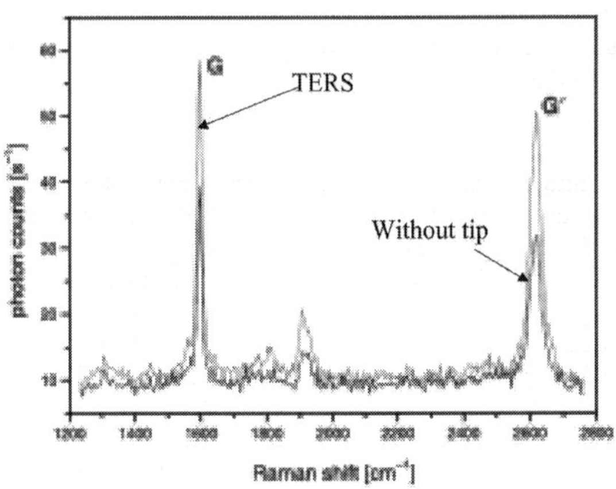

Fig. 3. Raman spectra detected with a sharp metal tip (green line) on top of the sample(distance; 1 nm) and with the tip retracted by 2 μm (black line). Note, the intensities of allRaman bands are increased with the tip close to the SWNTs (Hartschuh et al., 2003).

D. Mehtani et al. have introduced great potential of side illumination mode TERS for nanoscale chemical characterization and semiconductor (Mehtani et al., 2005). They demonstrated enhancement of Raman scattering signal of various molecular, polymeric and semiconducting materials as well as carbon nanotube (CNT) by comparing with far-field Raman signal. V. Poborchii et al. have introduced modified top illumination mode TERS using a silver particle on the top of quartz AFM cantilever probe immersed into glycerol

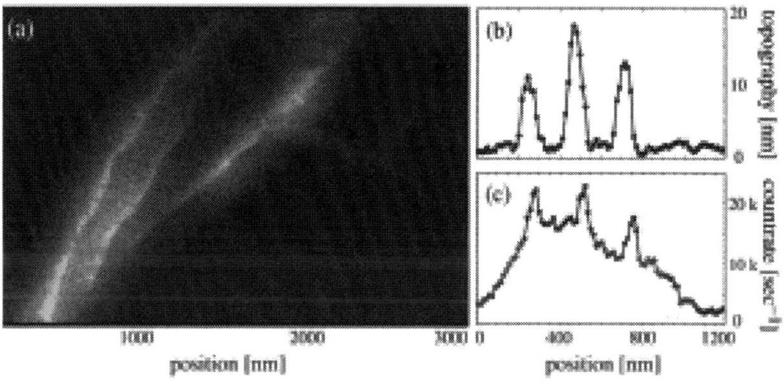

Fig. 4. (a) Raman image of SWNT bundles acquired by raster scanning a sharp metal tipand detecting the intensity of the G' band (scan area 3 × 1 μm, integration time 5 msper pixel). (b and c): Cross-sections taken along the white dotted line in (a) fortopography (b) and Raman signal (c) (Hartschuh et al., 2003).

droplet on Si surface (Poborchii et al.. 2005). As the experimental result, spatial resolution in a range of 100 nm was demonstrated. The system was used with depolarization optics without 364 nm primary light. As described above, TERS technology has tremendous potential to enhance Raman scattering signal compared with far-field Raman scattering signal, but the best spatial resolution was about 30 nm. It is not enough to apply the technology to measure the Si device.

2.3 SNOM

A research on this technology has been focused into improvement of near-field optical probe with very fine probe and high contrast against far-field light. At first, S. Hosaka et al. changed the optical fiber probe to the AFM cantilevered pyramidal optical probe because the fiber probe was broken by hard contact between the probe and sample. The cantilevered optical probe has a possibility to observe nanometer-sized pits formed by electron beam (EB) drawing. They succeeded in observing the 30 nm x 160 nm small pits by using the polarized near-field light as the illumination-collection mode SNOM. In the experiments, when they adopt an optical aperture on the cantilever, they had to focus illuminating laser beam into the aperture and to achieve another laser beam deflection optics for atomic force detection (optical lever). To achieve the requirements, they had to develop a through the lens (TTL) type optical lever. However, they have reported that illumination mode SNOM with an aperture on the top of metal probe has 2 components of near-field and far-field probes on the vicinity of the aperture (Fig. 5). The near-field probe power was very small rather than that of far-field probe. We needed to remove the far-field light and enhance the near-field probe power (high throughput) with small diameter. For the former, we adopted depolarization optics to detect only reflected near-field light without the far-field light reflected from the pyramidal probe. The illumination-collection mode SNOM optics was developed as shown in Fig. 6. The system obtained a spatial resolution of 300 nm from near-field Kerr effect image of perpendicular magnetic recorded bits (Fig. 7). The resolution was not enough to obtain nanometer resolution by using such optics and probe. The low

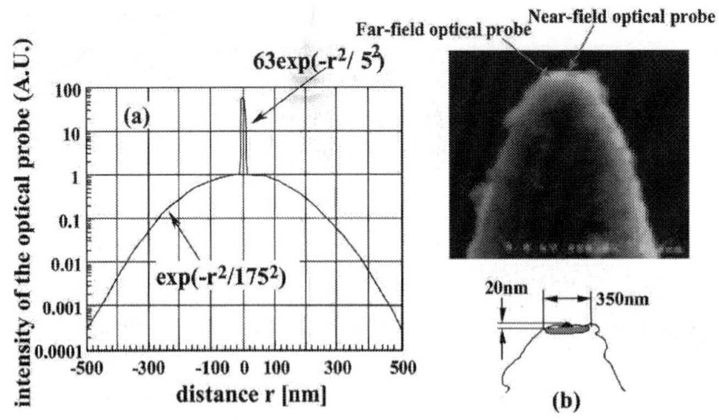

Fig. 5. Estimated optical probe profile (a) and the tip structure and its SEM image (b)(Hosaka et al., 1999).

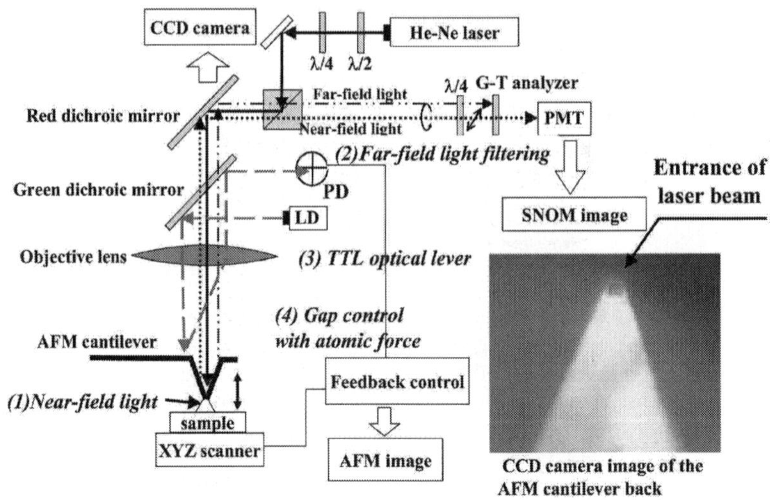

Fig. 6. Scheme of the illumination-collection SNOM system (Hosaka et al., 2007)

resolution was caused by the aperture formed on the top of the metal probe. In Raman spectroscopy, M. Yoshikawa et al. have reported that they developed tuning fork AFM cantilevered illumination-collection type SNOM for stress distribution in VLSI standard sample, which has been used as AFM check sample. The experimental result showed a resolution of about 250 nm from peak-frequency shift image of the sample around 520 cm^{-1} Si peak (Fig. 8). These data could not show high spatial resolution of less 50 nm. This might be caused by near-field optical probe with an aperture. Furthermore, the probe has a problem to make a contamination on Si device. Therefore, we have to solve the optical probe with no metal surface probe. We have proposed a plasmon effect near-field optical probe with Au inner film on outer pyramidal AFM conventional probe made of SiN with no aperture as described in next section.

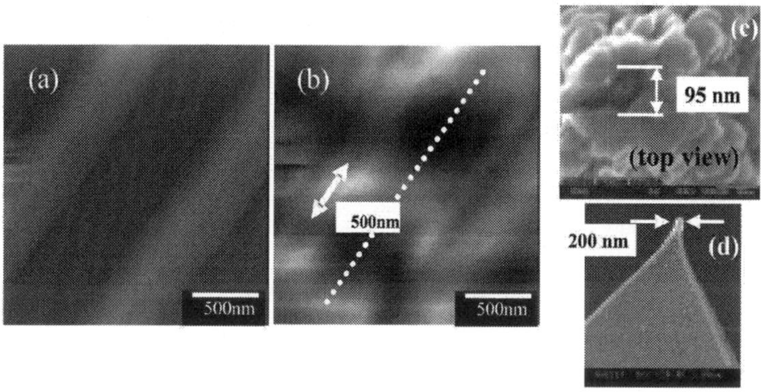

Fig. 7. SNOM observation of 640MB magneto-optical disc(2mmx2mm); (a) AFM imageand (b) SNOM image (Kerr effect image), (c) top view of the aperture and (d)pyramidal probe (Ono et al., 2005).

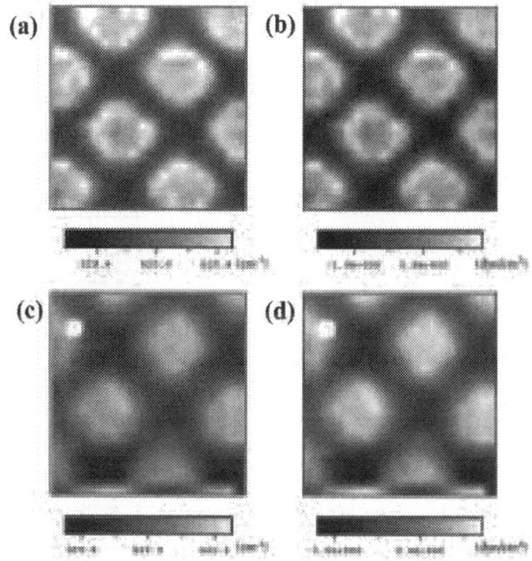

Fig. 8. The near-field Raman scattering Si images of (a) peak-frequency, and (b) stress inVLSI standards, measured by the pyramidical probe with a diameter of 100 nm. Theoptical microscope images of (c) peak-frequency, and (d) stress in VLSI standards,respectively (Yoshikawa et al., 2006),

3. Illumination-collection mode SNOM with aperture-less pyramidal probe

In order to consider whether we can obtain fine near-field optical probe from top of the aperture-less pyramidal AFM probe inside-coated with a metal film, we have studied light propagation through the top of probe when illuminating ultra-violet (UV) laser with a wave length of 363.8 nm using finite differential time domain (FDTD) method.

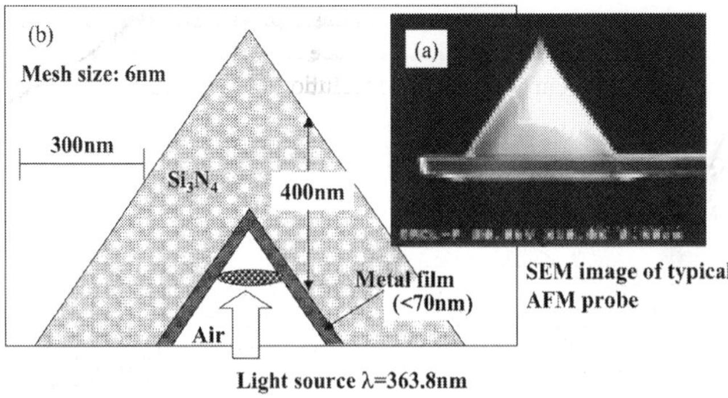

Fig. 9. Calculation model for near-field light emission from aperture-less cantileveredprobe with metal film using FDTD method.

Figure 9 shows the AFM cantilever image and scheme of the probe for SNOM probe. The cantilever is available for commercial pyramidal one, which is model of OMCL-TR400PSA-1 made by Olympus Inc. Figure 9(b) shows the calculation model of an enlarged image of the top of the pyramidal probe based on Fig. 9(a) when illuminating the UV laser into the probe. We executed FDTD calculation with very fine mesh with a size of 6 nm. As a result, we obtained near-field light profiles emitted from the top of the pyramidal probe for various metal films as shown in Fig. 10. The near-field light power becomes strong in order of aluminum, gold and silver. In the cases of Au and Ag, surface plasmon may occur on the inside metal film. In addition, the near-field light can be propagated along the pyramidal surfaces and emitted from the top. Figure 11 shows the light propagation from the top. The light can propagate to shallow region of <200 nm in FWHM of size far from the top.

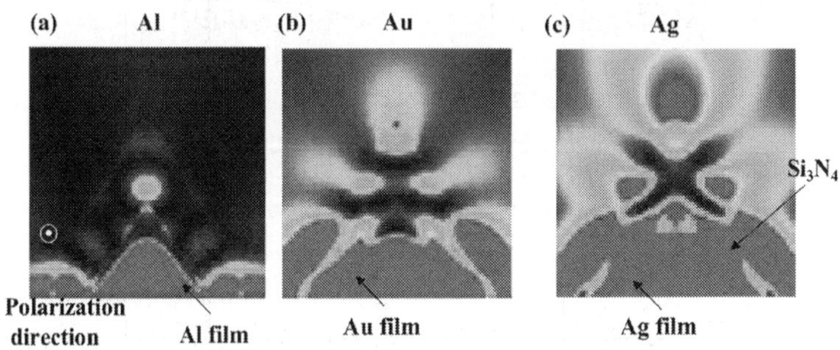

Fig. 10. Calculated images of near-field light emission from the top of aperture-less cantilevered probe with various metal films (50 nm) when illuminating an UV light witha wavelength of 363.8 nm (Hosaka et al., 2007).

Figure 12 shows calculated results of the near-field light propagation through the top for various Au film thicknesses when illuminating the UV laser. The figure shows that near-field light emits from the tip through the metal film and SiN dielectric material even by using thick metal film. The power decreases gradually with thickness. The profiles of the optical probe are almost same shape. As described above, we can use the near-field optical probe even when using the aperture-less pyramidal probe. The optical profiles, however, indicates that the estimated near-field optical probe size is too large to detect a detail of the light image or localized Raman signal with a resolution of <100 nm.

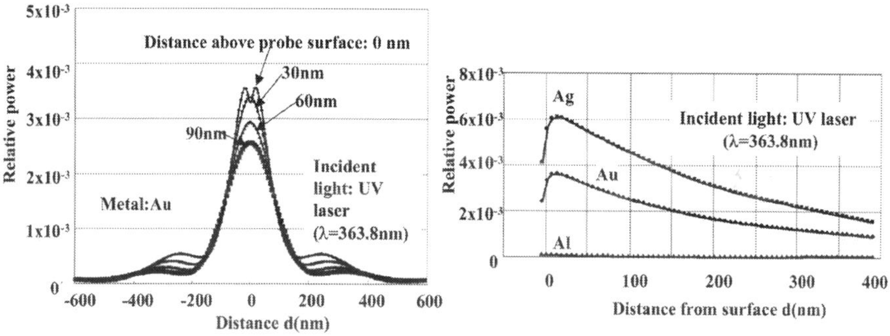

Fig. 11. Calculated results of near-field light propagation for the distances from thesurface of the aperture-less cantilevered probe with various metal film (thickness:50nm), (a) in-plain distribution n and (b).near-field light propagation above the sample surface (Hosaka et al., 2007).

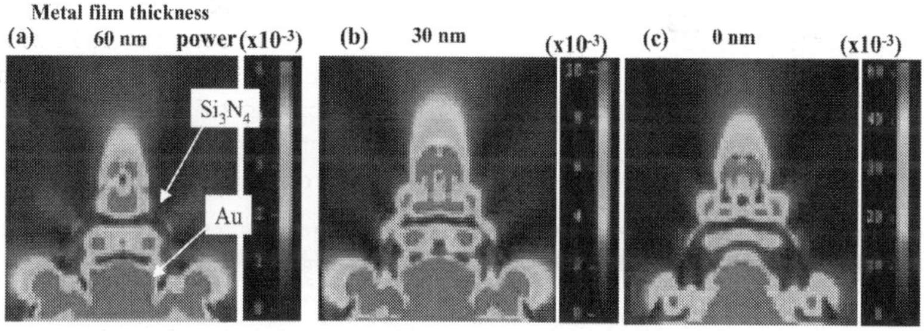

Fig. 12. Calculated images of near-field light emission from the top of aperture-lessmetal probe for various metal film thicknesses, (a) 60 nm-thick Au film, (b) 30 nm-thickAu film, and (c) without the metal film at incident light of UV light with 363.8 nm in

4. Prototype atomic force cantilevered SNOM system (Ono et al., 2005)

The prototyped SNOM system has 4 functions; (1) to keep the gap between the probe and the sample constant by controlling a Z-position of the sample using atomic force detected with the through the lens (TTL) type optical lever[6)] (AFM function), (2) to generate near-field light emitted from the top of the probe by illuminating the laser beam into the inside of the pyramidal probe (illumination-collection mode SNOM function), (3) to detect only near-field light reflected from the sample surface using a polarization (depolarization mode SNOM function) or Raman spectroscopy, and (4) to adjust the laser beams incident on the fixed positions of the cantilever using a charge-coupled device (CCD) camera for optical lever and near-field light adjustments. In order to achieve these functions, we have developed multi-beam optics. These optical axes coincide on the object lens. The main optics is shown in Fig. 6. In SNOM or Kerr effect detection, a He-Ne laser beam with a wavelength of 632.8 nm (red laser) and a semiconductor laser beam with 532 nm (green laser) for near-field light and atomic force detection of deflection optical lever, respectively, were used. In Raman spectroscopy, UV laser with a wavelength of 363.8 nm was used with Ar ion laser. The objective lens was used with a numerical aperture (NA) of around 0.5 in both cases.

The depolarization optics consists of a He-Ne laser source, a half wave ($\lambda/2$) plate, quarter wave ($\lambda/4$) plates, AFM cantilever with a small aperture, a G-T analyzer, and a photomultiplier tube (PMT). The linearly polarized light that emitted from the He-Ne laser is converted into circularly polarized light through the $\lambda/4$ plate, and is focused into the inside of the probe on the cantilever tip. After the polarized near-field light is reflected from the sample surface through or outside of the top, a plane of polarization is slightly rotated. By passing the reflected light through the $\lambda/4$ plate, the plane of the polarization is converted to the linearly polarized light with various angles. The G-T analyzer is adjusted to remove the far-field light reflected from inside wall of the probe. On the other hand, near-field light reflected from the sample surface has a little shift of polarization so that we can detect only polarization-rotated near-field light using G-T analyzer.

Figure 13 shows AFM and depolarization SNOM images of the sputtered Au film on the glass with a comparison of SEM image. The system observed not only many fine clacks in the film in AFM image, but also bright small lines with a width of <10 nm in SNOM image. The images indicate that both functions of AFM and SNOM have very fine resolution of less than 10 nm.

Fig. 13. Test sample observation of Au film; (a) SEM image, (b) AFM image and (c) SNOM image (Hosaka et al., 2007).

The optical system observed Kerr effect near field image of optical recording of conventional giga magneto-optical (MO) disc as shown in Fig. 14. From the rise-up at the signal edge, a resolution of Kerr effect image was less than 20 nm, considering a magnetic domain wall between switched bits. Comparing these data with previous image (Fig. 7), the aperture-less pyramidal SNOM has the potential to achieve fine spatial resolution in near-field light images.

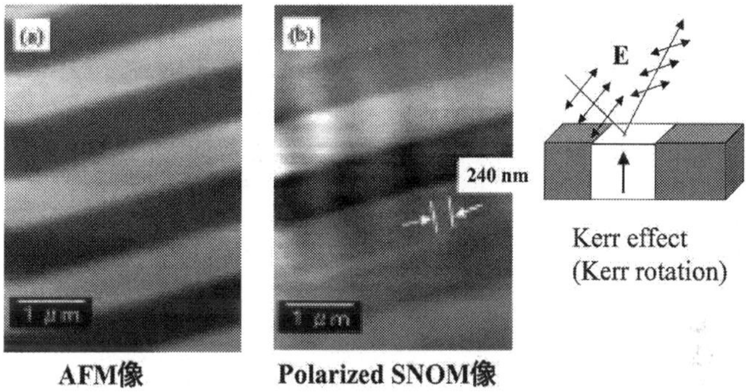

Fig. 14. AFM and SNOM images of 2.3 GB-MO recorded bits (MO disc spec.: min. bit length 233nm, track pitch: 670 nm (Hosaka et al., 2007).

5. Combination of SNOM and Raman spectroscopy (Hosaka et al. 2007)

We have prototyped the combination system with above system (Fig. 6) and Raman spectroscopy of Nanofinder30, which is made by Tokyo Instruments Inc., as shown in Fig. 15. The stimulated laser was used with UV line with a wavelength of 363.8 nm in Ar ion laser. Figure shows a scheme of optical functions of the system. In practice, the spectroscopy was inserted between the TTL type optical lever system and the depolarization optics. The sample with gates and shallow trench insulator (STI) on Si substrate was prepared for measuring its stress distribution.

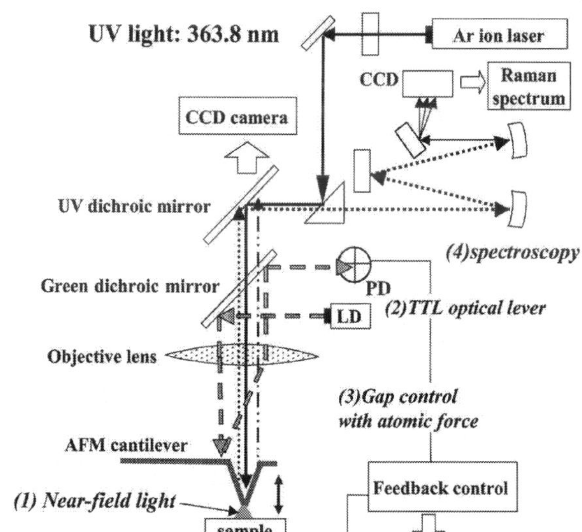

Fig. 15. Scheme of prototyped SNOM Raman spectroscopic microscopy (Hosaka et al., 2007).

Figure 16 shows Raman spectra change before and after AFM operation when illuminating the UV laser into the inside of aperture-less pyramidal probe. Figure 16(a) shows the spectrum under the sample was far from the probe. The Si Raman peak disappeared. Under controlling the system in the contact mode AFM, the peak at about 520 cm^{-1} was obtained. When adjusting the optics and using strong power of UV laser, strong signal was obtained as shown in Fig. 16(c).

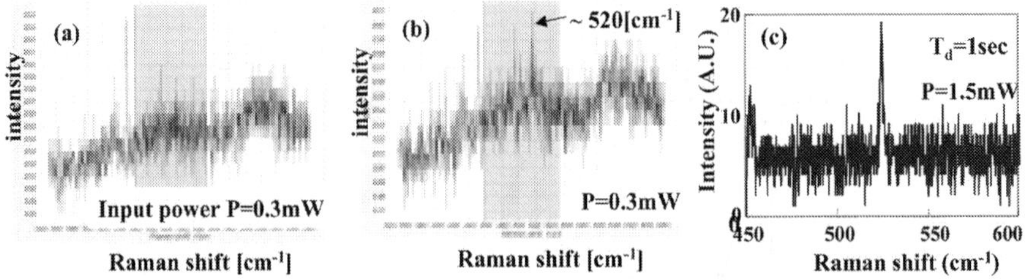

Fig. 16. Raman shift spectra of the silicon sample when illuminating UV light with a wavelength of 363.8 nm to inside of the pyramidal probe, (a) in separation between the probe and the sample, (b) under contact mode controlling at 20nN, and (c) incident laser power of 1.5 mW (Hosaka et al., 2007).

Figures 17(a) and 17(b) show spectra detected by the near-field optical Raman spectroscopy (NFRS) and optical microscopic Raman spectroscopy (OMRS) at an accumulation time of 16 sec, respectively. The OMRS can be easily achieved by removing the cantilever probe. Although the detected intensity of NFRS is weaker than that of OMRS, it is clarified that we detected Raman scattering peak of Si using the system and the aperture-less pyramidal probe.

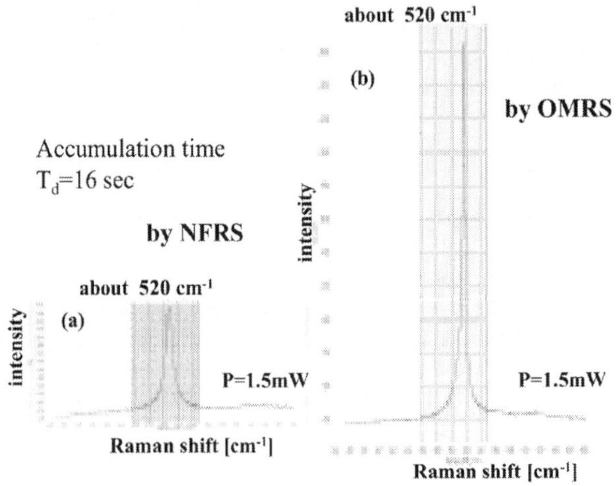

Fig. 17. Comparison of spectra in near-field optical Raman spectroscopy (NFRS) (a) andoptical microscopy Raman spectroscopy (OMRS) (Hosaka et al., 2007).

6. Raman spectroscopy of Si device test sample (Hosaka et al.)

The test sample structure and anticipated Raman peak shift of Si are shown in Fig. 18. The compression and tensile stresses distribution can be estimated in the sample. In the gate, compression stress occurs because of oxidation of both gate sides. In the Si area between the STI and the gate, tensile stress ocurrs because STI-SiO_2 volume is shrunk by annealing.

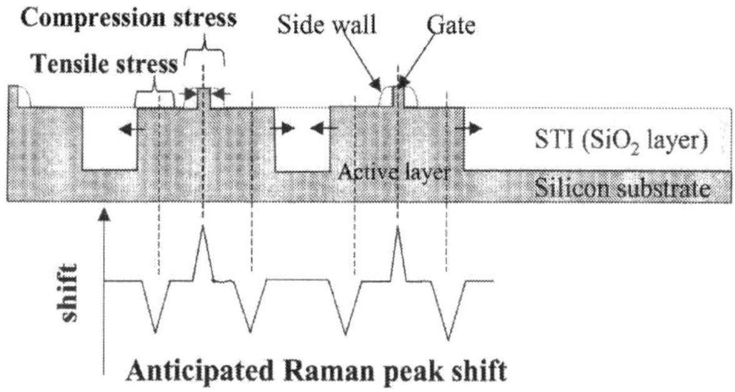

Fig. 18. A structure of the test sample, and anticipated stress model and Raman peak shift.

Using the sample, the surface structure was observed by AFM function in prototyped SNOM and scanning electron microscope (SEM). We can observe small dimension of the gate of about 25 nm as shown in Fig. 19. Then, we measured Raman spectrum around 520 cm^{-1} at each pixel on one line of the sample for peak-frequency shift and stress distribution.

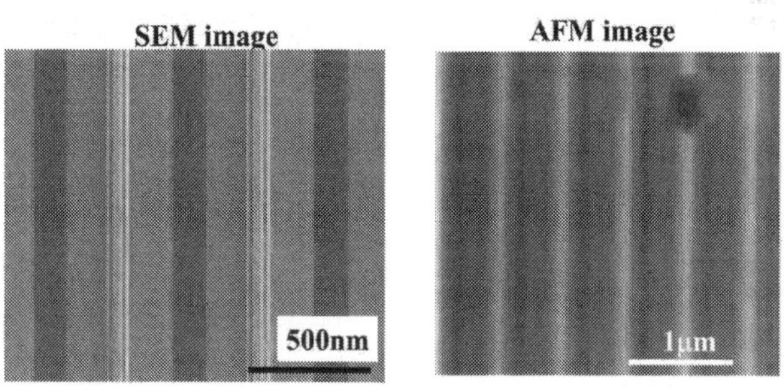

Fig. 19. SEM and AFM images of the sample, AFM image was taken by SNOM system. and stress model.

After the measurement, the peak-frequency shift was processed by Lorenz peak fitting of the spectrum for fine spatial resolution of less than 0.1 cm^{-1}. The peak-frequency shift profile is shown in Fig. 20. The profile is well agreed with the anticipated profile. The peak of compress stress appears at the gate part. The FWHM of the peak is less than 25 nm. The result was obtained when the focused laser beam with a power of 1.5 mW was illuminated into the pyramidal probe of AFM. The Au film was coated inside the probe with a thickness of 50 nm. The highest compression stress of 1-1.5x10^4 [Ncm^{-2}] was estimated, and is consistent with the result reported by M. Yoshikawa st al. Therefore, the technology has the potential to get near-field Raman scattering profile with fine spatial resolution of less than 25 nm.

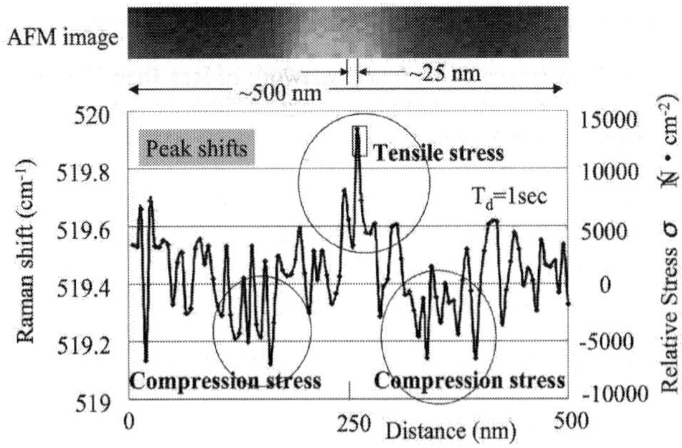

Fig. 20. AFM image of the gate and Raman spectrum peak shift around the gate.

7. Summary

Recent technologies of tip-enhanced Raman spectroscopy (TERS) and scanning near-field optical microscopy (SNOM) with Raman spectroscopy are reviewed.

In TERS technology, it has been developed based on surface-enhanced Raman spectroscopy (SERS). The some TERS are reviewed as follows:

1. Bottom illumination mode TERS has been described, and it has demonstrated fine spatial resolution and gigantic enhancement of Raman scattering signal using single wall carbon nano tube (SWNT).
2. Side- and modified top-illumination mode TERS have been described, and they have demonstrated enhancement of Raman scattering signal and resolution of subwavelength.
3. There are, however, some technical issues such as metal contamination, stress measurement of Si device, etc. This means that it is difficult to apply to an evaluation of semiconductor devices, etc.

In SNOM technologies, illumination-collection mode SNOM has been described with regard of aperture type SNOM probe and aperture-less pyramidal SNOM probe.

4. Using the aperture type SNOM probe, it is difficult to obtain fine spatial resolution because the aperture makes only near-field optical probe incompletely.
5. M. Yoshikawa et al. demonstrated spatial resolution of about 200 nm in peak-frequency shift image with Raman Si peak of VLSI standard sample using illumination-collection mode with the aperture type SNOM probe.
6. S. Hosaka et al. have proposed the aperture-less pyramidal probe in illumination-collection mode SNOM to improve near-field optical probe and to protect from the metal contamination to the device surface. The SNOM and Raman spectroscopy was combined with UV laser. The prototyped microscopy demonstrated following possibilities.
   a. By calculating near-field light propagations in the aperture-less pyramidal probe by FDTD method, strong near-field light propagation from top of the aperture-less

pyramidal probe occurs due to surface plasmon effect using Au and Ag as inside metal film.

b. Very fine SNOM image of the crack network of less than 10 nm in thin Au film using the aperture-less pyramidal probe SNOM was obtained with a spatial resolution of less than 10 nm.

c. The combination system of SNOM and Raman spectroscopy has the possibility to detect Raman scattering light using the aperture-less pyramidal probe.

d. The system measured the Si sample with STI and gate structure to get both AFM image and Si peak-frequency shift in stokes scattering.

e. The system observed compression and tensile stress of $1\text{-}1.5\times10^4$ [Ncm^{-2}] on the sample.

f. The spatial resolution of less than 25 nm was demonstrated in Raman spectroscopy.

As described above, scanning near-field Raman spectroscopic microscopy has the potential to measure a detail of the sample such as structure, magnetization, optical property, chemical characterization, stress distribution, etc with a fine spatial resolution of 25 nm. In the future, the technology is expected to be one of key technologies for evaluation of materials.

8. References

Durig, U., Pohl, D. W. and Rohner, F. (1986). *J. Appl. Phys.* Vol. 59, 3318.

Fischer, U. C. and Pohl D. W. (1989). *Phy. Rev. Lett.* Vol. 62, 458.

Hartschuh, A., Anderson, N. and Novotny, L. (2003). *J. Microscopy,* Vol. 210, 234-240.

Hayazawa N, Inouye Y, Sekkat Z, Kawata S. (2002). *J. Chem. Phys.* Vol. 117, 1296.

Hayazawa N, Inouye Y, Tarun A, Kawata S. (2002). *J. Appl. Phys.* Vol.92, 6983.

Hosaka, S, Shintani, T, Miyamoto, M, Kikukawa, A, Hirotsune, A, Terao, M, Yoshida, M, Fujita, K, Kammer, S. (1996). *J. Appl. Phys.,* Vol. 79, 8082.

Hosaka, S., Shintani, T., Kikukawa, A. and Itoh, K. (1999). *J. Microscopy,* Vol.194 , 369-373.

Hosaka, S., Sone, H., Takahashi, Y., Shintani, T., Kato, K., Saiki, T. (2003). *Microelectronic Engineering,* Vol. 67-68, 728.

Hosaka, S., Shimizu, T., Mine, K., Shimada, K., and Sone, H. (2007) . *J. Phys.: Conf. Ser.* Vol. 61, 425-429.

Hosaka, S., Koyabu, H., Mine, K., Aramomi, Y., Sone, H., Sato, E., Tochigi, K. (2007). *Proc. 2007 IEEE Int. Conf. Electron Devices and Solid-State Circuits* Tainan, Taiwan, pp. 297-300.

Hosaka, S., Koyabu, H., Mine, K., Aramomi, Y., Sone, H., Sato, E., Tochigi, K. to be submitted.

Nieman L.T., Krampert G.M., Martinez R.E. (2001). *Rev. Sci. Instrum.* Vol. 72, 1691.

Ono, M., Sone, H., and Hosaka, S. (2005). *Jpn. J. Appl. Phys.* Vol. 44, 5434.

Poborchii V., Tada T., Kanayama T. (2005). *Jpn. J. Appl. Phys.* Vol. 44, L202.

Pohl, D. W. (1986). *IBM J. Res. Develop.* Vol. 30, 417.

Rim, K., Hoyt J.L., Gibbons J.F. (2000). *IEEE Transactions on Electron Devices,* Vol. 45, 1406-1415.

Stockle R., Suh Y.D., Deckert V., Zenobi R. (2000) *Chem. Phys. Lett.* Vol. 318, 131.

Synge, E.H. (1932). *Philos. Mag.* Vol. 13: 297.

Tezuka, T., Sugiyama, N., and Takagi, S. (2001). *Appl. Phys. Lett.* Vol. 79, 1798-1800 .

Trautman J.K.,Macklin J.J., Brus L.E, Betzig E. (1994). *Nature,* Vol. 369, 40.

Welser, J., Hoyt, J.L., Gibbons, J.F. (1994). *Electron Device Letters*, Vol. 15, 100-102.

Wessel J. (1985).*J. Opt. Soc. Am. B,* Vol. 2, 1538.

Yoshikawa, M, Murakami, M, Matsuda, K, Sugie, R. Ishida, H, Shimizu, R, Jpn. J. Appl. Phys. 45, L486 (2006).